The Boundary Element Method for Engineers and Scientists

The Boundary Element Method for Engineers and Scientists

Theory and Applications

Second Edition

John T. Katsikadelis

School of Civil Engineering
National Technical University of Athens
Athens, Greece

ELSEVIER

AMSTERDAM • BOSTON • HEIDELBERG • LONDON
NEW YORK • OXFORD • PARIS • SAN DIEGO
SAN FRANCISCO • SINGAPORE • SYDNEY • TOKYO
Academic Press is an imprint of Elsevier

Academic Press is an imprint of Elsevier
125 London Wall, London EC2Y 5AS, United Kingdom
525 B Street, Suite 1800, San Diego, CA 92101-4495, United States
50 Hampshire Street, 5th Floor, Cambridge, MA 02139, United States
The Boulevard, Langford Lane, Kidlington, Oxford OX5 1GB, United Kingdom

British Library Cataloguing-in-Publication Data
A catalogue record for this book is available from the British Library.

Library of Congress Cataloging-in-Publication Data
A catalog record for this book is available from the Library of Congress.

ISBN: 978-0-12-804493-3

Publisher: Joe Hayton
Acquisition Editor: Brian Guerin
Editorial Project Manager: Edward Payne
Production Project Manager: Nicky Carter
Designer: Mark Rogers

Typeset by MPS Limited, Chennai, India

To my wife Efi
for her loving patience and support

Contents

Preface to the Second Edition

This second edition with the new title "The Boundary Element Method for Engineers and Scientists: Theory and Applications" reflects new developments that occurred after the first edition was written. Specifically, these are the *dual reciprocity method* (DRM) and the *analog equation method* (AEM), both established methods that have rendered the boundary element method (BEM) an efficient modern computational method for solving all linear and nonlinear problems, static as well as dynamic, in Engineering and Science. These developments are included in three new chapters, namely, in "The BEM for Potential Problems in Inhomogeneous Anisotropic Bodies," "The BEM for Time Dependent Problems," and "The BEM for Nonlinear Problems." Regarding the old chapters, the arrangement of the book remains the same, except for some small additions and emendations.

The first English edition under the title "Boundary Elements. Theory and Applications" appeared in 2002. It was published by Elsevier UK and was widely adopted internationally as a textbook for teaching the BEM as a formal course for boundary elements at the undergraduate or graduate level. This fact is confirmed by the translations of the book into Japanese by Prof. Masa Tanaka of Shinshu University, Nagano (Asakura, Tokyo, 2004), into Russian by Prof. Sergey Aleynikov of Voronezh State Architecture and Civil Engineering University (Издательство ABC, Publishing House of the Russian Civil Engineering Universities, Moscow, 2007), and into Serbian by Prof. Dragan Spasic of the University of Novi Sad (Gradjevinska Knjiga, Belgrade, 2011) (see Fig. 1). All the three professors, known for their research and writing about the boundary elements and mechanics in general, chose to translate the book as a textbook among several others of the international literature, after having checked its suitability as a textbook with their students.

The success of the first edition encouraged the author to prepare a new revised edition augmented by the three new chapters. They describe the established methods, that is, the DRM and the AEM, which overcome the inherent drawbacks of the BEM, namely, the inability to solve linear problems for which the fundamental solution cannot be established as well as nonlinear and time-dependent problems, using simple known static fundamental solutions. This is illustrated through the application of the DRM and the AEM to problems described by the complete second-order linear or nonlinear equation with variable coefficients (elliptic, parabolic, and hyperbolic).

FIGURE 1 Covers of the English, Japanese, Russian, and Serbian editions.

The material in these new chapters is presented in a systematic and comprehensive manner, as in the old chapters, so that the reader can understand the principles of DRM and AEM as well as their numerical implementation and computer programming. The material related to the AEM is the outcome of the author's long involvement with the method.

In closing, the author wishes to express, from this place too, his sincere thanks to his former student and coworker Dr. A.J. Yiotis for carefully reading the manuscript, his suggestions for constructive emendations, and his overall contribution to minimize the oversights in the text. Finally, warm thanks belong to Dr. Nikos G. Babouskos, also former student and coworker of the author, not only for his careful reading of the manuscript and his apposite suggestions for improvement of the book, but also for his assistance in developing the computer programs for the new chapters and in producing the numerical results of examples therein.

It is a pleasure to make grateful acknowledgment of many helpful suggestions which have been contributed by readers of the book.

John T. Katsikadelis
Athens
April 2016

Preface to the First Edition

The last three decades have been marked by the evolution of electronic computers and an enormous and widespread availability of computational power. This has boosted the development of computational methods and their application in engineering and in the analysis and design of *structures*, which extend from bridges to aircrafts and from machine elements to tunnels and the human body. New scientific *subfields* were generated in all engineering disciplines being described as "*Computational*," for example, *Computational Mechanics, Computational Fluid Mechanics, Computational Structural Analysis, Computational Structural Dynamics, etc*. The finite element method (FEM) and the boundary element method (BEM) are the most popular of the computational methods. While the FEM has been long established and is most well known in the engineering community, the BEM appeared later offering new computational capabilities with its effectiveness, accuracy, and low computational cost.

Although the BEM is taught as a regular course at an ever increasing number of universities, there is a noticeable lack of a textbook which could help students as well as professional engineers to understand the method, the underlying theory, and its application to engineering problems. An essential reason is that BEM courses are taught mainly as advanced graduate courses, and therefore much of the underlying fundamental knowledge of mathematics and mechanics is not covered in the respective undergraduate courses. Thus, the existing books on the BEM are addressed rather to academia and researchers who, somehow, have already been exposed to the BEM than to students following a BEM course for the first time and engineers who are using boundary element software in industry.

This observation stimulated the author to write the book at hand. His research in the development of the BEM during the last 25 years as well as the experience he acquired by teaching for many years the course of Boundary Elements at the Civil Engineering Department of the National Technical University of Athens, Greece, justify this endeavor. The author's ambition was to make the BEM accessible to the student as well to the professional engineer. For this reason, his main task was to organize and present the material in such a way so that the book becomes "user-friendly" and easy to comprehend, taking into account only the mathematics and mechanics to which students have been exposed during their undergraduate studies. This effort led to an innovative, in many aspects, way of presenting the BEM, including the derivation of fundamental solutions, the integral representation of the solutions and the boundary integral equations for various governing differential equations in a simple way minimizing a recourse to mathematics with which the student is not familiar. The indicial and tensorial notations, though they facilitate the authors' work and allow to borrow ready to use expressions from the literature, have been avoided in the present book. Nevertheless, all the necessary preliminary mathematical concepts have been included in order to make the book complete and self-sufficient.

In writing the book, topics requiring a detailed study for a deep and thorough understanding of the BEM, have been emphasized. These are:

i. The formulation of the physical problem.

ii. The formulation of the mathematical problem, which is expressed by the governing differential equations and the boundary conditions (boundary value problem).

iii. The conversion of the differential equations to boundary integral equations. This topic familiarizes the reader with special particular solutions, the so-called *fundamental solutions*, shows how they are utilized and helps to comprehend their singular behavior.

iv. The transformation of domain integrals to boundary line integrals or their elimination, in order to obtain pure boundary integral equations.

v. The numerical solution of the boundary integral equations. This topic, which covers a significant part of the book, deals with the numerical implementation of BEM rendering a powerful computational tool for solving realistic engineering problems. It contains the discretization of the boundary into elements, the modeling of its geometry, the approximation of the boundary quantities, as well as the techniques for the evaluation of regular and singular line integrals and in general the procedure for approximating the actual problem by a system of linear algebraic equations.

vi. A detailed description of the FORTRAN programs, which implement the numerical procedure for the various problems. The reader is provided with all the necessary information and the know-how so that he can write his own BEM-based computer programs for problems other than those included in the book.

vii. The use of the aforementioned computer programs for the solution of representative problems and the study of the behavior of the corresponding physical system.

Throughout the book, every concept is followed by example problems, which have been worked out in detail and with all the necessary clarifications. Furthermore, each chapter of the book is enriched with problems-to-solve. These problems serve a threefold purpose. Some of them are simple and aim at applying and better understanding the presented theory, some others are more difficult and aim at extending the theory to special cases requiring a deeper understanding of the concepts, and others are small projects which serve the purpose of familiarizing the student with BEM programming and the programs contained in the CD-ROM.

The latter class of problems is very important as it helps students to comprehend the usefulness and effectiveness of the method by solving real-life engineering problems. Through these problems students realize that the BEM is a powerful computational tool and not an alternative theoretical approach for dealing with physical problems. My experience in teaching the BEM shows that this is the students' most favorite type of problems. They are delighted to solve them, since they integrate their knowledge and make them feel confident in mastering the BEM.

The CD-ROM which accompanies the book contains the source codes of all the computer programs developed in the book, so that the student or the engineer can use them for the solution of a broad class of problems. Among them are general potential problems, problems of torsion, thermal conductivity, deflection of membranes and plates, flow of incompressible fluids, flow through porous media, in isotropic or anisotropic, homogeneous or composite bodies, as well as plane elastostatic problems in simply or multiply connected domains. As one can readily find out from the variety of the applications, the book is useful for engineers of all disciplines. The author is hopeful that the present book will introduce the reader to the BEM in an easy, smooth, and pleasant way and also contribute to its dissemination as a modern robust computational tool for solving engineering problems.

In closing, the author would like to express his sincere thanks to his former student and Visiting Assistant Professor at Texas A&M University Dr Filis Kokkinos for his carefully reading the manuscript and his suggestions for constructive changes. His critic and comments are greatly appreciated. Thanks also belong to my doctoral student, Mr G.C. Tsiatas, MSc, for checking the numerical results and the derivation of several expressions.

John T. Katsikadelis
Athens
January, 2002

Chapter | One

Introduction

CHAPTER OUTLINE

1.1 SCOPE OF THE BOOK

Since the boundary element method (BEM) became an appealing area of research over the last 55 years, its development was rapid. The related literature is enormous. Many books which describe the method and its numerical implementation have been published [1–20]. Some of them appeared after the publication of the first edition of this book. These books present the theoretical background and the numerical application of this modern tool of analysis. Hence, it would be fair to pose the question "what is the purpose of writing one more book on the topic?" The answer is quite simple. All the existing books, although they describe comprehensively the method, for the most part are written concisely. It could also be said that they are for academic use, and especially for the scientist that has already been exposed to the method and not for the student who studies the BEM for the first time. Moreover, since the BEM is a modern computational method for solving mainly engineering problems, it is intended for engineers. Therefore, the method must be presented in a way that can get across to them and bearing always in mind that extended utilization of advanced mathematics carries away authors in describing the method rather as a subject of applied mathematics than as a nice tool for solving engineering problems. For example, although the use of tensors provides a concise and elegant formulation, it puts engineering students off. For this purpose, the book at hand presents the BEM and provides derivation of all the necessary equations by incorporating only fundamental concepts and basic knowledge from differential and integral calculus, and numerical integration. Since, the scope of this book is

The Boundary Element Method for Engineers and Scientists.

to present the BEM in a comprehensive way and not to study in-depth all its potentials, the application of the method is illustrated by simple problems in the first instance. Some of them are boundary value problems governed by the Laplace or Poisson equation in two dimensions and plane elasticity problems. A considerable portion of the book is devoted to the numerical implementation of the method and its application to engineering problems. In all cases, computer programs are written in FORTRAN language. These programs, even though they solve important engineering problems, are not professional but educational. Mainly, they present the logical steps required for their construction and they familiarize students with the development of a BEM software.

The present book is not only a second revised edition of the previous book by this author "Boundary Elements. Theory and Applications," Elsevier 2002, but has been augmented by including three new chapters which present the BEM for the solution of steady state and time dependent problems (elliptic, parabolic, and hyperbolic) for which the fundamental solution either cannot be established or, even known, it is difficult to treat analytically and/or numerically in a systematic way. The last chapter includes the BEM for the solution of nonlinear problems. The new chapters present the most efficient methods which cope with the drawbacks of the conventional BEM, namely the DRM (dual reciprocity method) and the AEM (analog equation method). The latter, alleviated from any restrictions, renders the BEM an efficient computational tool for solving difficult problems in engineering and mathematical physics. Thus, the BEM offers a boundary method as an alternative to the domain type methods, such as finite difference method (FDM) and finite element method (FEM).

The author anticipates that the book at hand will help students as well as engineers and scientists to understand the BEM and apply it to problems they are faced with, either through the computer programs provided in the book or even their own. In addition, it is the author's strong belief that this book will contribute to a wider acceptance of the BEM as a modern efficient computational method.

1.2 BOUNDARY ELEMENTS AND FINITE ELEMENTS

The BEM constitutes a technique for analyzing the behavior of mechanical systems and especially of engineering structures subjected to external loading. The term *loading* is used here in the general sense, referring to the external source which produces a nonzero field function that describes the response of the system (temperature field, displacement field, stress field, etc.), and it may be heat, surface tractions, body forces, or even nonhomogeneous boundary conditions, for example, support settlement or support excitations.

Study of the behavior of structures is achieved today using computers. The reason is quite obvious, the low cost of the numerical versus the expensive experimental simulation. Numerical modeling can be used to study a wide variety of loadings and geometries of a structure and to determine the optimum design solution, before proceeding to its construction.

The most popular method used for the numerical analysis of structures during the last 45 years is the FEM. It is the method with which realistic problems of engineering are being solved, that is, the analysis of structural elements of arbitrary geometry, arbitrary loading, variety of constitutive relations, with linear or nonlinear behavior, in two or three dimensions. Justifiably, the FEM has been valued during the last 45 years as a modern computational tool.

A reasonable question to ask is "why do we need the BEM since we already have the FEM that solves engineering problems?" The answer is that modeling with finite elements can be ineffective and laborious for certain classes of problems. So the FEM, despite the generality of its application in engineering problems, is not free of drawbacks, the most important of which are:

i. Discretization is over the entire domain occupied by the body. Hence, generation and inspection of the finite element mesh exhibit difficulty and are both laborious and time-consuming, especially when the geometry of the body is not simple. For example, when there are holes, notches, or corners, mesh refinement and high element density is required at these critical regions of large solution gradients (Fig. 1.1a).

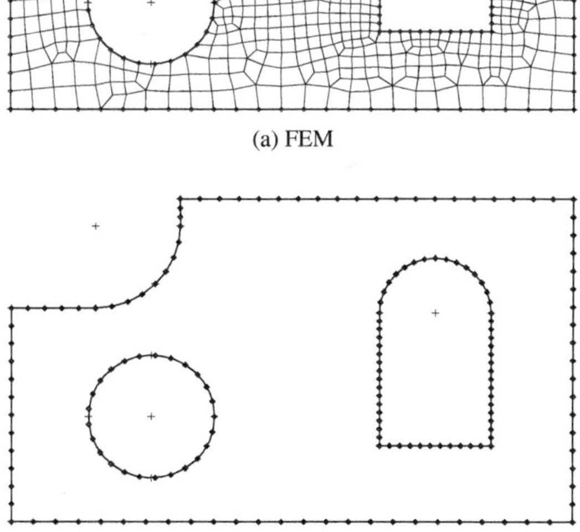

(a) FEM

(b) BEM

FIGURE 1.1 (a) Domain (FEM); (b) boundary (BEM) discretization.

ii. Modification of the discretized model to improve the accuracy of the solution or to reflect design changes can be difficult and requires a lot of effort and time.

iii. For infinite domains, for example, half-space or the complementary domain to a finite one, fabrication of fictitious closed boundaries is required in order to apply the FEM. This reduces the accuracy and sometimes may result in spurious or incorrect solutions.

iv. For problems described by differential equations of fourth or higher order (i.e., plate equations, or shell equations of sixth, eighth, or higher order), the conformity requirements demand such a tedious job that FEM may become impractical.

v. Although the FEM computes accurately the field function, which is the unknown of the problem, it is ineffective in determining its derivatives with the same accuracy. The accuracy drops considerably in areas of large gradients.

Drawbacks (i) and (ii) can be overcome by using updated releases of advanced professional finite element software, such as NASTRAN, ANSYS, ABACUS, etc., which are equipped with automatic and adaptable mesh generators. Essentially, the task of generating a finite element mesh is a hard geometric problem, and in several cases, it may prove to be far more difficult than the physical problem which is to be solved by the FEM. The new disadvantage, though, is that the codes for creating FEM meshes are closed, and any effort to produce them requires specialized knowledge from a different scientific area. An attempt, by the engineer, to learn the necessary material will only distract him from his original goal, which is to solve a physical problem.

On the contrary, the BEM possesses many advantages, the most important of which are:

i. Discretization is only over the boundary of the body, making the numerical modeling with the BEM easy (see Fig. 1.1b) and reducing the number of unknowns by one order. Thus, a remodeling to reflect design changes becomes simple.

ii. For infinite domains, the problem is formulated simply as an exterior one. Apparently, the fundamental solution has to satisfy some conditions at infinity, such as Sommerfeld's radiation condition for problems in dynamics. In this manner, computer programs developed for finite domains can be used, with just a few modifications, to solve problems in infinite domains. This is not possible with the FEM.

iii. The method is particularly effective in computing accurately the derivatives of the field function (e.g., fluxes, strains, stresses, moments). It can easily handle concentrated forces and moments, either inside the domain or on the boundary.

iv. The BEM allows evaluation of the solution and its derivatives at any point of the domain of the problem and at any instant in time. This is feasible because the method uses an integral representation of the

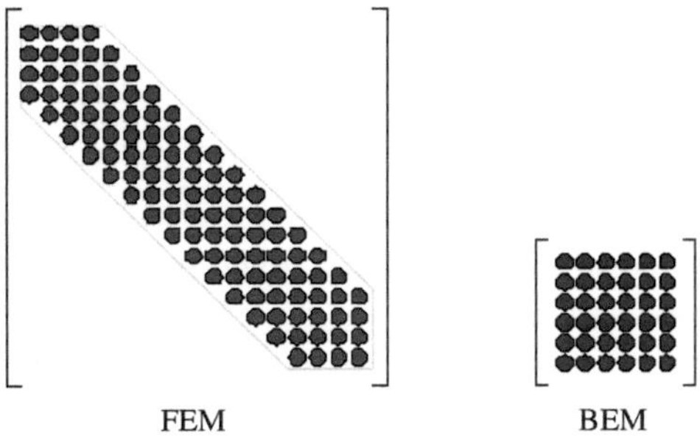

FIGURE 1.2 Coefficient matrices for FEM and BEM.

solution as a continuous mathematical expression, which can be differentiated and utilized as a mathematical formula. This is impossible with the FEM, since the solution is obtained only at the nodal points.

v. The method is well suited for solving problems in domains with geometric peculiarities, such as cracks.

On the other hand, the BEM exhibits the following inherent main disadvantages:

i. Application of the BEM requires the establishment of the integral representation of the solution. This is possible only if the problem is linear and its *fundamental solution* can be established. Hence the method cannot be used for problems whose fundamental solution is either unknown or cannot be determined. Such are, for example, problems described by differential equations with variable coefficients. The method is obviously not applicable to nonlinear problems for which a reciprocal identity cannot be established and hence the integral representations of the solution. During the last 25 years, intense research has been conducted and efficient methods have been developed that overcome the aforementioned disadvantages.

ii. The numerical implementation of the BEM results in systems of linear algebraic equations whose coefficient matrices are fully populated and nonsymmetrical. In a FEM model, however, the corresponding matrices have some very nice properties, they are banded and symmetric. This drawback of the BEM is counterbalanced by the much smaller dimensions of its matrices. The general format of the coefficient matrices for an FEM and BEM model is shown graphically in Fig. 1.2.

1.3 HISTORICAL DEVELOPMENT OF THE BEM

Until the beginning of the 1980s, the BEM was known as boundary integral equation method (BIEM). As a method for solving problems of mathematical physics, it has its origin in the work of G. Green [21]. He formulated, in 1828, the integral representation of the solution for the Dirichlet and Neumann problems of the Laplace equation by introducing the so-called Green's function for these problems. In 1872, Betti [22] presented a general method for integrating the equations of elasticity and deriving their solution in integral form. Basically, this may be regarded as a direct extension of Green's approach to the Navier equations of elasticity. In 1885, Somigliana [23] used Betti's reciprocal theorem to derive the integral representation of the solution for the elasticity problem, including in its expression the body forces, the boundary displacements and the tractions.

The fatherhood, however, of the BEM could be attributed to Fredholm. At the beginning of the 20th century, he was the first one to use singular boundary integral equations in order to find the unknown boundary quantities for problems of potential theory [24]. In fact, the method was employed as a mathematical tool to determine the necessary boundary conditions for a well-posed problem of mathematical physics, and not as a method to solve the problem. This is quite reasonable, because it was, and still is, not possible to find the analytic solution of the derived singular integral equations. In the aforementioned methods, the unknown boundary quantities have a direct physical or geometrical meaning and for this reason they are referred to as *direct* BEM. In addition to these methods, there were also other BEM formulations developed, in which the unknown boundary quantities have no direct physical or geometrical meaning, and thereafter they are given the name *indirect* BEM [25–27]. A detailed review of these methods may be found in [28]. Sherman [29,30], Mikhlin [31], and Muskhelishvili [32] used complex functions to develop BIEMs for the solution of plane elasticity problems. The reader can find some more details about the early history of the BEM in [33].

Closed form solutions of integral equations were only derived for some domains with very simple boundary geometry. Unfortunately, the work of Fredholm predated the computers that could make his ideas practical. For this reason, the BIEM was neglected until the end of the 1950s. Then, with the advent of computers, the method came back to the spotlight as an appealing numerical method for solving engineering problems. Numerical methods were developed for the solution of boundary integral equations and difficult physical problems of complex boundary geometry, which could not be tackled by other methods, were solved for the first time by the BIEM. The first works that laid the foundation of the BEM as a computational technique appeared in the early 1960s. Jaswon [34] and Symm [35] used Fredholm's equations to solve some two-dimensional problems of potential theory [36,37]. The merits of the BEM, which were listed in the previous section, attracted researchers and motivated them to further develop the method. Rizzo [38] and Cruse [39] applied the method to two- and three-dimensional elasticity problems, respectively. Rizzo and Shippy [40] extended the method to anisotropic elasticity, while Cruse and Rizzo [41] solved the elastodynamic problem. Ignaczak and Nowacki [42] expressed the integral equations of thermoelasticity and Mendelson [43] studied problems of elastoplastic torsion.

All the aforementioned problems are governed by second order partial differential equations. Another group of problems are those described by the biharmonic equation. In this case, the integral representation of the solution was derived from the Rayleigh-Green identity [44], and the approach was applied to plate bending and plane elasticity, with the latter being formulated in terms of Airy's stress function. The formulation consists of two boundary integral equations, one for each of the unknown boundary quantities. The first one arises from the boundary character of the integral representation of the field function, while the second is obtained from the integral representation either of the Laplacian of the field function or its normal derivative to the normal to the boundary. The second approach presented by Katsikadelis et al. [45] became the prevailing one and was adopted later by Bezine [46] and Stern [47] to solve the plate bending problem. An extended and detailed presentation of the plate bending problems that have been analyzed by the BEM can be found in Ref. [48,49]. Already in the late 1980s, one could find numerous publications in the literature, where the BEM was applied to a wide variety of engineering problems. Among them are static and dynamic, linear or nonlinear problems of elasticity, plates and shells, problems of elastodynamics, wave and earthquake engineering, geomechanics and foundation engineering, viscoelasticity, soil-structure interaction, fluid-structure interaction, fluid dynamics, unilateral contact, fracture mechanics, electricity and electromagnetism, heat conduction, acoustics, aerodynamics, corrosion, optimization, sensitivity analysis, inverse problems, problems of system identification, etc. It could be said that today the BEM has matured and become a powerful method for the analysis of engineering problems, and an alternative to the domain methods. The method has been established by the name BEM, which is attributed to the approach used to solve the boundary integral equations (i.e., discretization of the boundary into elements). Software based on the BEM has been developed for computers of simple or parallel architecture, along with professional high performance packages, like BEASY [50]. In 1978, C. Brebbia organized the first international conference on BEM, and since then conferences on BEM have been organized yearly by the international society for boundary elements and Wessex Institute of Technology (United Kingdom) in which the major developments in the field have been presented. Furthermore, all conferences on computational mechanics devote sessions to the BEM. A detailed review of this enormous work would occupy a lot of space in this book and, of course, it is beyond its scope. However, interested readers are referred to literature review articles [51,52], to the proceedings of the above BEM conferences, and to the numerous publications of the Computational Mechanics Publications, Southampton.

New developments in BEM aim at overcoming any drawbacks of the method. They deal with complicated time dependent problems, linear problems for which the fundamental solution is not known, as well as nonlinear problems. For all these types of problems the resulting integral solution involves domain integrals, which complicate the application of the method. The most efficient techniques that successfully overcome most of the difficulties and at the same time preserve the purely boundary character of the BEM, are the DRM [53], which, however, is subject to certain crucial restrictions, and the AEM [49,54,55]. The latter is general and applies without any limitations. A description of several well-known BEM variants developed over the past half century is presented with a critical review in [56].

1.4 STRUCTURE OF THE BOOK

As it was mentioned in Section 1.1, the scope of this book is to make the BEM comprehensible to the engineering student. For this purpose, its application is illustrated by the most representative problems described by second order partial differential equations. On these grounds, the book contains the introduction, ten chapters and three appendices. Every chapter is followed by the pertinent bibliography and recommended references for further study. Several exercises have been included at the end of each chapter in order to help the reader practice the concepts studied in the book.

Chapter 2 "Preliminary Mathematical Knowledge" presents some preliminary mathematical concepts that are necessary for developing the BEM. These concepts are the divergence theorem of Gauss, Green's reciprocal identity (Green's second theorem), and the definition of the delta function along with its properties. The chapter concludes with a section on calculus on variations, which provides the reader with an efficient mathematical tool to derive the governing differential equation as the Euler-Lagrange equation and the associated boundary conditions in complicated structural systems from stationary principles of mechanics.

In chapter 3 "The BEM for Potential Problems in Two Dimensions," the direct BEM is developed for some problems of potential theory, that is, boundary value problems for the two-dimensional Laplace and Poisson equations. It is extended also to the case of general second order partial differential equations with constant coefficients, which govern the homogeneous orthotropic, or generally, anisotropic bodies.

Chapter 4 "Numerical Implementation of the BEM" describes the numerical implementation of the BEM and the numerical solution of the singular boundary integral equations. For educational purposes, the solution is derived only for constant boundary elements and a computer program has been developed in FORTRAN language. The program is explained thoroughly and its structure is given in a systematic way, so that the student can become acquainted with the logic of writing BEM software. The method is also applied to domains containing holes and another computer program is provided for this case. Finally, one will find the method of subregions (subdivision of the original domain) as it applies to the BEM.

Chapter 5 "Boundary Element Technology" is devoted to the boundary element technology. The singular integral equations are integrated numerically using boundary elements. The elements may be subparametric, isoparametric, or superparametric and emphasis is put on the linear and parabolic elements. A considerable part of the chapter is devoted to the evaluation of the singular and near-singular integrals.

Chapter 6 "Applications" presents applications of the BEM to engineering problems that can be reduced to boundary value problems for the Laplace or Poisson equation. In particular, the BEM is applied to the Saint-Venant torsion problem for isotropic and anisotropic materials, to the bending of membranes and of simply supported plates, to heat conduction, and to the irrotational flow of incompressible fluids. For each of these problems, the reader is provided with a computer program and representative examples.

Chapter 7 "The BEM for Two-Dimensional Elastostatic Problems" studies the plane elasticity problem. The fundamental solution and the corresponding boundary integral equations are derived in a simple way so that the student will be able to follow all the steps and understand the subject. A computer program has been written also for this problem and several numerical applications are presented with the intention of familiarizing the student with the use of the program, on one hand, and of demonstrating the effectiveness of the method, on the other.

Chapter 8 "The BEM for Potential Problems in Inhomogeneous Anisotropic Bodies" presents the BEM for potential problems described by the general second order elliptic partial differential equation with variable coefficients. For this equation, while the reciprocal identity can be established, it is not possible to determine its fundamental solution. This is overcome by employing the DRM and the AEM, which use the simple known fundamental solution of the Laplace equation and are proven to be the most efficient BEM formulations. These two methods are presented in this chapter; first the DRM, with an emphasis on its capabilities and limitations, and then the AEM. The latter, alleviated from any restrictions, renders the BEM an efficient computational tool for solving all problems described by the complete second order differential equation and systems of them as well. The efficiency of these methods is illustrated by solving several representative problems.

Chapter 9 "The BEM for Time Dependent Problems" presents the BEM for problems described by the general second order hyperbolic and parabolic partial differential equations with variable coefficients. Though a reciprocal identity can be derived for these equations using time dependent reciprocal theorems, the establishment of the fundamental solution is out of the question. Here, the AEM, which uses the simple known fundamental solution of the Laplace equation, is presented as being the most suitable method to solve all problems described by the general second order hyperbolic and parabolic equations and systems of them as well, including problems described with fractional differential equations, such as the diffusion wave equation, which results in the hyperbolic and parabolic equations as special cases. The efficiency of the AEM is illustrated through well corroborated examples.

Chapter 10 "The BEM for Nonlinear Problems" presents the BEM for solving nonlinear steady state and time dependent potential problems described by the general second order nonlinear partial differential equation as well as systems of them. The DRM can treat such problems but only for specific forms of equations. Therefore, only the AEM is presented in this chapter as it proves to be the most appropriate method for solving these problems without any restrictions. Several example problems are presented, which validate the efficiency of the BEM to solve difficult nonlinear problems in engineering analysis and mathematical physics.

Appendix A lists useful relations, which facilitate the differentiation of the kernels found in the integral equations. Appendix B presents the Gaussian quadrature (numerical integration) for regular and singular integrals in one and two dimensions. Finally, Appendix C provides answers and/or hints to selected problems from those included at the end of the 10 chapters of the book.

1.5 THE COMPANION WEBSITE

The book is provided with a companion website containing the computer programs whose code is presented in the chapters of the present book as well as certain programs employed for the solution of the examples.

1.6 REFERENCES

[1] Brebbia CA. The boundary element method for engineers. London: Pentech Press; 1978.

[2] Banerjee PK, Butterfield R. Boundary element methods in engineering science. New York: McGraw-Hill; 1981.

[3] Brebbia CA, Telles JCF, Wrobel LC. Boundary element techniques. Southampton: Computational Mechanics Publications; 1984.

[4] Hartmann F. Introduction to boundary elements. Berlin: Springer-Verlag; 1989.

[5] Becker AA. The boundary element method in engineering: a complete course. New York: McGraw-Hill; 1992.

[6] Beer G, Watson JO. Introduction to finite and boundary element methods for engineers. New York: John Wiley and Sons; 1992.

[7] Chen G, Zhou J. Boundary element methods. London: Academic Press; 1992.

[8] El-Zafrany A. Techniques of the boundary element method. Upper Saddle, NJ: Prentice Hall PTR; 1993.

[9] Banerjee PK. The boundary element methods in engineering. London: McGraw-Hill; 1994.

[10] Hall WS. The boundary element method. Dordrecht: Kluwer Academic Publishers; 1994.

[11] Kane JH. Boundary element analysis in engineering continuum mechanics. Englewood Cliffs, New Jersey: Prentice Hall; 1994.

[12] Trevelyan J. Boundary elements for engineers: theory and applications. Southampton: Computational Mechanics Publications; 1994.

[13] Gaul L, Fiedler C. Methode der Randelemente in Statik und Dynamik. Braunschweig/Wiesbaden: Vieveg-Verlag; 1997.

[14] Paris F, Canas J. Boundary element method: Fundamentals and applications. Oxford: Oxford University Press; 1997.

[15] Brebbia CA, Dominguez J. Boundary elements: An introductory course. 2nd ed. Southampton: Computational Mechanics Publications; 2001.

[16] Constanda C. Direct and indirect boundary integral equations methods. Boca Raton, Fl: Chapman & Hall; 1999.

[17] Pozrikidis C. A Practical guide to boundary element methods with the software library BEMLIB. Boca Raton, Fl: Chapman & Hall; 2002.

[18] Aliabadi MH. The boundary element method. Chichester, England: John Wiley & Sons; 2002.

[19] Gaul L, Kögl M, Wagner M. Boundary elements for engineers and scientists. Berlin: Springer-Verlag; 2003.

[20] Beer G, Smith I, Duenser C. The boundary element method with programming: for engineers and scientists. Wien: Springer-Verlag; 2008.

[21] Green G. An essay on the application on mathematical analysis to the theories of electricity and magnetism, T. Wheelhouse: Nottingham; 1828.

[22] Betti E. Theoria dell' Elasticità. Il Nuovo Cimento 1872;6–10, Ser. 2

[23] Somigliana C. Sopra l' Equilibrio di' un Corpo Elastico Isotropo. Il Nuovo Cimento 1875;18(1):91–6 Ser. 3

[24] Fredholm I. Sur une classe d' equations fonctionelles. Acta Mathematica 1903;27:365–90.

[25] Kupradze VD. Potential methods in the theory of elasticity. Jerusalem: Israel Program for Scientific Translations; 1965.

[26] Kellog OD. Foundations of potential theory. Berlin: Springer-Verlag; 1967.

[27] Jaswon MA, Symm GT. Intergal equation methods in potential theory and elastostatics. London: Academic Press; 1977.

[28] Katsikadelis JT. The analysis of plates on elastic foundation by the boundary integral equation method. N.Y: Polytechnic University of New York; 1982, Ph.D. Dissertation

[29] Sherman DI. On the solution of the plane static problem of the theory of elasticity for displacements given on the boundary. Dokl Akad Nauk SSSR 1940;27:911–13.

[30] Sherman DI. On the solution of the plane static problem of the theory of elasticity for given external forces. Dokl Akad Nauk SSSR 1940;28:25–8.

[31] Mikhlin SG. Integral equations. London: Pergamon Press; 1957.

[32] Muskhelishvili NI. Some basic problems of the theory of elasticity. Holland: Noordhoff; 1963.

[33] Cheng AH-D, Cheng DT. Heritage and early history of the boundary element method. Eng Anal Bound Elem 2005;29:268–302.

[34] Jaswon MA. Integral equation methods in potential theory I. Proc R Soc Ser A 1963;275:23–32.

[35] Symm GT. Integral equation methods in potential theory II. Proc R Soc Ser A 1963;275:33–46.

[36] Jaswon MA, Ponter AR. An integral equation solution of the torsion problem. Proc R Soc Ser A 1963;275:237–46.

[37] Symm GT. An intergal equation solution in conformal mapping. Numer Math 1966;9:250–8.

[38] Rizzo FJ. An integral equation approach to boundary value problems of classical elastostatics. Q Appl Math 1967;Vol.25:83–95.

[39] Cruse T. Numerical solutions in three-dimensional elastostatics. Int J Solids Struct 1969;5:1259–74.

[40] Rizzo FJ, Shippy D. A merhod for stress determination in plane anisotropic elastic bodies. J Comp Mater 1970;Vol.4:36–61.

[41] Cruse T, Rizzo FJ. A direct formulation and numerical solution of the transient elastodynamic problem–I. Int J Math Anal Appl 1968;22:244–59.

[42] Ignaczak J, Nowacki W. Singular integral equations in thermoelasticity. Int J Eng Sci 1968;4:53–68.

[43] Mendelson A. 1975. Solution of the elastoplastic torsion problem by Boundary Integral Method, NASA, TN D-7882.

[44] Bergman S, Schiffer M. Kernel functions and elliptic differential equations in mathematical physics. New York: Academic Press; 1953.

[45] Katsikadelis JT, Massalas CV, Tzivanidis GJ. An integral equation solution of the plane problem of the theory of elasticity. Mech Res Commun 1977;4:199–208.

[46] Bezine G. Boundary integral equations for plate flexure with arbitrary boundary conditions. Mech Res Commun 1978;5:197–206.

[47] Stern M. A general boundary integral formulation for the numerical solution of plate bending problems. Int J Solids Struct 1979;15:769–82.

[48] Beskos DE, editor. Boundary element analysis of plates and shells. Berlin: Springer-Verlag; 1991.

[49] Katsikadelis JT. The boundary element method for plate analysis. Oxford: Academic Press, Elsevier; 2014.

[50] Trevelyan J. Use of the BEASY boundary element analysis software in engineering industry. In: Brebbia CA, Aliabadi MH, editors. Industrial Applications of the Boundary Element Method. Southampton: Computational Mechanics Publications; 1993 [chapter 5].

[51] Beskos DE. Boundary element methods in dynamic analysis. Part I. Appl Mech Rev 1987;40:1–23.

[52] Beskos DE. Boundary element methods in dynamic analysis. Part II. Appl Mech Rev 1997;50:149–97.

[53] Partridge PW, Brebbia CA, Wrobel LC. The dual reciprocity method. Southampton: Computational Mechanics Publications; 1992.

[54] Katsikadelis JT, Nerantzaki MS. A boundary-only BEM for linear and nonlinear problems. In: Kassab A, Brebbia CA, Chopra M, editors. Boundary elements, Vol. XX. Southampton: Computational Mechanics Publications; 1998. p. 309–20.

[55] Katsikadelis JT, Nerantzaki MS. The BEM for nonlinear problems. Eng Anal Bound Elem 1999;23:365–73.

[56] Kok Hwa Yu, KH, Kadarman AH, Djojodihardjo H. Development and implementation of some BEM variants-A critical review. Eng Anal Bound Elem 2010;34(10):884–99. Available from: http://dx.doi.org/10.1016/j.enganabound.2010.05.

Chapter | Two

Preliminary Mathematical Knowledge

CHAPTER OUTLINE

2.1 INTRODUCTION

In this chapter, some mathematical relations are presented which are required for the development and understanding of the boundary element method (BEM). Although these relations could have been included in an appendix, they are placed here to show the reader their important role in the theoretical foundation and development of the BEM. They will be used many a time and oft in the book and particularly for the transformation of the differential equations, which govern the response of physical systems within a domain, into integral equations on the boundary. The understanding of these mathematical concepts will give the reader a feeling of confidence for their subsequent use.

The Boundary Element Method for Engineers and Scientists.

2.2 THE GAUSS-GREEN THEOREM

The Gauss-Green theorem is a fundamental identity, which relates the integral of the derivative of a function over a domain Ω to the integral of that function on the boundary Γ. The domain may be two- or three-dimensional. For simplicity of presentation, this relationship is derived for the two-dimensional case. Consider the plane domain Ω bounded by the curve Γ. We shall work first with the derivative of a function $f = f(x, y)$ with respect to x. The integral over Ω may be written as a double integral, for which the integration is carried out first with respect to x and then with respect to y. Thus, we can write

$$\int_{\Omega} \frac{\partial f}{\partial x}\, d\Omega = \int_{y_1}^{y_2} \left(\int_{x_1}^{x_2} \frac{\partial f}{\partial x}\, dx \right) dy = \int_{y_1}^{y_2} \{ f(x_2, y) - f(x_1, y) \}\, dy \tag{2.1}$$

where

$$x_1 = x_1(y) \quad \text{and} \quad x_2 = x_2(y) \tag{2.2}$$

From the detail of Fig. 2.1, we have

$$\frac{dy}{ds} = \cos\alpha = n_x \Rightarrow dy = n_x\, ds \tag{2.3a}$$

$$-\frac{dx}{ds} = \sin\alpha = n_y \Rightarrow dx = -n_y\, ds \tag{2.3b}$$

where n_x and n_y are the components of the unit vector $\mathbf{n}$, which is normal to the boundary Γ. The negative sign in Eq. (2.3b) is due to the fact that the dx and the $\sin\alpha$ have opposite signs when the angle α is measured in the counter-clockwise sense with respect to the positive x-direction (see detail in Fig. 2.1).

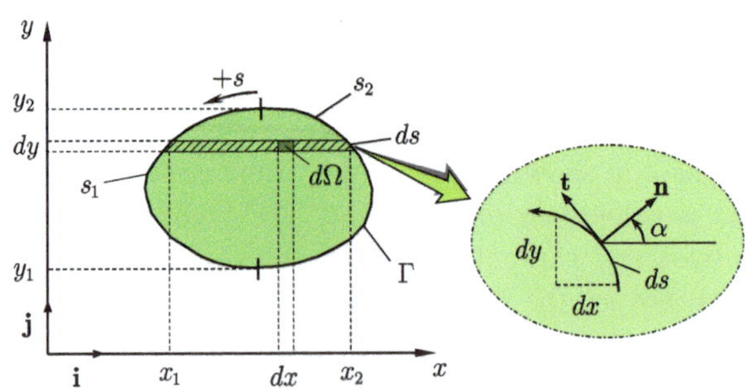

FIGURE 2.1 Integration over a plane domain Ω bounded by a curve Γ.

Consequently, Eq. (2.1) becomes

$$\int_{y_1}^{y_2} \{f(x_2, y) - f(x_1, y)\}\, dy = \int_{s_2} f(x_2, y)\; n_x\; ds - \int_{s_1} f(x_1, y)\; n_x\; ds \tag{2.4}$$

In the previous expression the integration on s_1 is performed in the negative direction (clockwise) when y varies from y_1 to y_2. Using uniform direction for the integration over s, both terms in Eq. (2.4) can be combined in a single expression

$$\int_\Omega \frac{\partial f}{\partial x}\; d\Omega = \int_\Gamma f\; n_x\; ds \tag{2.5}$$

Interchanging x with y in Eq. (2.5), we obtain

$$\int_\Omega \frac{\partial f}{\partial y}\; d\Omega = \int_\Gamma f\; n_y\; ds \tag{2.6}$$

If g is another function of x and y, then Eqs. (2.5) and (2.6) result in

$$\begin{gathered}\int_\Omega \frac{\partial (fg)}{\partial x}\, d\Omega = \int_\Gamma fg\; n_x\; ds = \int_\Omega g\frac{\partial f}{\partial x}\, d\Omega + \int_\Omega f\frac{\partial g}{\partial x}\, d\Omega \Rightarrow \\ \int_\Omega g\frac{\partial f}{\partial x}\, d\Omega = -\int_\Omega f\frac{\partial g}{\partial x}\, d\Omega + \int_\Gamma fg\; n_x\; ds\end{gathered} \tag{2.7}$$

$$\begin{gathered}\int_\Omega \frac{\partial (fg)}{\partial y}\, d\Omega = \int_\Gamma fg\; n_y\; ds = \int_\Omega g\frac{\partial f}{\partial y}\, d\Omega + \int_\Omega f\frac{\partial g}{\partial y}\, d\Omega \Rightarrow \\ \int_\Omega g\frac{\partial f}{\partial y}\, d\Omega = -\int_\Omega f\frac{\partial g}{\partial y}\, d\Omega + \int_\Gamma fg\; n_y\; ds\end{gathered} \tag{2.8}$$

Equations (2.7) and (2.8) state the integration by parts in two dimensions and are known as the *Gauss-Green theorem.*

2.3 THE DIVERGENCE THEOREM OF GAUSS

The divergence theorem results readily as an application of the Gauss-Green theorem. Consider the vector field $\mathbf{u} = u\,\mathbf{i} + v\,\mathbf{j}$, where $\mathbf{i}$, $\mathbf{j}$ denote the unit vectors along the x and y axes and $u = u(x, y)$, $v = v(x, y)$ its components. Applying Eq. (2.5) for $f = u$ and Eq. (2.6) for $f = v$ and adding, yields

$$\int_\Omega \left(\frac{\partial u}{\partial x} + \frac{\partial v}{\partial y}\right) d\Omega = \int_\Gamma (u\; n_x + v\; n_y)\; ds \tag{2.9}$$

If the coordinates x and y are represented by x_1 and x_2, respectively, then the components of the vector field $\mathbf{u}$ are denoted by u_i $(i=1,2)$ and those of the normal vector $\mathbf{n}$ by n_i. Therefore, Eq. (2.9) can be written as

$$\int_\Omega \left(\frac{\partial u_1}{\partial x_1} + \frac{\partial u_2}{\partial x_2}\right) d\Omega = \int_\Gamma (u_1\, n_1 + u_2\, n_2)\; ds \tag{2.10}$$

or using the summation convention

$$\int_\Omega \frac{\partial u_i}{\partial x_i}\, d\Omega = \int_\Gamma u_i\, n_i\; ds \quad (i=1,2) \tag{2.11}$$

Equations (2.9)–(2.11) can also be written using vector notation as

$$\int_\Omega \nabla \cdot \mathbf{u}\; d\Omega = \int_\Gamma \mathbf{u}\cdot\mathbf{n}\; ds \tag{2.12}$$

in which the symbolic vector ∇ is defined as

$$\nabla \equiv \mathbf{i}\frac{\partial}{\partial x} + \mathbf{j}\frac{\partial}{\partial y} = \mathbf{i}_1 \frac{\partial}{\partial x_1} + \mathbf{i}_2 \frac{\partial}{\partial x_2} \tag{2.13}$$

and represents the differential operator that produces the gradient of a scalar field.

The quantity $\nabla\cdot\mathbf{u}$, that is, the dot product of the vectors ∇ and $\mathbf{u}$, is referred to as the *divergence* of a vector field $\mathbf{u}$ at a point inside the domain Ω, whereas the quantity $\mathbf{u}\cdot\mathbf{n}$ is referred to as the *flux* of the vector field at a point on the boundary Γ. The latter dot product expresses the projection of $\mathbf{u}$ in the direction of $\mathbf{n}$. Eq. (2.12) relates the total divergence to the total flux of a vector field and it is known as the *divergence theorem of Gauss*. It is one of the most important theorems of integral calculus.

2.4 GREEN'S SECOND IDENTITY

Consider the functions $u=u(x,y)$ and $v=v(x,y)$, which are twice continuously differentiable in Ω and once on Γ. Applying Eq. (2.7) for $g=v$, $f=\frac{\partial u}{\partial x}$ and Eq. (2.8) for $g=v$, $f=\frac{\partial u}{\partial y}$ and adding the resulting equations, we arrive at the following

$$\begin{aligned}\int_\Omega v\left(\frac{\partial^2 u}{\partial x^2} + \frac{\partial^2 u}{\partial y^2}\right) d\Omega = &-\int_\Omega \left(\frac{\partial u}{\partial x}\frac{\partial v}{\partial x} + \frac{\partial u}{\partial y}\frac{\partial v}{\partial y}\right) d\Omega \\ &+ \int_\Gamma v\left(\frac{\partial u}{\partial x} n_x + \frac{\partial u}{\partial y} n_y\right) ds\end{aligned} \tag{2.14}$$

Similarly, applying Eq. (2.7) for $g = u$, $f = \frac{\partial v}{\partial x}$ and Eq. (2.8) for $g = u$, $f = \frac{\partial v}{\partial y}$ and adding the resulting equations, we obtain

$$\int_\Omega u\left(\frac{\partial^2 v}{\partial x^2} + \frac{\partial^2 v}{\partial y^2}\right) d\Omega = -\int_\Omega \left(\frac{\partial u}{\partial x}\frac{\partial v}{\partial x} + \frac{\partial u}{\partial y}\frac{\partial v}{\partial y}\right) d\Omega + \int_\Gamma u\left(\frac{\partial v}{\partial x} n_x + \frac{\partial v}{\partial y} n_y\right) ds \tag{2.15}$$

Subtracting Eq. (2.15) from Eq. (2.14) yields

$$\int_\Omega (v\nabla^2 u - u\nabla^2 v)\, d\Omega = \int_\Gamma \left(v\frac{\partial u}{\partial n} - u\frac{\partial v}{\partial n}\right) ds \tag{2.16}$$

where ∇^2 is known as the *Laplace operator* or *harmonic operator* and it is defined as

$$\nabla^2 \equiv \nabla\cdot\nabla = \left(\mathbf{i}\frac{\partial}{\partial x} + \mathbf{j}\frac{\partial}{\partial y}\right)\cdot\left(\mathbf{i}\frac{\partial}{\partial x} + \mathbf{j}\frac{\partial}{\partial y}\right) = \frac{\partial^2}{\partial x^2} + \frac{\partial^2}{\partial y^2} \tag{2.17}$$

while

$$\frac{\partial}{\partial n} \equiv \mathbf{n}\cdot\nabla = (n_x\,\mathbf{i} + n_y\,\mathbf{j})\cdot\left(\mathbf{i}\frac{\partial}{\partial x} + \mathbf{j}\frac{\partial}{\partial y}\right) = n_x\frac{\partial}{\partial x} + n_y\frac{\partial}{\partial y} \tag{2.18}$$

is the operator that produces the derivative of a scalar function in the direction of $\mathbf{n}$. Eq. (2.16) is known as *Green's second identity* for the harmonic operator or *Green's reciprocal identity*.

2.5 THE ADJOINT OPERATOR

Consider the complete second order differential equation with variable coefficients

$$L(u) = A\frac{\partial^2 u}{\partial x^2} + 2B\frac{\partial^2 u}{\partial x\,\partial y} + C\frac{\partial^2 u}{\partial y^2} + D\frac{\partial u}{\partial x} + E\frac{\partial u}{\partial y} + Fu = 0 \tag{2.19}$$

where $A, B, ..., F$ are given functions of x and y in Ω. Multiplying Eq. (2.19) by a function $v = v(x, y)$ and integrating over the domain, we have

$$\int_\Omega v\, L(u)\, d\Omega = 0 \tag{2.20}$$

Let us assume that v is twice continuously differentiable in Ω and once on Γ. Subsequently, integrating Eq. (2.20) by parts repeatedly until all derivatives of u are eliminated and incorporating Eqs. (2.7) and (2.8), we arrive at the reciprocal identity

$$\int_{\Omega}\{v\, L(u) - u\, L^*(v)\}\; d\Omega = \int_{\Gamma}(X\, n_x + Y\, n_y)\, ds \tag{2.21}$$

where

$$L^*(v) = \frac{\partial^2(Av)}{\partial x^2} + 2\frac{\partial^2(Bv)}{\partial x \partial y} + \frac{\partial^2(Cv)}{\partial y^2} - \frac{\partial(Dv)}{\partial x} - \frac{\partial(Ev)}{\partial y} + Fv \tag{2.22}$$

$$X = A\left(v\frac{\partial u}{\partial x} - u\frac{\partial v}{\partial x}\right) + B\left(v\frac{\partial u}{\partial y} - u\frac{\partial v}{\partial y}\right) + \left(D - \frac{\partial A}{\partial x} - \frac{\partial B}{\partial y}\right)uv \tag{2.23}$$

$$Y = B\left(v\frac{\partial u}{\partial x} - u\frac{\partial v}{\partial x}\right) + C\left(v\frac{\partial u}{\partial y} - u\frac{\partial v}{\partial y}\right) + \left(E - \frac{\partial B}{\partial x} - \frac{\partial C}{\partial y}\right)uv \tag{2.24}$$

The differential operator $L^*(\,)$ defined in Eq. (2.22) is referred to as the *adjoint* operator of $L(\,)$. Eq. (2.21) is the general form of the Green's second identity (2.16), which results readily as a special case of Eq. (2.21) for $A = C = 1$ and $B = D = E = 0$. The case $F \neq 0$ does not affect Eq. (2.16) as it is equivalent to adding and subtracting the term Fuv in the integral of the left hand side of this equation.

It can be shown that if the coefficients of operator (2.19) satisfy the conditions

$$\frac{\partial A}{\partial x} + \frac{\partial B}{\partial y} = D \tag{2.25a}$$

$$\frac{\partial B}{\partial x} + \frac{\partial C}{\partial y} = E \tag{2.25b}$$

Equation (2.22) becomes

$$L^*(v) = A\frac{\partial^2 v}{\partial x^2} + 2B\frac{\partial^2 v}{\partial x \partial y} + C\frac{\partial^2 v}{\partial y^2} + Fv \tag{2.26}$$

namely, the operator $L^*(\;)$ is identical to $L(\;)$ and in this case $L(\;)$ is called *self-adjoint*. The relations (2.25a,b) are referred to as the conditions of *self-adjointness*.

The character of the solution of Eq. (2.19), as well as the type of the problem to be solved, depends on the quantity $\Delta = B^2 - AC$. We distinguish three types of equations:

a. Elliptic type, for $\Delta < 0$
b. Parabolic type, for $\Delta = 0$
c. Hyperbolic type, for $\Delta > 0$

2.6 THE DIRAC DELTA FUNCTION

In problems of solid mechanics, we often come across *concentrated loads*, that is, actions that are applied to a very small region, theoretically at a point of space or at an instant of time.

For example, consider the plane elastic body Δ_1 having constant thickness h and occupying the half-plane $-\infty < x < +\infty$, $y \geq 0$ (Fig. 2.2). A circular disc Δ_2 of the same thickness h and radius R is in contact with the semi-infinite body at point $(x, y) = (0, 0)$ of its free boundary. The circular disc is loaded by a vertical concentrated unit load F as shown in Fig. 2.2. Because of the deformation of the elastic bodies the contact does not take place at a point, but it is extended over a small region. The function $f(x)$ represents the distribution per unit length of the force applied on the boundary of the body Δ_1. This function, which most probably has the shape shown in Fig. 2.3, is not *a priori* known. However, we know that it is sufficiently *concentrated* and

$$\int_{-\infty}^{+\infty} f(x)\, dx = 1 \tag{2.27}$$

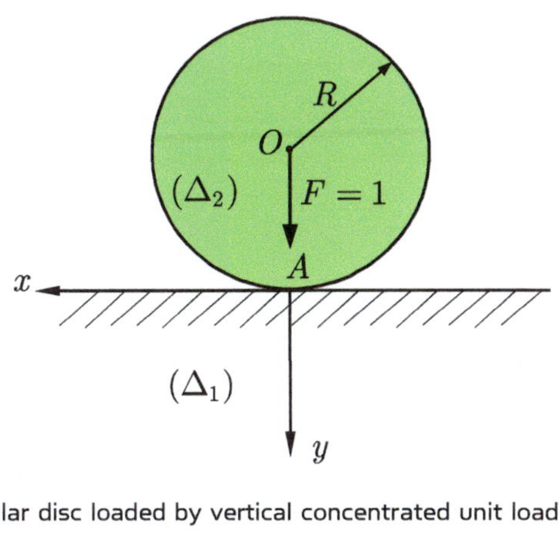

FIGURE 2.2 Circular disc loaded by vertical concentrated unit load *F*.

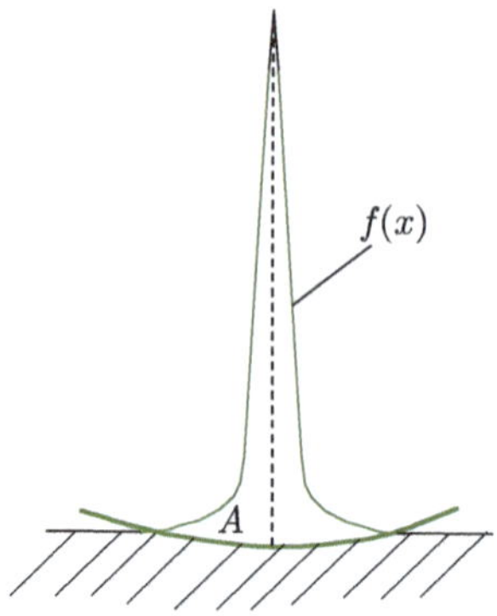

FIGURE 2.3 Force distribution on the boundary of the body.

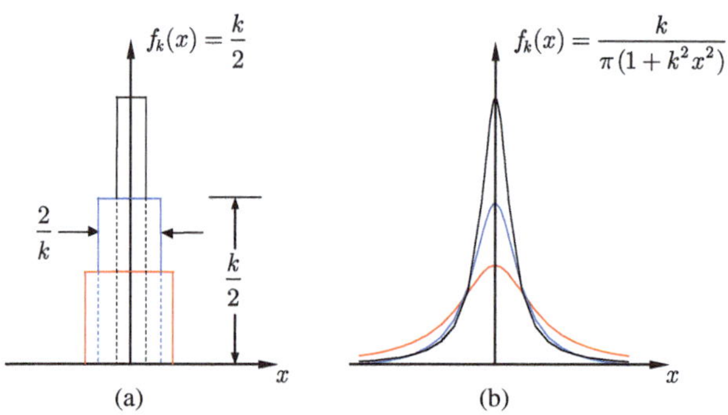

FIGURE 2.4 Functions f_k sufficiently *concentrated* for large values of k.

which expresses that the total force applied on the body Δ_1 is equal to unity. If we skip the problem of determining analytically the function f, we may assume *a priori* a prescribed shape for this function as

$$f_k(x) = \begin{cases} k/2, & |x| < \dfrac{1}{k} \\ 0, & |x| \geq \dfrac{1}{k} \end{cases} \tag{2.28a}$$

or

$$f_k(x) = \frac{k}{\pi(1 + k^2 x^2)} \tag{2.28b}$$

where k is a positive number.

Figures 2.4a and 2.4b show that both functions f_k defined in Eqs. (2.28a) and (2.28b), respectively, become sufficiently *concentrated* for large values of k. Moreover, they satisfy Eq. (2.27), which means that they are statically equivalent to f.

The distribution of $f(x)$ becomes more "*concentrated*" as the bodies become less deformable. This is expressed by increasing the values of k in Eqs. (2.28a) and (2.28b). The limiting case where the bodies are rigid is expressed by $k \to \infty$. This produces a fictitious distribution of the unit force per unit length, which is denoted by $\delta(x)$ and is defined as

$$\delta(x) = \lim_{k \to \infty} f_k(x) \tag{2.29}$$

where $f_k(x)$ is the function defined in Eq. (2.28a) or (2.28b). The function $\delta(x)$ is known as the *delta function* or *Dirac delta function*.

In mathematics, the delta function is treated in the theory of generalized functions. We give below the definition of the delta function as a generalized function of one and two variables, and we mention some of its properties used in developing the BEM.

The one-dimensional delta function is defined by the relation

$$\int_{-\infty}^{+\infty} \delta(x)\, h(x) dx = h(0) \tag{2.30}$$

for a point source applied at the position $x = 0$ or by the relation

$$\int_{-\infty}^{+\infty} \delta(x - x_o)\, h(x) dx = h(x_o) \tag{2.31}$$

for a point source applied at $x = x_o$. The function $h(x)$ is continuous in a finite interval containing the source point $x = 0$ or $x = x_o$. The one-dimensional delta function can also be described by the relations

$$\delta(x) = \begin{cases} 0, & x \neq 0 \\ \infty, & x = 0 \end{cases} \tag{2.32a}$$

$$\int_{-\infty}^{+\infty} \delta(x) dx = \int_{-\varepsilon}^{\varepsilon} \delta(x) dx = 1 \tag{2.32b}$$

where ε is a positive number. According to this definition, the function $\delta(x)$ has zero value everywhere except at point $x = 0$, where it becomes infinite, and satisfies Eq. (2.32b).

Equation (2.30) may be obtained from Eq. (2.32b) by applying the mean value theorem of integral calculus. Referring to Fig. 2.4a and choosing $\varepsilon = 1/k$, we obtain

$$\begin{aligned} \int_{-\infty}^{\infty} h(x)\, \delta(x)\, dx &= \lim_{\varepsilon \to 0} \int_{-\varepsilon}^{\varepsilon} h(x)\, f_k(x) dx = \lim_{\varepsilon \to 0} \left[h(x^*) \frac{1}{2\varepsilon} \right] 2\varepsilon \\ &= \lim_{\varepsilon \to 0} [h(x^*)] = h(0) \end{aligned}$$

The Dirac delta function $\delta(Q-Q_o)$ in two dimensions is defined as

$$\int_{\Omega}\delta(Q-Q_o)\,h(Q)\,d\Omega_Q=h(Q_o),\qquad Q(x,y),\quad Q_o(x_o,y_o)\in\Omega \tag{2.33}$$

for an arbitrary function $h(Q)$, which is continuous in the domain Ω containing the point $Q_o(x_o,y_o)$. The two-dimensional delta function may also be described by

$$\delta(Q-Q_o)=\begin{cases}0, & Q\neq Q_o\\ \infty, & Q=Q_o\end{cases} \tag{2.34a}$$

$$\int_{\Omega}\delta(Q-Q_o)\,d\Omega_Q=\int_{\Omega*}\delta(Q-Q_o)\,d\Omega_Q=1,\;Q_o(x_o,y_o)\in\Omega^*\subseteq\Omega \tag{2.34b}$$

In accordance to the one-dimensional delta function $\delta(x-x_o)$, the two-dimensional $\delta(Q-Q_o)$ may be defined as the limit of a set of functions. More specifically,

$$\delta(Q-Q_o)=\lim_{k\to\infty}f_k(r),\qquad Q(x,y),\,Q_o(x_o,y_o)\in\Omega \tag{2.35}$$

where

$$r=\sqrt{(x-x_o)^2+(y-x_o)^2}$$

For example the limit of the function

$$f_k(r)=\begin{cases}k^2/\pi, & r<\dfrac{1}{k}\\[2ex] 0, & r\geq\dfrac{1}{k}\end{cases} \tag{2.36a}$$

or

$$f_k(r)=\frac{ke^{-kr^2}}{\pi} \tag{2.36b}$$

for $k\to\infty$ is a two-dimensional delta function. We can also write $\delta(Q-Q_o)$ as the product of two one-dimensional delta functions as follows

$$\delta(Q-Q_o)=\delta(x-x_o)\,\delta(y-y_o) \tag{2.37}$$

Consider now the transformation

$$\left.\begin{aligned} x &= u(\xi, \eta) \\ y &= v(\xi, \eta) \end{aligned}\right\} \tag{2.38}$$

where u and v are assumed to be single-valued continuous differentiable functions of their arguments. The transformation of the function $\delta(Q - Q_o)$ from the Cartesian coordinates x and y to the curvilinear coordinates ξ and η can be performed in the following way.

It is supposed that under this transformation the point $x = x_o$, $y = y_o$ is mapped onto the point $\xi = \xi_o$, $\eta = \eta_o$. By changing coordinates, the integral in Eq. (2.33)

$$\int_\Omega \delta(x - x_o)\delta(y - y_o)h(x, y)\, dx\, dy = h(x_o, y_o)$$

becomes

$$\int_\Omega \delta[u(\xi, \eta) - x_o]\delta[v(\xi, \eta) - y_o]h(u, v)|J|\, d\xi\, d\eta = h(x_o, y_o) \tag{2.39}$$

where

$$J = \begin{vmatrix} \dfrac{\partial u}{\partial \xi} & \dfrac{\partial v}{\partial \xi} \\ \dfrac{\partial u}{\partial \eta} & \dfrac{\partial v}{\partial \eta} \end{vmatrix}$$

is the Jacobian of the transformation.

Equation (2.39) states that the symbolic function

$$\delta[u(\xi, \eta) - x_o]\ \delta[v(\xi, \eta) - y_o]\ |J|$$

assigns the value of the function $h(x, y)$ at the point where $u = x_o$, $v = y_o$, that is at the point where $\xi = \xi_o, \eta = \eta_o$. Consequently, we may write

$$\delta[u(\xi, \eta) - x_o]\ \delta[v(\xi, \eta) - y_o]\, |J| = \delta(\xi - \xi_o)\ \delta(\eta - \eta_o)$$

which may also be set in the form

$$\delta(x - x_o)\delta(y - y_o) = \frac{\delta(\xi - \xi_o)\delta(\eta - \eta_o)}{|J|} \tag{2.40}$$

provided that $|J| \neq 0$, a condition which implies that the transformation (2.38) is nonsingular (invertible).

We close this section by stating a property of the derivatives of the delta function.

i. For the mth order derivative of the one-dimensional delta function the following equation is valid

$$\int_a^b h(x)\,\frac{d^m\delta(x-x_o)}{dx^m}\,dx = (-1)^m\,\frac{d^m h(x_o)}{dx^m},\quad (a<x_o<b) \tag{2.41}$$

ii. For the $(m+n)^{th}$ order derivative of the two-dimensional delta function, the following equation is valid

$$\int_\Omega h(Q)\,\frac{\partial^{m+n}\,\delta(Q-Q_o)}{\partial x^m\,\partial y^n}\,d\Omega_Q = (-1)^{m+n}\,\frac{\partial^{m+n}\,h(Q_o)}{\partial x^m\,\partial y^n} \tag{2.42}$$

where $Q_o(x_o, y_o),\ \ Q(x, y) \in \Omega$.

2.7 CALCULUS OF VARIATIONS. EULER-LAGRANGE EQUATION

In mechanics, one of the applications of the calculus of variations is the derivation of the differential equation and the associated boundary conditions governing the response of complicated structural systems, especially when other approaches give doubtful results. In this context, we present the basics of the calculus of variations. The calculus of variations is a field of mathematical analysis that deals with maximizing or minimizing functionals. Functionals are mappings from a set of functions to the real numbers. They are often expressed as definite integrals, whose integrand depends on an unknown function or functions and/or certain of their derivatives. The basic problem of the calculus of variations is to determine the extremal functions, that is, the functions that make the functional attain a stationary value (minimum or maximum), in other words those functions where the rate of change of the functional is zero. The elementary part of the theory is concerned with the necessary condition which the required function must satisfy. This condition is generally in the form of a differential equation, the Euler-Lagrange equation, with boundary conditions, whose solution gives the required function.

A simple example of such a problem is to find the curve $u(x)$, that passes through the points $A(x_1, u_1)$ and $B(x_2, u_2)$ (Fig. 2.5) and produces a surface of revolution about the x axis with a minimum area. Thus, the function $u(x)$ is determined in such a way that the integral representing the area of the surface

$$I = 2\pi \int_{x_1}^{x_2} u(1+u'^2)^{1/2}\,dx \tag{2.43}$$

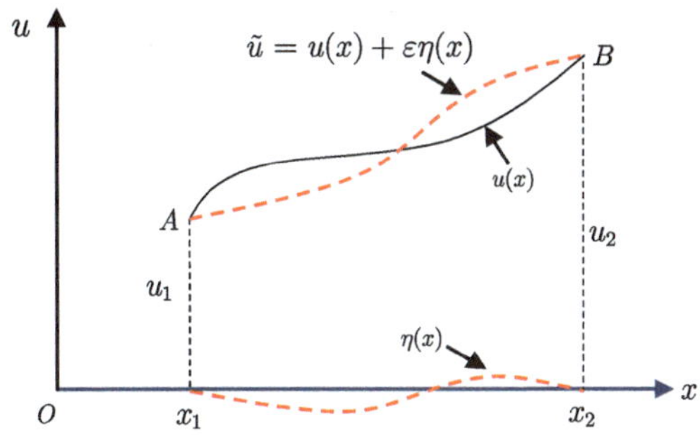

FIGURE 2.5 Function $u(x)$ and a variant $\tilde{u}(x)$.

is a minimum, and also that $u(x_1) = u_1$ and $u(x_2) = u_2$. Here it is assumed that u_1 and u_2 are nonnegative. We consider all continuous functions with continuous derivatives that satisfy the geometric or essential boundary conditions, that is, they pass through the points A and B. These functions constitute the set of admissible functions, where we will look for the minimizing function $u(x)$ of the integral (2.43). We suppose that $u(x)$ is the actual minimizing function, and we chose any continuously differentiable function $\eta(x)$ with $\eta(x_1) = \eta(x_2) = 0$, (Fig. 2.5). Then for any constant ε the function $\tilde{u}(x) = u(x) + \varepsilon\eta(x)$ is an admissible function. The change $\varepsilon\eta(x)$ is called the variation of $u(x)$ and is conventionally denoted by δu,

$$\delta u = \varepsilon\eta(x) \tag{2.44}$$

2.7.1 The Euler-Lagrange equation

We consider the integral

$$I(u) = \int_{x_1}^{x_2} F(x, u, u')dx \tag{2.45}$$

where we suppose that the integrand $F = F(x, u, u')$ has continuous second order derivatives with respect to its three arguments. The integral

$$I(\varepsilon) = \int_{x_1}^{x_2} \tilde{F}(x, u + \varepsilon\eta, u' + \varepsilon\eta')dx \tag{2.46}$$

obtained by replacing $u(x)$ with $u(x) + \varepsilon\eta(x)$ is a function of ε, once the functions $u(x)$ and $\eta(x)$ are specified, and takes on a minimum value when $\varepsilon = 0$. This is possible only if

$$\left.\frac{dI(\varepsilon)}{d\varepsilon}\right|_{\varepsilon=0} = 0 \tag{2.47}$$

Differentiating Eq. (2.46) with respect to ε and noting that the derivative passes under the integral sign we obtain

$$\frac{dI(\varepsilon)}{d\varepsilon} = \int_{x_1}^{x_2} \left(\frac{\partial \tilde{F}}{\partial u}\eta + \frac{\partial \tilde{F}}{\partial u'}\eta' \right) dx \tag{2.48}$$

which after integration by parts of the second term becomes

$$\frac{dI(\varepsilon)}{d\varepsilon} = \int_{x_1}^{x_2} \left(\frac{\partial \tilde{F}}{\partial u} - \frac{d}{dx}\frac{\partial \tilde{F}}{\partial u'} \right) \eta \, dx + \left[\frac{\partial \tilde{F}}{\partial u'}\eta(x) \right]_{x_1}^{x_2} \tag{2.49}$$

Noting that $\eta(x_1) = \eta(x_2) = 0$ and $\tilde{F} \to F$ when $\varepsilon \to 0$, the necessary condition (2.47) becomes

$$\int_{x_1}^{x_2} \left(\frac{\partial F}{\partial u} - \frac{d}{dx}\frac{\partial F}{\partial u'} \right) \eta \, dx = 0 \tag{2.50}$$

Since $\eta(x)$ is arbitrary we conclude that its coefficient in (2.50) must vanish in the interval (x_1, x_2)

$$\frac{\partial F}{\partial u} - \frac{d}{dx}\frac{\partial F}{\partial u'} = 0 \tag{2.51}$$

Thus, if $u(x)$ minimizes (or maximizes) the integral (2.45), it must satisfy the condition (2.51), which is called the Euler-Lagrange equation of the functional (2.45). When F is specified, the condition (2.51) yields the differential equation, whose solution in the interval (x_1, x_2) gives the extremal function.

The change in F corresponding to δu is

$$\Delta F = F(x, u + \delta u, u' + \delta u') - F(x, u, u') \tag{2.52}$$

Expanding the right-hand side in Taylor series yields

$$\Delta F = \frac{\partial F}{\partial u}\delta u + \frac{\partial F}{\partial u'}\delta u' + (\text{higher order terms of } \delta u \text{ and } \delta u') \tag{2.53}$$

In analogy with the differential the first two terms in the right hand of (2.53) are defined as the variation of F,

$$\delta F = \frac{\partial F}{\partial u}\delta u + \frac{\partial F}{\partial u'}\delta u' \tag{2.54}$$

The operator δ acts as the differential operator d, which for a function $F(x, y)$ of two independent variables reads

$$dF = \frac{\partial F}{\partial x} dx + \frac{\partial F}{\partial y} dy \tag{2.55}$$

There is, however, a significant difference between the two operators: the differential of a function represents a first-order approximation to the change in the function along a particular curve, while the variation of the functional is the first-order approximation to the change in the functional from curve to curve. Note that the complete analogy to the differential would be

$$\delta F = \frac{\partial F}{\partial x}\delta x + \frac{\partial F}{\partial u}\delta u + \frac{\partial F}{\partial u'}\delta u' \tag{2.56}$$

but as x is not varied, it is $\delta x = 0$ and the analogy is complete.

It can be easily verified that the laws for the variation of sums, product, ratios, powers, and so forth are completely analogous to differentiation, for example,

$$\delta(FG) = G\delta F + F\delta G \tag{2.57a}$$

$$\delta F^n = n\delta F^{n-1} \tag{2.57b}$$

$$\delta\left(\frac{F}{G}\right) = \frac{G\delta F - F\delta G}{G^2} \tag{2.57c}$$

Moreover, we have

$$\frac{d}{dx}(\delta u) = \varepsilon\frac{d\eta}{dx} = \delta\frac{du}{dx} \tag{2.58}$$

which shows that, if x is an independent variable, the operators d/dx and δ are commutative.

Using this notation, the variation of the integral (2.45) is written as

$$\begin{aligned}\delta I &= \int_{x_1}^{x_2} \delta F(x, u, u')dx \\ &= \int_{x_1}^{x_2} \left(\frac{\partial F}{\partial u}\delta u + \frac{\partial F}{\partial u'}\delta u'\right) dx\end{aligned} \tag{2.59}$$

From Eq. (2.58) we have $\delta u' = (\delta u)'$. Then integrating by parts the second term in the integrand yields

$$\delta I = \int_{x_1}^{x_2} \left(\frac{\partial F}{\partial u} - \frac{d}{dx}\frac{\partial F}{\partial u'}\right)\delta u \; dx + \left[\frac{\partial F}{\partial u'}\delta u(x)\right]_{x_1}^{x_2} \tag{2.60}$$

or taking into account that $\delta u(x_1) = \delta u(x_2) = 0$ we have

$$\delta I = \int_{x_1}^{x_2} \left(\frac{\partial F}{\partial u} - \frac{d}{dx}\frac{\partial F}{\partial u'} \right) \delta u \, dx \tag{2.61}$$

By virtue of (2.51) we conclude that $\delta I = 0$. Hence the necessary condition that a functional attains a minimum (or maximum) is the vanishing of its variation.

2.7.2 Natural boundary conditions

When the value of the unknown function $u(x)$ is not prescribed at one or both end points $x = x_1, x_2$, the variation $\delta u(x)$ need not vanish there. However, the right-hand side of Eq. (2.60) must vanish when $u(x)$ is the minimizing (or maximizing) function for all admissible variations $\delta u(x)$. Hence the second term in Eq. (2.60) must itself vanish

$$\left[\frac{\partial F}{\partial u'} \delta u(x) \right]_{x_1}^{x_2} = \left[\frac{\partial F}{\partial u'} \delta u(x) \right]_{x=x_2} - \left[\frac{\partial F}{\partial u'} \delta u(x) \right]_{x=x_1} = 0 \tag{2.62}$$

If $u(x)$ is not prescribed at either end point it is $\delta u(x_1) \neq 0$ or $\delta u(x_2) \neq 0$ and consequently Eq. (2.62) holds if

$$\left[\frac{\partial F}{\partial u'} \right]_{x=x_1} = 0 \quad \text{or} \quad \left[\frac{\partial F}{\partial u'} \right]_{x=x_2} = 0 \tag{2.63a, b}$$

The requirements that Eq. (2.63a) hold when $u(x_1)$ is not given, and that (2.63b) hold when $u(x_2)$ is not given, are called the natural boundary conditions.

On the basis of the above analysis, the function $u(x)$ that minimizes (or maximizes) the functional (2.45) results from the solution of the following boundary value problem

$$\frac{\partial F}{\partial u} - \frac{d}{dx}\frac{\partial F}{\partial u'} = 0 \quad \text{in} \quad x_1 \leq x \leq x_2 \tag{2.64}$$

$$u(x_1) = u_1 \quad \text{or} \quad \left[\frac{\partial F}{\partial u'} \right]_{x=x_1} = 0 \tag{2.65a}$$

and

$$u(x_2) = u_2 \quad \text{or} \quad \left[\frac{\partial F}{\partial u'} \right]_{x=x_2} = 0 \tag{2.65b}$$

2.7.3 Functional depending on a function of two variables

We consider the functional

$$I = \int_\Omega F(x, y, u, u_{,x}, u_{,y})\, dxdy \tag{2.66}$$

where $u = u(x, y)$ is a continuously differentiable function of x and y to be determined so that the functional attains a stationary value; Ω is a two-dimensional domain in the xy-plane with boundary Γ (Fig. 2.6). The procedure to derive the necessary conditions is here more complicated. However, it becomes simple if the required condition is obtained from the requirement that the variation $\delta I = 0$. This yields

$$\begin{aligned}\delta I &= \int_\Omega \delta F(x, y, u, u_{,x}, u_{,y})\, dxdy \\ &= \int_\Omega \left(\frac{\partial F}{\partial u}\delta u + \frac{\partial F}{\partial u_{,x}}\delta u_{,x} + \frac{\partial F}{\partial u_{,y}}\delta u_{,y} \right) dxdy\end{aligned} \tag{2.67}$$

We next get rid of the $\delta u_{,x}$ and $\delta u_{,y}$ from the two terms in the integrand. This can be achieved using integration by parts. Thus applying the Gauss-Green theorem, Eqs. (2.7) and (2.8) we obtain

$$\begin{aligned}\int_\Omega \frac{\partial F}{\partial u_{,x}}\delta u_{,x}\, dxdy &= \int_\Omega \frac{\partial F}{\partial u_{,x}}(\delta u)_{,x}\, dxdy \\ &= -\int_\Omega \frac{\partial}{\partial x}\frac{\partial F}{\partial u_{,x}}\delta u dxdy + \int_\Gamma \frac{\partial F}{\partial u_{,x}} n_x \delta u ds\end{aligned} \tag{2.68a}$$

$$\begin{aligned}\int_\Omega \frac{\partial F}{\partial u_{,y}}\delta u_{,y}\, dxdy &= \int_\Omega \frac{\partial F}{\partial u_{,y}}(\delta u)_{,y}\, dxdy \\ &= -\int_\Omega \frac{\partial}{\partial y}\frac{\partial F}{\partial u_{,y}}\delta u dxdy + \int_\Gamma \frac{\partial F}{\partial u_{,y}} n_y \delta u ds\end{aligned} \tag{2.68b}$$

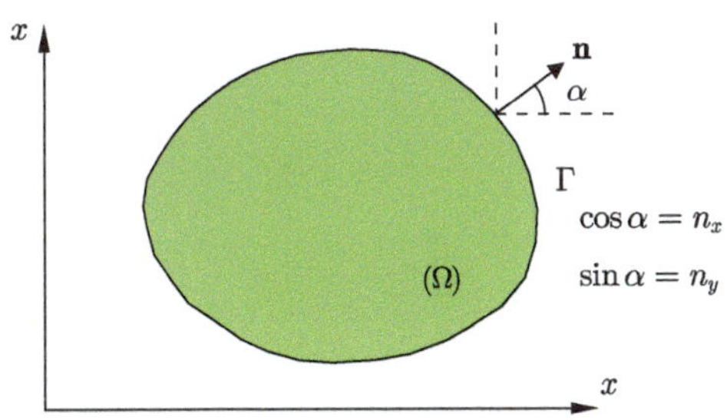

FIGURE 2.6 Domain Ω bounded by a curve Γ.

Introducing Eqs. (2.68a,b) in Eq. (2.67) gives

$$\begin{aligned}\delta I = &\int_{\Omega}\left(\frac{\partial F}{\partial u} - \frac{\partial}{\partial x}\frac{\partial F}{\partial u_{,x}} - \frac{\partial}{\partial y}\frac{\partial F}{\partial u_{,y}}\right)\delta u dx dy \\ &+ \int_{\Gamma}\left(\frac{\partial F}{\partial u_{,x}}n_x + \frac{\partial F}{\partial u_{,y}}n_y\right)\delta u ds\end{aligned} \tag{2.69}$$

If the function $u(x, y)$ is prescribed on Γ, then $\delta u = 0$ there and the boundary integral in Eq. (2.69) vanishes. Then, since δu is arbitrary in Ω its coefficient in the integrand must vanish. This yields the condition

$$\frac{\partial F}{\partial u} - \frac{\partial}{\partial x}\frac{\partial F}{\partial u_{,x}} - \frac{\partial}{\partial y}\frac{\partial F}{\partial u_{,y}} = 0 \tag{2.70}$$

which is the Euler-Lagrange equation of the functional.

When $u(x, y)$ is not prescribed on the boundary, then $\delta u \neq 0$ on Γ and the boundary integral vanishes only if the natural boundary condition is satisfied

$$\frac{\partial F}{\partial u_{,x}}n_x + \frac{\partial F}{\partial u_{,y}}n_y = 0 \ \text{ on } \Gamma \tag{2.71}$$

Thus the boundary condition on a part $\Gamma_1 \subseteq \Gamma$ reads

$$u = \overline{u} \quad \text{or} \quad \frac{\partial F}{\partial u_{,x}}n_x + \frac{\partial F}{\partial u_{,y}}n_y = 0 \tag{2.72}$$

The overbar designates a prescribed quantity.

2.7.4 Examples

EXAMPLE 2.1

Derive the differential equation of the function that passes through the points $A(x_1, u_1)$ and $B(x_2, u_2)$ (Fig. 2.5), and produces a surface of revolution about the x axis with a minimum area. The required function must minimize the functional (2.43). Here it is

$$F = u(1 + u'^2)^{1/2} \tag{a}$$

Hence

$$\frac{\partial F}{\partial u} = \sqrt{1 + u'^2}, \quad \frac{\partial F}{\partial u'} = \frac{uu'}{\sqrt{1 + u'^2}} \tag{b, c}$$

Inserting (b,c) in Eq. (2.51) gives

$$\sqrt{1+u'^2}-\frac{d}{dx}\frac{uu'}{\sqrt{1+u'^2}}=0 \tag{d}$$

which after reduction gives the differential equation

$$uu''-u'^2-1=0 \qquad \text{in } x_1<x<x_2 \tag{e}$$

with the boundary conditions

$$u(x_1)=u_1 \quad u(x_2)=u_2 \tag{f}$$

EXAMPLE 2.2

Derive the differential equation of the surface that passes through a space curve and has a minimum area. This is the well-known minimal surface problem or soap bubble problem. In this case the required minimizing function $u(x, y)$ has prescribed values on the boundary and minimizes the functional

$$I=\int_{\Omega}(1+u,_x^2+u,_y^2)^{1/2}\,dxdy \tag{a}$$

Hence, the functional is of the form (2.66) and the differential equation will result from Eq. (2.70) with

$$F=1+u,_x^2+u,_y^2 \tag{b}$$

for which we have

$$\frac{\partial F}{\partial u}=0,\quad \frac{\partial F}{\partial u,_x}=\frac{u,_x}{\sqrt{1+u,_x^2+u,_y^2}},\quad \frac{\partial F}{\partial u,_y}=\frac{u,_y}{\sqrt{1+u,_x^2+u,_y^2}}, \tag{c, d, e}$$

Inserting (c,d,e) in Eq. (2.70) gives

$$\frac{\partial}{\partial x}\frac{u,_x}{\sqrt{1+u,_x^2+u,_y^2}}+\frac{\partial}{\partial y}\frac{u,_y}{\sqrt{1+u,_x^2+u,_y^2}}=0 \tag{f}$$

and after reduction we obtain the differential equation

$$(1+u,_y^2)u,_{xx}-u,_x u,_y u,_{xy}+(1+u,_x^2)u,_{yy}=0 \ \text{ in } \Omega \tag{g}$$

with the boundary condition

$$u=\overline{u} \qquad \text{on } \Gamma \tag{h}$$

2.8 REFERENCES

This chapter has been designed to give a brief elementary description of the basic mathematical tools that will be employed throughout this book in developing the BEM formulations. The reader may look for more details about the Gauss-Green theorem and the Gauss divergence theorem in most books on calculus as well as in many books of engineering mathematics or mathematical physics, such as Smirnow [1], Hildebrand [2], Kreyszig [3], Sommerfeld [4], Tyn Myint and Debnath [5]. For a detailed discussion about the Dirac delta function, we refer to the book of Greenberg [6] or to the more advanced books of Roach [7] and Duff and Naylor [8]. For more information on the calculus of variation and the variational methods in mechanics the reader is advised to look in the books of Hildebrand [9] and Washizu [10].

[1] Smirnow WI. Lehrgang der höheren Mathematik, Teil II, 6ste Auflage, VEB Deutschder Verlag der Wissenschaften, Berlin; 1964.
[2] Hildebrand FB. Advanced calculus for applications. Englewood Cliffs, New Jersey: Prentice Hall; 1962.
[3] Kreyszig E. Advanced engineering mathematics. 4th ed. New York: John Wiley & Sons; 1979.
[4] Sommerfeld A. Partial differential equations in physics. New York and London: Academic Press; 1967.
[5] Tyn Myint U, Debnath L. Linear partial differential equations for scientists and engineers. 4th ed. Boston: Birkhäuser; 2007.
[6] Greenberg M. Application of Green's functions in science and engineering. Englewood Cliff, New Jersey: Prentice Hall; 1971.
[7] Roach GF. Green's functions. London: Van Nostrand Reinhold Company; 1970.
[8] Duff GFD, Naylor D. Differential equations of applied mathematics. New York: John Wiley & Sons; 1966.
[9] Hildebrand FB. Methods of applied mathematics. New York: Dover Publications, Inc; 1965.
[10] Washizu K. Variational methods in elasticity and plasticity. Oxford: Pergamon Press; 1975.

PROBLEMS

2.1. Convert the domain integral

$$\int_{\Omega} f \, d\Omega$$

to boundary line integral on Γ, when the function f is

i. $f = x$ **ii.** $f = y$
iii. $f = xy$ **iv.** $f = x^2$
v. $f = y^2$ **vi.** $f = x^2 + y^2$
vii. $f = \cos x$

2.2. Derive the operator ∇^2 in polar coordinates (r, θ) and convert the following domain integral to a boundary line integral on Γ

$$\int_{\Omega} \ln r \, d\Omega, \quad r = \sqrt{x^2 + y^2}$$

2.3. Evaluate the integrals

i. $\int_a^b \delta(x - x_o) dx, \ a < x_o < b$

ii. $\int_a^b \delta(k\,x)\, f(x) dx,\ a < 0 < b$

iii. $\int_a^b \delta(-\,x) dx,\ a < 0 < b$

iv. $\int_a^b \delta^{(n)}(x)\, \phi(x) dx,\ a < 0 < b$

2.4. Show that

i. $\delta(-\mathrm{x}) = \delta(x)$

ii. $\delta(ax + by)\delta(cx + dy) = \frac{\delta(x)\delta(y)}{|ad - bc|}$

2.5. Transform the delta function $\delta(P - P_o)$ for the points $P(x, y)$ and $P_o(x_o, y_o)$ into polar coordinates r and θ.

2.6. Derive Eq. (2.21) along with Eqs. (2.22) through (2.24).

2.7. Show that $\int_\Omega \nabla^2 u \ d\Omega = \int_\Gamma \frac{\partial u}{\partial n} ds$.

2.8. Given $L(u) = \nabla^2 u + \mathbf{a}\cdot\nabla u + cu$, where $\mathbf{a} = a_x\mathbf{i} + a_y\mathbf{j}$ is an arbitrary vector, show that

i. $L^*(u) = \nabla^2 u - \nabla\cdot(\mathbf{a}u) + cu$,

ii. $vL(u) - uL^*(v) = \nabla\cdot(v\nabla u - u\nabla v + \mathbf{a}uv)$

iii. derive Green's identity for the operator $L(u)$.

2.9. If $A,_x + B,_y = D$ and $B,_x + C,_y = E$, show that the reciprocal identity (2.21) can be written as

$$\int_\Omega \{v\ L(u) - u\ L(v)\}\ d\Omega = \int_\Gamma (v\nabla u\cdot\mathbf{m} - u\nabla v\cdot\mathbf{m})ds$$

where $\mathbf{m} = (An_x + Bn_y)\mathbf{i} + (Bn_x + Cn_y)\mathbf{j}$ is the vector in the direction of the *conormal* to the boundary.

2.10. Show that

$$\nabla u\cdot\mathbf{m} = (\mathbf{m}\cdot\mathbf{n})\nabla u\cdot\mathbf{n} + (\mathbf{m}\cdot\mathbf{t})\nabla u\cdot\mathbf{t}$$

where $\mathbf{m} = (An_x + Bn_y)\mathbf{i} + (Bn_x + Cn_y)\mathbf{j}$ is the vector in the direction of the *conormal* to the boundary.

2.11. The total potential energy of a string supported at its ends and subjected to a uniform tension S and a transverse load $f(x)$ is given as

$$\Pi = \int_0^l \left(\frac{S}{2}u'^2 - fu\right) dx$$

Derive the differential equation for the transverse displacement $u(x)$ by applying the principle of minimum potential energy.

2.12. The total potential energy of a membrane supported on its boundary and subjected to a uniform tension S and a transverse load $f(x, y)$ is given as

$$\Pi = \int_0^l \left[\frac{S}{2}(u,_x^2 + u,_y^2) - fu\right] dx$$

Derive the deferential equation for the transverse displacement $u(x, y)$ by applying the principle of minimum potential energy.

Chapter | Three

The BEM for Potential Problems in Two Dimensions

CHAPTER OUTLINE

3.1 INTRODUCTION

In this chapter the boundary element method (BEM) is developed for the solution of engineering problems described by the potential equation

$$\nabla^2 u = f(x, y) \quad (x, y \in \Omega) \tag{3.1}$$

This is the governing equation of potential theory in two dimensions, which for $f = 0$ is known as the Laplace equation, whereas for $f \neq 0$ is known as the Poisson equation. Its solution $u = u(x, y)$ represents the potential produced at a point (x, y) in the domain Ω due to a source $f(x, y)$ distributed over Ω. The potential Eq. (3.1) describes the response of many physical systems. It appears in steady state flow problems, such as fluid flow, thermal

The Boundary Element Method for Engineers and Scientists.

flow, electricity flow, as well as in torsion of prismatic bars, bending of membranes, etc. According to the definitions given in Section 2.5, Eq. (3.1) is of elliptic type, since $\Delta < 0$. Its solution is sought in a closed plane domain Ω having a boundary Γ on which either the function u or its derivative $\partial u/\partial n$ in the direction normal to Γ is prescribed. That is, the solution must satisfy the *boundary conditions* of the problem on the boundary Γ. *The boundary value problems* for the potential equation can be classified as follows:

i. *Dirichlet* problem

$$\nabla^2 u = f \ \text{ in } \Omega \tag{3.2a}$$

$$u = \overline{u} \ \text{ on } \Gamma \tag{3.2b}$$

ii. *Neumann* problem

$$\nabla^2 u = f \ \text{ in } \Omega \tag{3.3a}$$

$$\frac{\partial u}{\partial n} = \overline{u}_n \ \text{ on } \Gamma \tag{3.3b}$$

iii. *Mixed* problem

$$\nabla^2 u = f \ \text{ in } \Omega \tag{3.4a}$$

$$u = \overline{u} \ \text{ on } \Gamma_1 \tag{3.4b}$$

$$\frac{\partial u}{\partial n} = \overline{u}_n \ \text{ on } \Gamma_2 \tag{3.4c}$$

where $\Gamma_1 \cup \Gamma_2 = \Gamma$ and $\Gamma_1 \cap \Gamma_2 = \{\varnothing\}$.

iv. *Robin* problem

$$\nabla^2 u = f \ \text{ in } \Omega \tag{3.5a}$$

$$u + k(s)\frac{\partial u}{\partial n} = 0 \ \text{ on } \Gamma \tag{3.5b}$$

The quantities denoted by $\overline{u}$, $\overline{u}_n$, and $k(s)$ are known functions specified on the boundary.

All four problems can be expressed through a single formulation as

$$\nabla^2 u = f \ \text{ in } \Omega \tag{3.6a}$$

$$\alpha u + \beta \frac{\partial u}{\partial n} = \gamma \ \text{ on } \Gamma \tag{3.6b}$$

where $\alpha = \alpha(s)$, $\beta = \beta(s)$, and $\gamma = \gamma(s)$ are known functions specified on the boundary Γ. Apparently, each of the foregoing four boundary value problems (Eqs. 3.2–3.5) may be derived from Eq. (3.6) by appropriately specifying the functions α, β, and γ.

Two BEMs have been developed for solving the previously stated four boundary value problems: the *direct BEM* and the *indirect BEM* [12]. In this book, we present only the direct BEM.

3.2 FUNDAMENTAL SOLUTION

Let us consider a unit point source placed at point $P(x, y)$ of the xy-plane. Its density at $Q(\xi, \eta)$ may be expressed mathematically by the delta function as

$$f(Q) = \delta(Q - P) \tag{3.7}$$

hence the potential $v = v(Q, P)$ produced at point Q satisfies the equation

$$\nabla^2 v = \delta(Q - P) \tag{3.8}$$

A singular particular solution of Eq. (3.8) is called the *fundamental solution* of the potential Eq. (3.1) and it can be determined as follows.

We write Eq. (3.8) in polar coordinates with origin at the source point P. Since this solution is axisymmetric with respect to the source, it is independent of the polar angle θ, and thus Eq. (3.8) becomes

$$\frac{1}{r}\frac{d}{dr}\left(r\frac{dv}{dr}\right) = \delta(Q - P) \tag{3.9}$$

where

$$r = |Q - P| = \sqrt{(\xi - x)^2 + (\eta - y)^2} \tag{3.10}$$

that is, the Euclidean distance between points P and Q.

The right-hand side of Eq. (3.8) vanishes at all points of the plane, except at the origin $r = 0$, where it has an infinite value. Apart from point $r = 0$, Eq. (3.9) is written as

$$\frac{1}{r}\frac{d}{dr}\left(r\frac{dv}{dr}\right) = 0$$

which gives, after integrating twice

$$v = A \ln r + B$$

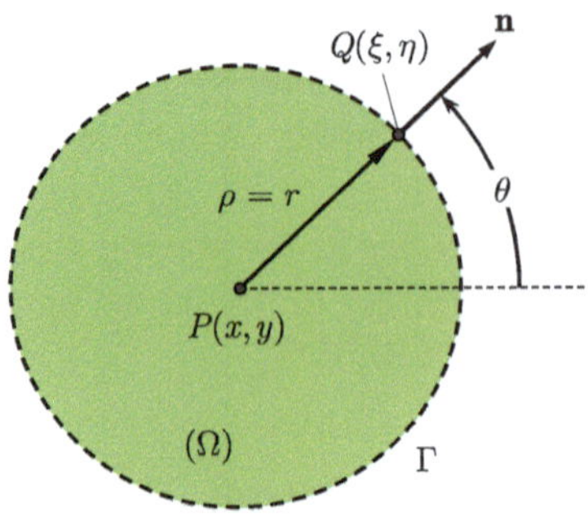

FIGURE 3.1 Circular domain Ω of radius ρ with a source P at its center.

where A and B are arbitrary constants. Since we look for a particular solution, we may set $B = 0$. The other constant, A, may be determined in the following way.

We consider the circular domain Ω with radius ρ centered at P (Fig. 3.1). Due to the axisymmetric nature of the problem, it is

$$\frac{\partial v}{\partial n} = \frac{\partial v}{\partial r} = A\,\frac{1}{r} \quad \text{and} \quad ds = r d\theta \tag{3.11}$$

Application of Green's identity (2.16) for $u = 1$ and $v = \ln r$ yields

$$-\int_{\Omega} \nabla^2 v \, d\Omega = -\int_{\Gamma} \frac{\partial v}{\partial n} \, ds$$

Using then Eqs. (3.8) and (3.11), and noting that for points on the boundary Γ it is $r = \rho$, the previous relation is written as

$$-\int_{\Omega} \delta(Q - P) \;\, d\Omega = -\int_{0}^{2\pi} A\,\frac{1}{\rho}\,\rho \; d\theta$$

or by applying Eq. (2.34b), it takes the form

$$1 = 2\pi\, A$$

from which we obtain

$$A = \frac{1}{2\pi} \tag{3.12}$$

Hence, the fundamental solution becomes

$$v = \frac{1}{2\pi}\,\ln r \tag{3.13}$$

From Eq. (3.10) it is evident that the fundamental solution does not change value, when points P and Q interchange their role. This means that v is symmetric with respect to these points, namely

$$v(Q, P) = v(P, Q) \tag{3.14}$$

The fundamental solution (3.13) is also known in the literature as the *free space Green's function*.

3.3 THE DIRECT BEM FOR THE LAPLACE EQUATION

In this section, we derive the solution of the Laplace equation

$$\nabla^2 u = 0 \ \text{ in } \Omega \tag{3.15}$$

with mixed boundary conditions (see Fig. 3.2)

$$u = \overline{u} \ \text{ on } \Gamma_1 \tag{3.16a}$$

$$\frac{\partial u}{\partial n} = \overline{u}_n \ \text{ on } \Gamma_2 \tag{3.16b}$$

where $\Gamma_1 \cup \Gamma_2 = \Gamma$, $\Gamma_1 \cap \Gamma_2 = \varnothing$. In the literature, condition (3.16a) is referred to as *essential* or *kinematic*, while (3.16b) is referred to as *natural*. Instead of the boundary conditions (3.16), we can use the general boundary condition (3.6b), but at the present time, we avoid it for the sake of simplicity.

Applying Green's identity (2.16), for the functions u and v that satisfy Eqs. (3.15) and (3.8), respectively, and assuming that the source lies at point P, we obtain

$$-\int_\Omega u(Q)\, \delta(Q-P)\, d\Omega_Q = \int_\Gamma \left[v(q,P)\frac{\partial u(q)}{\partial n_q} - u(q)\frac{\partial v(q,P)}{\partial n_q} \right] ds_q \tag{3.17}$$

where $P, Q \in \Omega$ and $q \in \Gamma$.

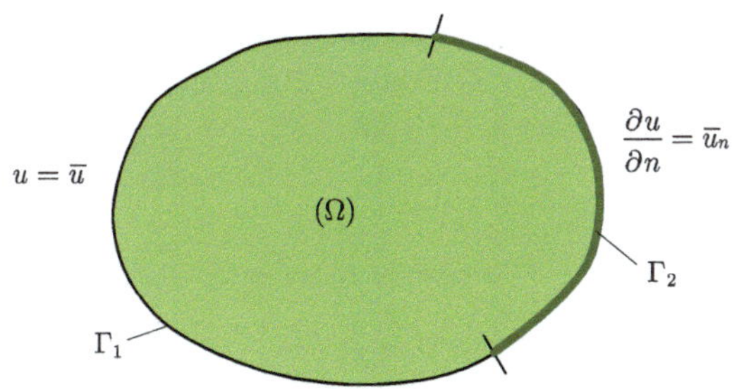

FIGURE 3.2 Domain Ω with mixed boundary conditions.

In the previous equations and hereafter, points inside the domain Ω are denoted by upper case letters, for example, P, Q, while points on the boundary Γ are denoted by lower case letters, for example, p, q. The subscripts of the differentials, for example, $d\Omega_Q$, ds_q, and the derivatives, for example, $\partial(\,)/\partial n_q$, denote the points that vary during integration or differentiation, respectively.

By virtue of Eq. (2.33), Eq. (3.17) is written as

$$u(P) = -\int_\Gamma \left[v(P,q)\frac{\partial u(q)}{\partial n_q} - u(q)\frac{\partial v(P,q)}{\partial n_q} \right] ds_q \tag{3.18}$$

The functions v and $\partial v/\partial n$ in the foregoing equation are both known quantities. They represent the fundamental solution of the Laplace equation and its normal derivative at point q of the boundary, which are given as

$$v = \frac{1}{2\pi}\ln r \tag{3.19}$$

$$\frac{\partial v}{\partial n} = \frac{1}{2\pi}\frac{\cos\phi}{r} \tag{3.20}$$

where $r = |q - P|$ and $\phi = \text{angle}(\mathbf{r}, \mathbf{n})$ (see Appendix A).

The expression (3.18) represents the solution of the differential Eq. (3.15) at any point P inside the domain Ω (*not* on the boundary Γ) in terms of the boundary values of u and its normal derivative $\partial u/\partial n$. The relation (3.18) is called the *integral representation of the solution* for the Laplace equation. It is apparent from the boundary conditions (3.16a) and (3.16b), that only one of the quantities u or $\partial u/\partial n$ is prescribed at a point $q(\xi, \eta)$ on the boundary. Consequently, it is not yet possible to determine the solution from the integral representation (3.18). Therefore, it is required to evaluate the boundary quantity, which is not prescribed by the boundary condition (either u or $\partial u/\partial n$). This is achieved by deriving the integral representation of u for points $P \equiv p$ lying on the boundary Γ.

We study the general case where the boundary is not smooth and P is a corner point (see Fig. 3.3). We consider the domain Ω^* which results from Ω after deleting a small circular section with center P, radius ε, and confined by the arcs PA and PB. The circular arc AB is denoted by Γ_ε and the sum of the arcs AP and PB by ℓ. The outward normal to Γ_ε coincides with the radius ε and is directed towards the center P. The angle between the tangents of the boundary at point P is denoted by α. Obviously, it is

$$\lim_{\varepsilon\to 0}(\theta_1 - \theta_2) = \alpha$$

$$\lim_{\varepsilon\to 0}\Gamma_\varepsilon = 0$$

$$\lim_{\varepsilon\to 0}(\Gamma - \ell) = \Gamma$$

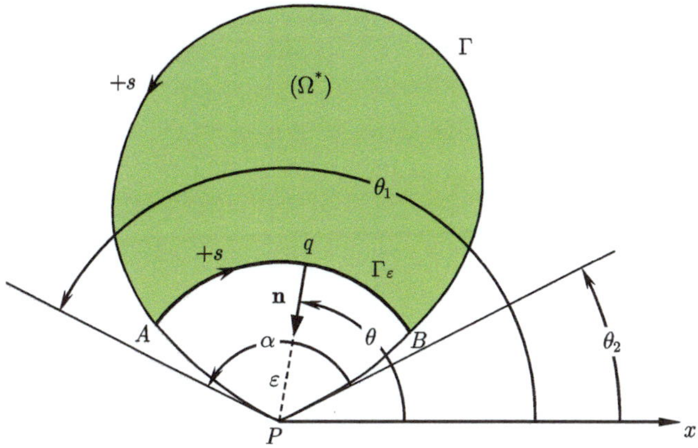

FIGURE 3.3 Geometric definitions related to a corner point P of a nonsmooth boundary.

Next we apply Green's identity (2.16) in the domain Ω^* for the functions u and v satisfying Eqs. (3.15) and (3.8), respectively. Since point P lies outside the domain Ω^*, where it is $\delta(Q-P)=0$, it follows that

$$\int_{\Omega^*} u\delta(Q-P)d\Omega = 0$$

and consequently Green's identity gives

$$0 = \int_{\Gamma-\ell}\left(v\frac{\partial u}{\partial n} - u\frac{\partial v}{\partial n}\right)ds + \int_{\Gamma_\varepsilon}\left(v\frac{\partial u}{\partial n} - u\frac{\partial v}{\partial n}\right)ds \tag{3.21}$$

We will examine next the behavior of the integrals in the above equation when $\varepsilon \to 0$. Apparently, the first integral becomes

$$\lim_{\varepsilon\to 0}\int_{\Gamma-\ell}\left(v\frac{\partial u}{\partial n} - u\frac{\partial v}{\partial n}\right)ds = \int_{\Gamma}\left(v\frac{\partial u}{\partial n} - u\frac{\partial v}{\partial n}\right)ds \tag{3.22}$$

The second one is written as (see Appendix A)

$$\begin{aligned}\int_{\Gamma_\varepsilon}\left(v\frac{\partial u}{\partial n} - u\frac{\partial v}{\partial n}\right) ds &= \int_{\Gamma_\varepsilon}\frac{1}{2\pi}\,\frac{\partial u}{\partial n}\,\ln r\; ds - \int_{\Gamma_\varepsilon}\frac{1}{2\pi}\;u\;\frac{\cos\phi}{r}\;ds \\ &= I_1 + I_2\end{aligned} \tag{3.23}$$

For the circular arc Γ_ε, it is $r=\varepsilon$ and $\phi=\pi$. Moreover, $ds=\varepsilon(-d\theta)$, because the angle θ is positive in the counter-clockwise sense, which is opposite to that of increasing s. Therefore, the first integral in Eq. (3.23) takes the form

$$I_1 = \int_{\Gamma_\varepsilon} \frac{1}{2\pi} \frac{\partial u}{\partial n} \ln r \, ds = \int_{\theta_1}^{\theta_2} \frac{1}{2\pi} \frac{\partial u}{\partial n} \varepsilon \ln \varepsilon \, d(-\theta) \tag{3.24}$$

According to the mean value theorem of integral calculus, the value of an integral is equal to the value of its integrand at some point O within the integration interval multiplied by the length of that interval. Hence,

$$I_1 = \frac{1}{2\pi} \left[\frac{\partial u}{\partial n} \right]_O \varepsilon \ln \varepsilon (\theta_1 - \theta_2)$$

When $\varepsilon \to 0$, the point O on the arc approaches point P. Of course, in this case the derivative $[\partial u / \partial n]_P$, though not defined, is bounded. Moreover, it is

$$\lim_{\varepsilon \to 0} (\varepsilon \ln \varepsilon) = 0$$

which implies that

$$\lim_{\varepsilon \to 0} I_1 = 0 \tag{3.25}$$

The second integral in the right-hand side of Eq. (3.23) may be written as

$$I_2 = -\int_{\Gamma_\varepsilon} \frac{1}{2\pi} u \frac{\cos\phi}{r} ds = -\int_{\theta_1}^{\theta_2} \frac{1}{2\pi} u \frac{-1}{\varepsilon} \varepsilon \, d(-\theta)$$

or by applying the mean value theorem

$$I_2 = -\frac{1}{2\pi} u_O (\theta_2 - \theta_1) = \frac{\theta_1 - \theta_2}{2\pi} u_O$$

and finally

$$\lim_{\varepsilon \to 0} I_2 = \frac{\alpha}{2\pi} u(p), \quad p \equiv P \in \Gamma \tag{3.26}$$

By virtue of Eqs. (3.25) and (3.26), Eq. (3.23) yields

$$\lim_{\varepsilon \to 0} \int_{\Gamma_\varepsilon} \left(v \frac{\partial u}{\partial n} - u \frac{\partial v}{\partial n} \right) ds = \frac{\alpha}{2\pi} u(p) \tag{3.27}$$

Incorporating now the findings of Eqs. (3.22) and (3.27) into Eq. (3.21), the latter gives for $\varepsilon \to 0$

$$\frac{\alpha}{2\pi} u(p) = -\int_{\Gamma} \left[v(p, q) \frac{\partial u(q)}{\partial n_q} - u(q) \frac{\partial v(p, q)}{\partial n_q} \right] ds_q \tag{3.28}$$

Equation (3.28) represents the integral representation of the solution for the Laplace equation (3.15) at points $p \in \Gamma$, where the boundary is not smooth. For points p, where the boundary is smooth, it is $\alpha = \pi$ and thus, Eq. (3.28) becomes

$$\frac{1}{2}u(p) = -\int_\Gamma \left[v(p,q)\frac{\partial u(q)}{\partial n_q} - u(q)\frac{\partial v(p,q)}{\partial n_q} \right] ds_q \tag{3.29}$$

A comparison between Eqs. (3.18) and (3.28) reveals that the function u is discontinuous when the point $P \in \Omega$ is on the boundary. It exhibits a jump equal to $(1 - \frac{\alpha}{2\pi})u(p)$ for corner points (Eq. 3.28), or $\frac{1}{2}u(p)$ for points on smooth parts of the boundary Γ (Eq. 3.29). When the point P is located outside the domain Ω, Green's identity (2.16) gives

$$0 = -\int_\Gamma \left[v(P,q)\frac{\partial u(q)}{\partial n_q} - u(q)\frac{\partial v(P,q)}{\partial n_q} \right] ds_q \tag{3.30}$$

Equations (3.18), (3.29), and (3.30) can be combined in a single general equation as

$$\varepsilon(P)\, u(P) = -\int_\Gamma \left[v(P,q)\frac{\partial u(q)}{\partial n_q} - u(q)\frac{\partial v(P,q)}{\partial n_q} \right] ds_q \tag{3.31}$$

where $\varepsilon(P)$ is a coefficient, referred to as the *free term coefficient*, which depends on the position of point P and is defined as

$$\varepsilon(P) = \begin{cases} 1 \ \text{for } P \text{ inside } \Omega \\ \dfrac{1}{2} \ \text{for } P \equiv p \text{ on the boundary } \Gamma \\ 0 \ \text{for } P \text{ outside } \Omega \end{cases}$$

Equation (3.29) constitutes a compatibility relation between the boundary values of u and $\partial u/\partial n$, meaning that only one of the quantities u and $\partial u/\partial n$ can be prescribed at each point of the boundary. This integral representation was employed by Fredholm [1] to prove that only one boundary condition is required for the well-posedness of the boundary value problem for the Laplace equation. At the same time, Eq. (3.29) can be viewed as an integral equation on the boundary Γ, that is a *boundary integral equation*, with respect to the not prescribed boundary quantity.

In what follows, we assume that boundary Γ is smooth. Thus, for the Dirichlet problem ($u = \overline{u}$ on Γ), Eq. (3.29) is written as

$$\frac{1}{2}\overline{u} = -\int_\Gamma \left(v\frac{\partial u}{\partial n} - \overline{u}\frac{\partial v}{\partial n} \right) ds \tag{3.32a}$$

in which the only unknown is the function $\partial u/\partial n$ on Γ. For the Neumann problem ($\partial u/\partial n = \overline{u}_n$), Eq. (3.29) becomes

$$\frac{1}{2}u = -\int_\Gamma \left(v\,\overline{u}_n - u\frac{\partial v}{\partial n} \right) ds \tag{3.32b}$$

with the only unknown the function u on Γ. For problems with mixed boundary conditions, Eq. (3.29) is treated as two separate equations (see Eq. 3.16), namely

$$\frac{1}{2}\overline{u} = -\int_\Gamma \left(v\frac{\partial u}{\partial n} - \overline{u}\frac{\partial v}{\partial n} \right) ds \ \text{ on } \Gamma_1 \tag{3.33a}$$

$$\frac{1}{2}u = -\int_\Gamma \left(v\,\overline{u}_n - u\frac{\partial v}{\partial n} \right) ds \ \text{ on } \Gamma_2 \tag{3.33b}$$

3.4 THE DIRECT BEM FOR THE POISSON EQUATION

In this case, we seek the solution of the boundary value problem which is governed by the Poisson equation

$$\nabla^2 u = f \ \text{ in } \Omega \tag{3.34}$$

and has mixed boundary conditions

$$u = \overline{u} \ \text{ on } \Gamma_1 \tag{3.35a}$$

$$\frac{\partial u}{\partial n} = \overline{u}_n \ \text{ on } \Gamma_2 \tag{3.35b}$$

The solution can be obtained in two different ways, which are presented in the following sections.

3.4.1 Application of Green's identity

The integral representation of the solution is obtained by applying Green's identity (2.16) for the functions u and v that satisfy Eqs. (3.34) and (3.8), respectively. This yields

$$\varepsilon(P)u = \int_\Omega vf\, d\Omega - \int_\Gamma \left(v\frac{\partial u}{\partial n} - u\frac{\partial v}{\partial n} \right) ds \tag{3.36}$$

The corresponding boundary integral equation for the smooth boundary is

$$\frac{1}{2}u = \int_{\Omega} vf \; d\Omega - \int_{\Gamma}\left(v\frac{\partial u}{\partial n} - u\frac{\partial v}{\partial n}\right) ds \tag{3.37}$$

3.4.2 Transformation of the Poisson equation to the Laplace equation

The solution of Eq. (3.34) can be obtained as a sum of two solutions

$$u = u_0 + u_1 \tag{3.38}$$

where u_0 is the solution of the homogeneous equation (Laplace equation) with boundary conditions $u_0 = \overline{u} - u_1$ on Γ_1 and $\partial u_0/\partial n = \overline{u}_n - \partial u_1/\partial n$ on Γ_2, and u_1 is a particular solution of the nonhomogeneous equation. Apparently, this method requires first the establishment of the particular solution.

i. *Particular solution* u_1

The particular solution of the nonhomogeneous equation is any function u_1 that satisfies only the governing equation

$$\nabla^2 u_1 = f \tag{3.39}$$

independently of boundary conditions. We will show next that

$$u_1 = \int_{\Omega} v\, f \; d\Omega \tag{3.40}$$

where v is the fundamental solution. Using the definition (2.33) of the Dirac delta function, Eq. (3.39) becomes

$$\begin{aligned}\nabla^2 u_1(P) &= f(P)\\ &= \textstyle\int_{\Omega}\delta(Q-P)\, f(Q) \;\, d\Omega_Q\end{aligned}$$

or taking into account Eq. (3.8), we have

$$\nabla^2 u_1(P) = \int_{\Omega} \nabla^2 v(Q,P)\, f(Q)\; d\Omega_Q = \nabla^2\left[\int_{\Omega} v(Q,P)\, f(Q)\; d\Omega_Q\right]$$

or

$$\nabla^2\left[u_1(P) - \int_{\Omega} v(Q,P)\, f(Q) \;\, d\Omega_Q\right] = 0$$

which is obviously satisfied by the particular solution

$$u_1(P) = \int_\Omega v(Q,P)\, f(Q)\ d\Omega_Q$$

It should be noted that the differentiations indicated by the ∇^2 operator are performed with respect to the coordinates of point P. Moreover, v is continuous with respect to this point. Based on these remarks, the operator ∇^2 was moved outside the integral in the previous equations.

Another method to establish a particular solution is by transforming Eq. (3.39) into the complex domain. The transformation is defined by

$$z = x + iy, \quad \overline{z} = x - iy \quad \left(i = \sqrt{-1}\right) \tag{3.41a}$$

and its inverse is

$$x = \frac{z + \overline{z}}{2}, \qquad y = \frac{z - \overline{z}}{2i} \tag{3.41b}$$

We can readily show that Eq. (3.39) is transformed to

$$4\,\frac{\partial^2 u_1}{\partial z\,\partial \overline{z}} = f(z, \overline{z}) \tag{3.42}$$

which, after two successive integrations, yields the particular solution $u_1(z, \overline{z})$. It is worth mentioning that any arbitrary integration functions can be omitted, because we look only for a particular solution. Subsequently, the transformation (3.41a) produces the particular solution $u_1(x, y)$ in the physical space.

EXAMPLE 3.1

Determine a particular solution $u_1(x, y)$ of the Poisson equation (3.34) when

$$f = x^2 + y^2 \tag{a}$$

Application of the transformation (3.41b) to equation (a) converts the function f into the complex domain,

$$f = z\,\overline{z} \tag{b}$$

Equation (3.42) may then be written as

$$\frac{\partial^2 u_1}{\partial z\,\partial \overline{z}} = \frac{1}{4}\, z\,\overline{z} \tag{c}$$

which, after successive integrations, yields

$$u_1 = \frac{1}{16} z^2 \bar{z}^2 \tag{d}$$

Substituting Eq. (3.41b) in the previous equation, we find the particular solution in the physical domain

$$u_1 = \frac{1}{16} (x^2 + y^2)^2 \tag{e}$$

ii. *Homogeneous solution* u_0

Once the particular solution u_1 has been established, the solution u_0 of the homogeneous equation will be obtained from the boundary value problem

$$\nabla^2 u_0 = 0 \text{ in } \Omega \tag{3.43}$$

$$u_0 = \bar{u} - u_1 \text{ on } \Gamma_1 \tag{3.44a}$$

$$\frac{\partial u_0}{\partial n} = \bar{u}_n - \frac{\partial u_1}{\partial n} \text{ on } \Gamma_2 \tag{3.44b}$$

This problem can be solved using the procedure presented in Section 3.3.

3.5 TRANSFORMATION OF THE DOMAIN INTEGRALS TO BOUNDARY INTEGRALS

In the process of solving the Poisson equation by BEM, domain integrals appear in the integral representation of the solution (3.36). These integrals are of the form

$$\int_\Omega v f \, d\Omega \tag{3.45}$$

Although the integrand vf is known, the fact that domain integrals need to be evaluated spoils the pure boundary character of the method, thus weakening the advantages of the BEM over domain methods. However, it is possible to overcome this drawback by converting the domain integral (3.45) to a boundary line integral on Γ, that is, on the boundary of the domain Ω. Next, two different approaches are presented for the conversion of integral (3.45).

i. *The function f is a polynomial of x and y.*

Let us assume that the function f is a first degree polynomial with respect to x and y,

$$f = \alpha_0 + \alpha_1\, x + \alpha_2\, y \tag{3.46}$$

where α_0, α_1, and α_2 are known constants. This function, apparently, satisfies the Laplace equation,

$$\nabla^2 f = 0 \tag{3.47}$$

Setting

$$\nabla^2 U = v \tag{3.48}$$

and applying Green's identity (2.16) for the functions f and U,

$$\int_\Omega (f\nabla^2 U - U\nabla^2 f)d\Omega = \int_\Gamma \left(f\frac{\partial U}{\partial n} - U\frac{\partial f}{\partial n} \right) ds$$

By virtue of Eqs. (3.47) and (3.48), the previous equation yields

$$\int_\Omega vf\, d\Omega = \int_\Gamma \left(f\frac{\partial U}{\partial n} - U\frac{\partial f}{\partial n} \right)\, ds \tag{3.49}$$

If the function f is defined in a subdomain $\Omega^* \subset \Omega$ with boundary Γ^* (see Fig. 3.4), then Eq. (3.49) becomes

$$\int_\Omega vf\, d\Omega = \int_{\Omega^*} vf\, d\Omega = \int_{\Gamma^*} \left(f\frac{\partial U}{\partial n} - U\frac{\partial f}{\partial n} \right)\, ds \tag{3.50}$$

The function $U = U(r)$ can be established as a particular solution of Eq. (3.48). For this purpose, we write the Laplace operator in polar coordinates,

$$\frac{1}{r}\,\frac{d}{dr}\left(r\,\frac{dU}{dr} \right) = \frac{1}{2\pi}\,\ln r$$

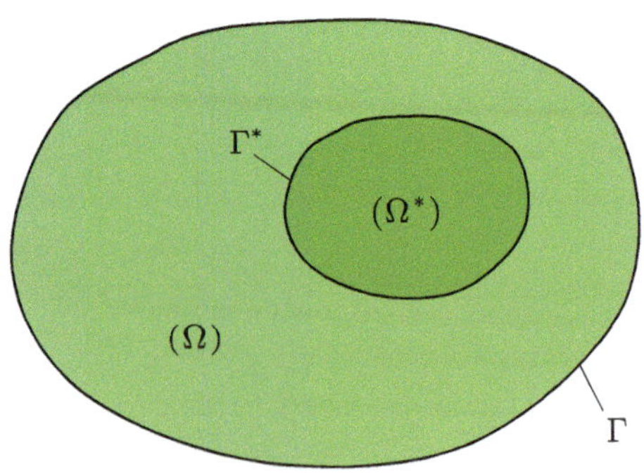

FIGURE 3.4 Subdomain $\Omega^* \subset \Omega$ with boundary Γ^*.

Integrating the above equation twice, we find the particular solution

$$U = \frac{1}{8\pi}\ r^2(\ln r - 1) \tag{3.51}$$

ii. *The function f is arbitrary.*

In this case f is an arbitrary function defined in $\Omega^* \subset \Omega$. We first establish another function F which satisfies the following Poisson equation

$$\nabla^2 F = f \tag{3.52}$$

This function F is determined as a particular solution of Eq. (3.52) using the procedure presented in Section 3.4.2. The Green identity (2.16) is then applied for the functions v and F in the domain Ω^*,

$$\int_{\Omega^*} (v\,\nabla^2 F - F\,\nabla^2 v)\ d\Omega = \int_{\Gamma^*} \left(v\frac{\partial F}{\partial n} - F\frac{\partial v}{\partial n} \right) ds$$

which becomes

$$\int_{\Omega^*} vf\ d\Omega = \int_{\Omega^*} F\ \delta(Q-P)\ d\Omega_Q + \int_{\Gamma^*} \left(v\frac{\partial F}{\partial n} - F\frac{\partial v}{\partial n} \right) ds$$

or on the basis of Eq. (3.36)

$$\int_{\Omega^*} vf\ d\Omega = \varepsilon(P)F + \int_{\Gamma^*} \left(v\frac{\partial F}{\partial n} - F\frac{\partial v}{\partial n} \right) ds \tag{3.53}$$

We distinguish two cases:

a. $\Omega^* \equiv \Omega$. The point P may be inside Ω or on Γ, then Eq. (3.53) yields

$$\int_{\Omega} vf\ d\Omega = \varepsilon(P)\ F + \int_{\Gamma} \left(v\frac{\partial F}{\partial n} - F\frac{\partial v}{\partial n} \right) ds \tag{3.54}$$

where $\varepsilon(P) = 1,\ 1/2$ depending on whether the point P is inside Ω or Γ, respectively.

b. $\Omega^* \subset \Omega$. The point P may be inside Ω^*, on Γ^*, or outside Ω^*. In this case Eq. (3.53) gives

$$\int_{\Omega^*} vf\ d\Omega = \varepsilon(P)F + \int_{\Gamma^*} \left(v\frac{\partial F}{\partial n} - F\frac{\partial v}{\partial n} \right) ds \tag{3.55}$$

where $\varepsilon(P) = 1, 1/2, 0$ depending on whether the point P is inside Ω^*, on Γ^* or outside Ω^*, respectively.

3.6 THE BEM FOR POTENTIAL PROBLEMS IN ANISOTROPIC BODIES

In this section, we develop the BEM solution for the following boundary value problem

$$k_{xx}\frac{\partial^2 u}{\partial x^2}+2\,k_{xy}\frac{\partial^2 u}{\partial x\,\partial y}+k_{yy}\frac{\partial^2 u}{\partial y^2}=f(x,y)\ \text{in}\ \Omega \tag{3.56a}$$

$$u=\overline{u}\ \text{on}\ \Gamma_1 \tag{3.56b}$$

$$\nabla u\cdot\mathbf{m}=\overline{q}_n\ \text{on}\ \Gamma_2 \tag{3.56c}$$

where

$$\mathbf{m}=(k_{xx}\,n_x+k_{xy}\,n_y)\,\mathbf{i}+(k_{xy}\,n_x+k_{yy}\,n_y)\,\mathbf{j}$$

is a vector in the direction of the *conormal* to the boundary [15] and k_{xx}, k_{xy}, k_{yy} are constants satisfying the *ellipticity condition* $k_{xy}^2-k_{xx}k_{yy}<0$. Obviously, when $k_{xx}=k_{yy}=1$ and $k_{xy}=0$, it is $\mathbf{m}=\mathbf{n}$ and the boundary value problem (3.56a,b,c) reduces to the mixed boundary value problem for the Poisson equation described by Eqs. (3.34) and (3.35). Eq. (3.56a) describes potential problems in homogeneous anisotropic bodies (see, e.g., Section 6.5).

3.6.1 Integral representation of the solution

Green's identity for the differential operator of Eq. (3.56a) can be obtained from Eq. (2.21) by setting $A=k_{xx}$, $B=k_{xy}$, $C=k_{yy}$, and $D=E=F=0$. Thus, one readily arrives at the following equation (see also Problem 2.9)

$$\int_\Omega [vL(u)-uL(v)]\,d\Omega=\int_\Gamma (v\nabla u\cdot\mathbf{m}-u\nabla v\cdot\mathbf{m})\,ds \tag{3.57}$$

where

$$L(\,)=k_{xx}\frac{\partial^2}{\partial x^2}+2\,k_{xy}\frac{\partial^2}{\partial x\,\partial y}+k_{yy}\frac{\partial^2}{\partial y^2} \tag{3.58}$$

The boundary condition (3.56c) may also be written, after rearranging its terms, as

$$\nabla u\cdot\mathbf{m}=\mathbf{q}\cdot\mathbf{n}=q_n \tag{3.59}$$

where

$$\mathbf{q} = q_x\,\mathbf{i} + q_y\,\mathbf{j} = \left(k_{xx}\frac{\partial u}{\partial x} + k_{xy}\frac{\partial u}{\partial y}\right)\mathbf{i} + \left(k_{xy}\frac{\partial u}{\partial x} + k_{yy}\frac{\partial u}{\partial y}\right)\mathbf{j} \qquad (3.60)$$

is a vector representing the *flux* of u, i.e., the flow of the potential u through the boundary Γ per unit length of the boundary, and q_n is its component (projection) in the direction of the normal to the boundary.

The resulting integral representation of the solution is obtained as

$$u(P) = \int_\Omega vf\,d\Omega - \int_\Gamma (v\nabla u\cdot\mathbf{m} - u\nabla v\cdot\mathbf{m})\,ds, \quad P\in\Omega \qquad (3.61)$$

where v is the fundamental solution of Eq. (3.56a), namely, a singular particular solution of the equation

$$L(v) = \delta(Q-P) \qquad (3.62)$$

3.6.2 Fundamental solution

The fundamental solution of Eq. (3.56a) can be established by transforming Eq. (3.62) in its canonical form [16]. For this purpose we consider the transformation

$$\left.\begin{aligned} \overline{x} &= y - a\,x \\ \overline{y} &= -b\,x \end{aligned}\right\} \qquad (3.63a)$$

and its inverse

$$\left.\begin{aligned} x &= -\frac{1}{b}\;\overline{y} \\ y &= \overline{x} - \frac{a}{b}\;\overline{y} \end{aligned}\right\} \qquad (3.63b)$$

where

$$a = \frac{k_{xy}}{k_{xx}} \quad \text{and} \quad b = \frac{\sqrt{D}}{k_{xx}} \qquad (3.64)$$

in which

$$D = \det(\mathbf{D}) = \begin{vmatrix} k_{xx} & k_{xy} \\ k_{xy} & k_{yy} \end{vmatrix} = k_{xx}\,k_{yy} - k_{xy}^2 > 0 \qquad (3.65a)$$

namely, the determinant of the matrix

$$\mathbf{D}=\begin{bmatrix} k_{xx} & k_{xy} \\ k_{xy} & k_{yy} \end{bmatrix} \tag{3.65b}$$

The derivatives involved in the operator (3.58) become under the transformation (3.63a)

$$\frac{\partial v}{\partial x}=-\left(a\frac{\partial v}{\partial \bar{x}}+b\frac{\partial v}{\partial \bar{y}}\right)$$

$$\frac{\partial v}{\partial y}=\frac{\partial v}{\partial \bar{x}}$$

$$\frac{\partial^2 v}{\partial x^2}=a^2\frac{\partial^2 v}{\partial \bar{x}^2}+2ab\frac{\partial^2 v}{\partial \bar{x}\partial \bar{y}}+b^2\frac{\partial^2 v}{\partial \bar{y}^2}$$

$$\frac{\partial^2 v}{\partial x\partial y}=-\left(a\frac{\partial^2 v}{\partial \bar{x}^2}+b\frac{\partial^2 v}{\partial \bar{x}\partial \bar{y}}\right)$$

$$\frac{\partial^2 v}{\partial y^2}=\frac{\partial^2 v}{\partial \bar{x}^2}$$

and, finally, the operator $L(v)$ takes the following concise form

$$L(v)=\frac{D}{k_{xx}}\left(\frac{\partial^2 v}{\partial \bar{x}^2}+\frac{\partial^2 v}{\partial \bar{y}^2}\right) \tag{3.66}$$

The function $\delta(Q-P)$ is transformed into the $\bar{x}\bar{y}$-plane using Eq. (2.40),

$$\delta(Q-P)=\frac{\delta(\overline{Q}-\overline{P})}{|J|}$$

where $Q(\xi,\eta)$, $P(x,y)$, $\overline{Q}(\bar{\xi},\bar{\eta})$, $\overline{P}(\bar{x},\bar{y})$ denote the coordinates of the points and

$$|J|=\begin{vmatrix} \dfrac{\partial x}{\partial \bar{x}} & \dfrac{\partial y}{\partial \bar{x}} \\ \dfrac{\partial x}{\partial \bar{y}} & \dfrac{\partial y}{\partial \bar{y}} \end{vmatrix}=\begin{vmatrix} 0 & 1 \\ -\dfrac{1}{b} & -\dfrac{a}{b} \end{vmatrix}=\frac{1}{b}$$

the Jacobian of the transformation (3.63a), which may be further written as

$$|J| = \frac{k_{xx}}{\sqrt{D}} \tag{3.67}$$

Consequently, the Dirac function is being transformed as

$$\delta(Q-P) = \frac{\sqrt{D}}{k_{xx}}\ \delta(\overline{Q}-\overline{P}) \tag{3.68}$$

Incorporating Eqs. (3.66) and (3.68) into Eq. (3.62), the latter becomes

$$\nabla^2 v = \frac{1}{\sqrt{D}}\delta(\overline{Q}-\overline{P}) \tag{3.69}$$

where the Laplace operator is expressed in terms of the $\overline{x}$ and $\overline{y}$ coordinates. A comparison between Eqs. (3.8) and (3.69) reveals that, according to Eq. (3.13), the fundamental solution of Eq. (3.69) in the $(\overline{x}, \overline{y})$ coordinate system is

$$v = \frac{1}{2\pi\sqrt{D}}\ln r \tag{3.70}$$

where

$$r = \sqrt{(\overline{\xi}-\overline{x})^2 + (\overline{\eta}-\overline{y})^2}$$

or back in the (x, y) system

$$r = \sqrt{(a^2+b^2)(\xi-x)^2 - 2a\,(\xi-x)(\eta-y) + (\eta-y)^2}$$

which by virtue of Eq. (3.64) becomes

$$r = \frac{\sqrt{k_{yy}\,(\xi-x)^2 - 2\,k_{xy}(\xi-x)\,(\eta-y) + k_{xx}\,(\eta-y)^2}}{\sqrt{k_{xx}}} \tag{3.71}$$

The fundamental solution is not altered if a constant is added to it. Thus, we can write

$$v = \frac{1}{2\pi\sqrt{D}}\ln r + \frac{1}{2\pi\sqrt{D}}\ln\left(\sqrt{\frac{k_{xx}}{D}}\right)$$

or

$$v = \frac{1}{2\pi\sqrt{D}} \ln r \tag{3.72}$$

where now

$$r = \sqrt{\frac{k_{yy}(\xi - x)^2 - 2\, k_{xy}(\xi - x)\,(\eta - y) + k_{xx}\,(\eta - y)^2}{D}} \tag{3.73}$$

It should be noted that the quantities k_{yy}/D, k_{xy}/D, and k_{xx}/D are the elements of the matrix $\mathbf{D}^{-1}$. If the material is orthotropic, the constants become $k_{xy} = 0$, $D = k_{xx}\,k_{yy}$, and Eq. (3.73) is reduced to

$$r = \sqrt{\frac{(\xi - x)^2}{k_{xx}} + \frac{(\eta - y)^2}{k_{yy}}} \tag{3.74}$$

For the established fundamental solution (3.72), the term $\nabla v \cdot \mathbf{m}$ appearing in the integral representation (3.61) becomes

$$\nabla v \cdot \mathbf{m} = \frac{1}{2\pi\sqrt{D}} \frac{\nabla r \cdot \mathbf{m}}{r} \tag{3.75}$$

where

$$\nabla r \cdot \mathbf{m} = \frac{\partial r}{\partial \xi} m_x + \frac{\partial r}{\partial \eta} m_y \tag{3.76}$$

in which the components of $\mathbf{m}$ are $m_x = k_{xx}\,n_x + k_{xy}\,n_y$ and $m_y = k_{xy}\,n_x + k_{yy}\,n_y$.

The derivatives in Eq. (3.76) are obtained by differentiating Eq. (3.73),

$$\frac{\partial r}{\partial \xi} = \frac{1}{D}\left(k_{yy}\frac{\xi - x}{r} - k_{xy}\frac{\eta - y}{r}\right) \tag{3.77}$$

and

$$\frac{\partial r}{\partial \eta} = \frac{1}{D}\left(-k_{xy}\frac{\xi - x}{r} + k_{xx}\frac{\eta - y}{r}\right) \tag{3.78}$$

3.6.3 Boundary integral equation

The boundary integral equation can be obtained from Eq. (3.61) when the point $P \in \Omega$ coincides with a point p on the boundary Γ. In this case, taking into account Eqs. (3.72) and (3.75) and the analysis presented in Section 3.3, for points p where the boundary is smooth, we have

$$\lim_{P\to p}\int_{\Gamma_\varepsilon} v\nabla u\cdot\mathbf{m}\, ds = 0$$

$$\lim_{P\to p}\int_{\Gamma_\varepsilon} u\nabla v\cdot\mathbf{m}\ ds = \frac{1}{2}u(p)$$

Consequently, the boundary integral equation takes the form

$$\frac{1}{2}u(p) = \frac{1}{2\pi\sqrt{D}}\int_\Omega f\ln r\, d\Omega\ - \frac{1}{2\pi\sqrt{D}}\int_\Gamma\left[(\nabla u\cdot\mathbf{m})\ln r - u\,\frac{\nabla r\cdot\mathbf{m}}{r}\right] ds \quad (3.79)$$

The kernels in the above equation are identical to those of Eqs. (3.28) or (3.37), except for the distance r, which is given by Eq. (3.73).

Equation (3.79) is used to determine the quantities u and $\nabla u\cdot\mathbf{m}$ on the parts of the boundary on which they are not prescribed. Subsequently, the solution at any point $P\in\Omega$ is evaluated from Eq. (3.61).

3.7 REFERENCES

The method of solving the potential equation by transforming it into a boundary integral equation is not new, but it goes back to the beginning of the last century. In 1900, Fredholm [1] employed it in the potential theory to determine the unknown boundary quantities from the prescribed ones. A detailed presentation of the method can be found in Kellogg's book [2] on potential theory. However, the researchers abandoned the boundary integral equation method as a solution method for potential problems, since it was practically impossible to solve the resulting singular boundary integral equations. The method was mainly limited to proving existence theorems for the solution of the differential equations.

Nevertheless, with the advent of modern computers the method came forth again and started slowly to be used as a computational method at the beginning of the 1960s. Jaswon [3] and Symm [4] presented a solution algorithm for the potential equation; Jaswon and Ponter [5] developed a numerical technique to solve the boundary integral equation for the classical Saint-Venant torsion problem of noncircular bars. They adopted a formulation in terms of the warping function and solved a Neumann problem for the Laplace equation. Numerical results were obtained for bars with various cross-sectional shapes, such as ellipses with or without holes, rectangles, equilateral triangles, and circles with notches. Mendelson [6] solved the same problem as a Dirichlet problem for the stress function. Later, Katsikadelis and Sapountzakis [7] formulated the boundary integral equations for the torsion problem of composite bars consisting of two or more materials. They developed the numerical technique for the solution of the boundary integral equations and produced numerical results for composite cross-sections of elliptical shape with elliptical or circular inclusions, for hollow box-shaped cross-sections and for a rectangular cross-section with an I-shaped inclusion. Symm [8] solved the problem of conformal mapping for simply connected domains having arbitrary shapes onto the unit circle $|w(z)| = 1$ of the complex domain. Christiansen [9] presented a complete collection of the integral equations for the Saint-Venant torsion problem.

In the last 30 years, the number of publications on BEM solutions for the potential equations has increased enormously. For a more detailed study of the boundary integral equation method, we refer to the books by Jaswon and Symm [10] and Gipson [11]. For the application of the BEM to the potential problems, we refer to the books by Banerjee and Butterfield [12], Brebbia and Dominguez [13], and Gaul et al. [14]. Rich sources for all aspects of the method applied to potential problems, that is, formulations, numerical implementation, programming, and applications to small and large scale problems of the real world in all fields of engineering and mathematical physics can be found in the proceedings of the series of international conferences on Boundary Elements/Mesh Reduction Methods (BEM/MRM), organized annually by Wessex Institute of Technology since 1978 as well as in the international Journal of Engineering Analysis with Boundary Elements. Finally, it is worth mentioning the books by Mikhlin [15] and Tyn Myint and Debnath [16], which may be useful to readers interested in the theory of partial differential equations of elliptic type.

[1] Fredholm I. Ofversight af Kongl, Svenska Vetenskaps—Akademiens Forhandligar 1900;57:39–46.
[2] Kellogg OD. Foundations of potential theory. Berlin: Springer-Verlag; 1967.
[3] Jaswon MA. Integral equation methods in potential theory, I. Proc R Soc London 1963;275:23–32 (A).
[4] Symm GT. Integral equation methods in potential theory, II. Proc R Soc London 1963;275:33–46 (A).
[5] Jaswon MA, Ponter AR. An integral equation solution of the torsion problem. Proc R Soc London 1963;275:237–46 (A).
[6] Mendelson A. 1973. Boundary integral methods in elasticity and plasticity, NASA, Technical Note, T.N. D-7418.
[7] Katsikadelis JT, Sapountzakis EI. Torsion of composite bars by the boundary element method. ASCE J Eng Mech 1985;111:1197–210.
[8] Symm GT. An integral equation method in conformal mapping. Numer Math 1966;9:250–8.
[9] Christiansen S. A review of some integral equations for solving the Saint-Venant torsion problem. J Elasticity 1978;8:1–20.
[10] Jaswon MA, Symm GT. Integral equation methods in potential theory and elastostatics. London: Academic Press, Inc; 1977.
[11] Gipson GS. Boundary element fundamentals—Basic concepts and recent developments in the Poisson equation. Southampton: Computational Mechanics Publications; 1987.
[12] Banerjee PK, Butterfield R. Boundary element methods in engineering science. U.K.: McGraw-Hill; 1981.
[13] Brebbia CA, Dominguez DJ. Boundary elements: An introductory course. 2nd ed. Southampton: Computational Mechanics Publications; 2001.
[14] Gaul L, Kögl M, Wagner M. Boundary elements for engineers and scientists. Berlin: Springer-Verlag; 2003.
[15] Mikhlin SG, editor. Linear equations of mathematical physics. New York: Holt. Rinehart and Winston; 1967.
[16] Tyn Myint U, Debnath L. Linear partial differential equations for scientists and engineers. 4th ed. Boston: Birkhäuser; 2007.

PROBLEMS

3.1. Determine a particular solution of the equation

$$\nabla^2 u = x\, y$$

3.2. Evaluate the following integral over a circle Ω having radius R and being centered at point $P(x_o, y_o)$

$$\int_\Omega v f \, d\Omega$$

where

$$v = \frac{1}{2\pi} \ln r, \quad r = |Q - P| = \sqrt{(x - x_o)^2 + (y - y_o)^2}$$

and

$$f(x, y) = \alpha_0 + \alpha_1\, x + \alpha_2\, y$$

3.3. Convert the domain integral

$$\int_\Omega \frac{\partial v}{\partial x} f \, d\Omega$$

where

$$v = \frac{1}{2\pi} \ln r, \quad r = |Q - P| = \sqrt{(x - x_o)^2 + (y - y_o)^2}$$

to a line integral on the boundary Γ, for the following two cases

i. $f = \alpha_0 + \alpha_1\, x + \alpha_2\, y$

ii. $f = x^2 + y^2$

3.4. Derive the integral representations of the derivatives $\partial u/\partial n_p$ and $\partial u/\partial t_p$, where u is the solution of the Laplace equation and n_p and t_p are the directions of the outward normal and the tangent to the boundary at point $p \in \Gamma$, respectively.

3.5. Derive the integral representations of the derivatives $\partial u/\partial x$ and $\partial u/\partial y$ at points: (1) $P \in \Omega$; and (2) $p \in \Gamma$.

Chapter | Four

Numerical Implementation of the BEM

CHAPTER OUTLINE

The Boundary Element Method for Engineers and Scientists.

4.1 INTRODUCTION

This chapter presents the numerical implementation of the boundary element method (BEM) for solving the potential problems discussed in the previous chapter. For realistic engineering applications, an analytical solution of the integral equation (3.29) is out of the question. However, a numerical solution of this equation is always feasible by employing the BEM.

Let us consider an arbitrary domain Ω with boundary Γ. The quintessence of the BEM is to discretize the boundary into a finite number of segments, not necessarily equal, which are called *boundary elements*. Two approximations are made over each of these elements. One concerns the geometry of the boundary, while the other has to do with the variation of the unknown boundary quantity over the element. The usually employed boundary elements are the *constant element*, the *linear element*, and the *parabolic* or *quadratic element*. On each element, we distinguish the *extreme points* or *end points* and the *nodes* or *nodal points*. The latter are the points at which values of the boundary quantities are assigned. In the case of constant elements the boundary segment is approximated by a straight line, which connects its extreme points. The node is placed at the mid-point of the straight line and the boundary quantity is assumed to be constant along the element and equal to its value at the nodal point (Fig. 4.1a). For linear elements, the boundary segment is approximated again by a straight line connecting its end points. The element has two nodes usually placed at the extreme points and the boundary quantity is assumed to vary linearly between the nodal values (Fig. 4.1b). Finally, for parabolic elements, the geometry is approximated by a parabolic arc. The element has three nodes, two of which are placed at the ends and the third somewhere in-between, usually at the mid-point (Fig. 4.1c).

For linear and parabolic elements, the geometry of the segment is depicted *isoparametrically*, that is, the geometry and the boundary quantity are approximated over the element by polynomials of the same degree. For constant elements, the geometry is depicted *superparametrically*, since it is represented with higher-order polynomial than that used to approximate the boundary quantity. Constant elements depict the boundary quantities discontinuously from element to element, in contrast to linear and parabolic elements, which depict them continuously. Although, the interelement continuity ensures a better approximation of the boundary quantity, it gives rise to complications at corner points or at points where the boundary conditions change type (mixed boundary value problems). These difficulties can be overcome by employing discontinuous linear and parabolic elements, which do not have nodal points at the ends of the element (see Chapter 5). The numerical solution of the integral equation (3.29) will be first presented by using constant boundary elements, because at this stage understanding the numerical implementation of the BEM overrides the need to incorporate more advanced numerical techniques, which improve the accuracy and efficiency of the BEM.

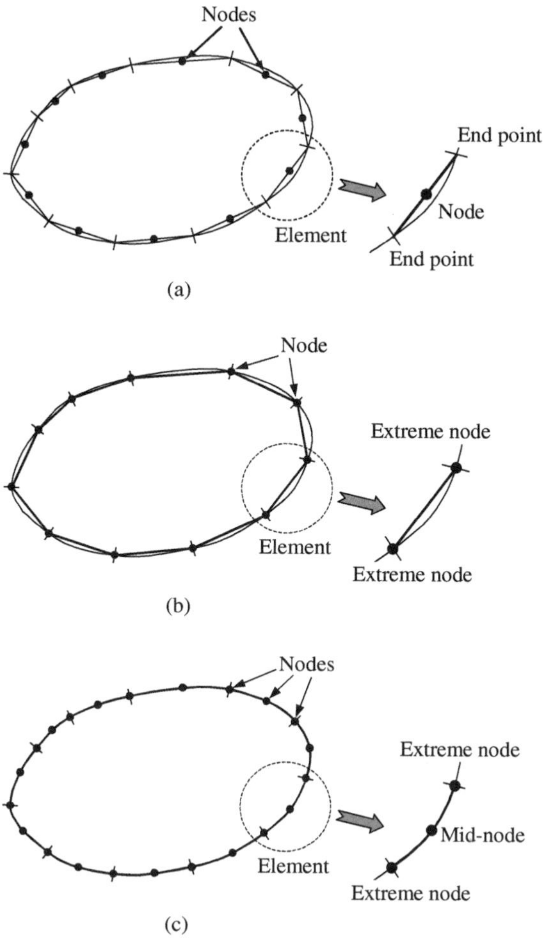

FIGURE 4.1 Various types of boundary elements. (a) Constant elements; (b) Linear elements; (c) Parabolic elements.

4.2 THE BEM WITH CONSTANT BOUNDARY ELEMENTS

The boundary Γ is discretized into N constant elements, which are numbered in the counter-clockwise sense. The values of the boundary quantity u and its normal derivative $\partial u/\partial n$ (denoted also as u_n) are assumed constant over each element and equal to their value at the mid-point of the element.

The discretized form of Eq. (3.29) is expressed for a given point p_i on Γ as

$$\frac{1}{2}u^i = -\sum_{j=1}^{N}\int_{\Gamma_j} v(p_i, q)\frac{\partial u(q)}{\partial n_q}\,ds_q + \sum_{j=1}^{N}\int_{\Gamma_j} u(q)\,\frac{\partial v(p_i, q)}{\partial n_q}\,ds_q \tag{4.1}$$

where Γ_j is the segment (straight line) on which the j-th node is located and over which integration is carried out, and p_i is the nodal point of the i-th element. For constant elements, the boundary is smooth at the nodal points, hence $\varepsilon(P)=1/2$. Moreover, the values of u and $\partial u/\partial n$ are constant on each element, so they can be moved outside the integral. Denoting by u^j and u_n^j the values of u and u_n, respectively, on the j-th element, Eq. (4.1) may be written as

$$-\frac{1}{2}u^i+\sum_{j=1}^{N}\left(\int_{\Gamma_j}\frac{\partial v}{\partial n}ds\right)u^j=\sum_{j=1}^{N}\left(\int_{\Gamma_j}vds\right)u_n^j \tag{4.2}$$

The integrals involved in the above equation relate the node p_i, where the fundamental solution is applied, to the node p_j ($j=1,2,\ldots,N$) (Fig. 4.2). Their values express the contribution of the nodal values u^j and u_n^j to the formation of the value $\frac{1}{2}u^i$. For this reason, they are often referred to as *influence coefficients*. These coefficients are denoted by $\hat{H}_{ij}$ and G_{ij}, which are defined as

$$\hat{H}_{ij}=\int_{\Gamma_j}\frac{\partial v(p_i,q)}{\partial n_q}ds \quad \text{and} \quad G_{ij}=\int_{\Gamma_j}v(p_i,q)\,ds \tag{4.3}$$

in which the point p_i remains constant (reference point), while the point q varies over the j-th element (integration point).

Introducing the notation (4.3) into Eq. (4.2), the discrete form of the solution becomes

$$-\frac{1}{2}u^i+\sum_{j=1}^{N}\hat{H}_{ij}\,u^j=\sum_{j=1}^{N}G_{ij}\,u_n^j \tag{4.4}$$

Moreover, setting

$$H_{ij}=\hat{H}_{ij}-\frac{1}{2}\delta_{ij} \tag{4.5}$$

where δ_{ij} is the *Kronecker delta*, which is defined as $\delta_{ij}=0$ for $i\neq j$ and $\delta_{ij}=1$ for $i=j$, Eq. (4.4) may further be written as

$$\sum_{j=1}^{N}H_{ij}\,u^j=\sum_{j=1}^{N}G_{ij}\,u_n^j \tag{4.6}$$

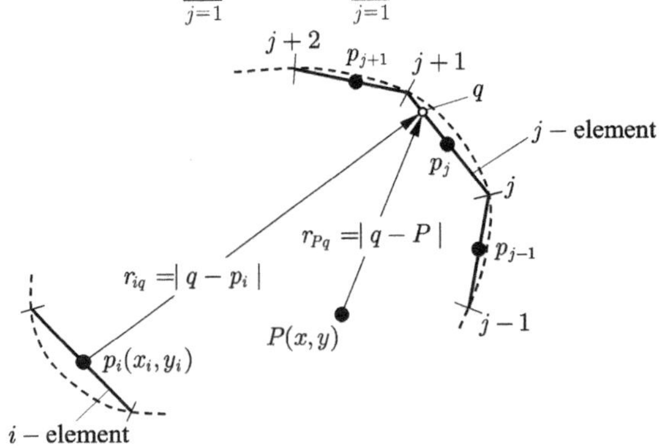

FIGURE 4.2 Nodal-point location and relative distances for constant element discretization.

Equation (4.6) is applied consecutively to all the nodes $p_i (i = 1, 2, \ldots, N)$ yielding a system of N linear algebraic equations, which are arranged in matrix form

$$[H]\{u\} = [G]\{u_n\} \tag{4.7}$$

where $[H]$ and $[G]$ are $N \times N$ square matrices, and $\{u\}$ and $\{u_n\}$ are $N \times 1$ vectors.

Let us assume mixed boundary conditions. In this case, the part Γ_1 of the boundary on which u is prescribed and the part Γ_2 on which u_n is prescribed, are discretized into N_1 and N_2 constant elements, respectively ($\Gamma_1 \cup \Gamma_2 = \Gamma$, $N_1 + N_2 = N$). Hence, Eq. (4.7) again contains N unknowns, that is $N - N_1$ values of u on Γ_2 and $N - N_2$ values of u_n on Γ_1. These N unknown quantities may be determined from the system of Eq. (4.7).

Prior to solving the system of the equations, it is necessary to separate the unknown from the known quantities. Eq. (4.7) may be written after partitioning of the matrices in the following way

$$\begin{bmatrix} [H_{11}] & [H_{12}] \end{bmatrix} \begin{Bmatrix} \{\overline{u}\}_1 \\ \{u\}_2 \end{Bmatrix} = \begin{bmatrix} [G_{11}] & [G_{12}] \end{bmatrix} \begin{Bmatrix} \{u_n\}_1 \\ \{\overline{u}_n\}_2 \end{Bmatrix} \tag{4.8}$$

where $\{\overline{u}\}_1$ and $\{\overline{u}_n\}_2$ denote the prescribed quantities on Γ_1 and Γ_2, respectively, while $\{u_n\}_1$ and $\{u\}_2$ denote the corresponding unknown quantities. Carrying out the multiplications and moving all the unknowns to the left hand side of the equation, we obtain

$$[A]\{X\} = \{B\} \tag{4.9}$$

where

$$[A] = \begin{bmatrix} [H_{12}] & -[G_{11}] \end{bmatrix} \tag{4.10a}$$

$$\{X\} = \begin{Bmatrix} \{u\}_2 \\ \{u_n\}_1 \end{Bmatrix} \tag{4.10b}$$

$$\{B\} = -[H_{11}]\{\overline{u}\}_1 + [G_{12}]\{\overline{u}_n\}_2 \tag{4.10c}$$

$[A]$ being a square matrix with dimensions $N \times N$, and $\{X\}$, $\{B\}$ vectors with dimension N.

The previous rearrangement of matrices is effective when the N_1 points where the values of u are prescribed, thus also the N_2 points where the values of u_n are prescribed, are consecutive. Otherwise, the partitioning of the matrices in Eq. (4.8) should be preceded by an appropriate rearrangement of columns in $[H]$ and $[G]$. Matrices $[A]$ and $\{B\}$ can also be constructed using another more straightforward procedure, which is based on the observation that matrix $[A]$ will eventually contain all the columns of $[H]$ and $[G]$ that correspond to the unknown boundary values of u and u_n, whereas vector $\{B\}$

will result as the sum of those columns of $[H]$ and $[G]$, which correspond to the known values u and u_n, after they have been multiplied by these values. It should be noted that a change of sign occurs, when the columns of $[G]$ or $[H]$ are moved to the other side of the equation. The aforementioned procedure is more suitable for writing the computer program.

The solution of the simultaneous equations (4.9) yields the unknown boundary quantities u and u_n. Therefore, knowing all the boundary quantities on Γ, the solution u can be computed at any point $P(x, y)$ in the domain Ω using Eq. (3.31) with $\varepsilon(P) = 1$. Applying the same discretization as in Eq. (4.1), we arrive at the following expression

$$u(P) = \sum_{j=1}^{N} \hat{H}_{ij}\, u^j - \sum_{j=1}^{N} G_{ij}\, u_n^j \tag{4.11}$$

The coefficients G_{ij} and $\hat{H}_{ij}$ are computed again from the integrals (4.3), but in this case the boundary point p_i is replaced in the expressions by the field point P in Ω (see Fig. 4.2).

The partial derivatives $\partial u/\partial x$ and $\partial u/\partial y$ can be evaluated at points within Ω by direct differentiation of Eq. (3.31) for $\varepsilon(P) = 1$. Since the fundamental solution and its derivatives are continuous functions of x and y, the differentiation passes under the integral sign giving

$$\frac{\partial u}{\partial x} = -\int_{\Gamma} \left[\frac{\partial v}{\partial x}\frac{\partial u}{\partial n} - u\frac{\partial}{\partial x}\left(\frac{\partial v}{\partial n}\right) \right] ds \tag{4.12}$$

$$\frac{\partial u}{\partial y} = -\int_{\Gamma} \left[\frac{\partial v}{\partial y}\frac{\partial u}{\partial n} - u\frac{\partial}{\partial y}\left(\frac{\partial v}{\partial n}\right) \right] ds \tag{4.13}$$

where the derivatives of the fundamental solution (3.13) are obtained as

$$\left.\begin{aligned}
&\frac{\partial v}{\partial x} = \frac{1}{2\pi}\frac{1}{r}\frac{\partial r}{\partial x}, \quad \frac{\partial v}{\partial y} = \frac{1}{2\pi}\frac{1}{r}\frac{\partial r}{\partial y}, \quad \frac{\partial v}{\partial n} = \frac{1}{2\pi}\frac{1}{r}\frac{\partial r}{\partial n}\\
&\frac{\partial}{\partial x}\left(\frac{\partial v}{\partial n}\right) = -\frac{1}{2\pi r^2}\left(\frac{\partial r}{\partial n}\frac{\partial r}{\partial x} - \frac{\partial r}{\partial t}\frac{\partial r}{\partial y}\right)\\
&\frac{\partial}{\partial y}\left(\frac{\partial v}{\partial n}\right) = -\frac{1}{2\pi r^2}\left(\frac{\partial r}{\partial n}\frac{\partial r}{\partial y} + \frac{\partial r}{\partial t}\frac{\partial r}{\partial x}\right)\\
&\frac{\partial r}{\partial x} = -\frac{\partial r}{\partial \xi} = -\frac{\xi - x}{r}, \quad \frac{\partial r}{\partial y} = -\frac{\partial r}{\partial \eta} = -\frac{\eta - y}{r}\\
&\frac{\partial r}{\partial n} = \nabla r\cdot\mathbf{n} = \frac{\partial r}{\partial \xi}n_x + \frac{\partial r}{\partial \eta}n_y, \quad \frac{\partial r}{\partial t} = \nabla r\cdot\mathbf{t} = -\frac{\partial r}{\partial \xi}n_y + \frac{\partial r}{\partial \eta}n_x
\end{aligned}\right\} \tag{4.14}$$

Expressions regarding partial derivatives of r along with their derivation are given in Appendix A. The last of Eq. (4.14) is obtained by noting that the components of the tangential unit vector **t** are $t_x = -n_y$ and $t_y = n_x$. Attention should be paid to the evaluation of the derivatives $\partial(\partial v/\partial n)/\partial x$ and $\partial(\partial v/\partial n)/\partial y$. The differentiation with respect to n is carried out at point $q(\xi,\eta)\in\Gamma$, while differentiation with respect to x or y is carried out at point $P(x, y)\in\Omega$.

Equations (4.12) and (4.13) are discretized in the same way as Eq. (4.1) and they yield the following expressions for the evaluation of the derivatives $u_{,x}$ and $u_{,y}$ at point $P(x, y)$

$$u_{,x}(P) = \left(\frac{\partial u}{\partial x}\right)_P = \sum_{j=1}^{N} (\hat{H}_{Pj})_{,x}\, u^j - \sum_{j=1}^{N} (G_{Pj})_{,x} u_n^j \tag{4.15}$$

$$u_{,y}(P) = \left(\frac{\partial u}{\partial y}\right)_P = \sum_{j=1}^{N} (\hat{H}_{Pj})_{,y}\, u^j - \sum_{j=1}^{N} (G_{Pj})_{,y} u_n^j \tag{4.16}$$

where the influence coefficients are given by the integrals

$$\left.\begin{aligned} (G_{Pj})_{,x} &= \int_{\Gamma_j} \frac{\partial v(P,q)}{\partial x}\, ds, & (\hat{H}_{Pj})_{,x} &= \int_{\Gamma_j} \frac{\partial}{\partial x}\left[\frac{\partial v(P,q)}{\partial n_q}\right] ds \\ (G_{Pj})_{,y} &= \int_{\Gamma_j} \frac{\partial v(P,q)}{\partial y}\, ds, & (\hat{H}_{Pj})_{,y} &= \int_{\Gamma_j} \frac{\partial}{\partial y}\left[\frac{\partial v(P,q)}{\partial n_q}\right] ds \end{aligned}\right\} \tag{4.17}$$

4.3 EVALUATION OF LINE INTEGRALS

The line integrals G_{ij} and $\hat{H}_{ij}$ defined in Eq. (4.3) are evaluated numerically using a standard Gaussian quadrature. Of course, these integrals can be evaluated analytically using symbolic languages, for example, *Mathematica* [1] or *Maple* [2], but the resulting expressions are very lengthy and may cover more than a page. Hence, the advantage of accuracy over the numerical integration is rather lost, due to the complexity of the mathematical expressions. For this reason, the Gaussian integration appears as the most suitable method for computing line integrals (see Appendix B). Two cases are distinguished for the integrals of the influence coefficients.

i. *Off-diagonal elements*, $i \neq j$

In this case the point $p_i\ (x_i, y_i)$ lies outside the j–element, which means that the distance $r = |q - p_i|$ does not vanish and, consequently, the integral is regular.

The Gaussian integration is performed over the interval $-1 \le \xi \le 1$,

$$\int_{-1}^{1} f(\xi)\, d\xi = \sum_{k=1}^{n} w_k\, f(\xi_k) \tag{4.18}$$

where n is the number of integration points (Gauss points), and ξ_k and w_k ($k = 1, 2, \ldots, n$) are the abscissas and weights of the Gaussian quadrature of order n, respectively.

Let us consider the element j over which the integration will be carried out. This element is defined by the coordinates (x_j, y_j) and (x_{j+1}, y_{j+1}) of its extreme points, which are expressed in a *global system* having axes x and y, and origin at point O (Fig. 4.3). Next, a *local system of axes* x' and y' is introduced at point p_j of the element. The local coordinates $(x', 0)$ of point q on the j-th element are related to the global coordinates of the xy-system through the expressions

$$x = \frac{x_{j+1} + x_j}{2} + \frac{x_{j+1} - x_j}{\ell_j} x', \qquad -\frac{\ell_j}{2} \leq x' \leq \frac{\ell_j}{2} \tag{4.19a}$$

$$y = \frac{y_{j+1} + y_j}{2} + \frac{y_{j+1} - y_j}{\ell_j} x', \qquad -\frac{\ell_j}{2} \leq x' \leq \frac{\ell_j}{2} \tag{4.19b}$$

where ℓ_j is the length of the j-th element, which is given in terms of the coordinates of the end points as

$$\ell_j = \sqrt{(x_{j+1} - x_j)^2 + (y_{j+1} - y_j)^2}$$

Expressions that map the global coordinates onto the integration interval $[-1, +1]$ are obtained by introducing in Eqs. (4.19a,b) the geometric relation

$$\frac{x'}{\ell_j/2} = \xi$$

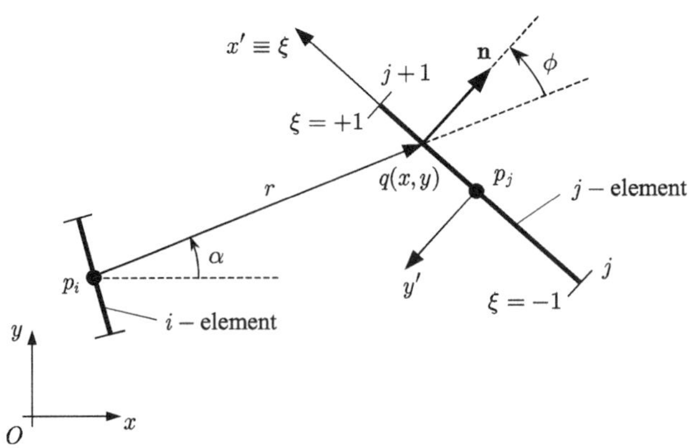

FIGURE 4.3 Global and local coordinate systems.

Thus, the resulting coordinate transformation becomes

$$x(\xi) = \frac{x_{j+1} + x_j}{2} + \frac{x_{j+1} - x_j}{2}\xi \tag{4.20a}$$

$$y(\xi) = \frac{y_{j+1} + y_j}{2} + \frac{y_{j+1} - y_j}{2}\xi \tag{4.20b}$$

Moreover, it is

$$\begin{aligned} ds &= \sqrt{dx^2 + dy^2} = \sqrt{\left(\frac{x_{j+1} - x_j}{2}\right)^2 + \left(\frac{y_{j+1} - y_j}{2}\right)^2}\, d\xi \\ &= \frac{\ell_j}{2}\, d\xi \end{aligned} \tag{4.21}$$

Hence, the Jacobian of the transformation is

$$\left|J(\xi)\right| = \frac{\ell_j}{2}$$

On the basis of the foregoing, the integrals of the influence coefficients are evaluated numerically in the following way:

a. The integral of G_{ij}

$$G_{ij} = \int_{\Gamma_j} v\, ds = \int_{-1}^{1} \frac{1}{2\pi} \ln[r(\xi)] \frac{\ell_j}{2}\, d\xi = \frac{\ell_j}{4\pi} \sum_{k=1}^{n} \ln\left[r(\xi_k)\right] w_k \tag{4.22}$$

where

$$r(\xi_k) = \sqrt{[x(\xi_k) - x_i]^2 + [y(\xi_k) - y_i]^2} \tag{4.23}$$

b. The integral of $\hat{H}_{ij}$

The Gauss integration can be employed to evaluate this integral. But it is simple to evaluate it analytically using the procedure described in Refs. [3,4].

Thus, referring to Fig. 4.4, we notice that

$$ds \cos\phi = r\, d\alpha$$

which can be used along with Eq. (4.14) and (A.7) to derive the expression

$$\hat{H}_{ij} = \int_{\Gamma_j} \frac{\partial v}{\partial n}\, ds = \int_{\Gamma_j} \frac{1}{2\pi} \frac{\cos\phi}{r}\, ds = \int_{\Gamma_j} \frac{1}{2\pi}\, da = \frac{a_{j+1} - a_j}{2\pi} \tag{4.24}$$

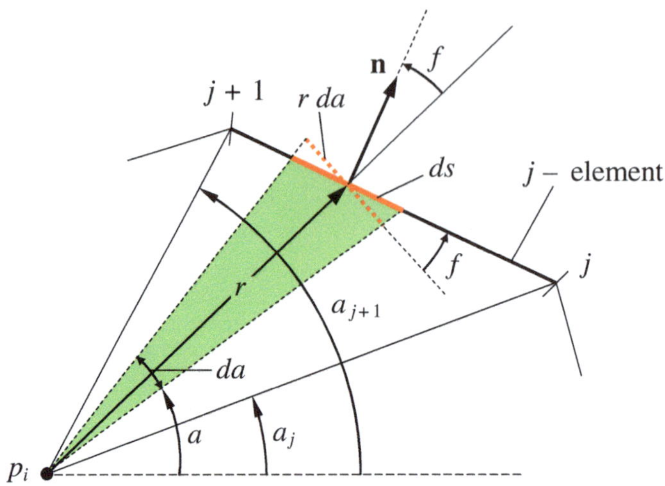

FIGURE 4.4 Definition of angles involved in the numerical integration over constant elements.

The angles a_{j+1} and a_j are computed from

$$\tan a_{j+1} = \frac{y_{j+1} - y_i}{x_{j+1} - x_i} \tag{4.25}$$

$$\tan a_j = \frac{y_j - y_i}{x_j - x_i} \tag{4.26}$$

where x_{j+1}, y_{j+1}, and x_j, y_j are the coordinates of the extreme points of the j-th element.

ii. *Diagonal elements,* $i = j$

In this case the node p_i coincides with node p_j, and r lies on the element. Consequently, it is $\phi = \pi/2$ or $\phi = 3\pi/2$, which yields $\cos\phi = 0$.

Moreover, we have

$$x_{p_j} = \frac{x_{j+1} + x_j}{2}, \qquad y_{p_j} = \frac{y_{j+1} + y_j}{2}$$

and

$$r(\xi) = \sqrt{[x(\xi) - x_{p_j}]^2 + [y(\xi) - y_{p_j}]^2} = \frac{\ell_j}{2}|\xi| \tag{4.27}$$

Hence,

$$\begin{aligned} G_{jj} &= \int_{\Gamma_j} \frac{1}{2\pi} \ln r \, ds = 2\int_0^{\ell_j/2} \frac{1}{2\pi} \ln r \, dr \\ &= \frac{1}{\pi}[r\ln r - r]_0^{\ell_j/2} = \frac{1}{\pi}\frac{\ell_j}{2}\left[\ln(\ell_j/2) - 1\right] \end{aligned} \tag{4.28a}$$

and

$$\hat{H}_{jj} = \frac{1}{2\pi}\int_{\Gamma_j} \frac{\cos\phi}{r}\, ds = \frac{1}{2\pi}\int_{-1}^{1} \frac{\cos\phi}{|\xi|}\, d\xi$$
$$= \frac{2}{2\pi}[\cos\phi \ln|\xi|]_0^1 = 0 \tag{4.28b}$$

It should be noted that for higher order elements (e.g. linear or parabolic) analytical integration is not applicable, and for this reason, other integration techniques are employed (see Chapter 5).

4.4 EVALUATION OF DOMAIN INTEGRALS

The integral representation of the solution of the Poisson equation (3.36) for points on the boundary, when discretized into N constant elements, may be written as

$$\varepsilon^i u^i = \int_\Omega vf\, d\Omega - \sum_{j=1}^{N} \int_{\Gamma_j} v\frac{\partial u}{\partial n}\, ds + \sum_{j=1}^{N} \int_{\Gamma_j} u\frac{\partial v}{\partial n}\, ds \tag{4.29}$$

In some cases the techniques presented in Section 3.5 for transforming the domain integral

$$F^i = \int_\Omega v(p_i, Q)\, f(Q)\, d\Omega_Q \tag{4.30}$$

to a boundary line integral are not suitable, and, if more advanced techniques are not utilized, such as the dual reciprocity method (see Section 8.3), the only recourse is a domain discretization. In this method, the domain Ω is divided into M two-dimensional elements or *cells*, for example, triangular or rectangular cells (see Fig. 4.5), over which numerical integration is performed. Thus, employing two-dimensional Gaussian integration, Eq. (4.30) becomes

$$F^i = \sum_{j=1}^{M} \left[\sum_{k=1}^{n} w_k\, v(p_i, Q_k)\, f(Q_k) \right] A_j \tag{4.31}$$

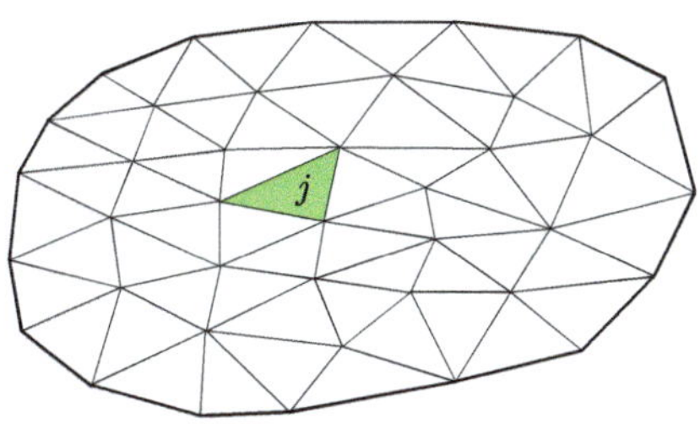

FIGURE 4.5 Discretization of domain Ω into triangular cells.

where Q_k and w_k ($k=1,2,\ldots,n$) are the k-th integration point and corresponding weight for the Gaussian quadrature on the j-th cell, and A_j ($j=1,2,\ldots,M$) is the area of the cell (see Appendix B).

By means of Eq. (4.31) and the notation introduced in Eqs. (4.3) and (4.5), Eq. (4.29) takes the following matrix form

$$[H]\{u\}+\{F\}=[G]\{u_n\} \tag{4.32}$$

Equation (4.32) is first reordered on the basis of the specified boundary conditions, and subsequently, they are solved for the unknown boundary quantities. The values of u at points inside Ω may then be computed from Eq. (4.29) for $\varepsilon^i=1$. It should be noted that, for a point i ($i=1,2,\ldots,M$) lying on the j-th cell, the domain integral exhibits a logarithmic singularity and special care must be taken for its evaluation (see Appendix B and also Ref. [5]).

4.5 PROGRAM LABECON FOR SOLVING THE LAPLACE EQUATION WITH CONSTANT BOUNDARY ELEMENTS

On the basis of the analysis presented in the previous sections, a computer program has been written in FORTRAN language [6] for the solution of boundary value problems described by the Laplace equation. This program employs constant elements to approximate the line integrals.

4.5.1 Main program

The structure of program LABECON is described by the macro flow chart shown in Fig. 4.6. The main program defines the maximum dimensions of the arrays and opens two files—the file containing the data and the file in which the results are written. Next, it calls the following eight subroutines:

INPUT	Reads the data from INPUTFILE
GMATR	Forms the matrix $[G]$ defined by Eq. (4.3)
HMATR	Forms the matrix $[H]$ defined by Eqs. (4.5) and (4.3)
ABMATR	Rearranges the matrices $[H]$ and $[G]$ according to the boundary conditions and forms the matrices $[A]$ and $\{B\}$ of Eq. (4.9)
SOLVEQ	Solves the system of linear equations $[A]\{X\}=\{B\}$ using Gauss elimination
REORDER	Rearranges the boundary values and forms the vectors $\{u\}$ and $\{u_n\}$
UINTER	Computes the values of u at the internal points
OUTPUT	Writes the results in OUTPUTFILE

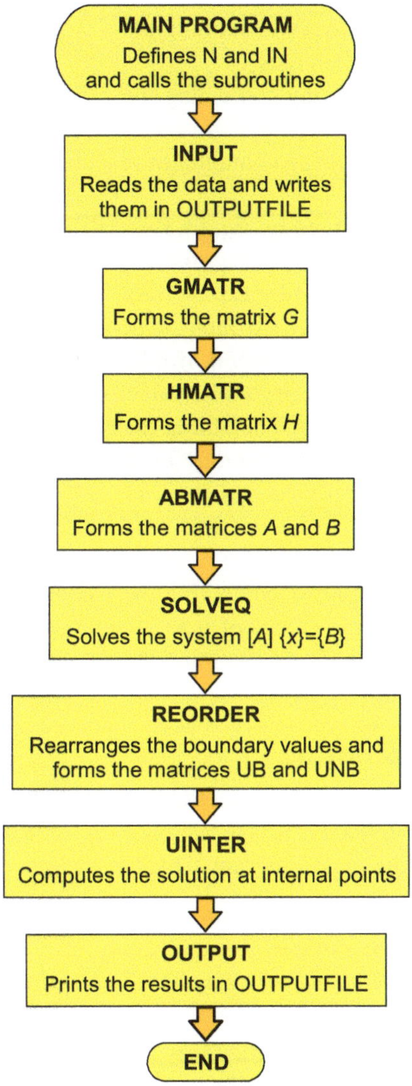

FIGURE 4.6 Macro flow chart of program LABECON.

The variables and arrays introduced in the program along with their meaning are given below:

N Number of boundary elements and boundary nodes

IN Number of internal points where the solution is computed

INDEX One-dimensional array in which a type of boundary condition is assigned to the nodes. It is INDEX(J) = 0 when u is prescribed, while INDEX(J) = 1 when $\partial u/\partial n$ is prescribed

XL One-dimensional array containing the x coordinates of the extreme points of all the elements

YL	One-dimensional array containing the y coordinates of the extreme points of all the elements
XM	One-dimensional array containing the x coordinates of all the boundary nodes
YM	One-dimensional array containing the y coordinates of all the boundary nodes
G	Square matrix defined by Eq. (4.3)
H	Square matrix defined by Eq. (4.5)
A	Square matrix defined by Eq. (4.10a)
UB	One-dimensional array. At input, it contains the boundary values of u, if INDEX(J) = 0, or $\partial u/\partial n$, if INDEX(J) = 1. At output it contains all the boundary nodal values of u
UNB	One-dimensional array containing the right-hand side vector of equation $[A]\{X\} = \{B\}$ given by Eq. (4.10c). At output it contains the boundary nodal values of $\partial u/\partial n$
XIN	One-dimensional array containing the x coordinates of the internal points at which the values of u are computed
YIN	One-dimensional array containing the y coordinates of the internal points at which the values of u are computed
UIN	One-dimensional array containing the computed values of u at the internal points

4.5.2 INPUT subroutine

The INPUT subroutine reads in free FORMAT all the required data. The data have been written in INPUTFILE, to which the user has assigned a specific name as required by the main program. This file contains the following data:

1. *User's name*. One line with the name of the user.
2. *Title*. One line with the title of the application.
3. *Extreme points of the boundary elements*. N pairs of values consisting of the coordinates XL and YL of the extreme points. They are given in the positive sense, which for a closed domain Ω is counter-clockwise (Fig. 4.7a), whereas for an open domain (external domain) is clockwise (Fig. 4.7b).

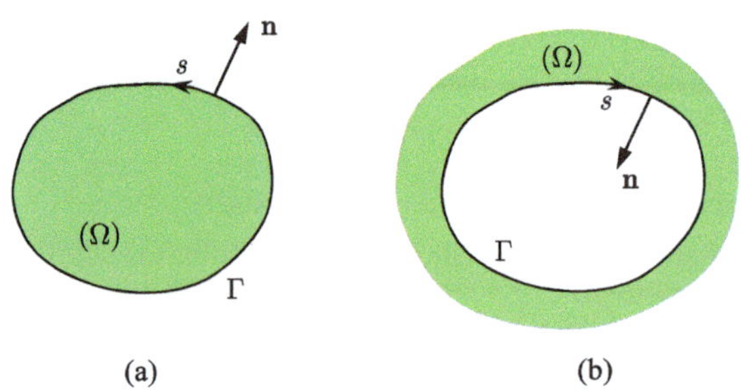

FIGURE 4.7 Positive direction and normal vector on the boundary of (a) closed domain; (b) open domain.

4. *Boundary conditions.* N pairs of prescribed values consisting either of INDEX = 0 and u, or INDEX = 1 and $\partial u/\partial n$.
5. *Internal points.* IN pairs of values consisting of the coordinates XIN and YIN of the internal points at which the values of u is computed.

Finally, the subroutine INPUT writes the data in OUTPUTFILE, to which the user assigns a specific name.

4.5.3 GMATR subroutine

The GMATR subroutine forms the matrix G defined in Eq. (4.3) by using the subroutines RLINTC and SLINTC. These subroutines perform the following tasks:

The RLINTC, (R)egular (L)ine (In)tegral for (C)onstant Elements, computes the off-diagonal elements of the G matrix.

The SLINTC, (S)ingular (L)ine (In)tegral for (C)onstant Elements, computes the diagonal elements of the G matrix.

4.5.4 RLINTC subroutine

The RLINTC subroutine computes regular line integrals on constant elements employing a four-point Gaussian integration scheme (see Appendix B). It uses the coordinates of the i-th nodal point and those of the extreme points of the j-th element (points j and $j+1$), it renames them to 0, 1, and 2 (see Fig. 4.8), respectively, and then it evaluates the integral of Eq. (4.22).

4.5.5 SLINTC subroutine

The SLINTC subroutine uses the coordinates of the extreme points of the j-th element (points j and $j+1$), it renames them to 1 and 2, respectively, and then calculates the integral on the basis of Eq. (4.28a).

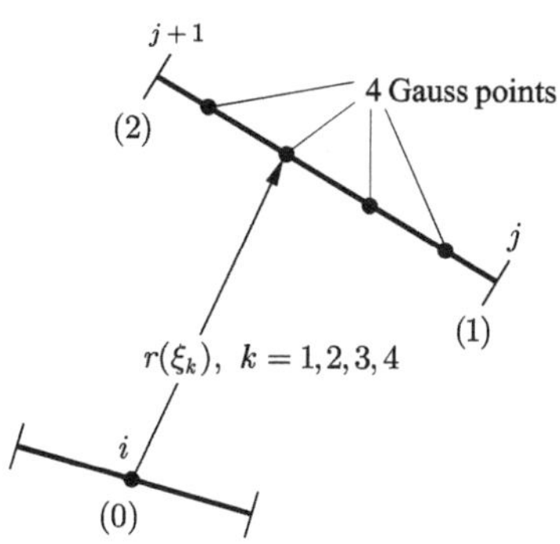

FIGURE 4.8 Four-point Gaussian quadrature and point numbering for subroutine RLINTC.

4.5.6 HMATR subroutine

The HMATR subroutine forms the matrix H defined in Eq. (4.5). The diagonal elements of the matrix $\hat{H}$ are $\hat{H}_{ii} = 0$ (see Eq. 4.28b), while the off-diagonal elements are calculated using the DALPHA subroutine.

4.5.7 DALPHA subroutine

The DALPHA subroutine uses the coordinates of the i-th nodal point and those of the extreme points of the j-th element (points j and $j+1$), it renames them to 0, 1, and 2 (see Fig. 4.8), respectively, and then calculates the off-diagonal elements H_{ij} according to Eq. (4.24).

4.5.8 ABMATR subroutine

The ABMATR subroutine generates the matrix A and the vector B of Eq. (4.9) from the columns of matrices G and H. The j-th column of matrix A consists of the corresponding column of H, if u is unknown (INDEX(J) = 1), or that of $-G$, if u_n is unknown (INDEX(J) = 0). The vector B results as the sum of the columns of $-H$ multiplied by the corresponding known values of u (INDEX(J) = 0) and the columns of G multiplied by the corresponding known values of u_n (INDEX(J) = 1). It should be noted that in the construction of A and B a change of sign occurs, when columns of H or G are shifted, respectively, to the left- or to the right-hand side of the equation (see Eq. 4.10).

4.5.9 SOLVEQ subroutine

The SOLVEQ subroutine employs the matrix A and the vector $B(=\text{UNB})$ and calls the subroutine LEQS, which solves the system of linear equations. The solution is stored in the vector UNB.

4.5.10 LEQS subroutine

The LEQS subroutine uses the matrix A, the vector B and the parameter N to solve the system of equations $AX = B$ by Gauss elimination. The solution is stored in the vector B. The output parameter LSING takes the value LSING = 0, when the matrix A is regular, or LSING = 1, when the matrix A is singular.

4.5.11 REORDER subroutine

The REORDER subroutine rearranges the vectors UB and UNB on the basis of the boundary condition vector INDEX. At output UB contains all the values of u and UNB all the values of its normal derivative $\partial u/\partial n$.

4.5.12 UINTER subroutine

The UINTER subroutine uses the boundary values of the vectors UB and UNB to compute the values of u at the specified internal points on the basis of Eq. (4.11).

The evaluation of derivatives $u_{,x}$ and $u_{,y}$ at the internal points is left as an exercise to the reader (see Problem 4.1).

4.5.13 OUTPUT subroutine

The OUTPUT subroutine writes all the results in the OUTPUTFILE to which the user gives a specific name. First, it lists the coordinates of the boundary nodal points along with the corresponding nodal values of u and $\partial u/\partial n$, and then the coordinates of the internal points and the computed values of u at those points.

In the sequel, we provide the listing of program LABECON. The electronic version of the code is available on this book's companion website.

```
C=====================================================================
C
      PROGRAM LABECON
C
C  This program solves the two dimensional (LA)place equation
C  using the (B)oundary (E)lement method with (CON)stant
C  boundary elements
C
      IMPLICIT REAL*8 (A-H,O-Z)
      CHARACTER*15 INPUTFILE,OUTPUTFILE
C
C  Set the maximum dimensions
C
      PARAMETER (N=16)
      PARAMETER (IN=9)
C
C   N= Number of boundary elements which is equal to the number
C      of boundary nodes
C  IN= Number of internal points where the function u is calculated
C
      DIMENSION INDEX(N)
      DIMENSION XL(N+1),YL(N+1),XM(N),YM(N),G(N,N),H(N,N)
      DIMENSION UB(N),A(N,N),UNB(N),XIN(IN),YIN(IN),UIN(IN)
C
C  Read the names and open the input and output files
C
      WRITE (*,'(A)')' Name of the INPUTFILE (max.15 characters)'
      READ  (*,'(A)') INPUTFILE
      WRITE (*,'(A)')' Name of the OUTPUTFILE (max.15 characters)'
      READ  (*,'(A)') OUTPUTFILE
      OPEN (UNIT=1, FILE=INPUTFILE)
      OPEN (UNIT=2, FILE=OUTPUTFILE)
C
C  Read the data from INPUTFILE
C
      CALL INPUT (XL,YL,XIN,YIN,INDEX,UB,N,IN)
```

```
C
C  Compute the G matrix
C
      CALL GMATR (XL,YL,XM,YM,G,N)
C
C  Compute the H matrix
C
      CALL HMATR (XL,YL,XM,YM,H,N)
C
C  Form the system of equations AX=B
C
      CALL ABMATR (G,H,A,UNB,UB,INDEX,N)
C
C  Solve the system of equations
C
      CALL SOLVEQ(A,UNB,N,LSING)
C
C  Form the vectors U and UN of all the boundary values
C
      CALL REORDER (UB,UNB,INDEX,N)
C
C  Compute the values UIN of u at the internal points
C
      CALL UINTER (XL,YL,XIN,YIN,UB,UNB,UIN,N,IN)
C
C  Print the results in the OUTPUTFILE

C
      CALL OUTPUT (XM,YM,UB,UNB,XIN,YIN,UIN,N,IN)
C
C  Close input and output files
C
      CLOSE(1)
      CLOSE(2)
C
      STOP
      END
C
C
C=====================================================================
C
      SUBROUTINE INPUT (XL,YL,XIN,YIN,INDEX,UB,N,IN)
C
C  This subroutine reads the data from the input file
C  and writes them in the output file
C
      IMPLICIT REAL*8 (A-H,O-Z)
      CHARACTER*80 NAME,TITLE
      DIMENSION XL(N+1),YL(N+1),XIN(IN),YIN(IN),INDEX(N),UB(N)
C
      WRITE(2,100)
  100 FORMAT('',69('*'))

C
C  Read user's name
C
      READ(1,'(A)')NAME
C
      WRITE(2,'(A)')NAME

C  Read the title of the program
C
      READ(1,'(A)')TITLE
C
      WRITE(2,'(A)')TITLE
C
```

```
      WRITE(2,200)N,IN
  200 FORMAT(//'BASIC PARAMETERS'//2X,'NUMBER OF BOUNDARY ELEMENTS='
     1,I3/2X,'NUMBER OF INTERNAL POINTS WHERE THE FUNCTION IS CALCULATED
     1=',I3)
C
C  Read the coordinates XL,YL of the extreme points of the boundary elements
C
      READ(1,*) (XL(I),YL(I),I=1,N)
C
C  Write the coordinates in the output file
C
      WRITE(2,300)
  300 FORMAT(//2X,'COORDINATES OF THE EXTREME POINTS OF THE BOUNDARY '
     1'ELEMENTS',//2X,'POINT',9X,'XL',15X,'YL')
      DO 20 I=1,N
   20 WRITE(2,400) I,XL(I),YL(I)
  400 FORMAT(2X,I3,2(3X,E14.5))
C
C  Read the boundary conditions and store them in UB(I).
C  If INDEX(I)=0, the value UB(I) is the prescribed value of u.
C  If INDEX(I)=1, the normal derivative un is prescribed.
C
      READ(1,*) (INDEX(I),UB(I),I=1,N)
C
C  Write the boundary conditions in the output file
C
      WRITE(2,500)

  500 FORMAT(//2X,'BOUNDARY CONDITIONS'//2X,'NODE',6X,'INDEX',
     1 7X,'PRESCRIBED VALUE')
      DO 30 I=1,N
   30 WRITE(2,600) I,INDEX(I),UB(I)
  600 FORMAT (2X,I3,9X,I1,8X,E14.5)
C
C  Read the coordinates of the internal points
C
      READ(1,*) (XIN(I),YIN(I),I=1,IN)
      RETURN
      END
C
C
C=================================================================
C
      SUBROUTINE GMATR (XL,YL,XM,YM,G,N)
C
C  This subroutine computes the elements of the G matrix
C
      IMPLICIT REAL*8 (A-H,O-Z)
      DIMENSION XL(N+1),YL(N+1),XM(N),YM(N)
      DIMENSION G(N,N)
C
C  Compute the nodal coordinates and store them in the arrays
C  XM and YM
C
      XL(N+1)=XL(1)
      YL(N+1)=YL(1)
      DO 10 I=1,N
      XM(I)=(XL(I)+XL(I+1))/2.
   10 YM(I)=(YL(I)+YL(I+1))/2.
C
C  Compute the elements of matrix G
C
      DO 20 I=1,N
      DO 20 J=1,N
      JP1=J+1
      IF(I.NE.J)THEN
      CALL RLINTC(XM(I),YM(I),XL(J),YL(J),XL(JP1),YL(JP1),RESULT)
```

```
      G(I,J)=RESULT
      ELSEIF(I.EQ.J)THEN
      CALL SLINTC(XL(J),YL(J),XL(JP1),YL(JP1),RESULT)
      G(I,J)=RESULT
      ENDIF
   20 CONTINUE
      RETURN
      END
C
C
C=============================================================
C
      SUBROUTINE RLINTC(X0,Y0,X1,Y1,X2,Y2,RESULT)
C
C  This subroutine computes the off-diagonal elements of the
C  matrix G
C
C  RA= The distance of point O from the Gauss integration point
C      on the boundary element
C
C  WG= The weights of the Gauss integration
C
C  XI= The coordinates of the Gauss integration points in the
C      interval [-1,1]
C
C  XC,YC= The global coordinates of the Gauss integration points
C
      IMPLICIT REAL*8 (A-H,O-Z)
      DIMENSION XC(4),YC(4),XI(4),WG(4)
      DATA XI/-0.86113631,-0.33998104,0.33998104,0.86113631/
      DATA WG/0.34785485,0.65214515,0.65214515,0.34785485/
      PI=ACOS(-1.)
      AX=(X2-X1)/2.
      AY=(Y2-Y1)/2.
      BX=(X2+X1)/2.
      BY=(Y2+Y1)/2.
C
C  Compute the line integral
C
      RESULT=0.
C
      DO 30 I=1,4
      XC(I)=AX*XI(I)+BX
      YC(I)=AY*XI(I)+BY
      RA=SQRT((XC(I)-X0)**2+(YC(I)-Y0)**2)
   30 RESULT=RESULT+DLOG(RA)*WG(I)
      SL=2.*SQRT(AX**2+AY**2)
      RESULT=RESULT*SL/(4.*PI)
      RETURN
      END
C
C
C=================================================================
C
```

```
      SUBROUTINE SLINTC(X1,Y1,X2,Y2,RESULT)
C
C  This subroutines computes the diagonal elements of the matrix G
C
      IMPLICIT REAL*8 (A-H,O-Z)
      PI=ACOS(-1.)
      AX=(X2-X1)/2.
      AY=(Y2-Y1)/2.
      SL=SQRT(AX**2+AY**2)
      RESULT=SL*(DLOG(SL)-1.)/PI
      RETURN
      END
C
C
C=====================================================================
C
      SUBROUTINE HMATR(XL,YL,XM,YM,H,N)
C

C  This subroutine computes the elements of the H matrix
C
      IMPLICIT REAL*8 (A-H,O-Z)
      DIMENSION XL(N+1),YL(N+1),XM(N),YM(N)
      DIMENSION H(N,N)
C
      PI=ACOS(-1.)
C
      XL(N+1)=XL(1)
      YL(N+1)=YL(1)
      DO 10  I=1,N
      XM(I)=(XL(I)+XL(I+1))/2.
   10 YM(I)=(YL(I)+YL(I+1))/2.
C
C  Compute the elements of the H matrix
C
      DO 20  I=1,N
      DO 20  J=1,N
      IF (I.NE.J) THEN
      CALL DALPHA (XM(I),YM(I),XL(J),YL(J),XL(J+1),YL(J+1),RESULT)
      H(I,J)=RESULT
      ELSEIF (I.EQ.J) THEN
      H(I,J)=-0.5
      ENDIF
   20 CONTINUE
      RETURN
      END
C
C
C=====================================================================
C
```

```
      SUBROUTINE DALPHA(X0,Y0,X1,Y1,X2,Y2,RESULT)
C
C  This subroutine computes the off diagonal elements of
C  the H matrix
C
      IMPLICIT REAL*8 (A-H,O-Z)
      PI=ACOS(-1.)
      DY1=Y1-Y0
      DX1=X1-X0
      DY2=Y2-Y0
      DX2=X2-X0
      DL1=SQRT(DX1**2+DY1**2)
      COS1=DX1/DL1
      SIN1=DY1/DL1
      DX2R=DX2*COS1+DY2*SIN1
      DY2R=-DX2*SIN1+DY2*COS1
      DA=ATAN2(DY2R,DX2R)
      RESULT=DA/(2.*PI)
      RETURN
      END
C
C
C===================================================================
C
      SUBROUTINE ABMATR (G,H,A,UNB,UB,INDEX,N)
C
C  This subroutine rearranges the matrices G and H and produces
C  the matrices A and B=UNB
C
      IMPLICIT REAL*8 (A-H,O-Z)
      DIMENSION G(N,N),H(N,N),A(N,N),UNB(N),UB(N),INDEX(N)
C
C  Reorder the columns of the system of equations and
C  store them in A
C
      DO 40 J=1,N
      IF(INDEX(J).EQ.0) THEN
      DO 20 I=1,N
   20 A(I,J)=-G(I,J)
      ELSEIF(INDEX(J).NE.0) THEN
      DO 30 I=1,N
   30 A(I,J)=H(I,J)
      END IF
   40 CONTINUE
C
C  Compute the right hand side vector and store it in UNB
C
      DO 50 I=1,N
      UNB(I)=0.
      DO 60 J=1,N
      IF(INDEX(J).EQ.0) THEN
      UNB(I)=UNB(I)-H(I,J)*UB(J)
      ELSEIF(INDEX(J).NE.0)THEN
```

```
      UNB(I)=UNB(I)+G(I,J)*UB(J)
      ENDIF
   60 CONTINUE
   50 CONTINUE
      RETURN
      END
C
C
C=================================================================
C
      SUBROUTINE SOLVEQ(A,UNB,N,LSING)
C
      IMPLICIT REAL*8 (A-H,O-Z)
      DIMENSION A(N,N),UNB(N)
      CALL LEQS(A,UNB,N,LSING)
      IF(LSING.EQ.0)THEN
      WRITE(2,150)
  150 FORMAT(/,'',69('*')//2X'The system has been solved regularly'/)
      ELSEIF(LSING.EQ.1)THEN
      WRITE(2,170)
  170 FORMAT(/,'',69('*')//2X'The system is singular'/)
      ENDIF
      RETURN
      END
C
C
C=================================================================
C
      SUBROUTINE LEQS(A,B,N,LSING)
C
C    This subroutine uses Gauss elimination to solve
C    a system of linear equations, [A]{X}={B}, where
C    A : One-dimensional array which contains the occasional row of
C        the two-dimensional array of the coefficients of the unknowns
C    B : One-dimensional array which contains the known coefficients
C    N : Integer denoting the number of the unknowns
C    LSING: Integer taking the values:
C          LSING = 0, if the system has been solved regularly
C          LSING = 1, if the system is singular
C
      IMPLICIT REAL*8 (A-H,O-Z)
      DIMENSION A(1),B(1)
C
      LSING=0
      JJ=-N
      DO 10 J=1,N
        JY=J+1
        JJ=JJ+N+1
        AMAX=0.0
        IHELP=JJ-J
      DO 20 I=J,N
        IJ=IHELP+I
        IF(ABS(AMAX)-ABS(A(IJ))) 30,20,20
   30     AMAX=A(IJ)
          IMAX=I
```

```
   20 CONTINUE
          IF(ABS(AMAX).EQ.0.) THEN
             LSING=1
             RETURN
          END IF
      I1=J+N*(J-2)
      IHELP=IMAX-J
      DO 40 K=J,N
        I1=I1+N
        I2=I1+IHELP
        ATEMP=A(I1)
        A(I1)=A(I2)
        A(I2)=ATEMP
   40 A(I1)=A(I1)/AMAX
      ATEMP=B(IMAX)
      B(IMAX)=B(J)
      B(J)=ATEMP/AMAX
          IF(J-N) 50,70,50
   50        IQS=N*(J-1)
             DO 10 IX=JY,N
               IXJ=IQS+IX
               IHELP=J-IX
               DO 60 JX=JY,N
                 IJREF=N*(JX-1)+IX
                 JJX=IJREF+IHELP
   60            A(IJREF)=A(IJREF)-(A(IXJ)*A(JJX))
   10 B(IX)=B(IX)-B(J)*A(IXJ)
   70     NY=N-1
      NN=N*N
      DO 80 J=1,NY
        I1=NN-J
        I2=N-J
        I3=N
      DO 80 K=1,J
        B(I2)=B(I2)-A(I1)*B(I3)
        I1=I1-N
   80   I3=I3-1
      RETURN
      END
C
C
C====================================================================
C
      SUBROUTINE REORDER(UB,UNB,INDEX,N)
C
C  This subroutine rearranges the arrays UB and UNB in such a way
C  that all values of the function u are stored in UB while all
C  the values of the normal derivative un are stored in UNB
C
      IMPLICIT REAL*8 (A-H,O-Z)
      DIMENSION UB(N),UNB(N),INDEX(N)
C
      DO 20  I=1,N
      IF(INDEX(I))20,20,10
   10 CH=UB(I)
      UB(I)=UNB(I)
      UNB(I)=CH
   20 CONTINUE
      RETURN
      END
```

```
C
C
C==================================================================
C
      SUBROUTINE UINTER(XL,YL,XIN,YIN,UB,UNB,UIN,N,IN)
C
C  This subroutine computes the values of u at the internal points
C
      IMPLICIT REAL*8 (A-H,O-Z)
      DIMENSION XL(N+1),YL(N+1),XIN(IN),YIN(IN),UB(N),UNB(N),UIN(IN)
C
C  Compute the values of u at the internal points
C
      DO 10  K=1,IN
      UIN(K)=0.
      DO 20  J=1,N
      JP1=J+1
      CALL DALPHA(XIN(K),YIN(K),XL(J),YL(J),XL(JP1),YL(JP1),RESH)
      CALL RLINTC(XIN(K),YIN(K),XL(J),YL(J),XL(JP1),YL(JP1),RESG)
   20 UIN(K)=UIN(K)+RESH*UB(J)-RESG*UNB(J)
   10 CONTINUE
      RETURN
      END
C
C
C==================================================================
C
      SUBROUTINE OUTPUT(XM,YM,UB,UNB,XIN,YIN,UIN,N,IN)
C
C  This subroutine prints the results in the output file.
C
      IMPLICIT REAL*8 (A-H,O-Z)
      DIMENSION XM(N),YM(N),UB(N),UNB(N)
      DIMENSION XIN(IN),YIN(IN),UIN(IN)
C
      WRITE(2,100)
  100 FORMAT('',69('*')///1X,'RESULTS'///2X,'BOUNDARY NODES'//
     1 11X,'X',15X,'Y',15X,'U',15X,'Un'/)
C
      DO 10 I=1,N
   10 WRITE(2,200) XM(I),YM(I),UB(I),UNB(I)
  200 FORMAT(4(2X,E14.5))
C
      WRITE (2,300)
  300 FORMAT(//,2X,'INTERNAL POINTS'//10X,'X',15X,'Y',11X,
     1 'SOLUTION U'/)
C
      DO 20 K=1,IN
   20 WRITE(2,400) XIN(K),YIN(K),UIN(K)
  400 FORMAT(3(2X,E14.5))
      WRITE(2,500)
  500 FORMAT(/,'',69('*'))
      RETURN
      END
C
C
C==================================================================
```

EXAMPLE 4.1

The scope of this example is to illustrate the use of program LABECON by solving a simple potential problem for the Laplace equation in the square domain Ω under mixed boundary conditions as shown in Fig. 4.9. The boundary is discretized into 16 constant elements and the solution is sought at 9 internal points (Fig. 4.10).

The exact solution is: $u(x, y) = 100\,(1 + x)$.

In this example the number of elements (N = 16) is relatively small and thus the data file may be created manually. However, if a rectangular domain is discretized into a large number of elements, one should form the data file by using the program RECT-1.FOR available on this book's companion website.

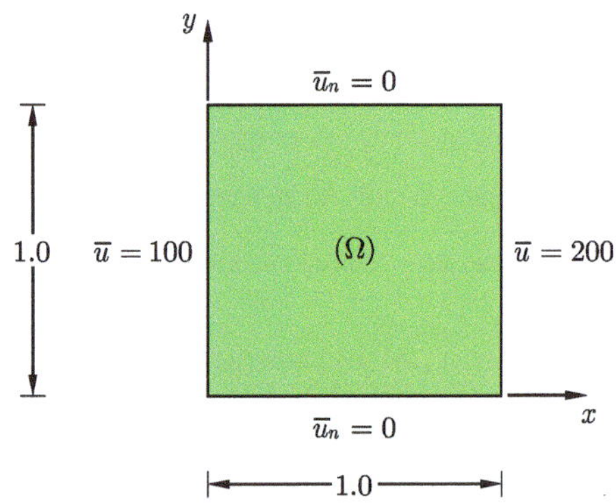

FIGURE 4.9 Square domain Ω and boundary conditions in Example 4.1.

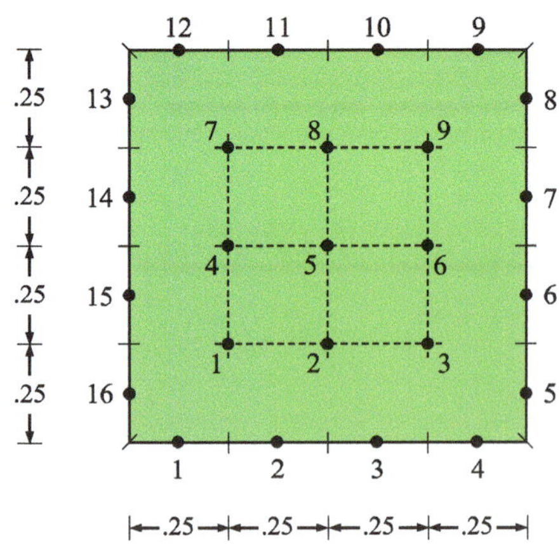

FIGURE 4.10 Boundary element discretization and internal points in Example 4.1.

EXAMPLE 4.1 (DATA)

```
J.T. KATSIKADELIS
Example 4.1

       .00000000         .00000000
       .25000000         .00000000
       .50000000         .00000000
       .75000000         .00000000
      1.00000000         .00000000
      1.00000000         .25000000
      1.00000000         .50000000
      1.00000000         .75000000
      1.00000000        1.00000000
       .75000000        1.00000000
       .50000000        1.00000000
       .25000000        1.00000000
       .00000000        1.00000000
       .00000000         .75000000
       .00000000         .50000000
       .00000000         .25000000

   1        .00000000
   1        .00000000
   1        .00000000
   1        .00000000
   0     200.00000000
   0     200.00000000
   0     200.00000000
   0     200.00000000
   1        .00000000
   1        .00000000
   1        .00000000
   1        .00000000
   0     100.00000000
   0     100.00000000
   0     100.00000000
   0     100.00000000

      0.25      0.25
      0.50      0.25
      0.75      0.25
      0.25      0.50
      0.50      0.50
      0.75      0.50
      0.25      0.75
      0.50      0.75
      0.75      0.75
```

EXAMPLE 4.1 (RESULTS)

```
**********************************************************************
J.T. KATSIKADELIS
Example 4.1


BASIC PARAMETERS

  NUMBER OF BOUNDARY ELEMENTS= 16
  NUMBER OF INTERNAL POINTS WHERE THE FUNCTION IS CALCULATED=  9


  COORDINATES OF THE EXTREME POINTS OF THE BOUNDARY ELEMENTS

  POINT          XL                YL
    1        .00000E+00        .00000E+00
    2        .25000E+00        .00000E+00
    3        .50000E+00        .00000E+00
    4        .75000E+00        .00000E+00
    5        .10000E+01        .00000E+00
    6        .10000E+01        .25000E+00
    7        .10000E+01        .50000E+00
    8        .10000E+01        .75000E+00
    9        .10000E+01        .10000E+01
   10        .75000E+00        .10000E+01
   11        .50000E+00        .10000E+01
   12        .25000E+00        .10000E+01
   13        .00000E+00        .10000E+01
   14        .00000E+00        .75000E+00
   15        .00000E+00        .50000E+00
   16        .00000E+00        .25000E+00


  BOUNDARY CONDITIONS

  NODE      INDEX       PRESCRIBED VALUE
    1         1           .00000E+00
    2         1           .00000E+00
    3         1           .00000E+00
    4         1           .00000E+00
    5         0           .20000E+03
    6         0           .20000E+03
    7         0           .20000E+03
    8         0           .20000E+03
    9         1           .00000E+00
   10         1           .00000E+00
   11         1           .00000E+00
   12         1           .00000E+00
   13         0           .10000E+03
   14         0           .10000E+03
   15         0           .10000E+03
   16         0           .10000E+03

**********************************************************************

  The system has been solved regularly

**********************************************************************
```

```
RESULTS

 BOUNDARY NODES

        X                 Y                 U                 Un

   .12500E+00        .00000E+00        .11188E+03        .00000E+00
   .37500E+00        .00000E+00        .13732E+03        .00000E+00
   .62500E+00        .00000E+00        .16268E+03        .00000E+00
   .87500E+00        .00000E+00        .18812E+03        .00000E+00
   .10000E+01        .12500E+00        .20000E+03        .10552E+03
   .10000E+01        .37500E+00        .20000E+03        .98417E+02
   .10000E+01        .62500E+00        .20000E+03        .98417E+02
   .10000E+01        .87500E+00        .20000E+03        .10552E+03
   .87500E+00        .10000E+01        .18812E+03        .00000E+00
   .62500E+00        .10000E+01        .16268E+03        .00000E+00
   .37500E+00        .10000E+01        .13732E+03        .00000E+00
   .12500E+00        .10000E+01        .11188E+03        .00000E+00
   .00000E+00        .87500E+00        .10000E+03       -.10552E+03
   .00000E+00        .62500E+00        .10000E+03       -.98417E+02
   .00000E+00        .37500E+00        .10000E+03       -.98417E+02
   .00000E+00        .12500E+00        .10000E+03       -.10552E+03

 INTERNAL POINTS

        X                 Y             SOLUTION U

   .25000E+00        .25000E+00        .12489E+03
   .50000E+00        .25000E+00        .15000E+03
   .75000E+00        .25000E+00        .17511E+03
   .25000E+00        .50000E+00        .12495E+03
   .50000E+00        .50000E+00        .15000E+03
   .75000E+00        .50000E+00        .17505E+03
   .25000E+00        .75000E+00        .12489E+03
   .50000E+00        .75000E+00        .15000E+03
   .75000E+00        .75000E+00        .17511E+03

**********************************************************************
```

As it was anticipated, the obtained solution is symmetric with respect to the axis passing through the center of the square and being parallel to the x-axis. Table 4.1 presents the computed values on the boundary and at the interior of the domain versus the number N of boundary elements. Comparing these results with the exact values, the computed boundary values of u and u_n converge rapidly to the exact ones. The accuracy at the internal points is even better, being attributed to the fact that these values are computed from Eq. (4.11), which is a weighted residual form for all the boundary values. The results have been obtained using the Microsoft Fortran PowerStation on a PC. The running time was a few seconds. Actually, the time required to solve the problem is the time needed to prepare the data file. For this reason, the user of LABECON is advised to write first a simple program that generates the coordinates XL, YL of the extreme points and XIN, YIN of the internal points, as well as the values of the INDEX vector and the boundary conditions. In this way, the tedious task of entering data by hand and the ensuing possible errors are avoided.

TABLE 4.1 Computed Boundary and Internal Values for Various Boundary Discretizations in Example 4.1

Point	16	48	80	112	144	Exact
	Number of Boundary Elements, *N*					
Values of u at the Boundary Nodes						
1	111.88	112.36	112.43	112.46	112.47	112.50
2	137.32	137.47	137.48	137.49	137.49	137.50
3	162.68	162.53	162.52	162.51	162.51	162.50
4	188.12	187.64	187.57	187.54	187.53	187.50
Values of u_n at the Boundary Nodes						
5	105.520	98.215	99.486	99.665	99.774	100.000
6	98.417	99.800	99.909	99.946	99.964	100.000
Values of u at the Internal Points						
1	124.89	124.98	124.99	124.99	125.00	125.00
5	150.00	150.00	150.00	150.00	150.00	150.00
9	175.11	175.02	175.01	175.01	175.00	175.00

EXAMPLE 4.2

The scope of this example is to demonstrate the efficiency of program LABECON in treating domains with curvilinear boundaries. Specifically, we want to compute u from the following Neumann problem:

$$\nabla^2 u = 0 \text{ in } \Omega$$

$$\frac{\partial u}{\partial n} = \overline{u}_n \text{ on } \Gamma$$

where the domain Ω is the ellipse shown in Fig. 4.11 and

$$\overline{u}_n = \frac{2(b^2x^2 - a^2y^2)}{\sqrt{b^4x^2 + a^4y^2}} \tag{a}$$

The exact solution is known to be

$$u = x^2 - y^2 + C \tag{b}$$

Practically, the discretization of the boundary into elements of equal length is not a simple problem. Besides, such a discretization would not give the best results, since the curvature of the boundary is more intense in the

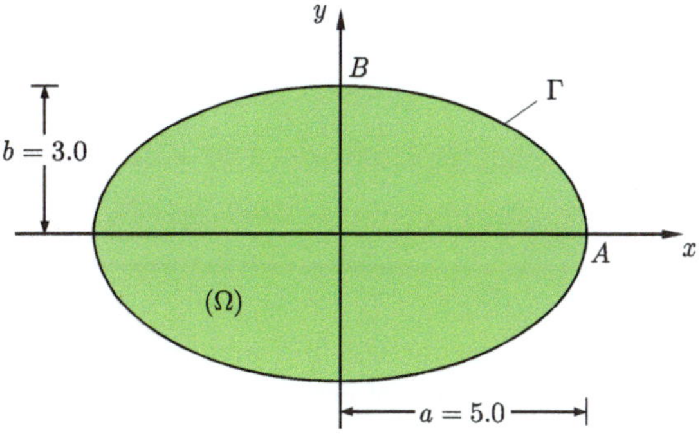

FIGURE 4.11 Elliptic domain Ω in Example 4.2.

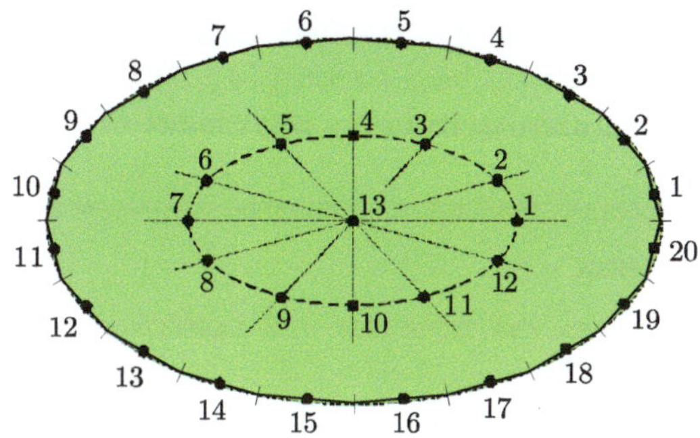

FIGURE 4.12 Boundary element discretization and internal points in Example 4.2.

neighborhood of point $A(5.0, 0.0)$ and decreases moving towards point $B(0.0, 3.0)$. A better discretization would have elements whose length is small near A and increases progressively towards B. This discretization can be achieved by establishing the extreme points of the element from the parametric equations of the ellipse for constant increment $\Delta\theta$ of the parameter. Namely, the coordinates are given as

$$x_i = a\cos\theta_i, \quad y_i = b\sin\theta_i \quad (i = 1, 2, \ldots, N)$$

where N is the total number of boundary elements, and

$$\theta_i = (i-1)\,\Delta\theta, \quad \Delta\theta = \frac{2\pi}{N}$$

For example, a discretization of the boundary into $N = 20$ elements (Fig. 4.12) produces the following lengths for elements 1 through 5 of the first quadrant

$$\ell_1 = 0.958, \quad \ell_2 = 1.097, \quad \ell_3 = 1.289, \quad \ell_4 = 1.457, \quad \ell_5 = 1.552$$

Apparently, the elements have the same lengths in each of the other three quadrants due to the symmetry with respect to the x and y axes.

The coordinates of the twelve internal points are computed similarly from the expressions

$$x_i = \frac{a}{2}\cos\theta_i, \quad y_i = \frac{b}{2}\sin\theta_i \quad (i = 1, 2, \ldots, 12)$$

$$\theta_i = (i-1)\Delta\theta, \quad \Delta\theta = \frac{2\pi}{12}$$

As it has already been mentioned, the preparation of the data file by keying in the data, besides the risk of error, is a tedious and time-consuming process. Since LABECON is not a commercial computer code, it does not offer a preprocessing interface for automatic preparation of the data. To facilitate this procedure and to reduce possible errors a FORTRAN program has been written that generates the required data and stores them in the INPUTFILE. The program has been named ELLIPSE-1 and it is available on this book's companion website. Its listing is given below.

```
C=====================================================================
C
      PROGRAM ELLIPSE1
C
C  This program creates the INPUTFILE for LABECON when the
C  domain is an ellipse
C
      IMPLICIT REAL*8 (A-H,O-Z)
      CHARACTER*15 INPUTFILE
      CHARACTER*80 NAME,TITLE
      PARAMETER (N=20)
      PARAMETER (IN=13)
      DIMENSION XL(N+1),YL(N+1),INDEX(N),XIN(IN),YIN(IN)
C
      WRITE (*,'(A)')' Name of the INPUTFILE (max.15 characters)'
      READ  (*,'(A)') INPUTFILE
      OPEN (1, FILE=INPUTFILE)
      WRITE (*,'(A)')' User NAME (max.15 characters)'
      READ(*,'(A)')NAME
      WRITE(1,'(A)')NAME
      WRITE (*,'(A)')' Program TITLE (max.15 characters)'
      READ(*,'(A)')TITLE
      WRITE(1,'(A)')TITLE
C
      WRITE(1,100)
  100 FORMAT(' ',79(' '))
C
      A=5.
      B=3.
      PI=ACOS(-1.)
      DTHETA=(2.*PI)/N
      DO 1 I=1,N
      THETA=(I-1)*DTHETA
      XL(I)=A*COS(THETA)
      YL(I)=B*SIN(THETA)
      WRITE(1,200)XL(I),YL(I)
  200 FORMAT(2(2X,F14.7))
    1 CONTINUE
```

```
C
      XL(N+1)=XL(1)
      YL(N+1)=YL(1)
      WRITE(1,100)
C
      DO 2 I=1,N-1
      INDEX(I)=1
      XM=(XL(I)+XL(I+1))/2.
      YM=(YL(I)+YL(I+1))/2.
      UNB=2.*((B*XM)**2-(A*YM)**2)/SQRT(B**2*(B*XM)**2+A**2*(A*YM)**2)
      WRITE(1,300) INDEX(I),UNB
    2 CONTINUE
  300 FORMAT (2X,I4,2X,F14.7)
      INDEX(N)=0
      XM=(XL(N)+XL(N+1))/2.
      YM=(YL(N)+YL(N+1))/2.
      UB=XM**2-YM**2
      WRITE (1,300) INDEX(N),UB
      WRITE (1,100)
C
      DA=A/2.
      DB=B/2.
      WRITE(1,100)
C
      DTHETA=(2.*PI)/12.
      DO 3 I=1,12
      THETA=(I-1)*DTHETA
      XIN(I)=DA*COS(THETA)
      YIN(I)=DB*SIN(THETA)
      WRITE(1,200)XIN(I),YIN(I)
    3 CONTINUE
C
      XIN(IN)=0.
      YIN(IN)=0.
      WRITE (1,200) XIN(IN),YIN(IN)
C
      STOP
      END
C
C
C=====================================================================
```

EXAMPLE 4.2 (DATA)

```
J.T. KATSIKADELIS
Example 4.2

     5.0000000          .0000000
     4.7552826          .9270510
     4.0450850         1.7633558
     2.9389263         2.4270510
     1.5450850         2.8531695
      .0000000         3.0000000
    -1.5450850         2.8531695
    -2.9389263         2.4270510
    -4.0450850         1.7633558
```

```
  -4.7552826           .9270510
  -5.0000000           .0000000
  -4.7552826          -.9270510
  -4.0450850         -1.7633558
  -2.9389263         -2.4270510
  -1.5450850         -2.8531695
    .0000000         -3.0000000
   1.5450850         -2.8531695
   2.9389263         -2.4270510
   4.0450850         -1.7633558
   4.7552826          -.9270510

1        9.1955756
1        4.9664676
1         .0000000
1       -3.7385842
1       -5.6807458
1       -5.6807458
1       -3.7385842
1         .0000000
1        4.9664676
1        9.1955756
1        9.1955756
1        4.9664676
1         .0000000
1       -3.7385842
1       -5.6807458
1       -5.6807458
1       -3.7385842
1         .0000000
1        4.9664676
0       23.5765287


   2.5000000           .0000000
   2.1650635           .7500000
   1.2500000          1.2990381
    .0000000          1.5000000
  -1.2500000          1.2990381
  -2.1650635           .7500000
  -2.5000000           .0000000
  -2.1650635          -.7500000
  -1.2500000         -1.2990381
    .0000000         -1.5000000
   1.2500000         -1.2990381
   2.1650635          -.7500000
    .0000000           .0000000
```

Referring to the Neumann problem of Example 4.2, the following should be taken into account, which are known from the theory of partial differential equations of the elliptic type [7].

a. In order for the problem to have a solution, it must be

$$\int_{\Gamma} \frac{\partial u}{\partial n} ds = 0$$

which can be readily proved by applying Eq. (2.16) for $v = 1$ and $\nabla^2 u = 0$.

TABLE 4.2 Computed Values of u at the Internal Points for Various Boundary Discretizations in Example 4.2

Point	Number of Boundary Elements, *N*						Exact
	20	40	100	300	700	1000	
1	7.1022	6.5060	6.2960	6.2556	6.2511	6.2506	6.2500
2	5.0959	4.4114	4.1759	4.1312	4.1262	4.1256	4.1250
3	1.0736	0.2201	−0.0646	−0.1177	−0.1236	−0.1243	−0.1250
4	−0.9207	−1.8715	−2.1842	−2.2422	−2.2485	−2.2492	−2.2500
13	1.2216	0.3514	0.0613	0.0073	0.0014	0.0007	0.0000

b. The solution u is not determined uniquely, but to the approximation of an arbitrary constant.

The first remark demands the following check:

$$\int_\Gamma \overline{u}_n \, ds \approx \sum_{i=1}^{N} \ell_i (\overline{u}_n)_i \approx 0$$

which can be shown that it is satisfied for the given data. The second remark produces difficulties in solving Eq. (4.9). The matrix $[A]=[H]$ is singular in this case and therefore, it can not be inverted. This difficulty can be overcome by prescribing arbitrarily the value of u at a node, say at the last node N, and then solving the problem with mixed boundary conditions. Without restriction of generality, we assign to u^N the exact value of u as it is computed from equation (b) with $C=0$, so that the obtained numerical results can be directly compared with the exact ones.

In Table 4.2, the computed values of u at internal points are presented for various values of N. Moreover, Fig. 4.13 shows the variation of error in u at the internal point 2 versus the number of boundary elements N. In this problem, the convergence is slower than in Example 4.1. This was anticipated, because the curved boundary of the ellipse is approximated by an inscribed polygon. A faster convergence can be achieved by employing another type of constant element which approximates the geometry with a parabolic arc [8].

4.6 DOMAINS WITH MULTIPLE BOUNDARIES

In many applications the domain Ω may contain holes. In that case the contours are more than one (Fig. 4.14). Namely, there is an outer boundary enclosing a finite number of nonintersecting inner contours. In mathematical

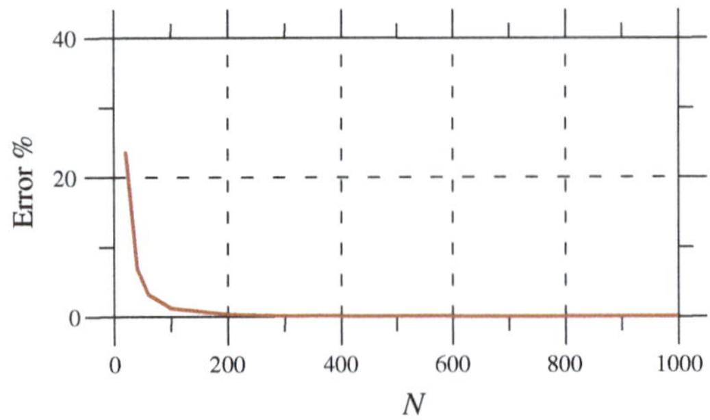

FIGURE 4.13 Variation of error in u at internal point 2 versus the number of boundary elements in Example 4.2.

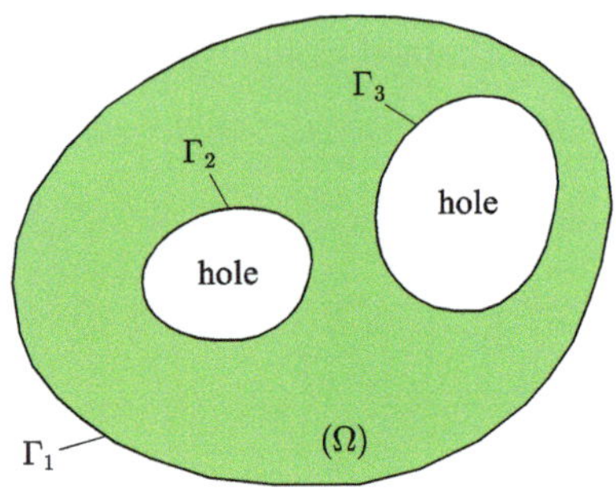

FIGURE 4.14 Multiply connected domain Ω.

terms, this type of domain is referred to as *multiply connected domain.* We come across these domains in many problems, such as torsion of bars with hollow cross-sections, fluid flow past obstacles, or heat conduction in pipes with thermal insulation.

It can be easily shown that Green's identity (2.16) is valid also for multiply connected domains, where the boundary Γ is the sum (union) of all the contours. Indeed, if we introduce the cuts AB and CD, which are arbitrary and not necessarily along straight lines (see Fig. 4.15), the domain Ω is converted to simply connected, that is one without holes. Green's identity (2.16) may then be written as

$$\begin{aligned}\int_{\Omega}(v\,\nabla^2 u-u\,\nabla^2 v)d\Omega=&\sum_{k=1}^{3}\int_{\Gamma_k}\left(v\frac{\partial u}{\partial n}-u\frac{\partial v}{\partial n}\right)ds+\int_{BA}\left(v\frac{\partial u}{\partial n}-u\frac{\partial v}{\partial n}\right)ds\\&+\int_{AB}\left(v\frac{\partial u}{\partial n}-u\frac{\partial v}{\partial n}\right)ds+\int_{DC}\left(v\frac{\partial u}{\partial n}-u\frac{\partial v}{\partial n}\right)ds\\&+\int_{CD}\left(v\frac{\partial u}{\partial n}-u\frac{\partial v}{\partial n}\right)ds\end{aligned}$$

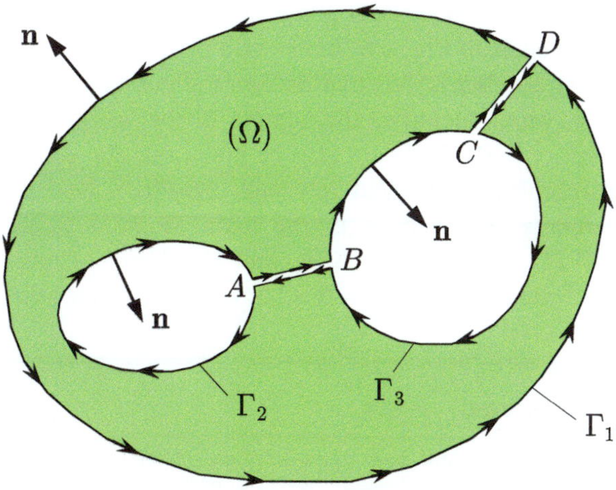

FIGURE 4.15 Positive direction and normal vector on the boundaries of a multiply connected domain.

Noting that

$$\int_{BA}\left(v\frac{\partial u}{\partial n}-u\frac{\partial v}{\partial n}\right)ds=-\int_{AB}\left(v\frac{\partial u}{\partial n}-u\frac{\partial v}{\partial n}\right)ds$$

$$\int_{DC}\left(v\frac{\partial u}{\partial n}-u\frac{\partial v}{\partial n}\right)ds=-\int_{CD}\left(v\frac{\partial u}{\partial n}-u\frac{\partial v}{\partial n}\right)ds$$

the foregoing equation becomes

$$\int_{\Omega}(v\,\nabla^2 u-u\,\nabla^2 v)d\Omega=\int_{\Gamma}\left(v\frac{\partial u}{\partial n}-u\frac{\partial v}{\partial n}\right)ds \tag{4.33}$$

where $\Gamma=\Gamma_1\cup\Gamma_2\cup\Gamma_3$.

Thus, Green's identity (2.16) applies also to multiply connected domains where the boundary integral is taken on all the contours. Consequently, relations resulting from Green's identity are valid for multiple boundaries as well. It should be noted that the positive sense on the inner contours is clockwise, which is opposite to that on the outer contour (Fig. 4.15).

4.7 PROGRAM LABECONMU FOR DOMAINS WITH MULTIPLE BOUNDARIES

The program LABECON can be readily modified to solve potential problems in domains with holes (multiply connected domains). The changes affect only the main program and the subroutines INPUT, GMATR, HMATR, and UINTER. The new program has been given the name LABECONMU to distinguish it from LABECON.

The structure of LABECONMU is the same as that of LABECON shown in the macro flow chart of Fig. 4.6. In its main program, two new parameters have been introduced, that is, NB that defines the number of boundaries, and the vector NL(I), which identifies the number of the last element on the *I*-th boundary (*I* = 1,2,. . .,NB). It should also be noted that the elements of all the boundaries are numbered consecutively and, therefore, N denotes the total number of elements. The listings of the main program as well as of the subroutines INPUT, GMATR, HMATR, and UINTER are presented below. Their electronic form is available on this book's companion website.

```
C=================================================================
C
      PROGRAM LABECONMU
C
C  This program solves the two dimensional (LA)place equation
C  using the (B)oundary (E)lement method with (CON)stant
C  boundary elements for domains with (MU)ltiple boundaries
C
      IMPLICIT REAL*8 (A-H,O-Z)
      CHARACTER*15 INPUTFILE,OUTPUTFILE
C
C  Set the maximum dimensions
C
      PARAMETER (N=24)
      PARAMETER (IN=8)
      PARAMETER (NB=2)
C
C   N= Number of boundary elements which is equal to the number
C      of boundary nodes
C  IN= Number of internal points where the function u is calculated
C  NB= Number of boundaries of the multiple boundary domain
C  NL= Array with entries the number of the last element on each
C      boundary
C
      DIMENSION INDEX(N),NL(NB)
      DIMENSION XL(N+1),YL(N+1),XM(N),YM(N),G(N,N),H(N,N)
      DIMENSION UB(N),A(N,N),UNB(N),XIN(IN),YIN(IN),UIN(IN)
C
C  Read the names and open the input and output files
C
      WRITE (*,'(A)')' Name of the INPUTFILE (max.15 characters)'
      READ  (*,'(A)') INPUTFILE
      WRITE (*,'(A)')' Name of the OUTPUTFILE (max.15 characters)'
      READ  (*,'(A)') OUTPUTFILE
      OPEN (UNIT=1, FILE=INPUTFILE)
      OPEN (UNIT=2, FILE=OUTPUTFILE)
C
C  Read data from INPUTFILE
C
      CALL INPUT (XL,YL,XIN,YIN,INDEX,UB,N,IN,NL,NB)
C
C  Compute the G matrix
C
      CALL GMATR (XL,YL,XM,YM,G,N,NL,NB)
C
C  Compute the H matrix
C
      CALL HMATR (XL,YL,XM,YM,H,N,NL,NB)
```

```
C
C  Form the system of equations AX=B
C
       CALL ABMATR (G,H,A,UNB,UB,INDEX,N)
C
C  Solve the system of equations
C
       CALL SOLVEQ(A,UNB,N,LSING)
C
C  Form the vectors U and UN of all the boundary values
C
       CALL REORDER (UB,UNB,INDEX,N)
C
C  Compute the values of U at the internal points
C
      CALL UINTER (XL,YL,XIN,YIN,UB,UNB,UIN,N,IN,NL,NB)
C
C  Print the results in the OUTPUTFILE
C
      CALL OUTPUT (XM,YM,UB,UNB,XIN,YIN,UIN,N,IN)
C
C  Close input and output files
C
      CLOSE(1)
      CLOSE(2)
C
      STOP
      END
C
C
C=====================================================================
C
      SUBROUTINE INPUT (XL,YL,XIN,YIN,INDEX,UB,N,IN,NL,NB)
C
C  This subroutine reads the input data from the input file
C  and writes them in the output file
C
      IMPLICIT REAL*8 (A-H,O-Z)
      CHARACTER*80 NAME,TITLE
      DIMENSION XL(N+1),YL(N+1),XIN(IN),YIN(IN),INDEX(N),UB(N),NL(NB)
C
      WRITE(2,100)
  100 FORMAT('',69('*'))
C
C  Read user's name
C
      READ(1,'(A)')NAME
C
      WRITE(2,'(A)')NAME
C
C  Read the title of the program
C
      READ(1,'(A)')TITLE
C
      WRITE(2,'(A)')TITLE
C
C  Read the number of the last element on each boundary
C
      READ(1,*) (NL(I),I=1,NB)
C
      WRITE(2,200)N,IN,NB
  200 FORMAT(//'BASIC PARAMETERS'//2X,'NUMBER OF BOUNDARY ELEMENTS='
     1,I3/2X,'NUMBER OF INTERNAL POINTS WHERE THE FUNCTION IS CALCULATED
     1=',I3//2X,'NUMBER OF BOUNDARIES =',I3)
      DO 250 I=1,NB
```

```
  250 WRITE(2,260)I,NL(I)
  260 FORMAT(2X,'BOUNDARY',I3,2X,'LAST BOUNDARY ELEMENT=',I3)
C
C  Read the coordinates XL,YL of the extreme points of the boundary elements
C
      READ(1,*) (XL(I),YL(I),I=1,N)
C
C  Write the coordinates in the output file
C
      WRITE(2,300)
  300 FORMAT(//2X,'COORDINATES OF THE EXTREME POINTS OF THE BOUNDARY '
     1'ELEMENTS',//2X,'POINT',9X,'XL',15X,'YL')
      DO 20 I=1,N
   20 WRITE(2,400) I,XL(I),YL(I)
  400 FORMAT(2X,I3,2(3X,E14.5))
C
C  Read the boundary conditions and store them in UB(I).
C  If INDEX(I)=0, the value UB(I) is the prescribed value of u.
C  If INDEX(I)=1, the normal derivative un is prescribed.
C
      READ(1,*) (INDEX(I),UB(I),I=1,N)
C
C  Write the boundary conditions in the output file
C
      WRITE(2,500)
  500 FORMAT(//2X,'BOUNDARY CONDITIONS'//2X,'NODE',6X,'INDEX',
     1 7X,'PRESCRIBED VALUE')
      DO 30 I=1,N
   30 WRITE(2,600) I,INDEX(I),UB(I)
  600 FORMAT (2X,I3,9X,I1,8X,E14.5)
C
C  Read the coordinates of the internal points
C
      READ(1,*) (XIN(I),YIN(I),I=1,IN)
      RETURN
      END
C
C
C=======================================================================
C
      SUBROUTINE GMATR (XL,YL,XM,YM,G,N,NL,NB)
C
C  This subroutine computes the elements of the G matrix
C
      IMPLICIT REAL*8 (A-H,O-Z)
      DIMENSION XL(N+1),YL(N+1),XM(N),YM(N),NL(NB)
      DIMENSION G(N,N)
C
C  Compute the nodal coordinates and store them in the arrays XM
C  and YM
C
      IF (NB.GT.1) GOTO 5
      XL(NL(1)+1)=XL(1)
      YL(NL(1)+1)=YL(1)
C
    5 DO 10  I=1,N
      XM(I)=(XL(I)+XL(I+1))/2.
   10 YM(I)=(YL(I)+YL(I+1))/2.
      IF(NB.LE.1) GO TO 40
      XM(NL(1))=(XL(NL(1))+XL(1))/2.
      YM(NL(1))=(YL(NL(1))+YL(1))/2.
      DO 15 K=2,NB
      XM(NL(K))=(XL(NL(K))+XL(NL(K-1)+1))/2.
   15 YM(NL(K))=(YL(NL(K))+YL(NL(K-1)+1))/2.
```

```
C
C  Compute the elements of matrix G
C
   40 DO 20 I=1,N
      X0=XM(I)
      Y0=YM(I)
      DO 20 J=1,N
      X1=XL(J)
      Y1=YL(J)
      X2=XL(J+1)
      Y2=YL(J+1)
      IF(NB.LE.1)GOTO 60
        IF(J.NE.NL(1))GOTO 50
          X2=XL(1)
          Y2=YL(1)
          GOTO 60
   50   DO 30 K=2,NB
        IF(J.NE.NL(K))GOTO 30
          X2=XL(NL(K-1)+1)
          Y2=YL(NL(K-1)+1)
          GOTO 60
   30   CONTINUE
   60 IF(I.NE.J)THEN
        CALL RLINTC(X0,Y0,X1,Y1,X2,Y2,RESULT)
        G(I,J)=RESULT
      ELSEIF(I.EQ.J)THEN
        CALL SLINTC(X1,Y1,X2,Y2,RESULT)
        G(I,J)=RESULT
      ENDIF
   20 CONTINUE
      RETURN
      END
C
C
C=====================================================================
C
      SUBROUTINE HMATR(XL,YL,XM,YM,H,N,NL,NB)
C
C  This subroutine computes the elements of the H matrix
C
      IMPLICIT REAL*8 (A-H,O-Z)
      DIMENSION XL(N+1),YL(N+1),XM(N),YM(N),NL(NB)
      DIMENSION H(N,N)
C
      PI=ACOS(-1.)
C
C  Compute the nodal coordinates and store them in the arrays XM
C  and YM
C
      IF (NB.GT.1) GOTO 5
      XL(NL(1)+1)=XL(1)
      YL(NL(1)+1)=YL(1)
```

```
C
    5 DO 10  I=1,N
      XM(I)=(XL(I)+XL(I+1))/2.
   10 YM(I)=(YL(I)+YL(I+1))/2.
      IF(NB.LE.1) GO TO 40
      XM(NL(1))=(XL(NL(1))+XL(1))/2.
      YM(NL(1))=(YL(NL(1))+YL(1))/2.
      DO 15 K=2,NB
      XM(NL(K))=(XL(NL(K))+XL(NL(K-1)+1))/2.
   15 YM(NL(K))=(YL(NL(K))+YL(NL(K-1)+1))/2.
C
C  Compute the elements of H matrix
C
   40 DO 20 I=1,N
      X0=XM(I)
      Y0=YM(I)
      DO 20 J=1,N
      X1=XL(J)
      Y1=YL(J)
      X2=XL(J+1)
      Y2=YL(J+1)
      IF(NB.LE.1)GOTO 60
        IF(J.NE.NL(1))GOTO 50
          X2=XL(1)
          Y2=YL(1)
          GOTO 60
   50   DO 30 K=2,NB
        IF(J.NE.NL(K))GOTO 30
          X2=XL(NL(K-1)+1)
          Y2=YL(NL(K-1)+1)
          GOTO 60
   30   CONTINUE
   60 IF (I.NE.J) THEN
        CALL DALPHA (X0,Y0,X1,Y1,X2,Y2,RESULT)
        H(I,J)=RESULT
      ELSEIF (I.EQ.J) THEN
        H(I,J)=-0.5
      ENDIF
   20 CONTINUE
      RETURN
      END
C
C
C=====================================================================
C
      SUBROUTINE UINTER(XL,YL,XIN,YIN,UB,UNB,UIN,N,IN,NL,NB)
C
C  This subroutine computes the values of u at the internal points
C
      IMPLICIT REAL*8 (A-H,O-Z)
      DIMENSION XL(N+1),YL(N+1),XIN(IN),YIN(IN),UB(N),UNB(N),UIN(IN)
      DIMENSION NL(NB)
C
C  Compute the values of u at the internal points
C
      IF (NB.GT.1) GOTO 5
      XL(NL(1)+1)=XL(1)
      YL(NL(1)+1)=YL(1)
```

```
C
    5 DO 10  KK=1,IN
      UIN(KK)=0.
      DO 20  J=1,N
      X0=XIN(KK)
      Y0=YIN(KK)
      X1=XL(J)
      Y1=YL(J)
      X2=XL(J+1)
      Y2=YL(J+1)
      IF(NB.LE.1)GOTO 60
        IF(J.NE.NL(1))GOTO 50
          X2=XL(1)
          Y2=YL(1)
          GOTO 60
   50   DO 30 K=2,NB
        IF(J.NE.NL(K))GOTO 30
          X2=XL(NL(K-1)+1)
          Y2=YL(NL(K-1)+1)
          GOTO 60
   30   CONTINUE
   60 CALL DALPHA(X0,Y0,X1,Y1,X2,Y2,RESH)
      CALL RLINTC(X0,Y0,X1,Y1,X2,Y2,RESG)
   20 UIN(KK)=UIN(KK)+RESH*UB(J)-RESG*UNB(J)
   10 CONTINUE
      RETURN
      END
C
C
C=====================================================================
```

EXAMPLE 4.3

This example demonstrates the use of the program LABECONMU for the solution of a simple potential problem with mixed boundary conditions. The square domain Ω is doubly connected, namely, it contains a hole. Its outer boundary has been discretized into 16 constant elements, while the inner one into 8 elements. The solution is sought at 8 internal points. The data and the boundary discretization are shown in Figs. 4.16 and 4.17. The data

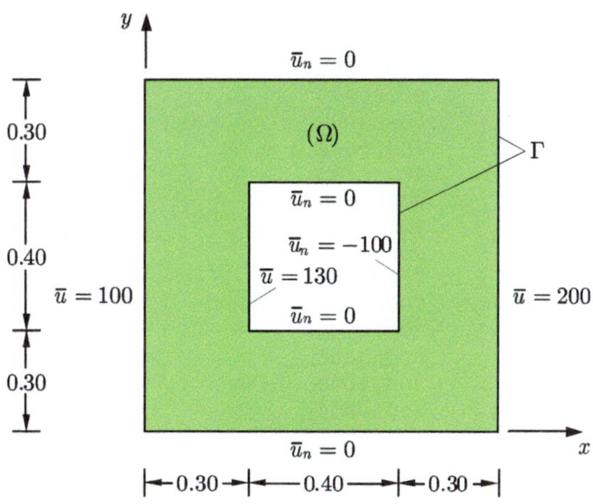

FIGURE 4.16 Doubly connected domain Ω and boundary conditions in Example 4.3.

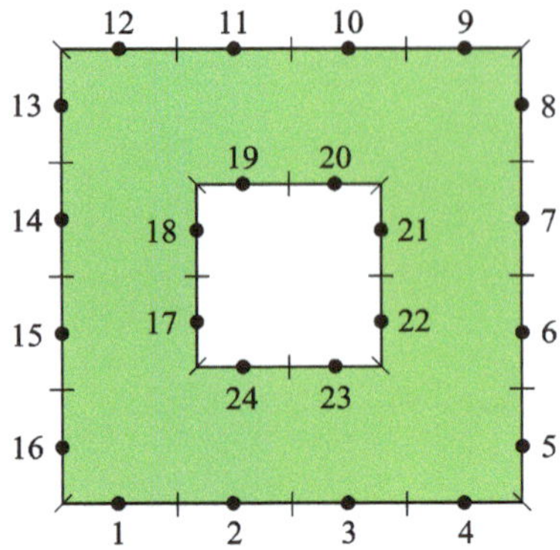

FIGURE 4.17 Boundary element discretization for the doubly connected domain in Example 4.3.

file has been created using program RECT-2.FOR, available on this book's companion website.

The exact solution is: $u(x, y) = 100(1 + x)$.

EXAMPLE 4.3 (DATA)

```
J.T. KATSIKADELIS
Example 4.3

  16
  24
        .00000000        .00000000
        .25000000        .00000000
        .50000000        .00000000
        .75000000        .00000000
       1.00000000        .00000000
       1.00000000        .25000000
       1.00000000        .50000000
       1.00000000        .75000000
       1.00000000       1.00000000
        .75000000       1.00000000
        .50000000       1.00000000
        .25000000       1.00000000
        .00000000       1.00000000
        .00000000        .75000000
        .00000000        .50000000
        .00000000        .25000000
        .30000000        .30000000
        .30000000        .50000000
        .30000000        .70000000
        .50000000        .70000000
        .70000000        .70000000
        .70000000        .50000000
        .70000000        .30000000
        .50000000        .30000000
```

```
1              .00000000
1              .00000000
1              .00000000
1              .00000000
0           200.00000000
0           200.00000000
0           200.00000000
0           200.00000000
1              .00000000
1              .00000000
1              .00000000
1              .00000000
0           100.00000000
0           100.00000000
0           100.00000000
0           100.00000000
0           130.00000000
0           130.00000000
1              .00000000
1              .00000000
1          -100.00000000
1          -100.00000000
1              .00000000
1              .00000000

    .15000000        .15000000
    .50000000        .15000000
    .85000000        .15000000
    .15000000        .50000000
    .85000000        .50000000
    .15000000        .85000000
    .50000000        .85000000
    .85000000        .85000000
```

EXAMPLE 4.3 (RESULTS)

```
**********************************************************************
J.T. KATSIKADELIS
Example 4.3

BASIC PARAMETERS

  NUMBER OF BOUNDARY ELEMENTS= 24
  NUMBER OF INTERNAL POINTS WHERE THE FUNCTION IS CALCULATED=  8

  NUMBER OF BOUNDARIES =  2
  BOUNDARY  1  LAST BOUNDARY ELEMENT= 16
  BOUNDARY  2  LAST BOUNDARY ELEMENT= 24
```

COORDINATES OF THE EXTREME POINTS OF THE BOUNDARY ELEMENTS

POINT	XL	YL
1	.00000E+00	.00000E+00
2	.25000E+00	.00000E+00
3	.50000E+00	.00000E+00
4	.75000E+00	.00000E+00
5	.10000E+01	.00000E+00
6	.10000E+01	.25000E+00
7	.10000E+01	.50000E+00
8	.10000E+01	.75000E+00
9	.10000E+01	.10000E+01
10	.75000E+00	.10000E+01
11	.50000E+00	.10000E+01
12	.25000E+00	.10000E+01
13	.00000E+00	.10000E+01
14	.00000E+00	.75000E+00
15	.00000E+00	.50000E+00
16	.00000E+00	.25000E+00
17	.30000E+00	.30000E+00
18	.30000E+00	.50000E+00
19	.30000E+00	.70000E+00
20	.50000E+00	.70000E+00
21	.70000E+00	.70000E+00
22	.70000E+00	.50000E+00
23	.70000E+00	.30000E+00
24	.50000E+00	.30000E+00

BOUNDARY CONDITIONS

NODE	INDEX	PRESCRIBED VALUE
1	1	.00000E+00
2	1	.00000E+00
3	1	.00000E+00
4	1	.00000E+00
5	0	.20000E+03
6	0	.20000E+03
7	0	.20000E+03
8	0	.20000E+03
9	1	.00000E+00
10	1	.00000E+00
11	1	.00000E+00
12	1	.00000E+00
13	0	.10000E+03
14	0	.10000E+03
15	0	.10000E+03
16	0	.10000E+03
17	0	.13000E+03
18	0	.13000E+03
19	1	.00000E+00
20	1	.00000E+00
21	1	-.10000E+03
22	1	-.10000E+03
23	1	.00000E+00
24	1	.00000E+00

```
*************************************************************************

 The system has been solved regularly

*************************************************************************
```

```
RESULTS

 BOUNDARY NODES

          X                Y                U                Un

     .12500E+00      .00000E+00      .11212E+03      .00000E+00
     .37500E+00      .00000E+00      .13776E+03      .00000E+00
     .62500E+00      .00000E+00      .16273E+03      .00000E+00
     .87500E+00      .00000E+00      .18811E+03      .00000E+00
     .10000E+01      .12500E+00      .20000E+03      .10536E+03
     .10000E+01      .37500E+00      .20000E+03      .95220E+02
     .10000E+01      .62500E+00      .20000E+03      .95220E+02
     .10000E+01      .87500E+00      .20000E+03      .10536E+03
     .87500E+00      .10000E+01      .18811E+03      .00000E+00
     .62500E+00      .10000E+01      .16273E+03      .00000E+00
     .37500E+00      .10000E+01      .13776E+03      .00000E+00
     .12500E+00      .10000E+01      .11212E+03      .00000E+00
     .00000E+00      .87500E+00      .10000E+03     -.10777E+03
     .00000E+00      .62500E+00      .10000E+03     -.99560E+02
     .00000E+00      .37500E+00      .10000E+03     -.99560E+02
     .00000E+00      .12500E+00      .10000E+03     -.10777E+03
     .30000E+00      .40000E+00      .13000E+03      .10833E+03
     .30000E+00      .60000E+00      .13000E+03      .10833E+03
     .40000E+00      .70000E+00      .14054E+03      .00000E+00
     .60000E+00      .70000E+00      .16002E+03      .00000E+00
     .70000E+00      .60000E+00      .17168E+03     -.10000E+03
     .70000E+00      .40000E+00      .17168E+03     -.10000E+03
     .60000E+00      .30000E+00      .16002E+03      .00000E+00
     .40000E+00      .30000E+00      .14054E+03      .00000E+00

 INTERNAL POINTS

          X                Y            SOLUTION U

     .15000E+00      .15000E+00      .11508E+03
     .50000E+00      .15000E+00      .15026E+03
     .85000E+00      .15000E+00      .18525E+03
     .15000E+00      .50000E+00      .11507E+03
     .85000E+00      .50000E+00      .18569E+03
     .15000E+00      .85000E+00      .11508E+03
     .50000E+00      .85000E+00      .15026E+03
     .85000E+00      .85000E+00      .18525E+03
************************************************************************
```

4.8 THE METHOD OF SUBDOMAINS

In certain problems the material properties of the body are piecewise continuous. Such, for example, is the torsion of a composite bar consisting of two or more materials of different shear moduli. Another example is the problem of heat conduction in a body having different coefficients of thermal conductivity in two or more subregions. In the literature such a body is referred to as a *multizone body* and the domain it occupies is referred to as a *composite domain*.

Potential problems in composite domains can be solved by applying the BEM separately to each of its subdomains. The reason is that the fundamental solution is valid only for homogeneous domains. Next, without restricting the

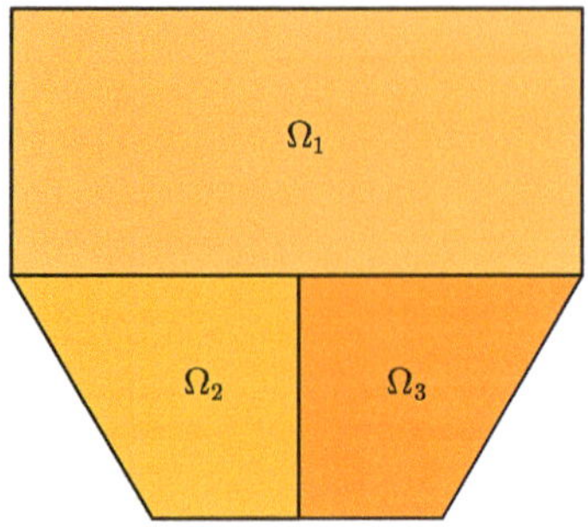

FIGURE 4.18 Composite domain $\Omega = \Omega_1 \cup \Omega_2 \cup \Omega_3$.

generality, we will illustrate the BEM by applying it to the composite domain of Fig. 4.18, which consists of the three subdomains Ω_1, Ω_2, and Ω_3. In discretizing the boundaries of these subdomains, special care is taken to have the same discretization on either side of each interface, namely, on common parts of the boundaries between subdomains (see Fig. 4.19).

In each subdomain the following vectors are defined:

Subdomain Ω_1

$\{u\}_1^1$, $\{u_n\}_1^1$	Nodal values on part Γ_1 of the external boundary
$\{u\}_{12}^1$, $\{u_n\}_{12}^1$	Nodal values on the interface Γ_{12}
$\{u\}_{13}^1$, $\{u_n\}_{13}^1$	Nodal values on the interface Γ_{13}

In the foregoing notation, the superscript denotes the subdomain, while the subscript denotes the neighboring subdomains or the corresponding free boundary. The number of the nodal points on Γ_1, Γ_{12}, and Γ_{13} are N_1, N_{12}, and N_{13}, respectively.

Subdomain Ω_2

$\{u\}_2^2$, $\{u_n\}_2^2$	Nodal values on part Γ_2 of the external boundary
$\{u\}_{12}^2$, $\{u_n\}_{12}^2$	Nodal values on the interface Γ_{12}
$\{u\}_{23}^2$, $\{u_n\}_{23}^2$	Nodal values on the interface Γ_{23}

The number of the nodal points on Γ_2, Γ_{12}, and Γ_{23} are denoted by N_2, N_{12}, and N_{23}, respectively.

Subdomain Ω_3

$\{u\}_3^3$, $\{u_n\}_3^3$	Nodal values on part Γ_3 of the external boundary
$\{u\}_{13}^3$, $\{u_n\}_{13}^3$	Nodal values on the interface Γ_{13}
$\{u\}_{23}^3$, $\{u_n\}_{23}^3$	Nodal values on the interface Γ_{23}

The number of the nodal points on Γ_3, Γ_{13}, and Γ_{23} are N_3, N_{13}, and N_{23}, respectively.

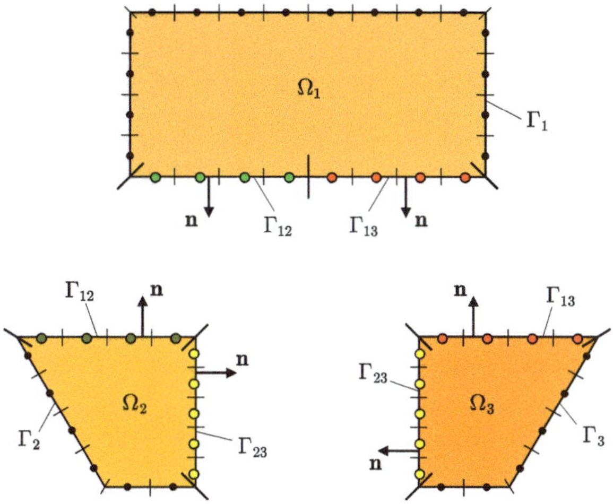

FIGURE 4.19 Boundary element discretization of composite domain ($\Gamma = \Gamma_1 \cup \Gamma_2 \cup \Gamma_3$). Subdomains and interfaces.

The boundary conditions are specified only on the external boundary Γ of Ω, that is, only on the parts Γ_1, Γ_2, Γ_3 ($\Gamma = \Gamma_1 \cup \Gamma_2 \cup \Gamma_3$). Both quantities u and u_n are unknown on either side of the interfaces Γ_{12}, Γ_{13}, Γ_{23}. Therefore, the total number of boundary unknowns in each subdomain is:

Subdomain Ω_1: N_1 on Γ_1, $2N_{12}$ on Γ_{12}, $2N_{13}$ on Γ_{13}, total $N_1 + 2N_{12} + 2N_{13}$
Subdomain Ω_2: N_2 on Γ_2, $2N_{12}$ on Γ_{12}, $2N_{23}$ on Γ_{23}, total $N_2 + 2N_{12} + 2N_{23}$
Subdomain Ω_3: N_3 on Γ_3, $2N_{13}$ on Γ_{13}, $2N_{23}$ on Γ_{23}, total $N_3 + 2N_{13} + 2N_{23}$

However, the available equations for the evaluation of the unknown boundary quantities are:
from subdomain Ω_1: $N_1 + N_{12} + N_{13}$
from subdomain Ω_2: $N_2 + N_{12} + N_{23}$
from subdomain Ω_3: $N_3 + N_{13} + N_{23}$

Therefore, the number of unknowns exceeds the number of available equations by $2(N_{12} + N_{13} + N_{23})$. The additional equations result from physical considerations, which are the so-called *continuity conditions* at the interfaces. These conditions express:

Continuity of the potential. The values of the potential on each side of the interface separating two subdomains are equal,

$$\left.\begin{array}{l}\{u\}_{12}^{1}=\{u\}_{12}^{2}\\ \{u\}_{13}^{1}=\{u\}_{13}^{3}\\ \{u\}_{23}^{2}=\{u\}_{23}^{3}\end{array}\right\} \tag{4.34}$$

Continuity of the flux. The flux q_n is a quantity related to the physical problem described by the potential equation. For example, in heat conduction problems the flux is given by Fourier's law $q_n = k\ u_n$, where k is the coefficient of thermal conductivity of the material. For the torsion problem the expression for the flux is more complicated. The outcoming flux from one subdomain is equal to the incoming flux in the adjacent subdomain. Thus, if q_n denotes the flux along the normal to the interface, its continuity across the interface requires

$$\left.\begin{array}{l}\{q_n\}_{12}^{2}=-\{q_n\}_{12}^{1}\\ \{q_n\}_{13}^{3}=-\{q_n\}_{13}^{1}\\ \{q_n\}_{23}^{3}=-\{q_n\}_{23}^{2}\end{array}\right\} \tag{4.35}$$

The minus sign in the right-hand side of Eq. (4.35) is explained by the fact that the positive direction for the flux coincides with the outward normal vector $\mathbf{n}$. This means that the two flux vectors at the common interface of adjacent subdomains are of opposite directions, because their corresponding normal vectors are opposite too (Fig. 4.19). Generally, the formulation of the continuity condition for the flux requires familiarity with the physical problem under consideration. For this reason, the above concepts will become more clear after studying Chapter 6.

Let us apply the flux continuity conditions to the heat flow problem. If the coefficients of thermal conductivity in the subdomains Ω_1, Ω_2, and Ω_3 are denoted by k_1, k_2, and k_3, respectively, then Eq. (4.35) may be written as

$$\left.\begin{array}{l}\{u_n\}_{12}^{2}=-[k_{12}]\{u_n\}_{12}^{1}\\ \{u_n\}_{13}^{3}=-[k_{13}]\{u_n\}_{13}^{1}\\ \{u_n\}_{23}^{3}=-[k_{23}]\{u_n\}_{23}^{2}\end{array}\right\} \tag{4.36}$$

where $[k_{12}]$, $[k_{13}]$, $[k_{23}]$ are square diagonal matrices with elements $k_{12} = k_1/k_2$, $k_{13} = k_1/k_3$, and $k_{23} = k_2/k_3$, respectively.

Equations (4.34) and (4.36) provide now the required $2(N_{12} + N_{13} + N_{23})$ additional equations for establishing all the unknowns of the problem.

On the basis of the foregoing, the matrix equations for the boundary of each subdomain become:

For the boundary of subdomain Ω_1

$$[H]^1\{u\}^1 = [G]^1\{u_n\}^1$$

or

$$\left[[H]_1^1 \quad [H]_{12}^1 \quad [H]_{13}^1\right]\begin{Bmatrix}\{u\}_1^1\\ \{u\}_{12}^1\\ \{u\}_{13}^1\end{Bmatrix} = \left[[G]_1^1 \quad [G]_{12}^1 \quad [G]_{13}^1\right]\begin{Bmatrix}\{u_n\}_1^1\\ \{u_n\}_{12}^1\\ \{u_n\}_{13}^1\end{Bmatrix}$$

The boundary conditions on part Γ_1 of the external boundary influence only the first term in each side of the above equation. Incorporating these conditions, we get

$$\left[[A]_1^1 \quad [H]_{12}^1 \quad [H]_{13}^1\right]\begin{Bmatrix}\{x\}_1^1\\ \{u\}_{12}^1\\ \{u\}_{13}^1\end{Bmatrix} = \{B\}_1^1 + \left[[G]_{12}^1 \quad [G]_{13}^1\right]\begin{Bmatrix}\{u_n\}_{12}^1\\ \{u_n\}_{13}^1\end{Bmatrix} \tag{4.37}$$

The vector $\{x\}_1^1$ contains all the unknown boundary quantities of Γ_1. The matrix $[A]_1^1$ and the vector $\{B\}_1^1$ are derived using the same procedure as for the case of Eq. (4.9).

For the boundary of subdomain Ω_2

$$[H]^2\{u\}^2 = [G]^2\{u_n\}^2$$

or

$$\left[[H]_2^2 \quad [H]_{12}^2 \quad [H]_{23}^2\right]\begin{Bmatrix}\{u\}_2^2\\ \{u\}_{12}^2\\ \{u\}_{23}^2\end{Bmatrix} = \left[[G]_2^2 \quad [G]_{12}^2 \quad [G]_{23}^2\right]\begin{Bmatrix}\{u_n\}_2^2\\ \{u_n\}_{12}^2\\ \{u_n\}_{23}^2\end{Bmatrix}$$

Using the continuity conditions (4.34) and (4.36), the above equation yields

$$\left[[H]_2^2 \ [H]_{12}^2 \ [H]_{23}^2\right]\begin{Bmatrix}\{u\}_2^2\\ \{u\}_{12}^1\\ \{u\}_{23}^2\end{Bmatrix} = \left[[G]_2^2 \quad -[G]_{12}^2[k_{12}] \ [G]_{23}^2\right]\begin{Bmatrix}\{u_n\}_2^2\\ \{u_n\}_{12}^1\\ \{u_n\}_{23}^2\end{Bmatrix}$$

Subsequently, applying the boundary conditions from part Γ_2 of the external boundary, we arrive at

$$\left[[A]_2^2 \quad [H]_{12}^2 \quad [H]_{23}^2\right]\begin{Bmatrix}\{x\}_2^2\\ \{u\}_{12}^1\\ \{u\}_{23}^2\end{Bmatrix} = \{B\}_2^2 + \left[-[G]_{12}^2[k_{12}] \quad [G]_{23}^2\right]\begin{Bmatrix}\{u_n\}_{12}^1\\ \{u_n\}_{23}^2\end{Bmatrix} \tag{4.38}$$

For the boundary of subdomain Ω_3

$$[H]^3\{u\}^3 = [G]^3\{u_n\}^3$$

or

$$\left[[H]_3^3 \quad [H]_{13}^3 \quad [H]_{23}^3\right]\begin{Bmatrix}\{u\}_3^3\\ \{u\}_{13}^3\\ \{u\}_{23}^3\end{Bmatrix} = \left[[G]_3^3 \quad [G]_{13}^3 \quad [G]_{23}^3\right]\begin{Bmatrix}\{u_n\}_3^3\\ \{u_n\}_{13}^3\\ \{u_n\}_{23}^3\end{Bmatrix}$$

and using again the continuity conditions (4.34) and (4.36), the above equation takes the form

$$\left[[H]_3^3 \; [H]_{13}^3 \; [H]_{23}^3\right]\begin{Bmatrix}\{u\}_3^3\\ \{u\}_{13}^1\\ \{u\}_{23}^2\end{Bmatrix} = \left[[G]_3^3 \; -[G]_{13}^3\,[k_{13}] \; -[G]_{23}^3\,[k_{23}]\right]\begin{Bmatrix}\{u_n\}_3^3\\ \{u_n\}_{13}^1\\ \{u_n\}_{23}^2\end{Bmatrix}$$

Finally, incorporating the boundary conditions for part Γ_3 of the external boundary, the last equation gives

$$\left[[A]_3^3 \; [H]_{13}^3 \; [H]_{23}^3\right]\begin{Bmatrix}\{x\}_3^3\\ \{u\}_{13}^1\\ \{u\}_{23}^2\end{Bmatrix} = \{B\}_3^3 + \left[-[G]_{13}^3\,[k_{13}] \; -[G]_{23}^3[k_{23}]\right]\begin{Bmatrix}\{u_n\}_{13}^1\\ \{u_n\}_{23}^2\end{Bmatrix} \tag{4.39}$$

Equations (4.37)–(4.39) of the three subdomains may then be combined in a single matrix equation as

$$[A]\{X\} = \{B\} \tag{4.40}$$

where
$\{X\}$: Vector consisting of all the unknown values on the external boundary and on the interfaces. Its dimension is: $N = N_1 + N_2 + N_3 + 2N_{12} + 2N_{13} + 2N_{23}$
$[A]$: Known square coefficient matrix of dimensions $N \times N$
$\{B\}$: Known vector of dimension N
The vectors $\{X\}$, $\{B\}$ and the matrix $[A]$ are defined by the equations

$$\{X\} = \begin{Bmatrix}\{x\}_1^1\\ \{x\}_2^2\\ \{x\}_3^3\\ \{u\}_{12}^1\\ \{u\}_{13}^1\\ \{u\}_{23}^2\\ \{u_n\}_{12}^1\\ \{u_n\}_{13}^1\\ \{u_n\}_{23}^2\end{Bmatrix}, \quad \{B\} = \begin{Bmatrix}\{B\}_1^1\\ \{B\}_2^2\\ \{B\}_3^3\end{Bmatrix} \tag{4.41}$$

FIGURE 4.20 Long and slender homogeneous domain divided in three subdomains.

$$[A] = \begin{bmatrix} [A]_1^1 & [0] & [0] & [H]_{12}^1 & [H]_{13}^1 & [0] & -[G]_{12}^1 & -[G]_{13}^1 & [0] \\ [0] & [A]_2^2 & [0] & [H]_{12}^2 & [0] & [H]_{23}^2 & [G]_{12}^2[k_{12}] & [0] & -[G]_{23}^2 \\ [0] & [0] & [A]_3^3 & [0] & [H]_{13}^3 & [H]_{23}^3 & [0] & [G]_{13}^3[k_{13}] & [G]_{23}^3[k_{23}] \end{bmatrix} \tag{4.42}$$

We notice that matrix $[A]$ is not fully populated. Its structure, therefore, allows the use of special techniques for the solution of Eq. (4.40), which reduce running time and disk space requirements. The method of subdomains may also be employed for long and slender homogeneous domains to overcome numerical problems associated with the integration of the fundamental solution over long distances. By splitting the domain into two or more subdomains (see Fig. 4.20), the aspect ratio of each subdomain is reduced, and the influence matrices $[H]$ and $[G]$ are computed more accurately.

4.9 REFERENCES

In the present chapter the BEM was presented as a numerical method suitable for solving problems described by the potential equation described by the Laplace or Poisson equation. The employed boundary discretization was the simplest one, that of constant elements. For further reading on the material of this chapter, one should look in the books by Jaswon and Symm [9], Brebbia [10], Banerjee and Butterfield [11], Brebbia and Dominguez [12], Kane [13], Beer et al. [14], and Pozrikidis [15]. The computer language employed for the programs was FORTRAN 77 of the Microsoft Fortran PowerStation (Professional). The FORTRAN 90, though more powerful, was not selected, because many readers of this book might not be familiar with it yet.

[1] Wolfram S. *Mathematica. a system for doing mathematics by computer.* Redwood City, California: Addison-Wesley Publishing Company; 1988.

[2] Robertson J. Engineering mathematics with MAPLE. New York: McGraw-Hill; 1996.

[3] Katsikadelis JT. The analysis of plates on elastic foundation by the boundary integral equation method. Polytechnic University of New York; 1982. Ph.D. Dissertation.

[4] Katsikadelis JT, Armenakas AE. Analysis of clamped plates on elastic foundation by the boundary integral equation method. ASME J Appl Mech 1984;51:574–80.

[5] Katsikadelis JT, Armenakas AE. Numerical evaluation of double integrals with a logarithmic or a cauchy-type singularity. ASME J Appl Mech 1983;50:682–4.

[6] Microsoft Fortran PowerSstation 4.0, Professional, Edition 1994–95.

[7] Tyn Myint U, Debnath L. Linear partial differential equations for scientists and engineers. 4th ed. Boston: Birkhäuser; 2007.

[8] Katsikadelis JT, Kallivokas LF. Clamped plates on Pasternak-type elastic foundation by the boundary element method. ASME J Appl Mech, 53. 1986. p. 909–16.
[9] Jaswon MA, Symm GT. Integral equation methods in potential theory and elastostatics. London: Academic Press, Inc; 1977.
[10] Brebbia CA. The boundary element method for engineers. Plymouth: Pentech Press; 1978.
[11] Banerjee PK, Butterfield R. Boundary element methods in engineering science. New York: McGraw-Hill; 1981.
[12] Brebbia CA, Dominguez J. Boundary elements: An introductory course. 2nd ed. Southampton: Computational Mechanics Publications; 2001.
[13] Kane JH. Boundary element analysis in engineering continuum mechanics. Englewood Cliffs, New Jersey: Prentice Hall; 1994.
[14] Beer G, Smith I, Duenser C. The boundary element method with programming. Wien: Springer-Verlag; 2008.
[15] Pozrikidis, C. A. C. Practical guide to boundary element methods with the software library BEMLIB. Boca Raton, Florida: CRC Press; 2002.

PROBLEMS

4.1. Write a FORTRAN subroutine that computes the derivatives $u_{,x}$ and $u_{,y}$ at internal points $P \in \Omega$.

4.2. Write a computer program to evaluate the domain integrals and embed it as a subroutine in LABECON and LABECOMU to solve the Poisson equation.

4.3. Modify appropriately the program LABECON so that it can be used to solve the anisotropic problem.

4.4. For the composite domain of the following figure and the indicated discretization, compute the matrix $[A]$ and the vector $\{B\}$ of Eq. (4.40).

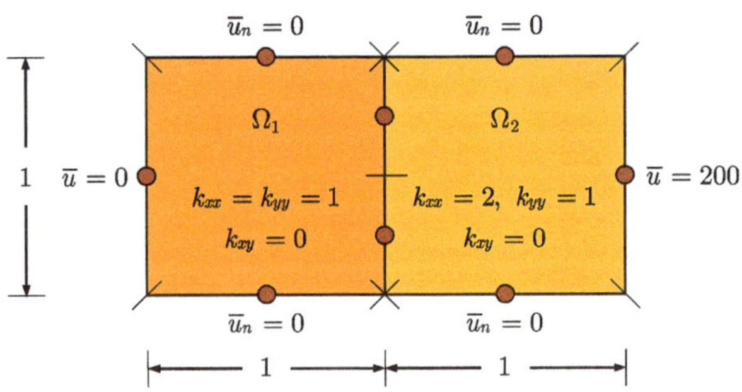

FIGURE P4.4

Chapter | Five

Boundary Element Technology

CHAPTER OUTLINE

5.1 INTRODUCTION

In the previous chapter, the boundary value problem described by the Poisson's differential equation (3.6a) and the boundary condition (3.6b) was modeled numerically by the boundary element method (BEM). This formulation originated from the discretization of the boundary integral equation (3.37) and resulted in the system of linear algebraic equation (4.9), which approximates the solution of the integral equation. The accuracy of the approximation and the efficiency of the BEM depends on the boundary discretization technique (i.e., the type of employed element), and on the method used for the integration of the kernel functions over the elements.

The constant element was presented in Chapter 4. This element approximates the actual geometry by a straight line, while the unknown boundary quantity is assumed to be constant on the element, resulting in a discontinuous distribution on the boundary. A better approximation of the boundary quantity can be achieved by adopting linear variation over the element.

The Boundary Element Method for Engineers and Scientists.

Even the linear element is not an ideal one as it cannot approximate accurately a curved boundary. For this reason, higher order elements have been developed, namely, elements that approximate both the boundary geometry and the boundary quantities by higher order interpolation polynomials, usually of second or third degree. Evidently, these elements model more accurately the curved boundaries and the distribution of the boundary quantities on the element. However, they have the weak point that the functions being integrated over each element become more complicated and the computer time is considerably increased.

Generally, the boundary elements may be classified in the following three categories [1]:

a. *Subparametric elements.* The polynomial approximating the geometry of the boundary is of a lower degree than that approximating the variation of the boundary quantity, for example, the element is a straight line, whereas the boundary quantity varies parabolically.

b. *Isoparametric elements.* The geometry and the boundary quantity are approximated by the same degree polynomials, for example, the linear and the parabolic elements defined in chapter "Numerical Implementation of the BEM" are isoparametric.

c. *Superparametric elements.* The geometry is approximated by a higher degree polynomial than that approximating the boundary quantity, for example, the element is modeled by a parabolic arc, whereas the boundary quantity is constant or varies linearly on it.

The use of the subparametric and superparametric elements is quite limited. Subparametric elements, however, result inevitably as a degenerate case of parabolic elements for rectilinear boundaries. Superparametric elements, in which the boundary geometry is modeled by a parabolic arc and the boundary quantity is constant, have been introduced and widely used by Katsikadelis and his coworkers [2–6]. This element has the advantages of the constant element regarding the simplicity of the assembly procedure for the influence matrices, whereas it approximates the curvilinear boundary with great accuracy. In general, the isoparametric elements are the most widely used elements, especially in commercial BEM codes.

The boundary elements are also distinguished in *continuous* and *discontinuous*. Continuous elements have nodes at their extreme points, and therefore they share nodes with the adjacent elements, while discontinuous elements have nodes located away from the extreme points. In the sections that follow, we will discuss the linear and the parabolic elements for continuous and discontinuous modeling.

5.2 LINEAR ELEMENTS

As it was mentioned before, linear elements approximate the geometry of the boundary by straight lines and the boundary quantity by a linear function on each element. In order to establish the expression for the variation of the boundary quantity over an element, its values at two nodal points are required. For this purpose, it is convenient to introduce a local coordinate system $Ox'y'$ on each element, where $-\ell/2 \leq x' \leq \ell/2$ with ℓ being the length of the element. For continuous elements the nodal points are placed at the extreme points (Fig. 5.1a), whereas for discontinuous elements the nodal points are placed between the end points (Fig. 5.1b).

Figure 5.2 shows the discretization of an elliptic boundary into 12 continuous linear elements, whereas Fig. 5.3 shows the discretization of the same boundary into 12 discontinuous linear elements. From these two figures, it becomes evident that the discretization of the boundary using discontinuous elements requires twice as many nodes compared to the continuous elements. However, a better approximation of the boundary quantity is achieved by employing the discontinuous elements (see Fig. 5.4). Nevertheless, the advantage of the discontinuous element is not the improved accuracy as compared to the continuous element, but rather its capability to overcome computational problems arising at points where the boundary quantity is discontinuous, for example, the normal derivative at corner points. *Hybrid* or *continuous–discontinuous* elements have also been invented, that is, elements having only one of the nodes placed at an extreme point.

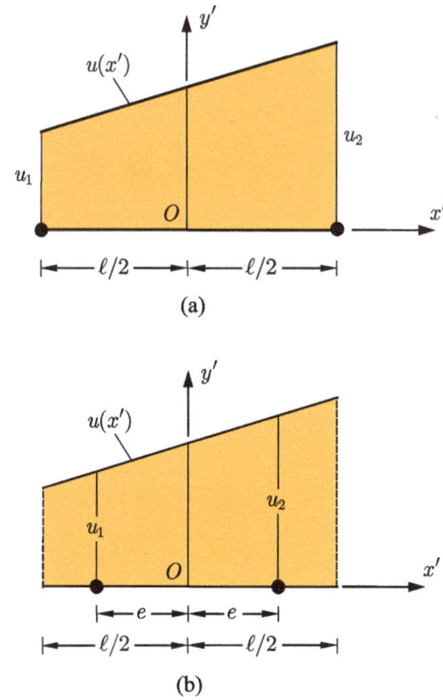

FIGURE 5.1 Continuous and discontinuous linear elements. (a) Continuous element; (b) Discontinuous element.

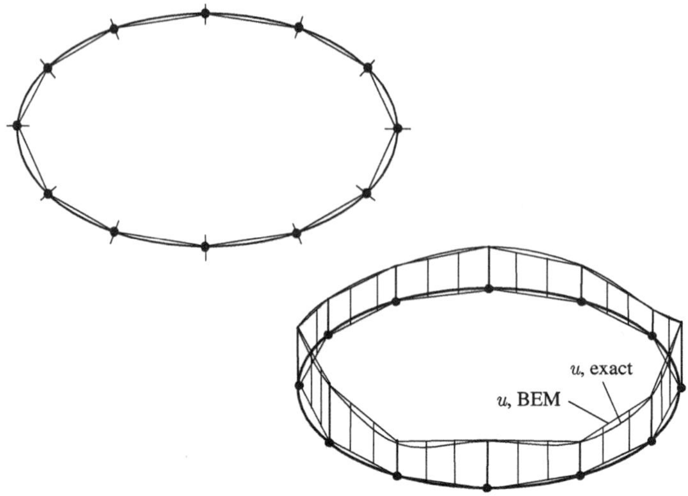

FIGURE 5.2 Elliptic boundary modeled by 12 continuous linear elements.

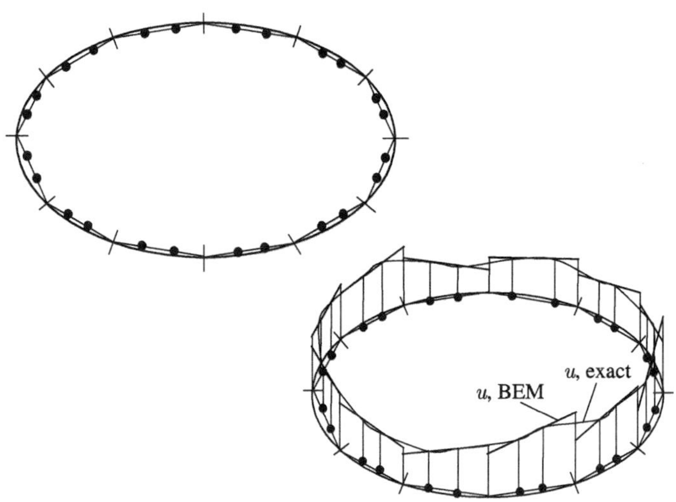

FIGURE 5.3 Elliptic boundary modeled by 12 discontinuous linear elements.

Let us consider the j-th element of the discretized boundary having end points j and $j+1$, and length ℓ_j (Fig. 5.5). In the local system of axes $O'x'y'$ its geometry is described by the equations

$$\begin{aligned} x' &= x' \quad (-\ell_j/2 \leq x' \leq \ell_j/2) \\ y' &= 0 \end{aligned}$$

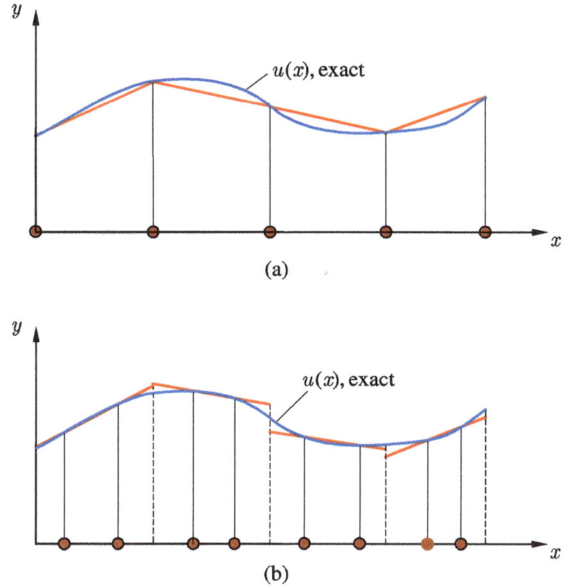

FIGURE 5.4 Approximation of the exact function $u(x)$ by linear continuous and discontinuous elements. (a) Continuous elements; (b) Discontinuous elements.

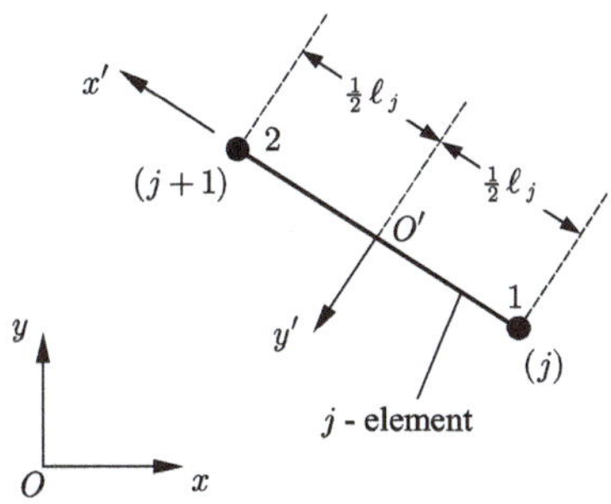

FIGURE 5.5 Global (Oxy) and local $(O'x'y')$ systems of axes for the j-th element.

whereas in the global system of axes Oxy by the equations

$$x = \frac{x_{j+1} + x_j}{2} + \frac{x_{j+1} - x_j}{\ell_j} x' \tag{5.1a}$$

$$y = \frac{y_{j+1} + y_j}{2} + \frac{y_{j+1} - y_j}{\ell_j} x' \tag{5.1b}$$

Recall that the positive direction of x' on the element is from point j towards point $j+1$, since the nodes and extreme points of the discretized boundary are numbered in the counter-clockwise sense. The interval $[-\ell_j/2, \ell_j/2]$ is normalized by setting

$$\xi = \frac{x'}{\ell_j/2} \tag{5.2}$$

and then Eq. (5.1) become

$$x = \frac{x_{j+1} + x_j}{2} + \frac{x_{j+1} - x_j}{2}\,\xi \tag{5.3a}$$

$$y = \frac{y_{j+1} + y_j}{2} + \frac{y_{j+1} - y_j}{2}\,\xi \tag{5.3b}$$

where $-1 \leq \xi \leq 1$.

Equation (5.3) may be rewritten as

$$x(\xi) = \frac{1}{2}(1-\xi)\,x_j + \frac{1}{2}(1+\xi)\,x_{j+1} \tag{5.4a}$$

$$y(\xi) = \frac{1}{2}(1-\xi)\,y_j + \frac{1}{2}(1+\xi)\,y_{j+1} \tag{5.4b}$$

The variation of the boundary quantity u (or $u_n = \partial u/\partial n$) is linear on the element. Hence, its distribution in the local system of axes is given by the expression

$$u = \frac{u^{j+1} + u^j}{2} + \frac{u^{j+1} - u^j}{\ell_j}\,x'$$

or

$$u = \frac{u^{j+1} + u^j}{2} + \frac{u^{j+1} - u^j}{2}\,\xi$$

or

$$u(\xi) = \frac{1}{2}(1-\xi)u^j + \frac{1}{2}(1+\xi)u^{j+1} \tag{5.5}$$

Using local numbering for the nodes, where j and $j+1$ are renamed to 1 and 2, respectively, Eqs. (5.4a), (5.4b), and (5.5) may all be expressed through the following general equation

$$f(\xi) = \psi_1(\xi)f_1 + \psi_2(\xi)f_2 \tag{5.6}$$

where f_1 and f_2 are the values of the function $f(x)$ at the nodes 1 and 2, and $f(\xi)$ represents any of the functions $x(\xi)$, $y(\xi)$, $u(\xi)$, or $u_n(\xi)$. The functions $\psi_1(\xi)$ and $\psi_2(\xi)$ are given as

$$\psi_1(\xi) = \frac{1}{2}(1 - \xi) \tag{5.7a}$$

$$\psi_2(\xi) = \frac{1}{2}(1 + \xi) \tag{5.7b}$$

and they express the influence of the nodal values f_j and f_{j+1} on the expression of $f(\xi)$ for the linear element. They are the functions of the linear interpolation and they are referred to as *linear shape functions*. From the above, one readily concludes that the linear element is isoparametric.

5.3 THE BEM WITH LINEAR BOUNDARY ELEMENTS

In this section, we will present the BEM for the potential equation using continuous linear elements. The number of the elements is equal to the number of nodes. The actual boundary is, thus, modeled by an inscribed polygon with the nodes placed at corner points.

In the case of constant elements, the substitute boundary which models the actual one is always smooth at the nodes and, therefore, the integral equation (3.29) is employed. In the case of linear elements, however, the nodes lie at the corners of the polygon and, consequently, the integral equation (3.28) should be applied instead with $\alpha = \alpha^i$, where α^i is the angle between the elements $(i-1)$ and i (Fig. 5.6). Thus, after discretizing the boundary into N linear elements, Eq. (3.28) becomes

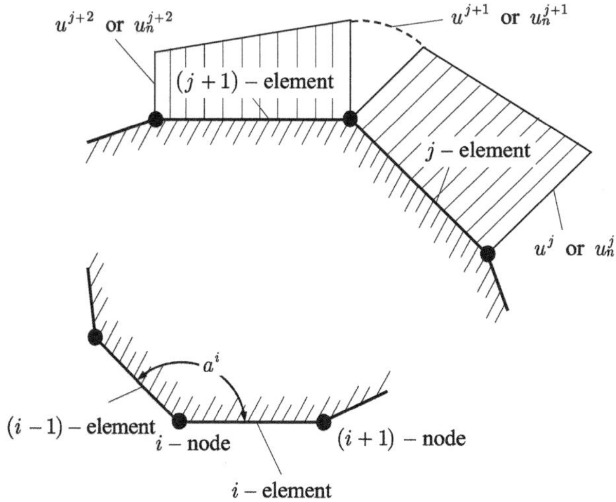

FIGURE 5.6 Modeling of the boundary with continuous linear elements.

$$\varepsilon^i\, u^i = -\sum_{j=1}^{N}\int_{\Gamma_j} v\, u_n\, ds + \sum_{j=1}^{N}\int_{\Gamma_j} u\, v_n\, ds \tag{5.8}$$

where $\varepsilon^i = \alpha^i/2\pi$ and $v_n = \partial v/\partial n$ denotes the derivative of the fundamental solution in the normal to the boundary direction.

We examine now the integrals over the j-th element. Using the linear approximation of Eq. (5.6) for the boundary quantities u and u_n, the line integral appearing in the first sum of Eq. (5.8) may be written as

$$\begin{aligned}\int_{\Gamma_j} v\, u_n\, ds &= \int_{-1}^{1} v\, \left[\psi_1(\xi)\, u_n^1 + \psi_2(\xi)\, u_n^2\right] \frac{\ell_j}{2}\, d\xi \\ &= u_n^1 \int_{-1}^{1} v\, \psi_1(\xi)\frac{\ell_j}{2}\, d\xi + u_n^2 \int_{-1}^{1} v\, \psi_2(\xi)\frac{\ell_j}{2}\, d\xi \\ &= g_1^{ij}\, u_n^1 + g_2^{ij}\, u_n^2\end{aligned} \tag{5.9}$$

where

$$g_1^{ij} = \frac{\ell_j}{2}\int_{-1}^{1} v\, \psi_1(\xi)\, d\xi \tag{5.10a}$$

$$g_2^{ij} = \frac{\ell_j}{2}\int_{-1}^{1} v\, \psi_2(\xi)\, d\xi \tag{5.10b}$$

and

$$v = \frac{1}{2\pi}\ln r \tag{5.11}$$

$$r = \sqrt{[x(\xi) - x_i]^2 + [y(\xi) - y_i]^2} \tag{5.12}$$

It should be made clear that the superscript i in the symbols g_1^{ij} and g_2^{ij} indicates the i-th node (p_i) where the source is applied, while the superscript j indicates the element over which the integration is carried out. Finally, the subscripts 1 and 2 denote in local numbering the points j and $(j+1)$, respectively.

In a similar fashion, the line integral appearing in the second sum of Eq. (5.8) may be written as

$$\int_{\Gamma_j} v_n\, u\, ds = h_1^{ij}\, u^1 + h_2^{ij}\, u^2 \tag{5.13}$$

where

$$h_1^{ij} = \frac{\ell_j}{2} \int_{-1}^{1} v_n \, \psi_1(\xi) \, d\xi \tag{5.14a}$$

$$h_2^{ij} = \frac{\ell_j}{2} \int_{-1}^{1} v_n \, \psi_2(\xi) \, d\xi \tag{5.14b}$$

and

$$v_n = \frac{\partial v}{\partial n} = \frac{1}{2\pi} \frac{\cos \phi}{r} \tag{5.15}$$

Substituting Eqs. (5.9) and (5.13) back into Eq. (5.8), the latter yields

$$-\varepsilon^i \, u^i + \sum_{j=1}^{N} \hat{H}_{ij} \, u^j = \sum_{j=1}^{N} G_{ij} \, u_n^j \tag{5.16}$$

where

$$\hat{H}_{ij} = \begin{cases} h_1^{i1} + h_2^{iN} & \text{for } j = 1 \\ h_1^{ij} + h_2^{i,j-1} & \text{for } j = 2, 3, \ldots, N \end{cases} \tag{5.17}$$

$$G_{ij} = \begin{cases} g_1^{i1} + g_2^{iN} & \text{for } j = 1 \\ g_1^{ij} + g_2^{i,j-1} & \text{for } j = 2, 3, \ldots, N \end{cases} \tag{5.18}$$

Equation (5.16) may be written in matrix form as

$$[H]\{u\} = [G]\{u_n\} \tag{5.19}$$

in which it has been set

$$[H] = -[\varepsilon] + [\hat{H}] \tag{5.20}$$

$[\varepsilon]$ is a diagonal matrix with elements of the coefficients ε^i.

5.3.1 Corner points and points of change of boundary conditions

In formulating Eq. (5.19), it was assumed that the quantities u and $u_n = \partial u / \partial n$ had a unique value. However, this is not always the case. For example, u_n is not continuous at the corner points of the actual boundary, since its value is generally different before and after the corner. Likewise, in mixed boundary conditions different values are prescribed at nodal points, where the boundary conditions change type. At corner points, we may distinguish the following cases for the boundary conditions [7]:

a. Known: u_n *before* and *after* the corner. Unknown: u at the corner.

b. Known: u at the corner and u_n *before* the corner. Unknown: u_n *after* the corner.

c. Known: u at the corner and u_n *after* the corner. Unknown: u_n *before* the corner.

d. Known: u at the corner. Unknown: u_n *before* and *after* the corner.

In all the above cases, u is assumed to be continuous at the corner points, having therefore a unique value at these points. The terms *before* or *after* refer to the value of the quantity just before or right after the corner point according to the positive sense on the boundary (see Section 4.7).

The unknown boundary quantities are determined under the assumption that u_n may be discontinuous at all nodal points, which means that we are dealing with $2N$ values of u_n. With this in mind, we can write Eq. (5.19) as

$$[H]\{u\} = [G^*]\{u_n^*\} \tag{5.21}$$

where $\{u_n^*\}$ is a vector containing the $2N$ values of the normal derivative (two at each node) and $[G^*]$ is a $N \times 2N$ matrix whose elements are defined as

$$\begin{aligned} G^*_{i,2j-1} &= g_1^{ij} \\ G^*_{i,2j} &= g_2^{ij} \quad (j = 1, 2, \ldots, N) \end{aligned} \tag{5.22}$$

For the first three cases of the corner boundary conditions—cases (a), (b), and (c)—only one boundary quantity is unknown. Therefore, rearranging the unknowns on the basis of the boundary conditions, Eq. (5.21) produces a system of N linear equations, which can be solved for the N unknown boundary quantities. The unknowns are rearranged by examining all the nodes. If the value of u is unknown at a node, then the respective column remains at the left-hand side of Eq. (5.21), otherwise this column is multiplied by the known value of u, its sign is switched, and it is shifted to the right-hand side of the equation. Similarly, if the value of u_n is unknown, then the respective column of $[G^*]$ is shifted with opposite sign to the left-hand side of the equation. The two consecutive columns corresponding to u_n are added together if the node is not a corner point of the *actual* boundary. After completing this process, the right-hand side of the equation contains only known quantities and, thus, the matrix multiplication results in a single vector. Corner points may also be treated using one-sided discontinuous elements before and after the corner (Fig. 5.7). Thus, two separate nodes appear in the equations at which two different values of u_n are computed.

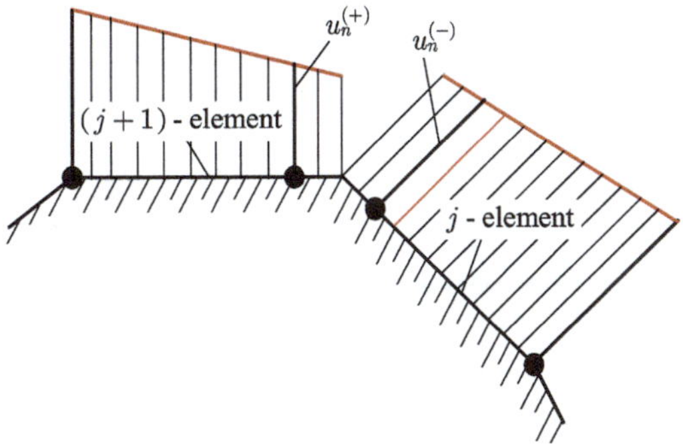

FIGURE 5.7 Discontinuous elements adjacent to a corner of the actual boundary ($u_n = \partial u/\partial n$).

It should be noted that an abrupt change in the boundary's slope (especially reentrant corners) or a change in the type of boundary conditions causes a local singularity in the behavior of the solution, which may even "*pollute*" the numerical results over the whole domain. A remedy to this problem is the refinement of the elements near the point of singularity. Nevertheless, this technique is not always successful in giving a reliable solution, especially for its derivative, and recourse to special techniques is unavoidable [8].

5.4 EVALUATION OF LINE INTEGRALS ON LINEAR ELEMENTS

The matrices $[G]$ and $[H]$ appearing in Eq. (5.19) require the computation of the line integrals (5.10) and (5.14) whose integrands are products of the fundamental solution v or its normal derivative $v_n = \partial v/\partial n$ and the linear shape functions $\psi_1(\xi)$ and $\psi_2(\xi)$. The integrations are carried out over the interval $[-1, +1]$. Two cases are considered for the linear elements as it was done for the constant elements. These cases are dictated by the behavior of the functions (5.11) and (5.15). Specifically, when the j-th element, over which the integration is performed, does not contain the source point i, that is, $i \neq j$, then it is always $r \neq 0$ and the integral is regular. On the other hand, when the source point lies on the j-th element, that is, $i = j$, then the distance r takes also the value $r = 0$ and the behavior of the integral is singular. The integration for the first case ($i \neq j$) will be referred to as *outside integration*, whereas for the second case ($i = j$) will be referred to as *inside integration*.

5.4.1 Outside integration

The integrals (5.10) and (5.14) may be evaluated analytically using symbolic languages (e.g., MAPLE). This process, however, yields very lengthy expressions, which in some cases may cover several pages, making them computationally impractical. A very practical and accurate approach is the numerical integration. Any integration rule may be utilized for this purpose, for example, trapezoidal rule, Simpson's rule, or Newton-Cotes integration formulae. But, the most suitable method for the numerical evaluation of the BEM integrals is the *Gaussian quadrature* (see Appendix B). This method approximates the integral with great accuracy using the least number of values of the integrand. The numerical integration should not be performed "blindly." The accuracy depends not only on the number of integration points, but also on how the integrand varies within the integration interval. A smooth variation of the integrand gives more accurate results. Therefore, the integration process requires a thoughtful consideration and special care is required when the integrand exhibits intense changes. In integrals (5.10) and (5.14), the shape functions vary smoothly and, consequently, the behavior of the integrand is dominated by the functions $\ln r$ and $1/r$.

In order to have a better insight into the variation of the integrand, we consider the domain of Fig. 5.8 and examine the function $g(\xi) = \psi_1(\xi)\,\ln r$ on the element with extreme points $(4,4)$ and $(3,4)$. Three locations for the source are considered:

i. location A, relatively far from the element,
ii. location B, at a relatively moderate distance from the element, and
iii. location C, relatively close to the element.

For the particular element, the transform equation (5.4a) and (5.4b) become

$$x(\xi) = \left[\frac{1}{2}(1-\xi)\right]4 + \left[\frac{1}{2}(1+\xi)\right]3 = 3.5 - 0.5\,\xi$$

$$y(\xi) = \left[\frac{1}{2}(1-\xi)\right]4 + \left[\frac{1}{2}(1+\xi)\right]4 = 4$$

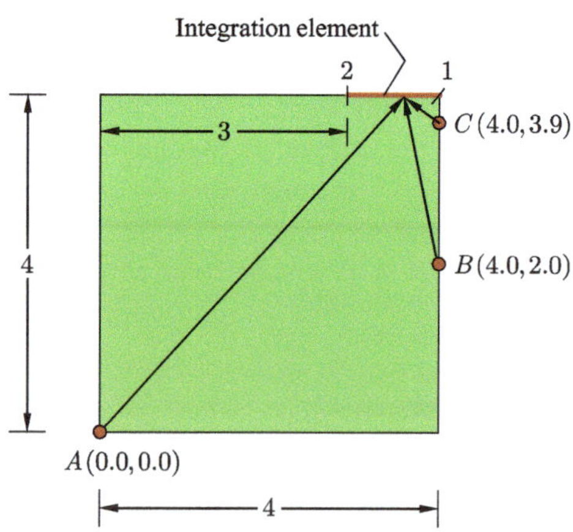

FIGURE 5.8 Points A, B, and C where the source is applied.

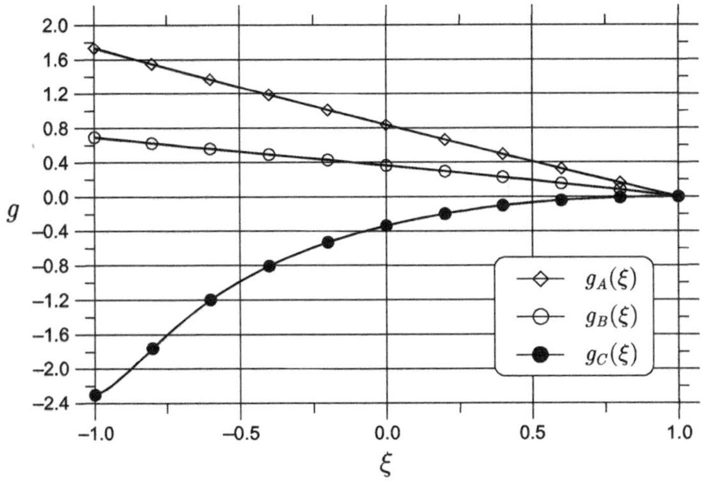

FIGURE 5.9 Behavior of the integrand $g(\xi) = \psi_1(\xi)\ln r$ for different locations of the source.

and according to Eq. (5.12), the relative distance from the source point is

$$r(\xi) = \sqrt{\left[x(\xi) - x_i\right]^2 + \left[y(\xi) - y_i\right]^2}$$

where x_i and y_i are the global coordinates of the source point.

The behavior of the integrand associated with points A, B, and C is illustrated graphically through the plots of $g(\xi)$ depicted in Fig. 5.9. It should be noticed that the variation of the function $g(\xi)$ is quite different when the source is located close to the element over which the integration is performed. Therefore, an efficient programming of the BEM should take into account such a behavior by using Gaussian quadratures with variable number of integration points (i.e., increasing number of points as the source is getting closer to the element). The number of integration points should be chosen in a way to ensure *sufficient accuracy*. However, an unnecessarily large number of integration points should be avoided, in order to keep the computation cost low.

5.4.2 Inside integration

In this case the source lies on the element over which the integration is performed. As the integration point runs along the whole element, it will coincide inevitably with the source point. There, the distance r vanishes and the integrands of Eqs. (5.10) and (5.14) exhibit a singular behavior, because the factors $\ln r$ and $\cos\phi/r$ become infinite for $r = 0$ (see Eqs. (5.11) and (5.15)). These integrals are known as *singular integrals*. Their value exists and is determined through special integration techniques, either analytical or numerical. Even an *indirect* method has been invented to circumvent the evaluation of the singular integrals by computing directly the coefficients H_{ii} and G_{ii} (see Section 5.4.3). In the sequel, we will first study the integrals with *logarithmic singularity*, and then those with *Cauchy type singularity* $(1/r)$.

INTEGRALS WITH LOGARITHMIC SINGULARITY

Analytical integration

We consider the general case of the discontinuous linear element shown in Fig. 5.10. The continuous element results as a special case, when the nodes are shifted to the extreme points. The two nodes of the linear element are designated locally by the numbers 1 and 2, while their global coordinates are denoted by (x_1, y_1) and (x_2, y_2). Using this notation, it can be easily proved that the coordinate transformation from the local to the global system is expressed by the equations

$$x(\xi) = \frac{\kappa_1 x_2 + \kappa_2 x_1}{\kappa} + \frac{x_2 - x_1}{\kappa}\,\xi \tag{5.23a}$$

$$y(\xi) = \frac{\kappa_1 y_2 + \kappa_2 y_1}{\kappa} + \frac{y_2 - y_1}{\kappa}\,\xi \tag{5.23b}$$

where $-1 \le \xi \le 1$ and

$$\kappa = \kappa_1 + \kappa_2, \quad \xi = \frac{x'}{\ell/2} \tag{5.24}$$

Furthermore, Eqs. (5.23) may be rewritten as

$$x(\xi) = \psi_1(\xi)\, x_1 + \psi_2(\xi)\, x_2$$
$$y(\xi) = \psi_1(\xi)\, y_1 + \psi_2(\xi)\, y_2$$

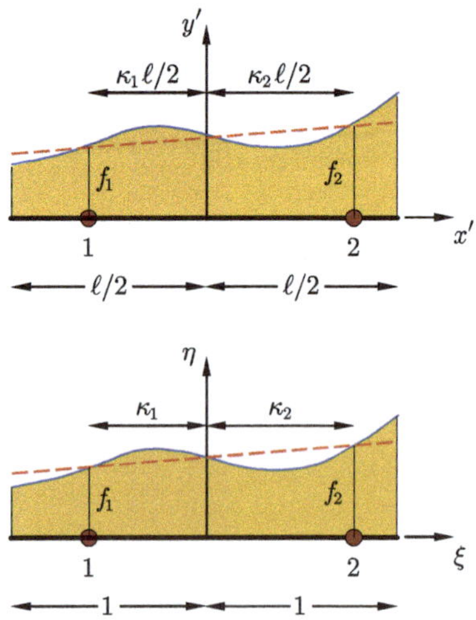

FIGURE 5.10 Discontinuous linear element in local coordinate systems.

in which the shape functions $\psi_1(\xi)$ and $\psi_2(\xi)$ are given by the expressions

$$\psi_1(\xi) = \frac{1}{\kappa}(\kappa_2 - \xi) \tag{5.25a}$$

$$\psi_2(\xi) = \frac{1}{\kappa}(\kappa_1 + \xi) \tag{5.25b}$$

with $0 \leq \kappa_1,\ \kappa_2 \leq 1$.

It is apparent that for $\kappa_1 = \kappa_2 = 1$, the shape functions, Eqs. (5.25), reduce to Eqs. (5.7), which represent the shape functions of the continuous linear element.

If the source lies on the J-th ($J = 1, 2$) local node of the element, its coordinates are going to be

$$x_J = \frac{\kappa_1\, x_2 + \kappa_2\, x_1}{\kappa} + \frac{x_2 - x_1}{\kappa}\xi_J$$

$$y_J = \frac{\kappa_1\, y_2 + \kappa_2\, y_1}{\kappa} + \frac{y_2 - y_1}{\kappa}\xi_J$$

where

$$\xi_1 = -\kappa_1, \quad \xi_2 = \kappa_2$$

and the relative distance of Eq. (5.12) becomes

$$r(\xi) = \frac{\ell_j}{2}|\xi - \xi_J| \tag{5.26}$$

where ℓ_j is the length of the j-th element.

Having described the discontinuous linear element, we shall study now the integrals of Eqs. (5.10) for the case where the source node i lies on the integration element j. The integral (5.10a) may then be written for a discontinuous element as

$$\begin{aligned}\frac{4\pi}{\ell_j} g_1^{ij} &= \int_{-1}^{1} \psi_1(\xi) \ln r\, d\xi \\ &= \int_{-1}^{\xi_J} \psi_1(\xi) \ln r\, d\xi + \int_{\xi_J}^{1} \psi_1(\xi) \ln r\, d\xi \\ &= I_1 + I_2\end{aligned} \tag{5.27}$$

where, the i-th node coincides with the J-th local node ($J = 1, 2$) of element j ($j = 1, 2, \ldots, N$). Note, that for discontinuous elements the number of nodes cannot be equal to the number of elements N. For example, if all the elements are discontinuous, the total number of nodes will be $2N$, which is the maximum possible number of nodes for a linear element discretization.

The transformation

$$\xi = -(1+\xi_J)\, z + \xi_J \tag{5.28}$$

maps the integration interval $[-1, \xi_J]$ of I_1 onto the interval [0, +1]. Substituting into the expression of I_1 which is given in Eq. (5.27), we have

$$\begin{aligned} I_1 &= \int_{-1}^{\xi_J} \psi_1(\xi) \ln r \, d\xi \\ &= \int_0^1 \frac{1}{\kappa} \left[(\kappa_2 - \xi_J) + (1+\xi_J)\, z\right] \ln\left[\frac{\ell_j}{2}(1+\xi_J)\, z\right](1+\xi_J)\, dz \end{aligned}$$

Introducing the quantities

$$\theta = \frac{\ell_j}{2}(1+\xi_J)\, z, \quad \theta_1 = \frac{\ell_j}{2}(1+\xi_J)$$

the above integral becomes

$$I_1 = \int_0^{\theta_1} \frac{1}{\kappa} \frac{2}{\ell_j} \left[(\kappa_2 - \xi_J) + \frac{2}{\ell_j}\theta\right] \ln\theta \, d\theta \tag{5.29}$$

which is of the form

$$I_1 = \int_0^{\theta_1} (a + b\theta) \ln\theta \, d\theta \tag{5.30}$$

where the two constants are

$$a = \frac{1}{\kappa} \frac{2}{\ell_j}(\kappa_2 - \xi_J) \quad \text{and} \quad b = \frac{1}{\kappa}\left(\frac{2}{\ell_j}\right)^2$$

The integral of Eq. (5.30) can be readily integrated by parts to yield

$$I_1 = \left[a\theta\,(\ln\theta - 1) + b\theta^2\left(\frac{1}{2}\ln\theta - \frac{1}{4}\right)\right]_0^{\theta_1} \tag{5.31}$$

or using the definitions of the constants a and b, it yields

$$\begin{aligned} I_1 = &\frac{1}{\kappa}(\kappa_2 - \xi_J)(1+\xi_J)\left\{\ln\left[\frac{\ell_j}{2}(1+\xi_J)\right] - 1\right\} \\ &+ \frac{1}{\kappa}(1+\xi_J)^2\left\{\frac{1}{2}\ln\left[\frac{\ell_j}{2}(1+\xi_J)\right] - \frac{1}{4}\right\} \end{aligned} \tag{5.32}$$

Referring now to the second integral of Eq. (5.27), the transformation

$$\xi = (1-\xi_J)\,z + \xi_J \tag{5.33}$$

maps the integration interval $[\xi_J, +1]$ of I_2 onto the interval $[0, +1]$. Thus, the new expression of the integral is

$$\begin{aligned} I_2 &= \int_{\xi_J}^{1} \psi_1(\xi) \ln r \, d\xi \\ &= \int_0^1 \frac{1}{\kappa}\left[(\kappa_2-\xi_J)-(1-\xi_J)\,z\right]\ln\left[\frac{\ell_j}{2}(1-\xi_J)\,z\right](1-\xi_J)\,dz \end{aligned}$$

Furthermore, by setting

$$\theta = \frac{\ell_j}{2}(1-\xi_J)\,z, \quad \theta_1 = \frac{\ell_j}{2}(1-\xi_J)$$

the above integral may also be written as

$$I_2 = \int_0^{\theta_1} \frac{1}{\kappa}\,\frac{2}{\ell_j}\left[(\kappa_2-\xi_J)-\frac{2}{\ell_j}\theta\right]\ln\theta \, d\theta \tag{5.34}$$

The integral of Eq. (5.34) has exactly the same form as that of Eq. (5.30), but with different values of the two constants, which are now defined as

$$a = \frac{1}{\kappa}\,\frac{2}{\ell_j}(\kappa_2-\xi_J) \quad \text{and} \quad b = -\frac{1}{\kappa}\left(\frac{2}{\ell_j}\right)^2$$

Substituting this set of constants into Eq. (5.31), we obtain the final expression for the second integral of Eq. (5.27)

$$\begin{aligned} I_2 = {} & \frac{1}{\kappa}(\kappa_2-\xi_J)(1-\xi_J)\left\{\ln\left[\frac{\ell_j}{2}(1-\xi_J)\right]-1\right\} \\ & -\frac{1}{\kappa}(1-\xi_J)^2\left\{\frac{1}{2}\ln\left[\frac{\ell_j}{2}(1-\xi_J)\right]-\frac{1}{4}\right\} \end{aligned} \tag{5.35}$$

The results of Eqs. (5.32) and (5.35) are combined into Eq. (5.27) to produce the analytical expression for the influence coefficient g_1^{ij} when the node i lies on the element j (i.e., singular case).

The other influence coefficient defined by Eq. (5.10b), is written for a discontinuous element as

$$\frac{4\pi}{\ell_j} g_2^{ij} = \int_{-1}^{1} \psi_2(\xi)\ln r \, d\xi$$
$$= \int_{-1}^{\xi_J} \psi_2(\xi)\ln r \, d\xi + \int_{\xi_J}^{1} \psi_2(\xi) \ln r \, d\xi = I_3 + I_4 \tag{5.36}$$

where the indices i, j, and J are defined as for the case of Eq. (5.27). The integrals I_3 and I_4 are evaluated following the procedure applied for I_1 and I_2. The resulting expressions for these two integrals are

$$I_3 = \frac{1}{\kappa}(\kappa_1 + \xi_J)(1+\xi_J)\left\{ \ln\left[\frac{\ell_j}{2}(1+\xi_J)\right] - 1 \right\}$$
$$-\frac{1}{\kappa}(1+\xi_J)^2 \left\{ \frac{1}{2}\ln\left[\frac{\ell_j}{2}(1+\xi_J)\right] - \frac{1}{4} \right\} \tag{5.37}$$

and

$$I_4 = \frac{1}{\kappa}(\kappa_1 + \xi_J)(1-\xi_J) \left\{ \ln\left[\frac{\ell_j}{2}(1-\xi_J)\right] - 1 \right\}$$
$$+\frac{1}{\kappa}(1-\xi_J)^2 \left\{ \frac{1}{2}\ln\left[\frac{\ell_j}{2}(1-\xi_J)\right] - \frac{1}{4} \right\} \tag{5.38}$$

which may be combined into Eq. (5.36) to give the analytical expression for the singular case of the influence coefficient g_2^{ij}.

The influence coefficients for continuous elements may be deduced from the foregoing equations by setting $\xi_1 = -1$ and $\xi_2 = 1$. In this case, Eqs. (5.27) and (5.36) give:

i. for $\xi_J = -1$, it is $\kappa_1 = \kappa_2 = 1$, $\kappa = 2$ and

$$g_1^{ij} = \frac{\ell_j}{4\pi}\left(\ln\ell_j - 1.5\right)$$
$$g_2^{ij} = \frac{\ell_j}{4\pi}\left(\ln\ell_j - 0.5\right) \tag{5.39a}$$

ii. for $\xi_J = +1$, it is $\kappa_1 = \kappa_2 = 1$, $\kappa = 2$ and

$$g_1^{ij} = \frac{\ell_j}{4\pi}\left(\ln\ell_j - 0.5\right)$$
$$g_2^{ij} = \frac{\ell_j}{4\pi}\left(\ln\ell_j - 1.5\right) \tag{5.39b}$$

Numerical integration

According to the previous discussion, integrals with logarithmic singularity, like those of Eqs. (5.27) and (5.36), can be set in the form

$$I = \int_0^1 f(x) \ln x \, dx \tag{5.40}$$

Special Gauss integration schemes have been developed for its numerical evaluation. Stroud and Secrest [9] approximated this type of integral as follows

$$\int_0^1 f(\xi) \ln\left(\frac{1}{\xi}\right) d\xi \approx \sum_{k=1}^{n} f(\xi_k)\, w_k \tag{5.41}$$

and produced tables with the integration points ξ_k and the corresponding weights w_k (see Appendix B and Refs. [9,10]). It should be emphasized here that the integration interval must always be reduced to [0, +1]. Therefore, when the source lies inside the element, the integration interval must be split into two intervals, $[-1, \xi_J]$ and $[\xi_J, +1]$. Consequently, the influence coefficients of Eq. (5.10) become

$$\begin{aligned} \frac{4\pi}{\ell_j} g_\alpha^{ij} &= \int_{-1}^{1} \psi_\alpha(\xi) \ln r \, d\xi \\ &= \int_{-1}^{\xi_J} \psi_\alpha(\xi) \ln r \, d\xi + \int_{\xi_J}^{1} \psi_\alpha(\xi) \ln r \, d\xi \quad (\alpha = 1, 2) \end{aligned} \tag{5.42}$$

where ξ_J $(J = 1, 2)$ denotes the node of the j-th element where the source node i is located and r is the relative distance between the source and the integration point given in Eq. (5.26).

The transformations (5.28) and (5.33) map the two integration intervals $[-1, \xi_J]$ and $[\xi_J, +1]$ of Eq. (5.42) onto the interval [0, +1], respectively. Eq. (5.42) is then written as

$$\begin{aligned} \frac{4\pi}{\ell_j} g_\alpha^{ij} &= \int_0^1 \psi_\alpha^{(1)}(z)(1 + \xi_J) \ln\left[\frac{\ell_j}{2}(1 + \xi_J) z\right] dz \\ &+ \int_0^1 \psi_\alpha^{(2)}(z)(1 - \xi_J) \ln\left[\frac{\ell_j}{2}(1 - \xi_J) z\right] dz \end{aligned}$$

or by expanding the logarithms

$$\begin{aligned}\frac{4\pi}{\ell_j} g_\alpha^{ij} = &(1+\xi_J)\ln\left[\frac{\ell_j}{2}(1+\xi_J)\right]\int_0^1 \psi_\alpha^{(1)}(z)\,dz \\ &+(1+\xi_J)\int_0^1 \psi_\alpha^{(1)}(z)\ln z\,dz \\ &+(1-\xi_J)\ln\left[\frac{\ell_j}{2}(1-\xi_J)\right]\int_0^1 \psi_\alpha^{(2)}(z)\,dz \\ &+(1-\xi_J)\int_0^1 \psi_\alpha^{(2)}(z)\ln z\,dz \\ &= I_1 + I_2 + I_3 + I_4\end{aligned} \tag{5.43}$$

where $\psi_\alpha^{(1)}(z)$ and $\psi_\alpha^{(2)}(z)$ $(\alpha = 1, 2)$ are transformed shape functions obtained from $\psi_\alpha(\xi)$ by expressing the variable ξ in terms of z according to Eqs. (5.28) and (5.33), respectively. The integrals I_1 and I_3 are regular and can be evaluated either analytically or by applying the conventional Gauss integration, while the I_2 and I_4 are singular and can be evaluated numerically by virtue of Eq. (5.41).

Integration by extracting the singularity

The integrals of Eq. (5.10) may also be written as

$$\begin{aligned}\frac{4\pi}{\ell_j} g_\alpha^{ij} &= \int_{-1}^{1} \psi_\alpha(\xi)\ln r\,d\xi \\ &= \int_{-1}^{1} [\psi_\alpha(\xi) - \psi_\alpha(\xi_J)]\ln r\,d\xi + \psi_\alpha(\xi_J)\int_{-1}^{1} \ln r\,d\xi\end{aligned} \tag{5.44}$$

The integrand of the first integral vanishes for $\xi = \xi_J$. Indeed, this integrand can be written as

$$\frac{\ln r(\xi)}{\left[\dfrac{1}{\psi_\alpha(\xi) - \psi_\alpha(\xi_J)}\right]}$$

It can be easily noticed that this expression takes for $\xi = \xi_J$ the indeterminate form $\frac{\infty}{\infty}$. Thus, applying consecutively L'Hôpital's rule and, taking into account that the derivatives $\psi'_\alpha(\xi) = d\psi_\alpha(\xi)/d\xi$ and $r'(\xi) = dr(\xi)/d\xi$ are finite and do not vanish, the integrand in question yields

$$\lim_{\xi\to\xi_J}\{[\psi_\alpha(\xi) - \psi_\alpha(\xi_J)]\ln r(\xi)\} = 0$$

This result shows that the first integral of Eq. (5.44) is regular and can be evaluated using conventional Gaussian quadrature. It is worth mentioning that the accuracy of this numerical integration is increased if it is carried out on two separate subintervals, $[-1, \xi_J]$ and $[\xi_J, +1]$. Apparently, this is dictated by the shape of the integrand as it is depicted in Fig. 5.11. The second integral of Eq. (5.44) exhibits a logarithmic singularity, but it can be readily evaluated analytically. In closing we could say that the method of *extracting the singularity* simplifies the evaluation of the integral (5.44), though it does not avoid the evaluation of singular integrals.

Table 5.1 presents the values of the integrals g_α^{ij}. They have been computed: (1) analytically from Eqs. (5.27) and (5.36); (2) using the special Gauss integration of Eq. (5.43) with 8 integration points; and (3) by extracting the singularity according to Eq. (5.44). In the last case, the regular integral was computed using eight Gauss points on each of the subintervals $[-1, \xi_J]$ and $[\xi_J, +1]$, while the singular one was evaluated analytically. For all cases the element data are $x_1 = 3.0$, $y_1 = 2.0$, $x_2 = 1.0$, $y_2 = 3.0$, and $\kappa_1 = \kappa_2 = 0.5$.

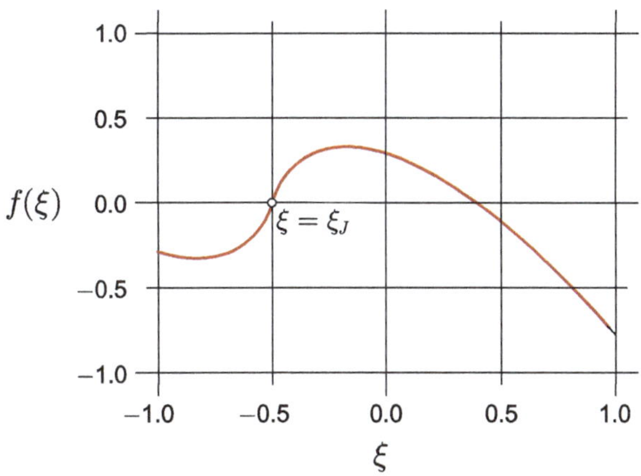

FIGURE 5.11 Variation of the function $f(\xi) = [\psi_1(\xi) - \psi_1(\xi_J)]\ln r(\xi)$ for $\xi_J = -0.5$.

TABLE 5.1 Values of the Integrals g_α^{ij}, when the Source i Lies on the Element

Influence Coefficient	Analytical Integration	Gauss Integration, Eq. (5.43)	Extracting the Singularity, Eq. (5.44)
g_1^{1j}	−0.2970889377	−0.2970889376	−0.2970889377
g_2^{1j}	0.0274675251	0.0274675250	0.0274675251
g_1^{2j}	0.0274675251	0.0274675250	0.0274675251
g_2^{2j}	−0.2970889377	−0.2970889376	−0.2970889377

Integrals with Cauchy type singularity

It has already been mentioned that the computation of the diagonal elements $\hat{H}_{ii}$ and G_{ii} of matrices $[H]$ and $[G]$ requires an inside integration. In the preceding subsection we presented techniques for the evaluation of the coefficients G_{ii} which are line integrals with logarithmic singularity. The coefficients $\hat{H}_{ii}$, however, are determined by evaluating integrals of the form (see Eqs. (5.14) and (5.15))

$$\int_{-1}^{1} \psi_\alpha(\xi)\frac{\cos\phi}{r}\,d\xi, \qquad r = \frac{\ell_j}{2}\,|\xi - \xi_J| \tag{5.45}$$

where $\psi_\alpha(\xi)$ $(\alpha = 1, 2)$ are given in Eq. (5.25) and r in Eq. (5.26).

Linear elements approximate the geometry by a straight line. Hence it is

$$\cos\phi = \cos\left(\pm\frac{\pi}{2}\right) = 0$$

along the whole element, since $\phi = \text{angle}\,(\mathbf{r}, \mathbf{n})$ (see Appendix A). Therefore, the integrand of Eq. (5.45) becomes

$$\psi_\alpha(\xi)\frac{\cos\phi}{r} = \begin{cases} 0, & \xi \neq \xi_J \\ \dfrac{0}{0}, & \xi = \xi_J \end{cases}$$

Applying L'Hôpital's rule, the above expression yields

$$\lim_{\xi\to\xi_J} \frac{\psi_\alpha(\xi)\cos\phi}{r(\xi)} = \lim_{\xi\to\xi_J} \frac{\psi'_\alpha(\xi)\cos\phi}{r'(\xi)} = 0 \tag{5.46}$$

because the derivative $\psi'_\alpha(\xi) = (-1)^\alpha/\kappa$ is constant (see Eq. 5.25) and finite in the interval $[-1, +1]$, and the same is true for the derivative $r'(\xi)$, which according to Eq. (5.26) is given as

$$r'(\xi) = \frac{\ell_j}{2}\,\text{sgn}(\xi) \tag{5.47}$$

where $\text{sgn}(\xi)$ (signum of ξ) is the function defined as

$$\text{sgn}(\xi) = \begin{cases} +1, & \xi > 0 \\ -1, & \xi < 0 \end{cases} \tag{5.48}$$

Consequently, the value of integral (5.45) is

$$\int_{-1}^{1} \psi_\alpha(\xi)\frac{\cos\phi}{r}\, d\xi = 0 \tag{5.49}$$

It should be noted that this result is not valid for higher order elements, since for these elements $\cos\phi(\xi) \neq 0$.

The integrand in (5.45) behaves as $1/r$ and becomes infinite when $r = 0$. This singularity is known as *Cauchy type singularity*. Analytical and numerical, as well as hybrid techniques have been developed for the evaluation of integrals with this type of singularity. Their use and programming require special care. In addition to the diagonal elements $\hat{H}_{ii}$, the coefficients ε^i must also be computed (see Eqs. 5.16 or 5.20), which, of course, increases the computational task. However, it is possible to evaluate directly the elements $H_{ii} = \hat{H}_{ii} - \varepsilon_i$ by an indirect method that avoids the evaluation of any singular integral.

5.4.3 Indirect evaluation of the diagonal influence coefficients

The matrices $[G]$ and $[H]$ in Eq. (4.7) for constant elements, or in Eq. (5.19) for linear elements, are affected only by the boundary geometry, its discretization and the employed type of element. Hence, these matrices do not depend on the boundary conditions, that is, they remain unchanged for given boundary geometry, boundary discretization, and type of boundary elements.

The indirect integration is based on the fact that the function $u = \alpha x + by + c$ is a solution to the Laplace equation, namely,

$$\nabla^2 u = 0 \ \text{ in } \Omega$$

with boundary conditions

$$u = \alpha\, x + b\, y + c \ \text{ on } \Gamma \tag{5.50a}$$

and

$$u_n = \nabla u \cdot \mathbf{n} = \alpha\, n_x + b\, n_y \ \text{ on } \Gamma \tag{5.50b}$$

a. *Evaluation of the elements H_{ii}*

Let us assume that $a = b = 0$ and $c = 1$. In this case, it will be $u = 1$ and $u_n = 0$ on the boundary. Obviously, these values must satisfy Eq. (5.19), which gives, after substitution,

$$[H]\{1\} = 0$$

where {1} is the vector whose elements are all equal to 1. The above equation may also be written as

$$\sum_{j=1}^{N} H_{ij} = 0$$

or

$$H_{ii} = -\sum_{\substack{j=1 \\ i \neq j}}^{N} H_{ij} \quad (i = 1, 2, \ldots, N) \tag{5.51}$$

Equation (5.51) states that the diagonal element in the i-th row of matrix $[H]$ is equal to the negative sum of the remaining elements in this row. It should be noted that Eq. (5.51) is valid only for a closed domain Ω. For infinite domains, it is not valid, because a constant value for u violates the regularity condition at the infinity. Nevertheless, it is possible even in this case to evaluate H_{ii} using an indirect approach (see Ref. [11]).

b. *Evaluation of the elements* G_{ii}

Once the matrix $[H]$ has been computed, the diagonal elements G_{ii} of matrix $[G]$ can also be computed using an indirect method, which avoids the evaluation of any integrals, either regular or singular. To this end, Eq. (5.19) is applied for the function

$$u = a\,x + b\,y$$

giving

$$[G]\{u_n\} = [H]\{u\}$$

or

$$\sum_{j=1}^{N} G_{ij}\, u_n^j = \sum_{j=1}^{N} H_{ij}\, u^j \quad (i = 1, 2, \ldots, N)$$

The above equation is solved for G_{ii}, yielding

$$G_{ii} = \frac{1}{u_n^i}\left(-\sum_{\substack{j=1 \\ j \neq i}}^{N} G_{ij}\, u_n^j + \sum_{j=1}^{N} H_{ij}\, u^j \right) \tag{5.52}$$

The boundary values u^j and u_n^j appearing in Eq. (5.52) are computed from the following expressions

$$u^j = a\,x_j + b\,y_j \tag{5.53a}$$

$$u_n^j = a\, n_x^j + b\, n_y^j \tag{5.53b}$$

where (x_j, y_j) are the coordinates of the j-th node, and (n_x^j, n_y^j) are the components of the unit vector normal to the j-th element. The constants a and b are chosen arbitrarily, but under the condition that $u_n^i \neq 0$. This means according to Eq. (5.53b), that the vector with components the coefficients a and b must not be normal to the vector $\mathbf{n}^j$ $(j = 1, 2, \ldots, N)$, or, in other words, must not be parallel to the j-th boundary element. This may be achieved by setting $a = 1$, $b = \lambda > 0$ and choosing λ so that

$$\lambda \neq \left| \frac{y_{j+1} - y_j}{x_{j+1} - x_j} \right| \tag{5.54}$$

5.5 HIGHER ORDER ELEMENTS

Constant and linear elements cannot approximate with sufficient accuracy the geometry of curvilinear boundaries. For this reason it is recommended to use curvilinear elements, for which the interpolating polynomials are of a degree higher than one. In general, their form in the normalized interval $[-1, +1]$ is going to be

$$f(\xi) = a_o + a_1 \xi + a_2 \xi^2 + a_3 \xi^3 + \cdots + a_n \xi^n \quad (-1 \leq \xi \leq 1) \tag{5.55}$$

For $n = 2$, Eq. (5.55) yields the interpolation function for the *quadratic* or *parabolic* element, for $n = 3$ that of the *cubic* element and so forth. In what follows we will limit our presentation to the parabolic element. For higher order elements or for a general theory on isoparametric elements the reader is advised to look in Ref. [12].

The boundary quantities u and u_n in the line integrals of Eq. (5.8) are functions of the arc length measured from some origin. When a parabolic variation is assumed the quantities u or u_n will be expressed by a polynomial of the form

$$f(s) = \alpha_o + \alpha_1 s + \alpha_2 s^2 \tag{5.56}$$

The coordinates of point $(x, y) \in \Gamma$, which varies during the integration, are also functions of s, that is, $x = x(s)$, $y = y(s)$. Hence

$$r = \sqrt{[x(s) - x_i]^2 + [y(s) - y_i]^2} = r(s)$$

and

$$v = v(s), \quad v_n = \frac{\partial v}{\partial n} = v_n(s)$$

Consequently, the integral to be evaluated is of the form

$$I = \int_{\Gamma_j} w(s)\, ds \tag{5.57}$$

Of course, the integration can be performed by first establishing the transformation $s = s(\xi)$ and then substituting it in the integral (5.57). Although this method is conceptually simple, its implementation requires the establishment of complicated expressions of s and for this reason it is not the most suitable. Instead, another method is presented below, which considerably simplifies the integration procedure.

The integration is performed over the normalized interval $[-1, +1]$ with respect to the intrinsic coordinate ξ, and thus, the integral (5.57) becomes

$$I = \int_{-1}^{1} w^*(\xi)|J(\xi)|\, d\xi \tag{5.58}$$

where $|J(\xi)|$ is the Jacobian of the transformation which maps the parabolic arc Γ_j of the xy-plane onto the straight line segment with $-1 \leq \xi \leq 1$ and $\eta = 0$ of the $\xi\eta$-plane (see Fig. 5.12).

The boundary quantity f (u or u_n) is approximated directly in the interval $[-1, +1]$ by a second order polynomial in ξ, namely,

$$f(\xi) = \alpha_0 + \alpha_1 \xi + \alpha_2 \xi^2 \tag{5.59}$$

The coefficients α_0, α_1, and α_2 are determined from the requirement that the function $f(\xi)$ takes the nodal values f_1, f_2, f_3 at points $\xi = -1, 0, 1$, respectively (see Fig. 5.12). Hence,

$$\begin{aligned} f(-1) &= f_1 \\ f(0) &= f_2 \\ f(1) &= f_3 \end{aligned} \tag{5.60}$$

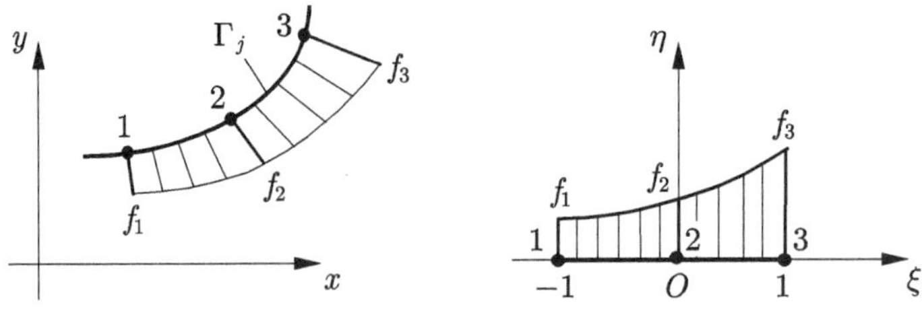

FIGURE 5.12 Parabolic element in global and local coordinate systems.

Applying conditions (5.60) to Eq. (5.59), we get the following system of equations for the unknown coefficients

$$\begin{aligned} \alpha_0 - \alpha_1 + \alpha_2 &= f_1 \\ \alpha_0 &= f_2 \\ \alpha_0 + \alpha_1 + \alpha_2 &= f_3 \end{aligned}$$

whose solution is

$$\begin{aligned} \alpha_0 &= f_2 \\ \alpha_1 &= \frac{f_3 - f_1}{2} \\ \alpha_2 &= \frac{f_1 - 2f_2 + f_3}{2} \end{aligned} \tag{5.61}$$

Introducing Eq. (5.61) into Eq. (5.59), we obtain the expression of the boundary quantity in terms of the three nodal values of element

$$f(\xi) = f_2 + \frac{f_3 - f_1}{2}\,\xi + \frac{f_1 - 2f_2 + f_3}{2}\,\xi^2 \tag{5.62}$$

Equation (5.62) may further be written in the form

$$f(\xi) = \psi_1(\xi)\, f_1 + \psi_2(\xi)\, f_2 + \psi_3(\xi)\, f_3 \tag{5.63}$$

or

$$f(\xi) = \sum_{\alpha=1}^{3} \psi_\alpha(\xi)\, f_\alpha \tag{5.64}$$

where

$$\begin{aligned} \psi_1(\xi) &= -\frac{1}{2}\xi\,(1 - \xi) \\ \psi_2(\xi) &= (1 - \xi)(1 + \xi) \\ \psi_3(\xi) &= \frac{1}{2}\xi\,(1 + \xi) \end{aligned} \tag{5.65}$$

The functions defined in Eq. (5.65) are the *shape functions* of the parabolic or quadratic element.

The mapping of the parabolic element from the xy-plane onto the interval $-1 \leq \xi \leq 1$ of the $\xi\eta$-plane is accomplished through the transformation

$$\begin{aligned} x(\xi) &= b_0 + b_1\xi + b_2\xi^2 \\ y(\xi) &= c_0 + c_1\xi + c_2\xi^2 \end{aligned} \tag{5.66}$$

We can readily conclude from Eqs. (5.59) and (5.66) that the parabolic element at hand is isoparametric, since both the geometry and the boundary quantity are approximated by polynomials of the same degree. The coefficients b_k and c_k $(k = 0, 1, 2)$ in Eq. (5.66) are evaluated from the requirement that the element arc should pass through the points (x_1, y_1), (x_2, y_2), and (x_3, y_3) for $\xi = -1, 0, 1$, respectively. These conditions are expressed mathematically as

$$\begin{aligned} x(-1) &= x_1, \quad x(0) = x_2, \quad x(1) = x_3 \\ y(-1) &= y_1, \quad y(0) = y_2, \quad y(1) = y_3 \end{aligned}$$

It is apparent that the above conditions yield expressions for $x(\xi)$ and $y(\xi)$, which are similar to those of Eq. (5.64). More specifically,

$$\begin{aligned} x(\xi) &= \sum_{\alpha=1}^{3} \psi_\alpha(\xi) x_\alpha \\ y(\xi) &= \sum_{\alpha=1}^{3} \psi_\alpha(\xi) y_\alpha \end{aligned} \tag{5.67}$$

where the shape functions $\psi_\alpha(\xi)$ are given in Eq. (5.65).

Equation (5.67) may be employed to express the distance r, as well as the kernels $v(r)$ and $v_n(r)$ of the integrals representing the influence coefficients, as functions of the variable ξ. Finally, the Jacobian of Eq. (5.58) is evaluated from the expression

$$ds = \sqrt{dx^2 + dy^2} = \sqrt{[x'(\xi)]^2 + [y'(\xi)]^2}\, d\xi$$

Hence

$$\begin{aligned} |J(\xi)| &= \left\{[x'(\xi)]^2 + [y'(\xi)]^2\right\}^{1/2} \\ &= \left[(b_1 + 2b_2\xi)^2 + (c_1 + 2c_2\xi)^2\right]^{1/2} \end{aligned} \tag{5.68}$$

where, on the basis of Eq. (5.61), it is

$$\begin{aligned} b_1 &= \frac{x_3 - x_1}{2}, \quad b_2 = \frac{x_1 - 2x_2 + x_3}{2} \\ c_1 &= \frac{y_3 - y_1}{2}, \quad c_2 = \frac{y_1 - 2y_2 + y_3}{2} \end{aligned} \tag{5.69}$$

After discretizing the boundary into N parabolic elements, the boundary integral equation (3.31) may be written as

$$\varepsilon^i u^i = - \sum_{j=1}^{N} \int_{\Gamma_j} v\, u_n\, ds + \sum_{j=1}^{N} \int_{\Gamma_j} u\, v_n\, ds \tag{5.70}$$

Taking into account the parabolic variation of the boundary quantities on the elements, the second line integral in the right-hand side of Eq. (5.70) yields

$$\begin{aligned} \int_{\Gamma_j} u(q)\, v_n(p_i, q)\, ds_q &= \int_{\Gamma_j} (\psi_1\, u^1 + \psi_2\, u^2 + \psi_3\, u^3)\, v_n\, ds \\ &= h_1^{ij}\, u^1 + h_2^{ij}\, u^2 + h_3^{ij}\, u^3 \end{aligned} \tag{5.71}$$

where it has been set

$$\begin{aligned} h_1^{ij} &= \int_{\Gamma_j} \psi_1\, v_n\, ds \\ h_2^{ij} &= \int_{\Gamma_j} \psi_2\, v_n\, ds \\ h_3^{ij} &= \int_{\Gamma_j} \psi_3\, v_n\, ds \end{aligned} \tag{5.72}$$

In order to evaluate the above integrals, their integrands are expressed in terms of the variable ξ

$$h_\alpha^{ij} = \int_{\Gamma_j} \psi_\alpha\, v_n\, ds = \int_{-1}^{1} \psi_\alpha(\xi) \frac{\cos\phi(\xi)}{2\pi r(\xi)}\, |J(\xi)|\, d\xi \quad (\alpha = 1, 2, 3) \tag{5.73}$$

Similarly, we obtain

$$\begin{aligned} \int_{\Gamma_j} u_n\, v\, ds &= \int_{\Gamma_j} (\psi_1\, u_n^1 + \psi_2\, u_n^2 + \psi_3\, u_n^3)\, v\, ds \\ &= g_1^{ij}\, u_n^1 + g_2^{ij}\, u_n^2 + g_3^{ij}\, u_n^3 \end{aligned} \tag{5.74}$$

where

$$g_\alpha^{ij} = \int_{\Gamma_j} \psi_\alpha \, v \, ds = \int_{-1}^{1} \psi_\alpha(\xi) \, \frac{\ln r(\xi)}{2\pi} \, |J(\xi)| d\xi \quad (\alpha = 1, 2, 3) \tag{5.75}$$

The computation of integrals (5.73) and (5.75) is performed numerically following a procedure similar to that employed for the linear element. Thus, for the outside integration ($i \neq j$) the conventional Gauss integration is utilized. However, for the inside integration ($i = j$), the analytical method for the computation of g_α^{ij} becomes too complicated and it is not recommended. Instead, the method of extracting the singularity turns out to be the most suitable. The elements H_{ii} are computed using the indirect method presented in Section 5.4.3, avoiding, thus, the evaluation of singular integrals.

The corner points or the points where there is a change in the boundary conditions are treated using one-sided discontinuous parabolic elements, for example, $\kappa_2 = 1$ (Fig. 5.13). For discontinuous parabolic elements, the coefficients of the polynomial (5.59) are determined from the conditions

$$\begin{aligned} f(-\kappa_1) &= f_1 \\ f(0) &= f_2 \\ f(\kappa_2) &= f_3 \end{aligned} \tag{5.76}$$

which yield

$$\begin{aligned} \alpha_0 - \alpha_1 \kappa_1 + \alpha_2 \kappa_1^2 &= f_1 \\ \alpha_0 &= f_2 \\ \alpha_0 + \alpha_1 \kappa_2 + \alpha_2 \kappa_2^2 &= f_3 \end{aligned} \tag{5.77}$$

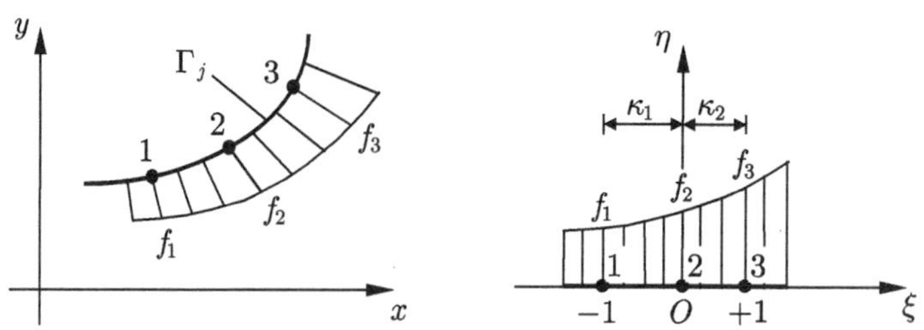

FIGURE 5.13 Discontinuous parabolic element in global and local coordinate systems.

Equation (5.77) are solved for the unknown coefficients α_0, α_1, and α_2 giving

$$\begin{aligned} \alpha_0 &= f_2 \\ \alpha_1 &= -\frac{\kappa_2^2}{K} f_1 + \frac{\kappa_2^2 - \kappa_1^2}{K} f_2 + \frac{\kappa_1^2}{K} f_3 \\ \alpha_2 &= \frac{\kappa_2}{K} f_1 - \frac{\kappa_1 + \kappa_2}{K} f_2 + \frac{\kappa_1}{K} f_3 \end{aligned} \tag{5.78}$$

where

$$K = \kappa_1\, \kappa_2 (\kappa_1 + \kappa_2)$$

Introducing the above values of the coefficients α_i in Eq. (5.59), the latter takes the known form

$$f(\xi) = \sum_{\alpha=1}^{3} \psi_\alpha(\xi)\, f_\alpha \quad (\alpha = 1, 2, 3) \tag{5.79}$$

where $\psi_\alpha(\xi)$ are shape functions given as

$$\begin{aligned} \psi_1(\xi) &= \frac{\kappa_2}{K}(-\kappa_2 + \xi)\xi \\ \psi_2(\xi) &= \frac{\kappa_1 + \kappa_2}{K}\left[\kappa_1\, \kappa_2 + (\kappa_2 - \kappa_1)\xi - \xi^2\right] \\ \psi_3(\xi) &= \frac{\kappa_1}{K}(\kappa_1 + \xi)\xi \end{aligned} \tag{5.80}$$

Evidently, Eq. (5.80) reduce to the shape functions of Eq. (5.65) for $\kappa_1 = \kappa_2 = 1$. Higher order elements can be derived by choosing interpolating polynomials of higher degree and following exactly the same procedure as for the quadratic element.

All continuous elements derived on the basis of the polynomial (5.55) for $n \geq 1$, produce a continuous variation of the boundary quantity on the whole boundary, that is, they do not exhibit jumps at the interelement nodes. This type of continuity is called C^0 *continuity*. These elements, however, do not ensure continuity of the derivative at the interelement nodes. This continuity, referred to as C^1 *continuity*, can be achieved using shape functions described by special third order polynomials known as *Hermite polynomials* or *Hermite interpolation functions*. In this case, the element has two nodes which are placed at its extreme points. The unknowns are the values f_1 and f_2 of the boundary quantity, along with the corresponding derivatives $\theta_1 = (df/d\xi)_1$ and $\theta_2 = (df/d\xi)_2$. Hence, the boundary quantity is expressed as

$$f(\xi) = \psi_1(\xi)\, f_1 + \psi_2(\xi)\, f_2 + \psi_3(\xi)\, \theta_1 + \psi_4(\xi)\, \theta_2 \tag{5.81}$$

in which the shape functions $\psi_\alpha(\xi)$ are given by the Hermite polynomials [1]

$$\begin{aligned} \psi_1(\xi) &= \frac{1}{4}(1-\xi)^2(2+\xi), \quad \psi_2(\xi) = \frac{1}{4}(1+\xi)^2(2-\xi) \\ \psi_3(\xi) &= \frac{1}{4}(1-\xi)^2(1+\xi), \quad \psi_4(\xi) = -\frac{1}{4}(1+\xi)^2(1-\xi) \end{aligned} \tag{5.82}$$

5.6 NEAR-SINGULAR INTEGRALS

Once the unknown boundary quantities have been established from the solution of the boundary integral equations, the values of the potential u and its derivatives $u_{,x} = \partial u/\partial x$ and $u_{,y} = \partial u/\partial y$ at internal points can be computed using the expressions (4.11), (4.15), and (4.16), or the corresponding ones for the linear and the parabolic element approximation. The influence coefficients involved in the aforementioned equations are expressed in terms of line integrals on the elements Γ_j whose integrands involve factors of the form

$$\ln r, \ \frac{1}{r}, \ \frac{1}{r^2} \tag{5.83}$$

where $r = |P - q|$ is the distance between the points $P \in \Omega$ and $q \in \Gamma_j$. Clearly, because point P lies inside the domain Ω, while point q lies on the boundary, it is always $r \neq 0$. Therefore, these integrals, at least theoretically, are regular since the value of their integrands is always finite. When point P lies *far* from the boundary, the functions (5.83) have a smooth variation and consequently the conventional Gauss integration gives accurate results. However, when point P lies *near* the boundary, the functions (5.83) may take a very large, though finite, value and, thus, their variation is not smooth any more.

In order to illustrate this behavior, we consider the element of Fig. 5.14. The internal point P lies near the boundary. Applying Eq. (5.3), one obtains the global coordinates in terms of the local coordinate ξ,

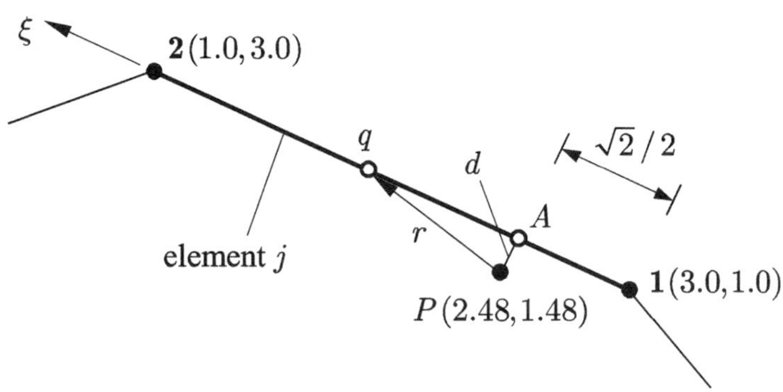

FIGURE 5.14 Linear element and internal point P near the boundary, $d = \min(r) = 0.02 \ll 1$.

$$x(\xi) = 2 - \xi$$
$$y(\xi) = 2 + \xi$$

and Eq. (5.12) gives the relative distance between points P and q as

$$\begin{aligned} r(\xi) &= \left\{[x(\xi)-2.48]^2+[y(\xi)-1.48]^2\right\}^{1/2} \\ &= \left[(0.48+\xi)^2+(0.52+\xi)^2\right]^{1/2} \end{aligned} \tag{5.84}$$

The variation of the functions $1/r(\xi)$ and $1/r(\xi)^2$ is shown in Figs. (5.15) and (5.16), respectively. We notice the large values of these functions at $\xi = -0.5$, that is at point A, which is the normal projection of point P on the element.

Consequently, the integrals of the functions (5.83) for points P near the boundary behave like singular integrals, although they are not. In the literature, these integrals are known as *near-singular integrals*. Their evaluation faces considerable difficulties, because neither the conventional Gauss integration nor those methods suitable for singular integrals can be employed. Nevertheless, other special techniques have been developed for their evaluation. Among them, the most popular are: the method of the *element subdivision*, and the method of the *coordinate transformation*. The first one has been discussed in detail by Lachat and Watson [13] and Kane et al. [14], while the second one has been presented by Telles [15]. Only the first method is presented below, while the other one can be found in the relevant literature.

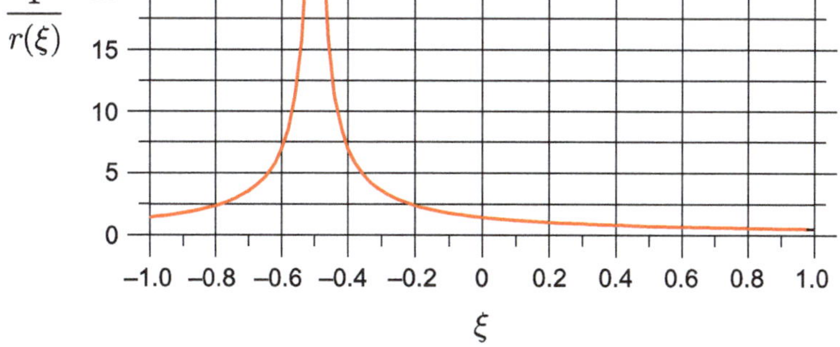

FIGURE 5.15 Variation of the function $1/r(\xi)$ along the element.

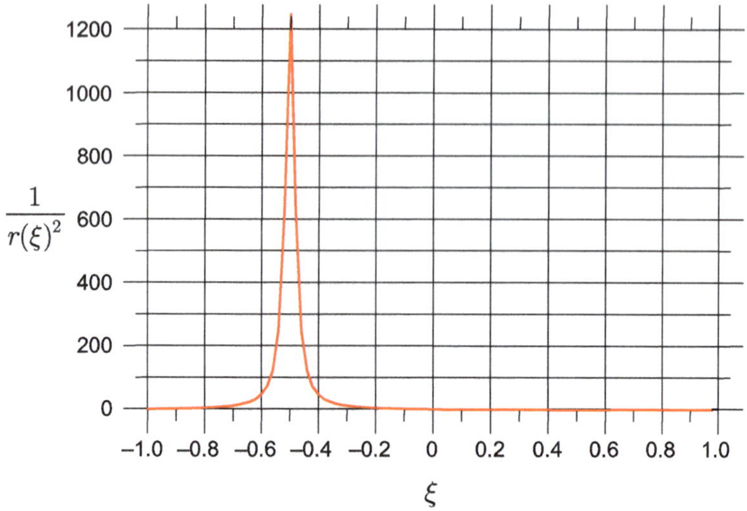

FIGURE 5.16 Variation of the function $1/r(\xi)^2$ along the element.

5.6.1 The method of element subdivision

This method has been successfully employed to achieve a uniform accuracy for the values of the line integrals. The success of this technique is based on the user's experience. This technique is illustrated by evaluating the following integral

$$\int_{-1}^{1} \frac{1}{r(\xi)} d\xi \tag{5.85}$$

The plot of its integrand, which is depicted in Fig. 5.15, suggests subdivision of the interval $[-1, +1]$ in four subintervals:

Subinterval 1: $[-1, -0.6]$
Subinterval 2: $[-0.6, -0.5]$
Subinterval 3: $[-0.5, -0.4]$
Subinterval 4: $[-0.4, +1]$

In each subinterval, the integral can be evaluated using any integration rule such as the trapezoidal rule, Simpson's rule, Newton-Cotes integration formulae, etc. The Gauss integration is also recommended for this case. Its application requires first transformation of each subinterval onto the interval $[-1, +1]$. If ξ_k and ξ_{k+1} represent the end points of the k-th subinterval, then the linear transformation

$$\xi = \frac{\xi_{k+1} + \xi_k}{2} + \frac{\xi_{k+1} - \xi_k}{2} \eta \; (-1 \leq \eta \leq 1) \tag{5.86}$$

TABLE 5.2 Values of the Near-Singular Integral (5.85) for Various Subdivisions of the Element

Number of Subintervals	Subintervals	Number of Gauss Points	$\int_{-1}^{1} \frac{1}{r} d\xi$
1	[−1.0, +1.0]	20	7.55905
2	[−1.0, −0.5] [−0.5, +1.0]	10 10	6.34858
2	[−1.0, −0.5] [−0.5, +1.0]	12 8	6.29535
2	[−1.0, −0.5] [−0.5, +1.0]	8 12	6.35520
4	[−1.0, −0.6] [−0.6, −0.5] [−0.5, −0.4] [−0.4, +1.0]	5 4 5 6	6.30650
4	[−1.0, −0.6] [−0.6, −0.5] [−0.5, −0.4] [−0.4, +1.0]	6 4 4 6	6.30934
Exact value			6.309586

serves the purpose. The value of the integral over the k-th subinterval is given as

$$I_k = \frac{\xi_{k+1} - \xi_k}{2} \sum_{i=1}^{n} \frac{1}{r\left(\frac{\xi_{k+1} + \xi_k}{2} + \frac{\xi_{k+1} - \xi_k}{2} \eta_i\right)} w_i \tag{5.87}$$

where η_i and w_i $(i = 1, 2, \ldots, n)$ are the abscissas and weights of the n-th order Gauss integration.

The integral (5.85) has been computed numerically using various subdivisions of the element shown in Fig. 5.14 and the results are given in Table 5.2 along with the exact value. These results reveal that the choice of element subdivision greatly affects the accuracy, even though the total number of integration points remains the same (20 points). Therefore, special care should be taken for the computation of near-singular integrals to avoid an uncontrolled error. A general rule for acceptable results is to choose two equal subintervals at both sides of point A, whose length is sufficiently smaller than the distance d (see Fig. 5.14).

5.7 REFERENCES

From the presentation of the BEM up to this point, it becomes evident that its evolution as a computational method for solving realistic engineering problems is based on its success in solving singular boundary integral equations. Therefore, the technology of the boundary element, namely, the construction of various types of elements, the effective integration over them, especially of the singular kernels, as well as their manipulation in order to treat discontinuities of the boundary quantities, is among the most important ingredients of the BEM and has been a field of intense research ever since the BEM appeared as a computational method. For more information about this subject, the reader is advised to look in books by Brebbia and Dominguez [7], Kane [16], and Banerjee and Butterfield [12]. Regarding the evaluation of singular integrals, a rich technical literature is available. Suggestively, we mention the work of Hall [17] and Doblaré [18], as well as the book of Sladek and Sladek [19], who presented methods for the evaluation of singular integrals. Hayami and Brebbia [20], apart from the singular integrals, treated the near-singular ones. A method for the computation of line integrals with logarithmic singularity has also been presented by Katsikadelis and Armenakas [21]. Theocaris [22,23] has published extended work on the integration of singular integrals. Considerable work, especially on the evaluation of hypersingular integrals, has been published by Guiggiani and his coworkers [24,25]. The reader can also find extended literature related to this subject in the chapters *Computational Aspects* of the proceedings of the international conferences on *Boundary Elements* (BEM/MRM), which have been organized uninterruptedly by Wessex Institute of technology since 1978.

[1] Reddy JN. An introduction to the finite element method. New York: McGraw-Hill International Editions; 1993.
[2] Katsikadelis JT, Sapountzakis E. Torsion of composite bars by the boundary element method. ASCE J Eng Mech 1985;111:1197–210.
[3] Katsikadelis JT, Kallivokas L. Plates on biparametric elastic foundation. ASCE J Eng Mech 1988;114:547–875.
[4] Katsikadelis JT, Armenakas AE. A new boundary equation solution to the plate problem. ASME J Appl Mech 1989;56:264–374.
[5] Katsikadelis JT. Special methods for plate analysis. In: Beskos D, editor. Boundary element analysis of plates and shells. Berlin: Springer-Verlag; 1991. p. 221–311.
[6] Katsikadelis JT, Kokkinos FT. Static and dynamic analysis of composite shear walls by the boundary element method. Acta Mech 1987;68:231–50.
[7] Brebbia CA, Dominguez J. Boundary elements: an introductory course. 2nd ed. Southampton: Computational Mechanics Publications; 2001.
[8] Lefeber D. Solving problems with singularities using boundary elements. In: Brebbia CA, Connor JJ, editors. Topics in engineering, Vol. 6. Southampton: Computational Mechanics Publications; 1989.
[9] Stroud AH, Secrest D. Gaussian quadrature formulas. Englewood Cliffs, New Jersey: Prentice Hall; 1966.
[10] Abramowitz M, Stegun I, editors. Handbook of mathematical functions. 10th ed. New York: Dover Publications; 1972.
[11] Brebbia CA, Telles JCF, Wrobel LC. Boundary element techniques. Berlin: Springer-Verlag; 1984.
[12] Banerjee PK, Butterfield R. Boundary element methods in engineering science. U.K: McGraw-Hill; 1981.
[13] Lachat JC, Watson JO. Effective numerical treatment of boundary integral equations: a formulation for three-dimensional elastostatics. Int J Numer Methods Eng 1976;10: 991–1005.
[14] Kane JH, Gupta A, Saigal S. Reusable Intrinsic Sample Point (RISP) algorithm for efficient numerical integration of three dimensional curved boundary elements. Int J Numer Methods Eng 1989;28:1661–76.
[15] Telles JCF. A self-adaptive coordinate transformation for efficient numerical evaluations of general boundary element integrals. Int J Numer Methods Eng 1987;24:959–73.

[16] Kane JH. Boundary elements analysis in engineering continuum mechanics. Englewood Cliffs, New Jersey: Prentice Hall; 1994.
[17] Hall WS. Integration methods for singular boundary element integrands. In: Brebbia CA, editor. Boundary Elements X, vol. 1, Mathematical and Computational Aspects. Southampton: Computational Mechanics Publications; 1988. p. 219–36.
[18] Doblaré M. Computational aspects of the boundary element method. In: Brebbia CA, editor. Topics in boundary element research, Vol. 3. Berlin: Springer-Verlag; 1987. p. 51–131.
[19] Sladek V, Sladek J, editors. Singular integrals in boundary element methods. Southampton: Computational Mechanics Publications; 1998.
[20] Hayami K, Brebbia CA. Quadrature methods for singular and nearly singular integrals in 3-D boundary element method. In: Brebbia CA, editor. Boundary elements X, computational aspects, vol. 1. Southampton: Computational Mechanics Publications; 1988.
[21] Katsikadelis JT, Armenakas AE. Numerical evaluation of line integrals with a logarithmic singularity. AIAA J 1983;23:1135–7.
[22] Theocaris PS. Numerical solution of singular integral equations: methods. ASCE J Eng Mech 1981;107:733–51.
[23] Theocaris PS. Numerical solution of singular integral equations: applications. ASCE J Eng Mech 1981;107:753–71.
[24] Guiggiani M. Direct evaluation of hypersingular integrals in 2D BEM. In: Hackbusch W, editor. Notes in numerical field mechanics, vol. 33. Braunschweig: Vieweg; 1992. p. 23–4.
[25] Guiggiani M, Krishnasamy G, Rudolphi TJ, Rizzo FJ. A general algorithm for the numerical equations. ASME J Appl Mech 1992;59:604–14.

PROBLEMS

5.1. Derive the shape functions for the cubic element when the interior nodal points are placed at

i. $\xi_2 = -1/3$ and $\xi_3 = 1/3$,

ii. $\xi_2 = -1/2$ and $\xi_3 = 1/2$.

5.2. Given the circular sector of radius $R = 3$ and angle $\theta_0 = \pi/12$, compute its area by approximating the circular arc with (1) linear, (2) quadratic, and (3) cubic elements. For each case determine the error.

5.3. Compute the near-singular integral using Gauss integration

$$I = \int_{-1}^{+1} \frac{dx}{\left[(x-0.25)^2+0.05\right]^4}$$

5.4. Compute the integral

$$\int_{\Gamma_j} \psi_1(\xi) \ln r \, ds$$

when Γ_j is a quadratic element passing through the points **1**(4.30, 2.50), **2**(4.10, 2.90), **3**(3.80, 3.20) and the source lies at point P(4.15, 2.65).

5.5. Compute the integrals g_α^{ij} ($\alpha = 1, 2$) for the linear element with nodal points **1**(1, 2) and **2**(1.5, 2.3), when $\kappa_1 = \kappa_2 = 0.5$ and the source lies consecutively at point **1** and point **2**.

Chapter | Six

Applications

CHAPTER OUTLINE

6.1 INTRODUCTION

As it was mentioned in Chapter 3, the Laplace and Poisson equations describe the behavior of many physical systems. In this chapter the boundary element method (BEM) will be employed to solve several problems, such as torsion of noncircular prismatic bars, deflection of membranes, bending of simply supported plates, heat conduction, and fluid flow. All these problems are governed either by the Laplace or the Poisson equation.

6.2 TORSION OF NONCIRCULAR BARS

6.2.1 The warping function

Let us consider a bar of arbitrary cross-section twisted by moments M_t applied at its ends (Fig. 6.1). The cross-section is constant along the length of the bar. According to Saint-Venant's torsion theory [1,2], the deformation

The Boundary Element Method for Engineers and Scientists.

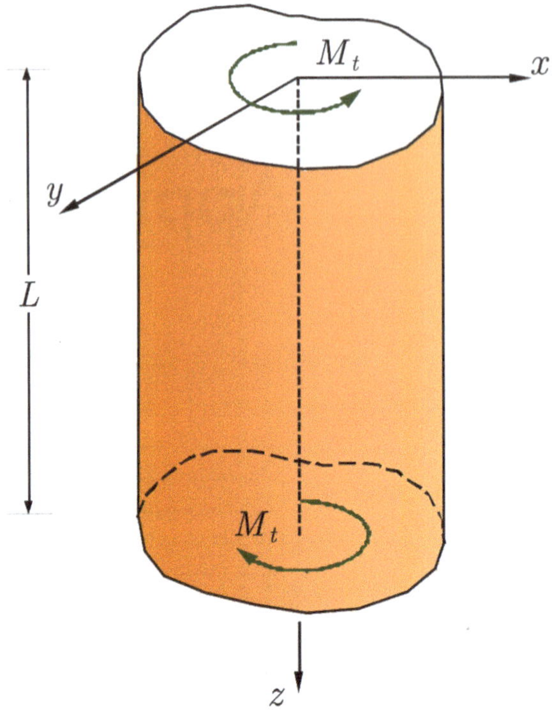

FIGURE 6.1 Bar of arbitrary cross-section twisted by moments end M_t.

of the bar consists of (1) rotations of the cross-sections about an axis passing through the *twist center* of the bar, and (2) warping of the cross-sections, which is the same for all sections. Choosing the origin of the coordinate system at the twist center of an end section (Fig. 6.1), the rotation at a distance z is θz, where θ is a constant expressing the rotation of a cross-section per unit length. Referring to Fig. 6.2 and assuming that this rotation is small, the displacements u and v of point $A(x, y, z)$ due to the rotation are determined as

$$u = -(AA')\sin\alpha = -r\,\theta\,z\frac{y}{r} = -\theta\,z\,y \tag{6.1a}$$

$$v = (AA')\cos\alpha = r\,\theta\,z\frac{x}{r} = \theta\,z\,x \tag{6.1b}$$

The warping of the cross-section is defined as

$$w = \theta\,\phi(x, y) \tag{6.2}$$

where $\phi(x, y)$ is the *warping function*.

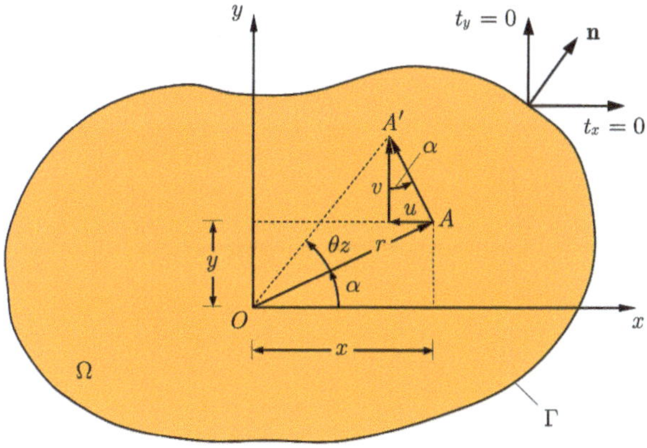

FIGURE 6.2 Displacement components in a cross-section of a twisted bar.

The displacements given by Eqs. (6.1a,b) and (6.2) yield the following strain components

$$\left.\begin{aligned} \varepsilon_x &= \frac{\partial u}{\partial x} = 0, \quad \varepsilon_y = \frac{\partial v}{\partial y} = 0 \\ \varepsilon_z &= \frac{\partial w}{\partial z} = 0, \quad \gamma_{xy} = \frac{\partial u}{\partial y} + \frac{\partial v}{\partial x} = 0 \\ \gamma_{xz} &= \frac{\partial w}{\partial x} + \frac{\partial u}{\partial z} = \theta\left(\frac{\partial \phi}{\partial x} - y\right) \\ \gamma_{yz} &= \frac{\partial w}{\partial y} + \frac{\partial v}{\partial z} = \theta\left(\frac{\partial \phi}{\partial y} + x\right) \end{aligned}\right\} \tag{6.3}$$

Hence, for homogeneous linear elastic material the corresponding stress components resulting from the above strain components are

$$\left.\begin{aligned} &\sigma_x = \sigma_y = \sigma_z = \tau_{xy} = 0 \\ &\tau_{xz} = G\theta\left(\frac{\partial \phi}{\partial x} - y\right) \\ &\tau_{yz} = G\theta\left(\frac{\partial \phi}{\partial y} + x\right) \end{aligned}\right\} \tag{6.4}$$

The equilibrium equations for the three-dimensional state of stress in the absence of body forces, are

$$\frac{\partial \sigma_x}{\partial x}+\frac{\partial \tau_{xy}}{\partial y}+\frac{\partial \tau_{xz}}{\partial z}=0 \tag{6.5a}$$

$$\frac{\partial \tau_{xy}}{\partial x}+\frac{\partial \sigma_y}{\partial y}+\frac{\partial \tau_{yz}}{\partial z}=0 \tag{6.5b}$$

$$\frac{\partial \tau_{xz}}{\partial x}+\frac{\partial \tau_{yz}}{\partial y}+\frac{\partial \sigma_z}{\partial z}=0 \tag{6.5c}$$

Introducing the constitutive relations (6.4) into Eqs. (6.5a–c) we obtain

$$\frac{\partial \tau_{xz}}{\partial z}=0 \tag{6.6a}$$

$$\frac{\partial \tau_{yz}}{\partial z}=0 \tag{6.6b}$$

$$\frac{\partial^2 \phi}{\partial x^2}+\frac{\partial^2 \phi}{\partial y^2}=0 \tag{6.6c}$$

The first two of Eqs. (6.6a–c) always hold, since the stress components τ_{xz} and τ_{yz} are independent of z. The last one, Eq. (6.6c), expresses the condition which must be satisfied by the warping function $\phi(x, y)$.

In addition, the stress components should satisfy the traction boundary conditions on the surface of the bar, namely

$$\left.\begin{aligned} \sigma_x n_x+\tau_{xy} n_y+\tau_{xz} n_z&=t_x \\ \tau_{xy} n_x+\sigma_y n_y+\tau_{yz} n_z&=t_y \\ \tau_{xz} n_x+\tau_{yz} n_y+\sigma_z n_z&=t_z \end{aligned}\right\} \tag{6.7}$$

where n_x, n_y, n_z are the direction cosines of the outward normal vector and t_x, t_y, t_z are the traction components on the surface of the bar.

(1) We examine first the boundary conditions on the cylindrical surface of the bar. This surface is traction free, that is, $t_x = t_y = t_z = 0$. Moreover, it is $n_z = 0$. Taking into account Eq. (6.4), we can readily show that the first two of the boundary conditions (6.7) are identically satisfied, while the third one yields

$$\left(\frac{\partial \phi}{\partial x}-y\right) n_x+\left(\frac{\partial \phi}{\partial y}+x\right) n_y=0$$

which may also be written as

$$\frac{\partial \phi}{\partial x} n_x + \frac{\partial \phi}{\partial y} n_y = y\, n_x - x\, n_y$$

or

$$\frac{\partial \phi}{\partial n} = y\, n_x - x\, n_y \tag{6.8}$$

Evidently, the warping function ϕ may be determined from Eq. (6.6c) and (6.8) by solving a Neumann problem for the Laplace equation, provided that the function

$$\phi_n(s) = y\, n_x - x\, n_y \tag{6.9}$$

satisfies the existence condition for the solution of the Neumann problem, that is,

$$\int_\Gamma \phi_n \, ds = 0 \tag{6.10}$$

This condition results from Green's identity, Eqs. (2.16), for $v=1$ and $u=\phi$. Indeed, using Eq. (2.3), Eq. (6.9) is written as

$$\begin{aligned} \phi_n &= y\frac{dy}{ds} + x\frac{dx}{ds} \\ &= \frac{1}{2}\frac{d}{ds}(x^2 + y^2) \end{aligned}$$

Consequently, noting that the function $\frac{1}{2}(x^2+y^2)$ is continuous on the whole boundary, we obtain

$$\begin{aligned} \int_\Gamma \phi_n \, ds &= \frac{1}{2}\int_\Gamma \frac{d}{ds}(x^2+y^2)\, ds \\ &= \frac{1}{2}\left[x^2+y^2\right]_B^B = 0 \end{aligned}$$

where B is any point on the boundary taken as the origin for the variable s.

(2) On the end cross-sections $z=0$ and $z=L$, it is $n_x = n_y = 0$ and $n_z = 1$. Thus, the boundary conditions (6.7) become

$$\tau_{xz} = t_x, \quad \tau_{yz} = t_y, \quad \sigma_z = t_z = 0 \tag{6.11}$$

which state that the end cross-sections are subjected only to tangential tractions.

We can readily prove that the stress resultants of these tractions vanish. Namely,

$$\int_{\Omega} \tau_{xz}\, d\Omega = 0 \quad \text{and} \quad \int_{\Omega} \tau_{yz}\, d\Omega = 0 \tag{6.12}$$

Indeed, Eq. (6.5c) becomes

$$\frac{\partial \tau_{xz}}{\partial x} + \frac{\partial \tau_{yz}}{\partial y} = 0 \tag{6.13}$$

Moreover, the first of Eq. (6.12) may be written as

$$\begin{aligned} \int_{\Omega} \tau_{xz}\, d\Omega &= \int_{\Omega} \left[\tau_{xz} + x \left(\frac{\partial \tau_{xz}}{\partial x} + \frac{\partial \tau_{yz}}{\partial y} \right) \right] d\Omega \\ &= \int_{\Omega} \left[\frac{\partial (x\tau_{xz})}{\partial x} + \frac{\partial (x\tau_{yz})}{\partial y} \right] d\Omega \end{aligned}$$

Next applying Gauss's divergence theorem (2.9) and using the last of Eq. (6.7), the above equation yields

$$\int_{\Omega} \tau_{xz}\, d\Omega = \int_{\Gamma} x(\tau_{xz}\, n_x + \tau_{yz}\, n_y)\, ds = 0 \tag{6.14}$$

since $n_z = 0$ and $t_z = 0$.

In a similar way, we can prove the second of Eq. (6.12).

The moment resultant on the cross-section $z = 0$ is going to be

$$M_t = \int_{\Omega} (x\, \tau_{yz} - y\, \tau_{xz})\, d\Omega = G\theta \int_{\Omega} \left(x^2 + y^2 + x \frac{\partial \phi}{\partial y} - y \frac{\partial \phi}{\partial x} \right) d\Omega \tag{6.15}$$

Setting

$$I_t = \int_{\Omega} \left(x^2 + y^2 + x \frac{\partial \phi}{\partial y} - y \frac{\partial \phi}{\partial x} \right) d\Omega \tag{6.16}$$

we arrive at

$$M_t = G I_t\, \theta \tag{6.17}$$

The constant quantity I_t, which depends only on the shape of the cross-section, is referred to as *torsional constant*. The quantity GI_t is called the *torsional rigidity* of the cross-section. Denoting by $\overline{\theta} = \theta L$ the relative rotation of the end cross-sections, Eq. (6.17) may also be written as

$$M_t = \frac{GI_t}{L}\overline{\theta} \tag{6.18}$$

The quantity GI_t/L expresses the *torsional stiffness coefficient* of a bar having length L. This coefficient appears in the stiffness matrix of grid elements or three-dimensional beam elements.

From the foregoing analysis, we conclude that the determination of the torsional constant of bars as well as of the shear stresses due to torsion, require the establishment of the warping function ϕ of the cross-section. For simple cross-sectional geometries, (e.g., elliptical, rectangular, triangular) the warping function ϕ can be determined using exact or approximate analytical solutions. However, for cross-sections of complex shape, as happens in realistic engineering problems, it is necessary to solve a Neumann problem for the Laplace equation in an arbitrary domain Ω. Hence, the warping function ϕ is established as the solution of the following boundary value problem

$$\nabla^2 \phi = 0 \quad \text{in } \Omega \tag{6.19a}$$

$$\frac{\partial \phi}{\partial n} = y\, n_x - x\, n_y \quad \text{on } \Gamma \tag{6.19b}$$

The function ϕ is determined exactly apart from an arbitrary constant term. That is, a function ϕ^* is obtained in the form

$$\phi^*(x, y) = \phi(x, y) + C \tag{6.20}$$

where $\phi(x, y)$ is the exact solution and C is a constant.

Apparently, the stress components and the torsional constant are not influenced by this arbitrary constant, because, according to Eqs. (6.4) and (6.16), only the derivatives of ϕ are required for the evaluation of these quantities. With respect to the displacement w, the arbitrary constant introduces a rigid body motion in the direction of the bar axis (see Eq. (6.2)), which, however, does not influence the deformation of the cross-section. The constant C can be determined by specifying the displacement w at a point of the cross-section, for example, $w=0$, which yields $\phi^*=0$ at this point and then $\phi(x, y) = \phi^*(x, y) - C$. It is advisable to choose the *twist center* of the cross-section as the point of zero axial displacement. Therefore, the establishment of this point should come first, if it is not *a priori* known.

DETERMINATION OF THE TWIST CENTER

The solution of the boundary value problem (6.19) gives the warping surface, if the origin of the coordinate axes is taken at the twist center of the cross-section, that is, the point which does not undergo any displacement during the rotation of the cross-section. In axisymmetric cross-sections (e.g., rectangle, equilateral triangle, ellipse, etc.) the twist center can be readily established by inspection since it coincides with the geometric center of the cross-section. However, for cross-sections of arbitrary shape, the twist center is not known and it should be determined in order to establish the warping function. This can be achieved by working as follows.

When the origin O does not coincide with the twist center with coordinates (x_0, y_0), Eqs. (6.1a,b), (6.2)–(6.4), (6.6c), and (6.8) are written respectively as

$$u = -\theta\, z\,(y - y_0) \tag{6.21a}$$

$$v = \theta\, z\,(x - x_0) \tag{6.21b}$$

$$w = \theta\, \phi(x, y) \tag{6.21c}$$

$$\left.\begin{aligned} &\varepsilon_x = 0, \quad \varepsilon_y = 0, \quad \varepsilon_z = 0, \quad \gamma_{xy} = 0 \\ &\gamma_{xz} = \theta\left[\frac{\partial \phi}{\partial x} - (y - y_0)\right] \\ &\gamma_{yz} = \theta\left[\frac{\partial \phi}{\partial y} + (x - x_0)\right] \end{aligned}\right\} \tag{6.22}$$

$$\left.\begin{aligned} &\sigma_x = \sigma_y = \sigma_z = \tau_{xy} = 0 \\ &\tau_{xz} = G\theta\left[\frac{\partial \phi}{\partial x} - (y - y_0)\right] \\ &\tau_{yz} = G\theta\left[\frac{\partial \phi}{\partial y} + (x - x_0)\right] \end{aligned}\right\} \tag{6.23}$$

$$\frac{\partial^2 \phi}{\partial x^2} + \frac{\partial^2 \phi}{\partial y^2} = 0 \tag{6.24}$$

$$\frac{\partial \phi}{\partial n} = (y - y_0)\, n_x - (x - x_0)\, n_y \tag{6.25}$$

Equation (6.25) may further be written as

$$\frac{\partial}{\partial n}(\phi + y_0\, x - x_0\, y) = y n_x - x\, n_y \tag{6.26}$$

or

$$\frac{\partial \phi^*}{\partial n} = y\, n_x - x\, n_y \tag{6.27}$$

where

$$\phi^* = \phi + y_0\, x - x_0\, y + C \tag{6.28a}$$

and consequently

$$\phi = \phi^* - y_0\, x + x_0\, y - C \tag{6.28b}$$

Since $\nabla^2\phi^* = \nabla^2\phi$, it is apparent that the Neumann problem being solved, is actually

$$\left.\begin{aligned} \nabla^2\phi^* &= 0 \\ \frac{\partial\phi^*}{\partial n} &= y\, n_x - x\, n_y \end{aligned}\right\} \tag{6.29}$$

which yields the function ϕ^*. As a result, the stress components and the torsional constant should be expressed in terms of the function ϕ^* and not ϕ. This can be accomplished by introducing ϕ from Eq. (6.28b) into the expressions (6.23), which gives

$$\tau_{xz} = G\theta\left(\frac{\partial\phi^*}{\partial x} - y\right) \tag{6.30a}$$

$$\tau_{yz} = G\theta\left(\frac{\partial\phi^*}{\partial y} + x\right) \tag{6.30b}$$

Using the above relations in Eq. (6.15) to express the moment M_t with respect to the origin of the coordinate axes, we obtain

$$I_t = \int_\Omega \left(x^2 + y^2 + x\,\frac{\partial\phi^*}{\partial y} - y\,\frac{\partial\phi^*}{\partial x}\right) d\Omega \tag{6.31}$$

On the basis of Eqs. (6.28a), (6.29), (6.30a,b), and (6.31), it can be concluded that.

a. If the origin of the coordinates does not coincide with the twist center of the cross-section, then the warping function ϕ^*, obtained as the solution of the Neumann problem, has undergone a rigid body rotation in the plane of the cross-section and a displacement parallel to the axis of the bar (see Ref. [1]).

b. The stress components τ_{xz}, τ_{yz} and the torsion constant I_t do not depend on the position of the origin of the coordinate axes.

The actual warping function ϕ is determined by establishing first the quantities x_0, y_0 and C. These quantities can be obtained from the minimization of the strain energy produced by axial normal warping stresses, which are ignored by the Saint-Venant's theory. This energy is given as [3]

$$\begin{aligned}\Pi(x_0, y_0, C) &= \frac{1}{2} E\theta^2 \int_\Omega \phi^2 \, d\Omega \\ &= \frac{1}{2} E\theta^2 \int_\Omega (\phi^* - y_0 x + x_0 y - C)^2 \, d\Omega\end{aligned} \tag{6.32}$$

The minimization conditions require

$$\left.\begin{aligned}\frac{\partial \Pi}{\partial x_0} &= 0 \\ \frac{\partial \Pi}{\partial y_0} &= 0 \\ \frac{\partial \Pi}{\partial C} &= 0\end{aligned}\right\} \tag{6.33}$$

Equation (6.32) is differentiated with respect to each of the three quantities to yield

$$\left.\begin{aligned}I_x\, x_0 - I_{xy}\, y_0 - S_x\, C &= -I_1 \\ I_{xy}\, x_0 - I_y\, y_0 - S_y\, C &= -I_2 \\ S_x\, x_0 - S_y\, y_0 - A\, C &= -I_3\end{aligned}\right\} \tag{6.34}$$

where it was set

$$\left.\begin{aligned}A &= \int_\Omega d\Omega, & S_x &= \int_\Omega y\, d\Omega, & S_y &= \int_\Omega x\, d\Omega \\ I_x &= \int_\Omega y^2\, d\Omega, & I_{xy} &= \int_\Omega xy\, d\Omega, & I_y &= \int_\Omega x^2\, d\Omega \\ I_1 &= \int_\Omega y\, \phi^*\, d\Omega, & I_2 &= \int_\Omega x\, \phi^*\, d\Omega, & I_3 &= \int_\Omega \phi^*\, d\Omega\end{aligned}\right\} \tag{6.35}$$

Since the torsion problem is solved by the BEM, the domain integrals in Eq. (6.35) should be converted to boundary line integrals in order to maintain the pure boundary character of the method. This can be achieved using Eqs. (2.5), (2.6), and (2.9). Thus we can write

$$A = \int_\Omega d\Omega = \frac{1}{2} \int_\Omega \left(\frac{\partial x}{\partial x} + \frac{\partial y}{\partial y} \right) d\Omega = \frac{1}{2} \int_\Gamma (x\, n_x + y\, n_y)\, ds \tag{6.36}$$

$$S_x = \int_\Omega y\, d\Omega = \frac{1}{2} \int_\Omega \frac{\partial}{\partial y}(y^2) d\Omega = \frac{1}{2} \int_\Gamma y^2\, n_y\, ds \tag{6.37}$$

$$S_y = \int_\Omega x \, d\Omega = \frac{1}{2}\int_\Omega \frac{\partial}{\partial x}(x^2) d\Omega = \frac{1}{2}\int_\Gamma x^2 \, n_x \, ds \tag{6.38}$$

$$I_x = \int_\Omega y^2 \, d\Omega = \frac{1}{3}\int_\Omega \frac{\partial}{\partial y}(y^3) d\Omega = \frac{1}{3}\int_\Gamma y^3 \, n_y \, ds \tag{6.39}$$

$$I_y = \int_\Omega x^2 \, d\Omega = \frac{1}{3}\int_\Omega \frac{\partial}{\partial x}(x^3) d\Omega = \frac{1}{3}\int_\Gamma x^3 n_x \, ds \tag{6.40}$$

$$\begin{aligned} I_{xy} &= \int_\Omega x\,y \, d\Omega = \frac{1}{4}\int_\Omega \left[\frac{\partial}{\partial x}(x^2\,y) + \frac{\partial}{\partial y}(x\,y^2)\right] d\Omega \\ &= \frac{1}{4}\int_\Gamma x\,y\,(xn_x + yn_y)\, ds \end{aligned} \tag{6.41}$$

To treat the integrals involving the function ϕ^*, Green's second identity (2.16) is applied consecutively for the functions

$$u_1 = \frac{yr^2}{8}, \quad u_2 = \frac{xr^2}{8}, \quad u_3 = \frac{r^2}{4}, \quad (r^2 = x^2 + y^2) \tag{6.42}$$

and for $v = \phi^*$. Noting that $\nabla^2 u_1 = y$, $\nabla^2 u_2 = x$, and $\nabla^2 u_3 = 1$, we obtain

$$I_1 = \int_\Omega y\phi^* \, d\Omega = \int_\Gamma \left(\phi^* \frac{\partial u_1}{\partial n} - u_1 \frac{\partial \phi^*}{\partial n}\right) ds \tag{6.43}$$

$$I_2 = \int_\Omega x\phi^* \, d\Omega = \int_\Gamma \left(\phi^* \frac{\partial u_2}{\partial n} - u_2 \frac{\partial \phi^*}{\partial n}\right) ds \tag{6.44}$$

$$I_3 = \int_\Omega \phi^* \, d\Omega = \int_\Gamma \left(\phi^* \frac{\partial u_3}{\partial n} - u_3 \frac{\partial \phi^*}{\partial n}\right) ds \tag{6.45}$$

The foregoing integrals are computed using the BEM discretization with constant elements.

The steps required in order to solve the torsion problem for bars with an arbitrary cross-section can be summarized as:

a. The function ϕ^* is determined as the solution of the Neumann problem described by Eqs. (6.29) with respect to the arbitrarily chosen xy-system of axes. Special care should be taken to ensure the existence of a solution for the Neumann problem (see Example 4.2).

b. The coordinates x_0, y_0 of the twist center and the constant C are computed from the solution of Eq. (6.34).

c. The warping function ϕ is evaluated using Eq. (6.28b).

d. The boundary stress τ_{tz} is computed using the procedure described in Section 6.2.2, while the torsion constant I_t is given by Eq. (6.31) or even better by its boundary integral form (6.47), which is derived right below.

The domain integral (6.31) is converted into a boundary line integral to avoid not only the domain integration but also the evaluation of the derivatives of ϕ^*, which are involved in the integrand. This is achieved following by working as follows:

Equation (6.31) may be written as

$$I_t = \int_\Omega \left[\frac{\partial}{\partial x}(xy^2 - y\phi^*) + \frac{\partial}{\partial y}(yx^2 + x\phi^*) \right] d\Omega \tag{6.46}$$

which by virtue of the Gauss divergence theorem (2.9) it is transformed to the following boundary line integral

$$I_t = \int_\Gamma \left[(xy^2 - y\phi^*)n_x + (yx^2 + x\phi^*)n_y \right] ds \tag{6.47}$$

REMARKS

The solution of the Saint-Venant torsion problem can also be obtained by determining the following functions:

a. The function $\psi(x, y)$, which is the conjugate of $\phi(x, y)$, by solving the Dirichlet problem

$$\nabla^2 \psi = 0 \quad \text{in } \Omega$$

$$\psi = \frac{1}{2}(x^2 + y^2) + C \quad \text{on } \Gamma$$

where C is an arbitrary constant. In this case, the shear stresses and the torsional constant are expressed as

$$\tau_{xz} = G\theta \left(\frac{\partial \psi}{\partial y} - y \right)$$

$$\tau_{yz} = G\theta \left(\frac{\partial \psi}{\partial x} + x \right)$$

$$I_t = \int_\Omega \left(x^2 + y^2 - x\frac{\partial \psi}{\partial x} - y\frac{\partial \psi}{\partial y} \right) d\Omega$$

b. The Prandtl's stress function $F(x, y)$ which is the solution of the Dirichlet problem for the Poisson equation

$$\nabla^2 F = -2 \quad \text{in } \Omega$$

$$F = C \quad \text{on } \Gamma$$

where C is an arbitrary constant. The shear stresses and the torsional constant are given as

$$\tau_{xz} = G\theta \frac{\partial F}{\partial y}$$

$$\tau_{yz} = -G\theta \frac{\partial F}{\partial x}$$

$$I_t = -\int_\Omega \left(x \frac{\partial F}{\partial x} + y \frac{\partial F}{\partial y} \right) d\Omega$$

The formulation in terms of the warping function ϕ is preferred over the formulations in terms of the foregoing two functions, because:

i. Once the function ψ or F is established, the determination of the warping function, hence of the axial displacements, from these functions requires the solution of an additional potential problem which is equally difficult as the original one.

ii. When the cross-section does not have holes, that is, the domain Ω is simply connected, the arbitrary constant C appearing in the boundary condition can be given an arbitrary value, for example, $C = 0$. However, when the cross-section has holes (hollow cross-section), that is, the domain Ω is multiply connected, then the constant C does not, generally, take the same value on all the contours and its exact values should be determined. This can be done by imposing additional conditions which ensure the uniqueness of the displacements. Therefore, the solution of the problem becomes even more complicated.

6.2.2 Evaluation of stresses

The stress components τ_{xz} and τ_{yz} are evaluated using Eq. (6.4). Apparently, these are determined by evaluating first the derivatives of the function ϕ. For points inside the domain of the cross-section, the derivatives can be computed by applying Eqs. (4.12) and (4.13).

The maximum values of the stresses appear on the boundary. The stress τ_{nz} is zero, whereas the stress τ_{tz} is given by the relation

$$\tau_{tz} = -\tau_{xz}\, n_y + \tau_{yz}\, n_x$$

which by virtue of Eq. (6.4) becomes

$$\tau_{tz} = G\theta \left(\frac{\partial \phi}{\partial t} + x\, n_x + y\, n_y \right) \tag{6.48}$$

The derivative $\partial\phi/\partial t = \partial\phi/\partial s$ can be computed from its integral representation on the boundary (see Problem 3.4). But it is simpler to evaluate it using numerical differentiation of ϕ along the boundary as it is described next.

Let us consider three consecutive nodal points $i-1,\ i,\ i+1$ on the boundary. The values ϕ_{i-1} and ϕ_{i+1} can be expressed in terms of the value of ϕ and its derivatives at point i using Taylor series expansions

$$\phi_{i-1}=\phi_i-(\phi_{,s})_i\ s_1+\frac{1}{2}(\phi_{,ss})_i\ s_1^2-\frac{1}{6}(\phi_{,sss})_i\ s_1^3+\ldots$$

$$\phi_{i+1}=\phi_i+(\phi_{,s})_i\ s_2+\frac{1}{2}(\phi_{,ss})_i\ s_2^2+\frac{1}{6}(\phi_{,sss})_i\ s_2^3+\ldots$$

where

$$s_1=\frac{\ell_{i-1}+\ell_i}{2}\quad\text{and}\quad s_2=\frac{\ell_i+\ell_{i+1}}{2}$$

with ℓ_{i-1},ℓ_i, and ℓ_{i+1} being the lengths of the respective elements.

In the above expressions, we neglect terms of order higher than the second and subsequently we eliminate the second derivatives. This yields the *central difference* approximation for the first derivative

$$(\phi_{,s})_i=\left(\frac{\partial\phi}{\partial t}\right)_i=\alpha_1\ \phi_{i-1}+\alpha_2\ \phi_i+\alpha_3\ \phi_{i+1} \tag{6.49}$$

where

$$\alpha_1=-\frac{s_2}{s_1(s_1+s_2)},\quad \alpha_2=-\frac{s_1-s_2}{s_1\,s_2},\quad \alpha_3=\frac{s_1}{s_2(s_1+s_2)} \tag{6.50}$$

If $s_1=s_2=\Delta s$, we obtain the known expression

$$\left(\frac{\partial\phi}{\partial t}\right)_i=\frac{\phi_{i+1}-\phi_{i-1}}{2\,\Delta s} \tag{6.51}$$

At points near the corners the derivative $\partial\phi/\partial t$ is discontinuous. For this reason *forward (backward) differences* should be employed when the corner precedes (follows) the point i. The finite difference expressions for the evaluation of $\partial\phi/\partial t$ for forward and backward differences are derived next.

For forward differences we consider the values ϕ_i, ϕ_{i+1}, and ϕ_{i+2}. The Taylor series approximations of ϕ_{i+1} and ϕ_{i+2} in terms of the value of ϕ and its derivatives at point i give

$$\phi_{i+1}=\phi_i+(\phi_{,s})_i\ s_1+\frac{1}{2}(\phi_{,ss})_i\ s_1^2+\frac{1}{6}(\phi_{,sss})_i\ s_1^3+\ldots$$

$$\phi_{i+2}=\phi_i+(\phi_{,s})_i\ (s_1+s_2)+\frac{1}{2}(\phi_{,ss})_i\ (s_1+s_2)^2+\frac{1}{6}(\phi_{,sss})_i\ (s_1+s_2)^3+\ldots$$

Neglecting the terms of order higher than the second, the above equations yield

$$(\phi_{,s})_i = \left(\frac{\partial \phi}{\partial t}\right)_i = \alpha_1\, \phi_i + \alpha_2\, \phi_{i+1} + \alpha_3\, \phi_{i+2} \tag{6.52}$$

where

$$\alpha_1 = -\frac{2s_1 + s_2}{s_1(s_1+s_2)}, \quad \alpha_2 = \frac{s_1+s_2}{s_1 s_2}, \quad \alpha_3 = -\frac{s_1}{s_2(s_1+s_2)} \tag{6.53}$$

$$s_1 = \frac{\ell_i + \ell_{i+1}}{2} \quad \text{and} \quad s_2 = \frac{\ell_{i+1}+\ell_{i+2}}{2}$$

For backward differences we consider the values ϕ_{i-2}, ϕ_{i-1}, ϕ_i and employing an analogous procedure, we obtain

$$(\phi_{,s})_i = \left(\frac{\partial \phi}{\partial t}\right)_i = \alpha_1\, \phi_i + \alpha_2\, \phi_{i-1} + \alpha_3\, \phi_{i-2} \tag{6.54}$$

$$\alpha_1 = \frac{2s_1 + s_2}{s_1(s_1+s_2)}, \quad \alpha_2 = -\frac{s_1+s_2}{s_1 s_2}, \quad \alpha_3 = \frac{s_1}{s_2(s_1+s_2)} \tag{6.55}$$

$$s_1 = \frac{\ell_i + \ell_{i-1}}{2} \quad \text{and} \quad s_2 = \frac{\ell_{i-1}+\ell_{i-2}}{2}$$

6.2.3 Program TORSCON for solving the torsion problem with constant elements

The program LABECON can be readily modified to solve the torsion problem. Besides the main program, the subroutines that are modified are: INPUT, UINTER, and OUTPUT. Moreover, three new subroutines are added, namely, TORCENTER, TORSTIF, and TORSTRESS, which compute the center of twist of the cross-section, the torsion constant I_t, and the boundary stress τ_{tz}, respectively. The listings of the main program and the modified subroutines as well as the new subroutines are given below. The electronic form of the code TORSCON is available on this book's companion website.

```
C=================================================================
C
      PROGRAM TORSCON
C
C
C  This program solves the Saint-Venant TORSION problem
C  as a Neumann problem for the Laplace equation using
C  the boundary element method with (CON)stant
C  boundary elements
C
```

```
      IMPLICIT REAL*8 (A-H,O-Z)
      CHARACTER*15 INPUTFILE,OUTPUTFILE
C
C  Set the maximum dimensions
C
      PARAMETER (N=20)
      PARAMETER (IN=13)
C
C  N= Number of boundary elelments equal to number of boundary
C     nodes
C IN= Number of internal points where the function u is calculated
C
      DIMENSION INDEX(N)
      DIMENSION XL(N+1),YL(N+1),XM(N),YM(N),G(N,N),H(N,N),UB(N)
      DIMENSION A(N,N),UNB(N),XIN(IN+1),YIN(IN+1),UIN(IN+1),SL(N),TTZ(N)
      DIMENSION AA(3,3),BB(3)
C
C  Read the names and open the input and output files
C
      WRITE (*,'(A)')' Name of the INPUTFILE (max.15 characters)'
      READ  (*,'(A)') INPUTFILE
      WRITE (*,'(A)')' Name of the OUTPUTFILE (max.15 characters)'
      READ  (*,'(A)') OUTPUTFILE
      OPEN (1, FILE=INPUTFILE)
      OPEN (2, FILE=OUTPUTFILE)
C
C  Read data from INPUTFILE
C
      CALL INPUT(XL,YL,XIN,YIN,INDEX,UB,N,IN)
C
C  Compute the G matrix
C
      CALL GMATR(XL,YL,XM,YM,G,N)
C
C  Compute the H matrix
C
      CALL HMATR(XL,YL,XM,YM,H,N)
C
C  Form the system of equations AX=B
C
      CALL ABMATR(G,H,A,UNB,UB,INDEX,N)
C
C  Solve the system of equations
C
      CALL SOLVEQ(A,UNB,N,LSING)
C
C  Form the vectors U and UN of all the bounday values
C
      CALL REORDER(UB,UNB,INDEX,N)
C
C  Compute the coordinates of the twist center of the

C  cross-section and the Neumann's problem constant
C
      CALL TORCENTER (XL,YL,N,UB,UNB,AA,BB,XTC,YTC,CT)
C
C  Compute the values of U at the internal points
C
      CALL UINTER(XL,YL,XM,YM,XIN,YIN,UB,UNB,UIN,N,IN,XTC,YTC,CT)
C
C  Compute the the torsion constant D
C
      CALL TORSTIF(XL,YL,N,UB,XTC,YTC,D)
C
C  Compute the boundary stress Ttz
C
      CALL TORSTRESS(XL,YL,XM,YM,UB,XTC,YTC,TTZ,SL,N)
```

```
C
C  Print the results in the OUTPUTFILE
C
      CALL OUTPUT (XM,YM,UB,UNB,XIN,YIN,UIN,D,TTZ,N,IN,XTC,YTC,CT)
C
C  Close input and output files
C
      CLOSE(1)
      CLOSE(2)
      STOP
      END
C
C
C=====================================================================
C
      SUBROUTINE INPUT (XL,YL,XIN,YIN,INDEX,UB,N,IN)
C
C
C  This subroutine reads the data from the input file
C  and writes them in the output file
C
      IMPLICIT REAL*8 (A-H,O-Z)
      CHARACTER*80 NAME,TITLE
      DIMENSION XL(N+1),YL(N+1),XIN(IN),YIN(IN),INDEX(N),UB(N)
C
      WRITE(2,100)
  100 FORMAT('',69('*'))
C
C  Read user's name
C
      READ(1,'(A)')NAME
C
      WRITE(2,'(A)')NAME
C
C  Read the title of the program
C
      READ(1,'(A)')TITLE
C
      WRITE(2,'(A)')TITLE
C
      WRITE(2,200)N,IN
  200 FORMAT(//'BASIC PARAMETERS'//2X,'NUMBER OF BOUNDARY ELEMENTS='
     1,I3/2X,'NUMBER OF INTERNAL POINTS WHERE THE FUNCTION IS CALCULATED
     1=',I3)
C
C  Read the coordinates XL,YL of the extreme points of the boundary elements
C
      READ(1,*)  (XL(I),YL(I),I=1,N)
C
C  Write the coordinates in the output file

C
      WRITE(2,300)
  300 FORMAT(//2X,'COORDINATES OF THE EXTREME POINTS OF THE BOUNDARY ELE
     1MENTS',//2X,'POINT',9X,'XL',15X,'YL')
      DO 20 I=1,N
   20 WRITE(2,400) I,XL(I),YL(I)
  400 FORMAT(2X,I3,2(3X,E14.5))
C
C  Compute the boundary values of Un and store in UB(I) (I=1,N-1),UB(N)=0.
C
      DO 10 I=1,N-1
      DX=XL(I+1)-XL(I)
      DY=YL(I+1)-YL(I)
      SL=DSQRT(DX**2+DY**2)
      ENX=DY/SL
      ENY=-DX/SL
```

```
      XM=(XL(I)+XL(I+1))/2.
      YM=(YL(I)+YL(I+1))/2.
      UB(I)=YM*ENX-XM*ENY
   10 INDEX(I)=1
      UB(N)=0.
      INDEX(N)=0
C
C  Write the boundary conditions in the output file
C
      WRITE(2,500)
  500 FORMAT(//2X,'BOUNDARY CONDITIONS'//2X,'NODE',6X,'INDEX',
     1 7X,'PRESCRIBED VALUE')
      DO 30 I=1,N
   30 WRITE(2,600) I,INDEX(I),UB(I)
  600 FORMAT (2X,I3,9X,I1,8X,E14.5)
C
C  Read the coordinates of the internal points
C
      READ(1,*) (XIN(I),YIN(I),I=1,IN)
      RETURN
      END
C
C
C=====================================================================
C
      SUBROUTINE TORCENTER (XL,YL,N,UB,UNB,AA,BB,XTC,YTC,CT)
C
C
C  This subroutine computes the coordinates of the twist
C  center of the cross-section and the arbitrary constant
C
      IMPLICIT REAL*8(A-H,O-Z)
      DIMENSION XL(N+1),YL(N+1),UB(N),UNB(N)
      DIMENSION XC(4),YC(4),XI(4),WG(4)
      DIMENSION AA(3,3),BB(3)
C
C        N = Number of boundary elements
C    XL,YL = Coordinates of the extreme points of the
C             boundary elements
C  XTC,YTC = Coordinates of the twist center
C       CT = Arbitrary constant of the Neumann problem
C       WG = Weights of the Gauss integration
C       XI = Coordinates of the Gauss integration points
C             in the interval [-1,1]
C    XC,YC = Global coordinates of the Gauss integration points
C
      DATA XI/-0.86113631,-0.33998104,0.33998104,0.86113631/
      DATA WG/0.34785485,0.65214515,0.65214515,0.34785485/
      PI=ACOS(-1.)

      XL(N+1)=XL(1)
      YL(N+1)=YL(1)
C
      AREA=0.
      SX=0.
      SY=0.
      AIX=0.
      AIY=0.
      AIXY=0.
      AI1=0.
      AI2=0.
      AI3=0.
C
      DO 10 I=1,N
      AX=(XL(I+1)-XL(I))/2.
      AY=(YL(I+1)-YL(I))/2.
      BX=(XL(I+1)+XL(I))/2.
```

```
      BY=(YL(I+1)+YL(I))/2.
      SL=DSQRT(AX**2+AY**2)
      ENX=AY/SL
      ENY=-AX/SL
      TERMA=0.
      TERMSX=0.
      TERMSY=0.
      TERMIX=0.
      TERMIY=0.
      TERMIXY=0.
      TERMI1=0.
      TERMI2=0.
      TERMI3=0.
        DO 40 K=1,4
        XC(K)=AX*XI(K)+BX
        YC(K)=AY*XI(K)+BY
        TERMA=TERMA+0.5*(XC(K)*ENX+YC(K)*ENY)*WG(K)*SL
        TERMSX=TERMSX+0.5*YC(K)**2*ENY*WG(K)*SL
        TERMSY=TERMSY+0.5*XC(K)**2*ENX*WG(K)*SL
        TERMIX=TERMIX+1./3.*YC(K)**3*ENY*WG(K)*SL
        TERMIY=TERMIY+1./3.*XC(K)**3*ENX*WG(K)*SL
        TERMIXY=TERMIXY+0.25*XC(K)*YC(K)*(XC(K)*ENX+YC(K)*ENY)
     1                  *WG(K)*SL
      U1=(XC(K)**2*YC(K)+YC(K)**3)/8.
      U1N=(2.*XC(K)*YC(K)*ENX+(XC(K)**2+3.*YC(K)**2)*ENY)/8.
        TERMI1=TERMI1+(UB(I)*U1N-UNB(I)*U1)*WG(K)*SL
      U2=(XC(K)**3+XC(K)*YC(K)**2)/8.
      U2N=((3.*XC(K)**2+YC(K)**2)*ENX+2.*XC(K)*YC(K)*ENY)/8.
        TERMI2=TERMI2+(UB(I)*U2N-UNB(I)*U2)*WG(K)*SL
      U3=(XC(K)**2+YC(K)**2)/4.
      U3N=(XC(K)*ENX+YC(K)*ENY)/2.
        TERMI3=TERMI3+(UB(I)*U3N-UNB(I)*U3)*WG(K)*SL
   40   CONTINUE
      AREA=AREA+TERMA
      SX=SX+TERMSX
      SY=SY+TERMSY
      AIX=AIX+TERMIX
      AIY=AIY+TERMIY
      AIXY=AIXY+TERMIXY
      AI1=AI1+TERMI1
      AI2=AI2+TERMI2
      AI3=AI3+TERMI3
   10 CONTINUE
C
C  Coordinates of the twist center and Neumann's constant
C
      AA(1,1)=AIX
      AA(1,2)=-AIXY
      AA(1,3)=-SX
      AA(2,1)=-AIXY
      AA(2,2)=AIY
      AA(2,3)=SY
      AA(3,1)=SX
      AA(3,2)=-SY
      AA(3,3)=-AREA
      BB(1)=-AI1
      BB(2)=AI2
      BB(3)=-AI3
      CALL LEQS(AA,BB,3,KS)
      XTC=BB(1)
      YTC=BB(2)
      CT=BB(3)
      WRITE(*,*)XTC,YTC,CT
C
      RETURN
      END
C
```

```
C=================================================================
C
      SUBROUTINE UINTER(XL,YL,XM,YM,XIN,YIN,UB,UNB,UIN,N,IN,XTC,YTC,CT)
C
C
C  This subroutine computes the values of u at the internal points
C
      IMPLICIT REAL*8 (A-H,O-Z)
      DIMENSION XL(N+1),YL(N+1),XIN(IN+1),YIN(IN+1)
      DIMENSION UB(N),UNB(N),UIN(IN+1),XM(N),YM(N)
C
      XIN(IN+1)=XTC
      YIN(IN+1)=YTC
C
C  Compute the values of u at the internal points
C
      DO 10  K=1,IN+1
      UIN(K)=0.
      DO 20  J=1,N
      JP1=J+1
      CALL DALPHA(XIN(K),YIN(K),XL(J),YL(J),XL(JP1),YL(JP1),RESH)
      CALL RLINTC(XIN(K),YIN(K),XL(J),YL(J),XL(JP1),YL(JP1),RESG)
   20 UIN(K)=UIN(K)+RESH*UB(J)-RESG*UNB(J)
   10 CONTINUE
C
      DO 30 K=1,N
      DX=XL(K+1)-XL(K)
      DY=YL(K+1)-YL(K)
      SL=DSQRT(DX**2+DY**2)
      ENX=DY/SL
      ENY=-DX/SL
      UB(K)=UB(K)-YTC*XM(K)+XTC*YM(K)-CT
      UNB(K)=UNB(K)-YTC*ENX+XTC*ENY
   30 CONTINUE
C
      DO 40 K=1,IN+1
      UIN(K)=UIN(K)-YTC*XIN(K)+XTC*YIN(K)-CT
   40 CONTINUE
      RETURN
      END
C
C
C=================================================================
C
      SUBROUTINE TORSTIF (XL,YL,N,UB,XTC,YTC,D)
C
C
C  This subroutine computes the torsion constant
C
      IMPLICIT REAL*8 (A-H, O-Z)
      DIMENSION XL(N+1),YL(N+1),UB(N)
      DIMENSION XC(4),YC(4),XI(4),WG(4)
C
C      N = Number of boundary elements
C  XL,YL = Coordinates of the extreme points of the boundary elements
C      D = Torsion constant
C     WG = Weights of the Gauss integration
C     XI = Coordinates of the Gauss integration points in the
C          interval [-1, 1]
C  XC,YC = Global coordinates of the Gauss integration points
C
      DATA XI/-0.86113631,-0.33998104,0.33998104,0.86113631/
      DATA WG/0.34785485,0.65214515,0.65214515,0.34785485/
      XL(N+1)=XL(1)
      YL(N+1)=YL(1)
      D=0.
```

```
C
      DO 10 I=1,N
      AX=(XL(I+1)-XL(I))/2.
      AY=(YL(I+1)-YL(I))/2.
      BX=(XL(I+1)+XL(I))/2.
      BY=(YL(I+1)+YL(I))/2.
      SL=DSQRT(AX**2+AY**2)
      ENX=AY/SL
      ENY=-AX/SL
C
      TERM=0.
        DO 40 K=1,4
        XC(K)=AX*XI(K)+BX-XTC
        YC(K)=AY*XI(K)+BY-YTC
        TERM=TERM+WG(K)*((XC(K)*YC(K)**2-YC(K)*UB(I))*ENX
     1            +(YC(K)*XC(K)**2+XC(K)*UB(I))*ENY)*SL
   40   CONTINUE
      D=D+TERM
   10 CONTINUE
      RETURN
      END
C
C
C=====================================================================
C
      SUBROUTINE TORSTRESS(XL,YL,XM,YM,UB,XTC,YTC,TTZ,SL,N)
C
C
C  This subroutine computers the boundary shear stress τtz
C  in the tangential direction
C
C  TTZ = Shear stresses at the boundary nodal points
C   SL = Distances between the boundary nodal points
C
      IMPLICIT REAL*8 (A-H, O-Z)
      DIMENSION XL(N+1),YL(N+1),XM(N),YM(N)
      DIMENSION TTZ(N),SL(N),UB(N)
C
      XL(N+1)=XL(1)
      YL(N+1)=YL(1)
      DO 10 I=1,N
      AX=(XL(I+1)-XL(I))/2.
      AY=(YL(I+1)-YL(I))/2.
      SL(I)=DSQRT(AX**2+AY**2)
   10 CONTINUE

C
      DO 20 I=1,N
      AX=(XL(I+1)-XL(I))/2.
      AY=(YL(I+1)-YL(I))/2
      SSL=DSQRT(AX**2+AY**2)
      ENX=AY/SSL
      ENY=-AX/SSL
C
      IF (I.EQ.1) THEN
      S1=SL(N)+SL(1)
      S2=SL(I)+SL(I+1)
      B1=UB(N)
      B2=UB(1)
      B3=UB(2)
      ELSE IF (I.EQ.N) THEN
      S1=SL(N-1)+SL(N)
      S2=SL(N)+SL(1)
      B1=UB(N-1)
      B2=UB(N)
      B3=UB(1)
      ELSE
```

```
      S1=SL(I-1)+SL(I)
      S2=SL(I)+SL(I+1)
      B1=UB(I-1)
      B2=UB(I)
      B3=UB(I+1)
      ENDIF
      UBT=(S1**2*B3-S2**2*B1+(S2**2-S1**2)*B2)
     1     /(S1*S2*(S1+S2))
      TTZ(I)=UBT+(XM(I)-XTC)*ENX+(YM(I)-YTC)*ENY
   20 CONTINUE
      RETURN
      END
C
C
C=====================================================================
C
      SUBROUTINE OUTPUT(XM,YM,UB,UNB,XIN,YIN,UIN,D,TTZ,N,IN,XTC,YTC,CT)
C
C
C  This subroutine prints the results in the outputfile.
C
      IMPLICIT REAL*8 (A-H,O-Z)
      DIMENSION XM(N),YM(N),UB(N),UNB(N)
      DIMENSION XIN(IN+1),YIN(IN+1),UIN(IN+1),TTZ(N)
C
      WRITE(2,100)
  100 FORMAT('',69('*')//1X,'RESULTS'//2X,'BOUNDARY NODES'//
     1         11X,'X',15X,'Y',15X,'U',14X,'Un'/)
C
      DO 10 I=1,N
   10 WRITE(2,200) XM(I),YM(I),UB(I),UNB(I)
  200 FORMAT(4(2X,E14.5))
C
      WRITE(2,900)
  900 FORMAT(/'',69('*')//3X,'COORDINATES OF THE TWIST CENTER',
     1       '   AND   ARBITRARY CONSTANT'/)
      WRITE(2,950)XTC,YTC,CT
  950 FORMAT(3X,'XTC=',E11.5,2X,'YTC=',E11.5,11X,'C=',E11.5)
      WRITE(2,500)
C
      WRITE (2,300)
  300 FORMAT(/,2X,'INTERNAL POINTS'//10X,'X',15X,'Y',11X,
     1        'SOLUTION U'/)
      DO 20 K=1,IN

   20 WRITE(2,400)  XIN(K),YIN(K),UIN(K)
  400 FORMAT(3(2X,E14.5))
      WRITE(2,600) D
  600 FORMAT(/,'',69('*')//2X,'TORSION CONSTANT    D=',E11.5/)
      WRITE(2,700)
  700 FORMAT('',69('*')//2X,'BOUNDARY STRESS Ttz'//
     1         11X,'X',15X,'Y',13X,'Ttz'/)
      DO 30 I=1,N
   30 WRITE(2,800) XM(I),YM(I),TTZ(I)
  800 FORMAT(3(2X,E14.5))
      WRITE(2,500)
  500 FORMAT(/,'',69('*'))
      RETURN
      END
C
C
C=====================================================================
```

EXAMPLE 6.1

In this example the program TORSCON is employed to solve the torsion problem for a bar of elliptic cross-section with semi-axes $a=5.0$ and $b=3.0$. The boundary is discretized into N unequal constant elements as in

Example 4.2. The coordinates of the extreme points are computed from the relations

$$x_i = a \cos \theta_i \quad \text{and} \quad y_i = b \sin \theta_i$$

where

$$\theta_i = -\Delta\theta/2 + (i-1)\,\Delta\theta, \quad \Delta\theta = 2\pi/\mathrm{N}\ (i = 1, 2, \ldots, \mathrm{N})$$

The coordinates of the internal points, which are located on concentric ellipses, are computed from the relations

$$x_k = a_i \cos \theta_j \quad \text{and} \quad y_k = b_i \sin \theta_j$$

$$k = (i-1)\,\mathrm{N2} + j$$

$$a_i = i\,\Delta a,\ \Delta a = a/(\mathrm{N1}+1)\ (i = 1, 2, \ldots, \mathrm{N1})$$

$$b_i = i\,\Delta b,\ \Delta b = b/(\mathrm{N1}+1)\ (i = 1, 2, \ldots, \mathrm{N1})$$

$$\theta_j = (j-1)\,\Delta\theta,\ \Delta\theta = 2\pi/\mathrm{N2}\ (j = 1, 2, \ldots, \mathrm{N2})$$

in which N1 denotes the number of internal concentric ellipses and N2 the number of points on each of the ellipses.

The analytical solution yields [1]

$$\phi = \frac{b^2 - a^2}{a^2 + b^2}\,xy, \qquad I_t = \frac{\pi a^3 b^3}{a^2 + b^2}, \qquad \frac{\tau_{tz}}{G\theta} = \frac{2\sqrt{b^4 x^2 + a^4 y^2}}{a^2 + b^2}$$

The data file, which has been produced using program ELLIPSE-3.FOR, available also on the companion website, and the results of program TORSCON for N = 20, N1 = 1, and N2 = 12 (IN = 12) are presented below.

EXAMPLE 6.1 (DATA)

```
J.T. Katsikadelis
Example 6.1

        4.9384417        -.4693034
        4.9384417         .4693034
        4.4550326        1.3619715
        3.5355339        2.1213203
        2.2699525        2.6730196
         .7821723        2.9630650
        -.7821723        2.9630650
       -2.2699525        2.6730196
       -3.5355339        2.1213203
       -4.4550326        1.3619715
       -4.9384417         .4693034
       -4.9384417        -.4693034
       -4.4550326       -1.3619715
```

```
-3.5355339      -2.1213203
-2.2699525      -2.6730196
 -.7821723      -2.9630650
  .7821723      -2.9630650
 2.2699525      -2.6730196
 3.5355339      -2.1213203
 4.4550326      -1.3619715


 2.5000000        .0000000
 2.1650635        .7500000
 1.2500000       1.2990381
  .0000000       1.5000000
-1.2500000       1.2990381
-2.1650635        .7500000
-2.5000000        .0000000
-2.1650635       -.7500000
-1.2500000      -1.2990381
  .0000000      -1.5000000
 1.2500000      -1.2990381
 2.1650635       -.7500000
  .0000000        .0000000
```

EXAMPLE 6.1 (RESULTS)

```
**********************************************************************
J.T. Katsikadelis
Example 6.1


BASIC PARAMETERS

  NUMBER OF BOUNDARY ELEMENTS= 20
  NUMBER OF INTERNAL POINTS WHERE THE FUNCTION IS CALCULATED= 13


  COORDINATES OF THE EXTREME POINTS OF THE BOUNDARY ELEMENTS

  POINT          XL                YL
    1        .49384E+01       -.46930E+00
    2        .49384E+01        .46930E+00
    3        .44550E+01        .13620E+01
    4        .35355E+01        .21213E+01
    5        .22700E+01        .26730E+01
    6        .78217E+00        .29631E+01
    7       -.78217E+00        .29631E+01
    8       -.22700E+01        .26730E+01
    9       -.35355E+01        .21213E+01
   10       -.44550E+01        .13620E+01
   11       -.49384E+01        .46930E+00
   12       -.49384E+01       -.46930E+00
   13       -.44550E+01       -.13620E+01
   14       -.35355E+01       -.21213E+01
   15       -.22700E+01       -.26730E+01
   16       -.78217E+00       -.29631E+01
   17        .78217E+00       -.29631E+01
   18        .22700E+01       -.26730E+01
   19        .35355E+01       -.21213E+01
   20        .44550E+01       -.13620E+01
```

BOUNDARY CONDITIONS

NODE	INDEX	PRESCRIBED VALUE
1	1	.00000E+00
2	1	-.14314E+01
3	1	-.19716E+01
4	1	-.17030E+01
5	1	-.95863E+00
6	1	.00000E+00
7	1	.95863E+00
8	1	.17030E+01
9	1	.19716E+01
10	1	.14314E+01
11	1	.00000E+00
12	1	-.14314E+01
13	1	-.19716E+01
14	1	-.17030E+01
15	1	-.95863E+00
16	1	.00000E+00
17	1	.95863E+00
18	1	.17030E+01
19	1	.19716E+01
20	0	.00000E+00

****

The system has been solved regularly

****

RESULTS

BOUNDARY NODES

X	Y	U	Un
.49384E+01	.00000E+00	.44409E-15	-.24373E-16
.46967E+01	.91564E+00	-.19551E+01	-.14314E+01
.39953E+01	.17416E+01	-.31820E+01	-.19716E+01
.29027E+01	.23972E+01	-.31957E+01	-.17030E+01
.15261E+01	.28180E+01	-.19795E+01	-.95863E+00
.00000E+00	.29631E+01	-.11102E-14	.12682E-15
-.15261E+01	.28180E+01	.19795E+01	.95863E+00
-.29027E+01	.23972E+01	.31957E+01	.17030E+01
-.39953E+01	.17416E+01	.31820E+01	.19716E+01
-.46967E+01	.91564E+00	.19551E+01	.14314E+01
-.49384E+01	.00000E+00	-.11102E-14	.24373E-16
-.46967E+01	-.91564E+00	-.19551E+01	-.14314E+01
-.39953E+01	-.17416E+01	-.31820E+01	-.19716E+01
-.29027E+01	-.23972E+01	-.31957E+01	-.17030E+01
-.15261E+01	-.28180E+01	-.19795E+01	-.95863E+00
.00000E+00	-.29631E+01	-.24425E-14	-.12682E-15
.15261E+01	-.28180E+01	.19795E+01	.95863E+00
.29027E+01	-.23972E+01	.31957E+01	.17030E+01
.39953E+01	-.17416E+01	.31820E+01	.19716E+01
.46967E+01	-.91564E+00	.19551E+01	.14314E+01

```
*********************************************************************

  COORDINATES OF THE TWIST CENTER    AND    ARBITRARY CONSTANT

  XTC= .12682E-15   YTC= .24373E-16              C=-.19551E+01

*********************************************************************

 INTERNAL POINTS

         X                 Y               SOLUTION U

     .25000E+01        .00000E+00        -.44409E-15
     .21651E+01        .75000E+00        -.73446E+00
     .12500E+01        .12990E+01        -.73517E+00
     .00000E+00        .15000E+01        -.22204E-15
    -.12500E+01        .12990E+01         .73517E+00
    -.21651E+01        .75000E+00         .73446E+00
    -.25000E+01        .00000E+00        -.88818E-15
    -.21651E+01       -.75000E+00        -.73446E+00
    -.12500E+01       -.12990E+01        -.73517E+00
     .00000E+00       -.15000E+01        -.88818E-15
     .12500E+01       -.12990E+01         .73517E+00
     .21651E+01       -.75000E+00         .73446E+00
     .00000E+00        .00000E+00        -.88818E-15

*********************************************************************

 TORSION CONSTANT     D= .30472E+03

*********************************************************************

 BOUNDARY STRESS Ttz

         X                 Y                  Ttz

     .49384E+01        .00000E+00         .29371E+01
     .46967E+01        .91564E+00         .29825E+01
     .39953E+01        .17416E+01         .32838E+01
     .29027E+01        .23972E+01         .37469E+01
     .15261E+01        .28180E+01         .41137E+01
     .00000E+00        .29631E+01         .42484E+01
    -.15261E+01        .28180E+01         .41137E+01
    -.29027E+01        .23972E+01         .37469E+01
    -.39953E+01        .17416E+01         .32838E+01
    -.46967E+01        .91564E+00         .29825E+01
    -.49384E+01        .00000E+00         .29371E+01
    -.46967E+01       -.91564E+00         .29825E+01
    -.39953E+01       -.17416E+01         .32838E+01
    -.29027E+01       -.23972E+01         .37469E+01
    -.15261E+01       -.28180E+01         .41137E+01
     .00000E+00       -.29631E+01         .42484E+01
     .15261E+01       -.28180E+01         .41137E+01
     .29027E+01       -.23972E+01         .37469E+01
     .39953E+01       -.17416E+01         .32838E+01
     .46967E+01       -.91564E+00         .29825E+01
```

Table 6.1 presents the computed values of the warping function, boundary shear stress, and torsion constant of the elliptic cross-section for various values of N and also helps to draw conclusions about the accuracy of the BEM. Moreover, Fig. 6.3 shows the contours of the warping surface $\phi = w/\theta$.

TABLE 6.1 Computed Values of ϕ, $\tau_{tz}/G\theta$, and I_t in a Bar of Elliptic Cross-Section for Various Values of N

Point x, y	Number of Boundary Elements, N						Exact
	20	60	100	160	220	300	
Values of ϕ at Internal Points							
2.1651, 0.7500	−0.7345	−0.7607	−0.7629	−0.7636	−0.7639	−0.7640	−0.7641
1.2500, 1.2990	−0.7352	−0.7607	−0.7629	−0.7637	−0.7639	−0.7640	−0.7641
Values of $\tau_{tz}/G\theta$ at Boundary Nodes							
5.0000, 0.0000	2.9371	2.6846	2.6608	2.6525	2.6499	2.6486	2.6471
0.0000, 3.0000	4.2484	4.3930	4.4050	4.4091	4.4104	4.4110	4.4118
Values of I_t							
	304.72	311.08	311.57	311.74	311.79	311.82	311.84

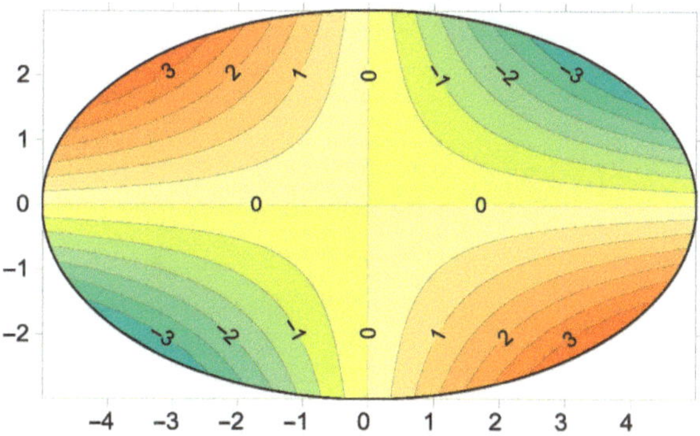

FIGURE 6.3 Contours of the warping function in a bar of elliptic cross-section in Example 6.1.

EXAMPLE 6.2

The program TORSCON is used to solve the torsion problem for a bar with square cross-section of side $a = 4.0$. The values of I_t and $\max\tau_{tz}/G\theta$ are computed for various values of N and are listed in Table 6.2. The data file is constructed for each case of discretization with program RECT-3.FOR, available on the companion website. The warping function $\phi = w/\theta$ is shown in Figs. 6.4 and 6.5. The exact values have been computed from the analytical expressions derived for the rectangular cross-section $a \times b$ [1].

TABLE 6.2 Computed Values of I_t and $\max \tau_{tz}/G\theta$ in a Bar of Square Cross-Section for Various Values of N in Example 6.2

Boundary Elements, N	I_t	$\max \tau_{tz}/G\theta$
20	35.718	2.6186
60	35.979	2.6938
100	35.988	2.6988
140	35.989	2.7010
220	35.989	2.7010
Exact value	35.990	2.7010

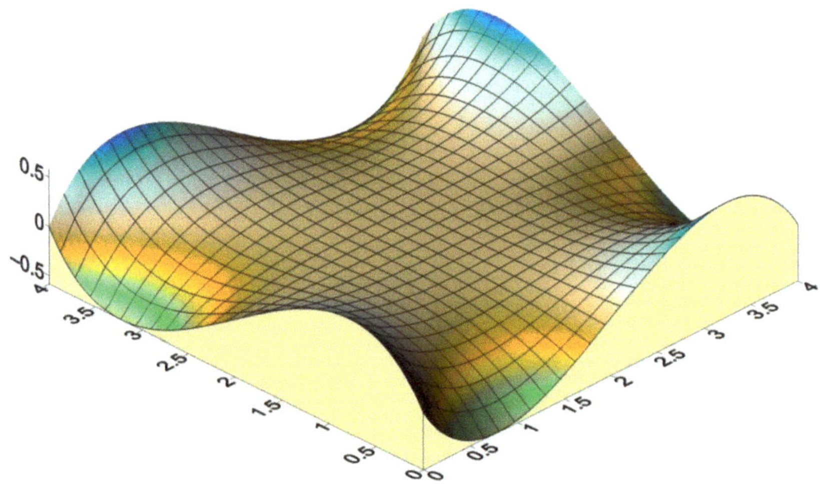

FIGURE 6.4 Warping function $\phi = w/\theta$ for a bar of square cross-section in Example 6.2.

$$I_t = \frac{1}{3}a^3 b\left[1 - \frac{192}{\pi^5}\ \frac{a}{b} \sum_{n=1,3,5,\ldots}^{\infty} \frac{1}{n^5}\tanh\left(\frac{n\pi b}{2a}\right)\right]$$

$$\frac{\max\tau_{tz}}{G\theta} = a\left[1 - \frac{8}{\pi^2} \sum_{n=1,3,5,\ldots}^{\infty} \frac{1}{n^2\cosh\left(\dfrac{n\pi b}{2a}\right)}\right]$$

EXAMPLE 6.3

In this example the torsion problem is solved for the bridge cross-section shown in Fig. 6.6. The solution was obtained using the program TORSCONMU, which results from the TORSCON by replacing the LABECON components

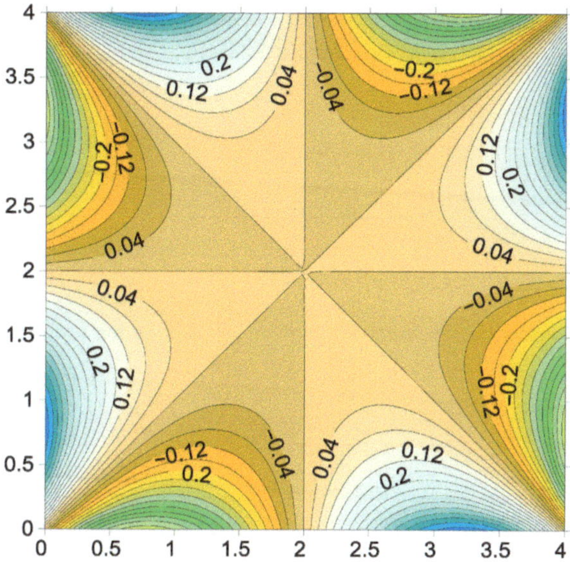

FIGURE 6.5 Contours of the warping function $\phi = w/\theta$ in a bar of square cross-section in Example 6.2.

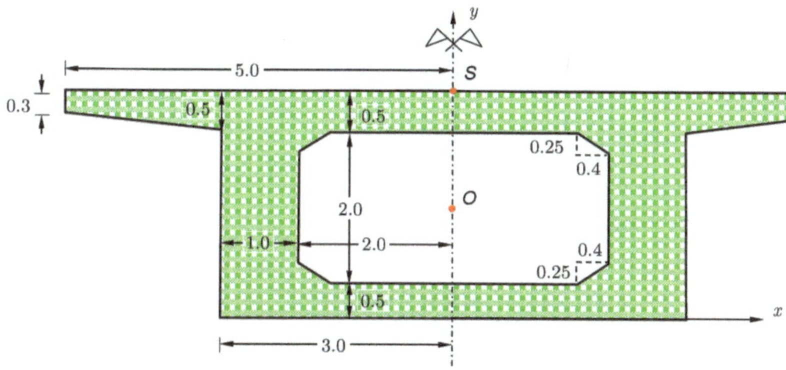

FIGURE 6.6 Bridge cross-section in Example 6.3.

with those of LABECONMU. This program is available on this book's companion website. The results have been obtained with $N_1 = 512$ and $N_2 = 456$ boundary elements on the external and internal boundaries, respectively. The data file is constructed with program BRIDGE.FOR, available also on the companion website. The computed coordinates of the twist center O are $x_0 = 0, y_0 = 1.449$. The computed torsional constant is $I_t = 29.203$ and the maximum shear stress $\max\tau_{tz}/G\theta = 2.75$ occurring at point S of the external boundary. Fig. 6.7 shows the contours of the warping function, while Fig. 6.8 shows the distribution of the shear stress $\tau_{tz}/G\theta$ along the boundaries.

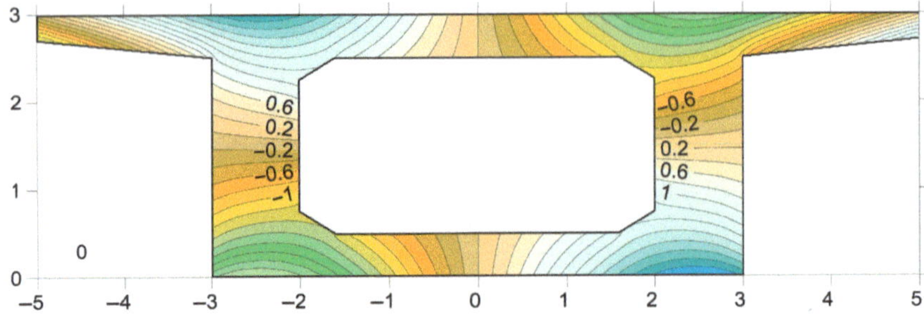

FIGURE 6.7 Contours of the warping function $\phi = w/\theta$ on the bridge cross-section in Example 6.3.

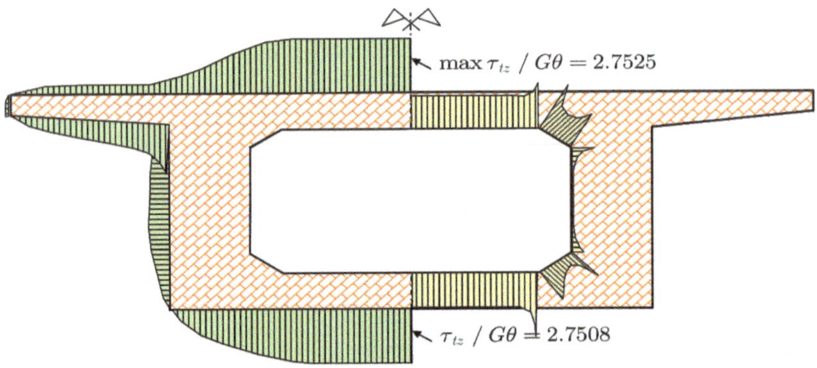

FIGURE 6.8 Distribution of the shear stress $\tau_{tz}/G\theta$ along the boundaries of the bridge cross-section in Example 6.3.

6.2.4 Torsion of anisotropic bars

Let us consider a bar of noncircular cross-section. It consists of an anisotropic material having one plane of material property symmetry, which is perpendicular to the z-axis of the bar (see Fig. 6.1). Such a material is termed *monoclinic* and is described by 13 independent elastic constants for this material instead of 21. In this case, the generalized Hooke's law valid for the general anisotropic body may be simplified and be expressed by the following six relations [4]

$$\left.\begin{aligned}
\varepsilon_x &= \alpha_{11}\,\sigma_x + \alpha_{12}\,\sigma_y + \alpha_{13}\,\sigma_z + \alpha_{16}\,\tau_{xy} \\
\varepsilon_y &= \alpha_{12}\,\sigma_x + \alpha_{22}\,\sigma_y + \alpha_{23}\,\sigma_z + \alpha_{26}\,\tau_{xy} \\
\varepsilon_z &= \alpha_{13}\,\sigma_x + \alpha_{23}\,\sigma_y + \alpha_{33}\,\sigma_z + \alpha_{36}\,\tau_{xy} \\
\gamma_{yz} &= \alpha_{44}\,\tau_{yz} + \alpha_{45}\,\tau_{xz} \\
\gamma_{xz} &= \alpha_{45}\,\tau_{yz} + \alpha_{55}\,\tau_{xz} \\
\gamma_{xy} &= \alpha_{16}\,\sigma_x + \alpha_{26}\,\sigma_y + \alpha_{36}\,\sigma_z + \alpha_{66}\,\tau_{xy}
\end{aligned}\right\} \tag{6.56}$$

where α_{ij} represent the material constants.

Equations (6.56) by virtue of Eqs. (6.3), yield

$$\sigma_x = \sigma_y = \sigma_z = \tau_{xy} = 0 \tag{6.57}$$

and

$$\left.\begin{aligned} \alpha_{55}\,\tau_{xz} + \alpha_{45}\,\tau_{yz} &= \theta\left(\frac{\partial\phi}{\partial x} - y\right) \\ \alpha_{45}\,\tau_{xz} + \alpha_{44}\,\tau_{yz} &= \theta\left(\frac{\partial\phi}{\partial y} + x\right) \end{aligned}\right\} \tag{6.58}$$

Equations (6.58) are solved for τ_{xz} and τ_{yz} yielding

$$\tau_{xz} = \frac{\theta}{|\alpha|}\left[\alpha_{44}\left(\frac{\partial\phi}{\partial x} - y\right) - \alpha_{45}\left(\frac{\partial\phi}{\partial y} + x\right)\right] \tag{6.59a}$$

$$\tau_{yz} = \frac{\theta}{|\alpha|}\left[-\alpha_{45}\left(\frac{\partial\phi}{\partial x} - y\right) + \alpha_{55}\left(\frac{\partial\phi}{\partial y} + x\right)\right] \tag{6.59b}$$

where

$$|\alpha| = \det[\alpha] = \begin{vmatrix} \alpha_{55} & \alpha_{45} \\ \alpha_{45} & \alpha_{44} \end{vmatrix} = \alpha_{44}\,\alpha_{55} - \alpha_{45}^2 \tag{6.60}$$

Introducing Eqs. (6.57) and (6.59) into the equilibrium equations (6.5a,b,c) and the boundary conditions (6.7), we arrive at following boundary value problem for the warping function ϕ

$$\overline{\alpha}_{44}\,\frac{\partial^2\phi}{\partial x^2} - 2\,\overline{\alpha}_{45}\,\frac{\partial^2\phi}{\partial x\partial y} + \overline{\alpha}_{55}\,\frac{\partial^2\phi}{\partial y^2} = 0 \quad \text{in } \Omega \tag{6.61}$$

$$\nabla\phi\cdot\mathbf{m} = y\,m_x - x\,m_y \quad \text{on } \Gamma \tag{6.62}$$

where

$$\mathbf{m} = (\overline{\alpha}_{44}\,n_x - \overline{\alpha}_{45}\,n_y)\;\mathbf{i} + (-\overline{\alpha}_{45}\,n_x + \overline{\alpha}_{55}\,n_y)\;\mathbf{j} \tag{6.63}$$

$$\overline{a}_{44} = \frac{a_{44}}{\sqrt{|\alpha|}}, \quad \overline{a}_{45} = \frac{a_{45}}{\sqrt{|\alpha|}}, \quad \overline{a}_{55} = \frac{a_{55}}{\sqrt{|\alpha|}}$$

$\mathbf{m}$ is a vector in the direction of the *conormal* to the boundary. Equation (6.61) along with the boundary condition (6.62) allow the determination of the warping function $\phi(x, y)$. Equation (6.61) is of the form of Eq. (3.56a) and can be solved using the BEM as it was presented in Section 3.6.

The twisting moment at the end of the cross-sections is

$$M_t = \int_\Omega (x\tau_{yz} - y\tau_{xz})d\Omega = GI_t\theta \tag{6.64}$$

where

$$G = \frac{1}{\sqrt{|\alpha|}} \tag{6.65}$$

is a constant having the same dimensions as the shear modulus, and

$$I_t = \int_\Omega \left[\overline{\alpha}_{55}\, x^2 + 2\overline{\alpha}_{45}\, xy + \overline{\alpha}_{44}\, y^2 - (\overline{\alpha}_{45}\, x + \overline{\alpha}_{44}\, y)\frac{\partial\phi}{\partial x} + (\overline{\alpha}_{55}\, x + \overline{\alpha}_{45}\, y)\frac{\partial\phi}{\partial y}\right] d\Omega \tag{6.66}$$

The latter may be converted to a line integral on the boundary taking the form

$$\begin{aligned} I_t = \int_\Gamma \Bigg\{ & \left[\overline{\alpha}_{44}\left(xy^2 - y\phi\right) + \overline{\alpha}_{45}\left(\frac{1}{2}x^2 y - x\phi\right)\right] n_x \\ & + \left[\overline{\alpha}_{55}\left(x^2 y + x\phi\right) + \overline{\alpha}_{45}\left(\frac{1}{2}xy^2 + y\phi\right)\right] n_y \Bigg\} ds \end{aligned} \tag{6.67}$$

For orthotropic materials, it is $a_{16} = a_{26} = a_{36} = \alpha_{45} = 0$ and the foregoing equations become

$$\left.\begin{aligned} \tau_{xz} &= \theta G_{xz}\left(\frac{\partial\phi}{\partial x} - y\right) \\ \tau_{yz} &= \theta G_{yz}\left(\frac{\partial\phi}{\partial y} + x\right) \end{aligned}\right\} \tag{6.68}$$

$$\overline{G}_{xz}\frac{\partial^2\phi}{\partial x^2} + \overline{G}_{yz}\frac{\partial^2\phi}{\partial y^2} = 0 \tag{6.69}$$

$$\mathbf{m} = \overline{G}_{xz}\, n_x\, \mathbf{i} + \overline{G}_{yz}\, n_y\, \mathbf{j} \tag{6.70}$$

$$\begin{aligned} I_t &= \int_\Omega \left[\overline{G}_{xz}\left(y^2 - y\frac{\partial\phi}{\partial x}\right) + \overline{G}_{yz}\left(x^2 + x\frac{\partial\phi}{\partial y}\right)\right] d\Omega \\ &= \int_\Gamma \left[\overline{G}_{xz}(xy^2 - y\phi)n_x + \overline{G}_{yz}(yx^2 + x\phi)n_y\right] ds \end{aligned} \tag{6.71}$$

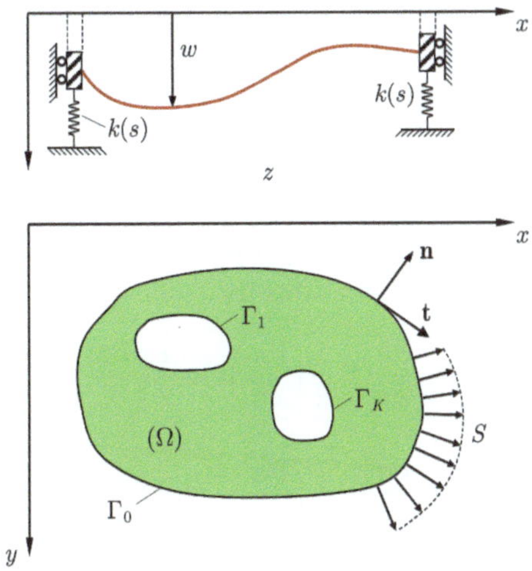

FIGURE 6.9 Elastic membrane occupying the multiply connected domain Ω.

where

$$G_{xz} = \frac{1}{\alpha_{55}}, \quad G_{yz} = \frac{1}{\alpha_{44}} \tag{6.72}$$

$$G = \sqrt{G_{xz} G_{yz}} \tag{6.73}$$

$$\overline{G}_{xz} = \overline{a}_{44} = \frac{G_{xz}}{G}, \quad \overline{G}_{yz} = \overline{a}_{55} = \frac{G_{yz}}{G} \tag{6.74}$$

are the nondimensional shear moduli in the xz and yz planes, respectively.

6.3 DEFLECTION OF ELASTIC MEMBRANES

We consider a flat elastic membrane of uniform thickness h occupying the two-dimensional multiply connected domain Ω of the xy-plane bounded by $K+1$ curves (see Fig. 6.9). The membrane is fixed or elastically supported along its boundary $\Gamma = \cup_{i=0}^{i=K} \Gamma_i$ and subjected to a uniform tension S, which is large enough so that it is not appreciably altered when the membrane is deflected by a distributed load $f(x, y)$.

The equilibrium equation of the deflected membrane can be derived by considering the equilibrium of a deflected element $d\Omega = dx\,dy$. Here, however, it will be derived using an energy approach, because it allows a better understanding of the linearization of the problem.

The initial flat membrane is deflected to a surface $w(x, y)$ when subjected to a transverse load of density $f(x, y)$. Due to the lateral defection the prestressed membrane is further stretched and additional strains are produced in its middle surface. The linear deflection theory of membranes is based on the following assumptions:

a. The prestress of the membrane is large enough so that the tension S remains unchanged during the deflection.

b. The additional strains of the middle surface of the membrane due to its elastic in-plane deformation ($u(x, y)$ and $v(x, y)$) are negligible compared to those due to the deflection $w(x, y)$ of the middle surface.

The second assumption implies that the strain components are given as

$$\varepsilon_x = \frac{\partial u}{\partial x} + \frac{1}{2}\left(\frac{\partial w}{\partial x}\right)^2 \approx \frac{1}{2}\left(\frac{\partial w}{\partial x}\right)^2 \tag{6.75a}$$

$$\varepsilon_y = \frac{\partial v}{\partial y} + \frac{1}{2}\left(\frac{\partial w}{\partial y}\right)^2 \approx \frac{1}{2}\left(\frac{\partial w}{\partial y}\right)^2 \tag{6.75b}$$

$$\gamma_{xy} = \frac{\partial v}{\partial x} + \frac{\partial u}{\partial y} + \frac{\partial w}{\partial x}\frac{\partial w}{\partial y} \approx \frac{\partial w}{\partial x}\frac{\partial w}{\partial y} \tag{6.75c}$$

The strain energy of the deflected membrane is written as

$$U = \frac{h}{2}\int_\Omega (\sigma_x\varepsilon_x + 2\,\tau_{xy}\gamma_{xy} + \sigma_y\varepsilon_y)d\Omega \tag{6.76}$$

or taking into account that $h\sigma_x = h\sigma_y = S$ and $\tau_{xy} = 0$, and using Eqs. (6.75), we arrive at

$$U = \frac{S}{2}\int_\Omega \left[\left(\frac{\partial w}{\partial x}\right)^2 + \left(\frac{\partial w}{\partial y}\right)^2\right] d\Omega \tag{6.77}$$

Thus, the total potential energy of the deflected membrane becomes

$$\begin{aligned}\Pi(w) = &\int_\Omega \left\{\frac{S}{2}\left[\left(\frac{\partial w}{\partial x}\right)^2 + \left(\frac{\partial w}{\partial y}\right)^2\right] - fw\right\} d\Omega \\ &+ \int_\Gamma \left[\frac{1}{2}k(s)\,w^2 - R(s)\,w\right] ds\end{aligned} \tag{6.78}$$

where $k(s)$ is the stiffness modulus of the elastic support and $R(s)$ is the density of the externally applied transverse load along the boundary.

The equilibrium of the membrane requires that the first variation of the total potential, $\delta\Pi$, vanishes. Thus, using the procedure described in Section 2.7.3 we can readily derive the differential equation of the membrane and the associated boundary condition, which formulate the following boundary value problem

$$S\,\nabla^2 w = -f \quad \text{in } \Omega \tag{6.79a}$$

$$\beta_1\, w + \beta_2\, q = \beta_3 \quad \text{on } \Gamma \tag{6.79b}$$

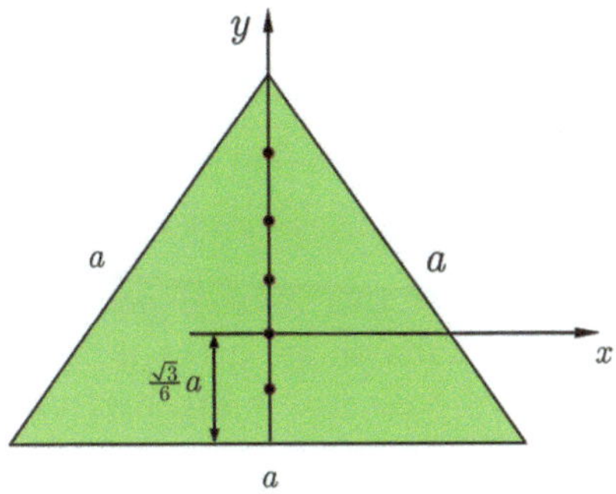

FIGURE 6.10 Triangular membrane in Example 6.4.

where $\beta_1 = k(s)$, $\beta_2 = S$, $\beta_3 = R(s)$; $q = \partial w / \partial n$ is the derivative of w along the direction normal to the boundary. Consequently, the problem of determining the deflection surface of an elastic membrane is reduced to the problem of solving Poisson's equation under Robin's boundary condition. Note that for $k(s) \to \infty$, the boundary condition (6.79b) yields $w = 0$, namely, the Dirichlet boundary condition.

EXAMPLE 6.4

Determine the deflection surface of an elastic membrane having the shape of an equilateral triangle with side length $a = 5.0$ m, Fig. 6.10. The membrane is fixed along its boundary and is subjected to a uniformly distributed load $f = 10\ \text{kN/m}^2$ and a tension $S = 1\ \text{kN/m}$. The coordinate axes are taken as shown in the figure.

The deflection $w(x, y)$ of the membrane is sought in the form

$$w = w_0 + w_1$$

where w_0 is the solution of the homogeneous equation and w_1 is a particular solution.

i. Particular solution w_1

The particular solution is established using the procedure described in Section 3.4.2 by transforming Eq. (6.79a) into the complex domain using Eq. (3.42). Thus, we find

$$w_1 = -\frac{10}{4}\left(x^2 + y^2\right)$$

ii. Homogeneous solution w_0

The homogeneous solution will be obtained from the following boundary value problem

$$\nabla^2 w_0 = 0 \quad \text{in } \Omega$$

$$w_0 = \frac{10}{4}(x^2 + y^2), \quad (x, y) \in \Gamma$$

using the program LABECON.

The problem admits the analytical solution [1]

$$w = -\frac{f}{2S}\left[\frac{1}{2}(x^2 + y^2) - \frac{1}{a\sqrt{3}}(y^3 - 3x^2 y) - \frac{1}{18}a^2\right]$$

The computed deflections at five internal points along with the corresponding values of the exact solution are given in Table 6.3. Moreover, the computed deflection surface is presented in Fig. 6.11 using contours.

TABLE 6.3 Deflections $w(0, y)$ of the Triangular Membrane on the Section $x = 0$ for Various Values of N in Example 6.4

y	Number of Boundary Elements, N						Exact
	30	60	90	150	210	300	
−0.7217	5.4550	5.4323	5.4284	5.4264	5.4259	5.4256	5.4253
0.0000	6.9743	6.9514	6.9475	6.9455	6.9450	6.9447	6.9444
0.7217	5.8895	5.8663	5.8624	5.8604	5.8599	5.8596	5.8524
1.4434	3.5043	3.4793	4.4753	3.4733	3.4728	3.4725	3.4722
2.1651	1.1159	1.0933	1.0883	1.0862	1.0856	1.0853	1.0851

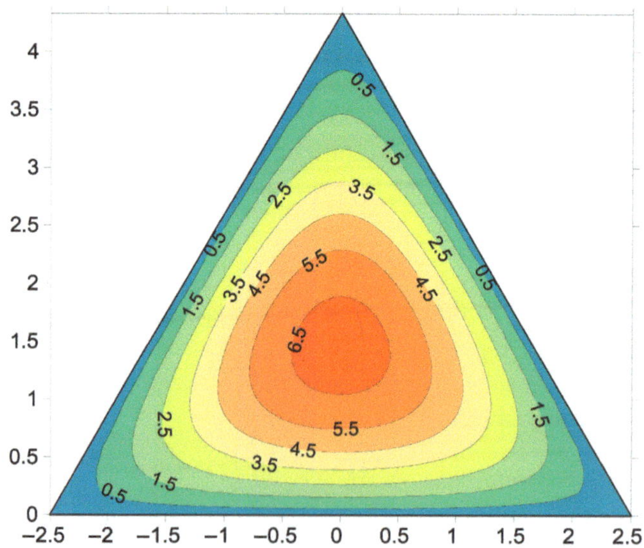

FIGURE 6.11 Deflection surface of the triangular membrane in Example 6.4.

6.4 BENDING OF SIMPLY SUPPORTED PLATES

The deflection $w(x, y)$ of a thin elastic plate occupying the two-dimensional domain Ω of the xy-plane satisfies the equation [5,6]

$$\nabla^4 w = \frac{f}{D} \tag{6.80}$$

where

$D = \frac{Eh^3}{12\,(1-\nu^2)}$ is the flexural rigidity of the plate,
ν is the Poisson's ratio,
h is the constant thickness of the plate,
$f = f(x, y)$ is the distributed transverse load, and
$\nabla^4 = \nabla^2\,\nabla^2 = \left(\frac{\partial^2}{\partial x^2} + \frac{\partial^2}{\partial y^2}\right)^2 = \frac{\partial^4}{\partial x^4} + 2\frac{\partial^4}{\partial x^2 \partial y^2} + \frac{\partial^4}{\partial y^4}$ is the biharmonic operator.

The bending and twisting moments are given by the expressions

$$M_x = -D\left(\frac{\partial^2 w}{\partial x^2} + \nu\frac{\partial^2 w}{\partial y^2}\right) \tag{6.81a}$$

$$M_y = -D\left(\frac{\partial^2 w}{\partial y^2} + \nu\frac{\partial^2 w}{\partial x^2}\right) \tag{6.81b}$$

$$M_{xy} = -M_{yx} = D\,(1-\nu)\frac{\partial^2 w}{\partial x\,\partial y} \tag{6.81c}$$

For a simply supported plate the deflection must satisfy the following boundary conditions on the plate boundary Γ

$$w = 0 \tag{6.82a}$$

$$M_n = -D\left(\frac{\partial^2 w}{\partial n^2} + \nu\frac{\partial^2 w}{\partial t^2}\right) = 0 \tag{6.82b}$$

where M_n is the bending moment in the direction n normal to the boundary and t denotes the tangential to the boundary direction.

Note that for a curvilinear boundary it is [6,7]

$$\frac{\partial^2 w}{\partial t^2} = \frac{\partial^2 w}{\partial s^2} + \kappa\frac{\partial w}{\partial n} \tag{6.83}$$

where $\kappa = \kappa(s)$ is the curvature of the boundary. When the boundary of the simply supported plate consists of rectilinear segments, obviously it is

$$\kappa = 0, \quad w = 0, \quad \frac{\partial w}{\partial s} = 0, \quad \frac{\partial^2 w}{\partial s^2} = 0$$

In this case, Eq. (6.83) results in

$$\frac{\partial^2 w}{\partial t^2}=0$$

while Eq. (6.82b) reduces to

$$\frac{\partial^2 w}{\partial n^2}=0$$

Consequently, on the basis of the above two equations, the deflection should satisfy the following equation on the boundary of the plate

$$\nabla^2 w = \frac{\partial^2 w}{\partial n^2} + \frac{\partial^2 w}{\partial t^2} = \frac{\partial^2 w}{\partial x^2} + \frac{\partial^2 w}{\partial y^2} = 0 \quad \text{on } \Gamma \tag{6.84}$$

For points inside the domain Ω, Eqs. (6.81a,b) give

$$M_x + M_y = -D(1+\nu)\nabla^2 w$$

So, by setting

$$M = \frac{M_x + M_y}{1+\nu} = -D\,\nabla^2 w \tag{6.85}$$

we can write Eq. (6.80) as

$$\nabla^2(-D\,\nabla^2 w) = -f$$

which by virtue of Eq. (6.85) may be split into the following two Poisson's equations

$$\nabla^2 M = -f \tag{6.86}$$

$$\nabla^2 w = -\frac{M}{D} \tag{6.87}$$

Equation (6.84) yields $M=0$ on the boundary of the plate. Therefore, the solution of Eq. (6.80) for a simply supported plate with a polygonal boundary can be obtained from the following two Dirichlet problems:

$$\left.\begin{aligned} \nabla^2 M &= -f \quad && \text{in } \Omega \\ M &= 0 && \text{on } \Gamma \end{aligned}\right\} \tag{6.88}$$

and

$$\left.\begin{aligned} \nabla^2 w &= -\frac{M}{D} \quad && \text{in } \Omega \\ w &= 0 && \text{on } \Gamma \end{aligned}\right\} \tag{6.89}$$

The solution of the plate equation by separating it into two potential equations is attributed to Marcus [8]. Its use is limited, because it can treat only simply supported plates with a polygonal boundary. For the extension of the method to plates with a curvilinear boundary under any type of boundary conditions, the reader is referred to the work of Paris and De Leon [9] or Katsikadelis [10]. Nevertheless, the solution of the general plate problem can be obtained by the BEM according to the procedure developed for the biharmonic operator [6,10].

The solution of Eqs. (6.88) and (6.89) using the BEM requires the evaluation of the domain integrals

$$\int_{\Omega} vf \, d\Omega \quad \text{and} \quad \int_{\Omega} vM \, d\Omega$$

where v is the fundamental solution of the Laplace equation. The function f is known and as a result the first of the above integrals can be evaluated using any of the methods presented in Section (3.5). The function M, however, is not given by an analytical expression, but by its numerical values at any desired points. These values can be used to convert the domain integral into a boundary line integral employing the dual reciprocity method (see Section 8.3).

6.5 HEAT TRANSFER PROBLEMS

The heat transfer equation is derived from the energy conservation principle. Its general form is given as [11]

$$-\nabla \cdot \mathbf{q}(\mathbf{x}, t) + f(\mathbf{x}, t) = \frac{\partial(u\rho)}{\partial t}, \quad \mathbf{x}(x, y, z) \in \Omega \tag{6.90}$$

where

t: time

$\mathbf{q}$: heat flux

f: rate of internal heat generation due to heat sources

ρ: material density

u: specific internal energy

Ω: domain occupied by the body.

For two-dimensional problems, the heat may flow in any direction of the xy-plane. This flow is described by the *flux vector* $\mathbf{q}$, whose direction is that of the heat flow and its magnitude expresses the heat passing per unit time through the unit surface normal to the direction of the heat flow.

According to the generalized Fourier's law, the thermal flux density depends linearly on the gradient of the temperature field.

Namely, denoting by $T = T(x, y, t)$ the temperature, the flux is expressed as

$$\mathbf{q} = \begin{Bmatrix} q_x \\ q_y \end{Bmatrix} = -\begin{bmatrix} k_{xx} & k_{xy} \\ k_{yx} & k_{yy} \end{bmatrix} \begin{Bmatrix} \dfrac{\partial T}{\partial x} \\ \dfrac{\partial T}{\partial y} \end{Bmatrix} = -\begin{Bmatrix} k_{xx}\dfrac{\partial T}{\partial x} & + & k_{xy}\dfrac{\partial T}{\partial y} \\ k_{yx}\dfrac{\partial T}{\partial x} & + & k_{yy}\dfrac{\partial T}{\partial y} \end{Bmatrix} \tag{6.91}$$

or

$$\mathbf{q} = -\mathbf{D} \cdot \nabla T \tag{6.92}$$

The matrix

$$\mathbf{D} = \begin{bmatrix} k_{xx} & k_{xy} \\ k_{yx} & k_{yy} \end{bmatrix} \tag{6.93}$$

provides information about the heat transfer in any direction and it is referred to as the *conductivity matrix*. It is the *constitutive matrix* for the heat transfer problem. For nonhomogeneous materials, the constitutive matrix depends on the position of point $\mathbf{x}(x, y)$, that is, $\mathbf{D} = \mathbf{D}(x, y)$. The determinant of the matrix $\mathbf{D}$ does not vanish, that is,

$$|D| \neq 0 \tag{6.94}$$

In general $\mathbf{D}$ is not a symmetric matrix. However, for the simplicity of the expressions it is assumed here to be symmetric, thus $k_{xy} = k_{yx}$.

If the material is orthotropic, it is $k_{xy} = k_{yx} = 0$ and the matrix $\mathbf{D}$ can be simplified taking the form

$$\mathbf{D} = \begin{bmatrix} k_{xx} & 0 \\ 0 & k_{yy} \end{bmatrix} \tag{6.95}$$

Moreover, for an isotropic material it is $k_{xx} = k_{yy} = k$ and consequently we may write

$$\mathbf{D} = k \begin{bmatrix} 1 & 0 \\ 0 & 1 \end{bmatrix} \tag{6.96}$$

The negative sign in Eq. (6.92) is due to the fact the heat flows from higher to lower temperature regions, while the gradient ∇T is directed towards regions of higher temperature.

It is known from thermodynamics that the internal energy depends linearly on the temperature

$$u = c\, T(\mathbf{x}, t) \tag{6.97}$$

where c is the *specific heat*.

Taking into account Eqs. (6.92) and (6.97), Eq. (6.90) becomes

$$\nabla \cdot (\mathbf{D} \cdot \nabla T) + f = \frac{\partial (cT\rho)}{\partial t} \tag{6.98}$$

If the material properties, that is, heat conductivity, specific heat, and density, do not depend on the temperature, the differential equation (6.98) becomes linear.

In steady-state heat transfer, that is when the thermal equilibrium has been reached, the temperature distribution within the body does not depend anymore on time and Eq. (6.98) is simplified to

$$\nabla \cdot (\mathbf{D} \cdot \nabla T) + f = 0 \tag{6.99}$$

Further, when the conductivity matrix is constant, that is, it does not depend on point $\mathbf{x}$, the material is thermally homogeneous and Eq. (6.99) takes the form

$$k_{xx}\frac{\partial^2 T}{\partial x^2} + 2k_{xy}\frac{\partial^2 T}{\partial x \partial y} + k_{yy}\frac{\partial^2 T}{\partial y^2} + f = 0 \tag{6.100}$$

To solve the heat transfer problem, Eq. (6.100) needs to be subject to appropriate prescribed boundary conditions. These boundary conditions can be of the following three types [12,13]

$$\left.\begin{aligned} &T = \overline{T} \quad \text{on } \Gamma_1 \\ &q_n = \overline{q}_n \quad \text{on } \Gamma_2 \\ &q_n = -h_o(T - T_\alpha) \quad \text{on } \Gamma_3 \end{aligned}\right\} \tag{6.101}$$

where $\Gamma_1 \cup \Gamma_2 \cup \Gamma_3 = \Gamma$; $q_n = \mathbf{q} \cdot \mathbf{n}$ denotes the projection of the heat flux vector on the normal to the boundary, h_o represents the heat transfer coefficient, and T_α the ambient temperature outside the body. The first boundary condition is a Dirichlet condition or otherwise known as an *essential* condition. The second is a Neumann condition or otherwise known as a *natural* condition. The third condition, otherwise known as a Robin condition, is a linear relationship between the flux and the temperature on the boundary. It is particularly important in heat transfer applications, as it represents the convection condition.

If the material is orthotropic, $k_{xy} = 0$, Eq. (6.100) becomes

$$k_{xx}\frac{\partial^2 T}{\partial x^2} + k_{yy}\frac{\partial^2 T}{\partial y^2} + f = 0 \tag{6.102}$$

Finally, if the material is isotropic, that is, $k_{xx} = k_{yy} = k$, Eq. (6.100) and the boundary conditions (6.101) become

$$k\,\nabla^2 T + f = 0 \tag{6.103}$$

$$T = \overline{T} \quad \text{on } \Gamma_1 \tag{6.104a}$$

$$-k\frac{\partial T}{\partial n} = \overline{q}_n \quad \text{on } \Gamma_2 \tag{6.104b}$$

$$-k\frac{\partial T}{\partial n} = -\,h_o(T - T_\alpha) \quad \text{on } \Gamma_3 \tag{6.104c}$$

As it was previously mentioned, the first two boundary conditions are typical Dirichlet and Neumann boundary conditions, respectively, while the third one has the form of Eq. (3.6b), namely

$$\alpha\, T + \beta\,\frac{\partial T}{\partial n} = \gamma \tag{6.105}$$

where $\alpha = h_o$, $\beta = -\,k$, and $\gamma = h_o T_\alpha$.

For a numerical solution of the problem with the BEM, the third boundary condition, Eq. (6.104c), is treated as follows. Equation (6.105) is applied to all nodal points on part Γ_3 of the boundary, which yields

$$[\alpha]\{T\} + [\beta]\{T_n\} = \{\gamma\} \tag{6.106}$$

where

$$[\alpha] = h_o[I], \quad [\beta] = -\,k[I], \quad \{\gamma\} = h_o\, T_\alpha \begin{Bmatrix} 1 \\ 1 \\ \vdots \\ 1 \\ 1 \end{Bmatrix} \tag{6.107}$$

in which $[I]$ is the unit matrix of dimensions $N_3 \times N_3$, where N_3 is the number of nodal points on Γ_3.

It is apparent that all the types of boundary conditions can be written in the form of Eq. (6.106) by specifying appropriately the coefficients α, β and γ, for example, for the Dirichlet condition it is $\alpha = 1$, $\beta = 0$, $\gamma = \overline{T}$. Thus, Eq. (6.106) along with Eq. (4.7), which for $u = T$ becomes

$$[H]\{T\} = [G]\{T_n\} \tag{6.108}$$

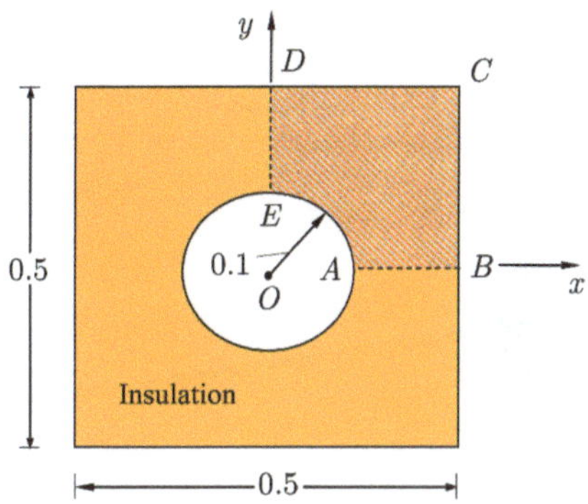

FIGURE 6.12 Insulated circular heating duct. Temperature inside $T = 300°$ C, outside $T = -15°$ C in Example 6.5.

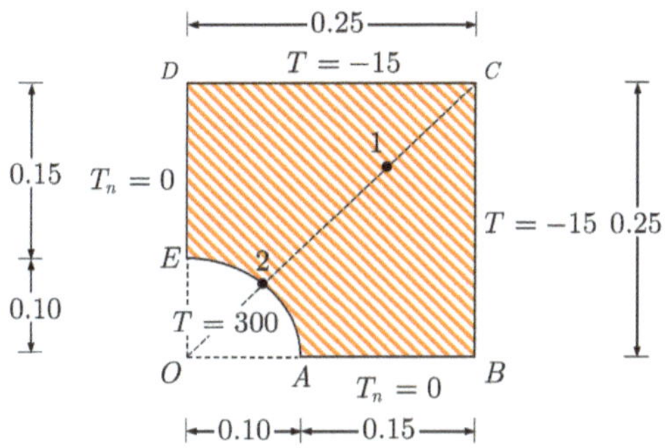

FIGURE 6.13 Quadrant $ABCDE$ and boundary conditions ($T_n = \partial T/\partial n$) in Example 6.5.

can be combined into a single equation forming a system of 2N linear algebraic equations, which allow the determination of the unknown boundary quantities.

EXAMPLE 6.5

We consider the circular insulated metal heating duct of Fig. 6.12, in which a liquid of temperature 300° C flows. We have to determine the temperature distribution inside the insulation when the outside temperature is −15° C. The thermal conductivity k is constant.

The problem can be solved using the BEM for domains with multiple boundaries (see Section 4.6). For the problem at hand, however, we can take advantage of the symmetry of the domain with respect to both the x and y axes and confine the solution to the quadrant $ABCDE$ (Fig. 6.13). The symmetry implies zero flux in the directions normal to the cuts AB and ED. All the boundary conditions related to the reduced domain are shown in Fig. 6.13.

TABLE 6.4 Computed Values of the Temperature T and Its Derivatives $\partial T/\partial x$ and $\partial T/\partial n$ for Various Values of N in Example 6.5

Point	Number of Boundary Elements, N 21	41	71	141	211	421
	T					
1	33.973	34.299	34.278	34.269	34.266	34.263
	$\partial T/\partial x$					
1	−656.78	−660.77	−661.20	−661.10	−661.07	−661.03
	$\partial T/\partial n$					
2	3063.9	3102.6	3116.1	3123.7	3125.4	3126.8

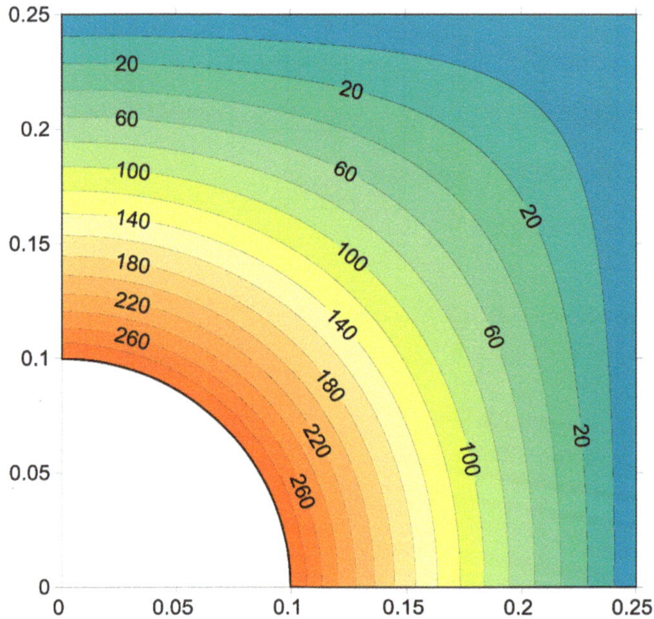

FIGURE 6.14 Temperature contours in the quadrant $ABCDE$ in Example 6.5.

The computed values of the temperature T and its derivative $\partial T/\partial x$ at the internal point 1 $(0.175, 0.175)$, and of the normal derivative $\partial T/\partial n$ at the boundary point 2 $(0.1\sqrt{2}/2, 0.1\sqrt{2}/2)$ are given in Table 6.4 for various values of the number of constant boundary elements (nodal points) N. Finally, the contour plots of the temperature distribution in the insulation are depicted in Fig. 6.14.

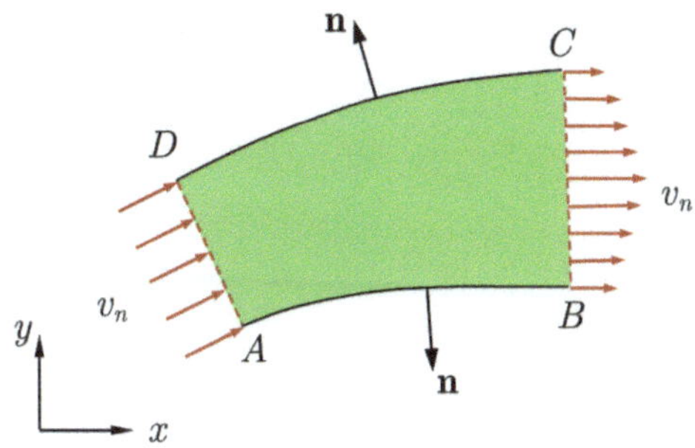

FIGURE 6.15 Two-dimensional fluid flow.

6.6 FLUID FLOW PROBLEMS

An ideal or perfect fluid is one that has zero viscosity and is incompressible. A good approximation to the solution of the inviscid (nonviscous) fluid flow problem is achieved by satisfying only the continuity equation, which for a two-dimensional flow is written as [14]

$$\nabla \cdot (\rho \mathbf{v}) + \frac{\partial \rho}{\partial t} + f = 0 \tag{6.109}$$

where ρ is the density of the fluid, $\mathbf{v}(v_x, v_y)$ the velocity of the fluid at point (x, y), and $f = f(x, y)$ the distribution of a possible internal source. When the density is constant (incompressible fluid), Eq. (6.109) becomes

$$\nabla \cdot \mathbf{v} + f/\rho = 0 \tag{6.110}$$

Consider the two-dimensional fluid flow of Fig. 6.15. If the flow is irrotational, the vorticity is zero, and the velocity satisfies the equation

$$\nabla \times \mathbf{v} = \left(\frac{\partial v_y}{\partial x} - \frac{\partial v_x}{\partial y} \right) \mathbf{k} = 0$$

which yields

$$\frac{\partial v_y}{\partial x} - \frac{\partial v_x}{\partial y} = 0 \tag{6.111}$$

Therefore, a potential function ϕ exists, which produces the velocity field as

$$\mathbf{v} = \nabla \phi = \frac{\partial \phi}{\partial x} \, \mathbf{i} + \frac{\partial \phi}{\partial y} \, \mathbf{j} \tag{6.112}$$

Thus the velocity components are

$$v_x = \frac{\partial \phi}{\partial x} \quad \text{and} \quad v_y = \frac{\partial \phi}{\partial y} \tag{6.113}$$

Obviously, Eq. (6.111) is satisfied identically and Eq. (6.110) becomes

$$\nabla \cdot \nabla \phi + f/\rho = 0 \tag{6.114}$$

or

$$\nabla^2 \phi + f/\rho = 0 \tag{6.115}$$

The solution of the differential equation (6.115) under boundary conditions, which will be presented next, allows the determination of the potential ϕ. Thereafter, the velocity components are obtained from Eq. (6.113).

The boundary conditions on the velocity potential may be derived from physical considerations. For this purpose, we consider the irrotational and inviscid flow inside a part of a pipe, as shown in Fig. 6.15.

As there is no penetration through the rigid walls AB and DC of the pipe, the normal component v_n of the velocity is zero,

$$v_n = \frac{\partial \phi}{\partial n} = 0 \tag{6.116}$$

Along the cross-section AD (inlet) the distribution of v_n may be given, namely,

$$v_n = \frac{\partial \phi}{\partial n} = \overline{v}_n \tag{6.117}$$

while at the cross-section BC (outlet) the conditions are:

i. The velocity component v_n may be prescribed, but its distribution should satisfy along with Eqs. (6.116) and (6.117) the mass conservation principle in the given domain, namely,

$$\int_\Gamma v_n \, ds = -\int_{AD} v_n \, ds + \int_{BC} v_n \, ds = -\int_\Omega f \, d\Omega \tag{6.118}$$

ii. The condition of tranquility of flow (fully developed laminar flow) may be imposed, provided that the cross-section BC is placed sufficiently far from regions of intense variation in the velocity field. This condition is expressed mathematically as

$$\frac{\partial v_n}{\partial n} = \frac{\partial^2 \phi}{\partial n^2} = 0 \tag{6.119}$$

The latter condition was not included in those discussed in Chapter 3 and thus it requires special treatment.

If section BC is placed away from regions of intense variation of the velocity, as for the laminar flow of case (ii), and its shape is such that the velocity is everywhere normal to it (e.g., straight line), the tangential component of the velocity vanishes. Therefore, it is

$$v_t = \frac{\partial \phi}{\partial s} = 0$$

which implies

$$\phi = C \tag{6.120}$$

where C is an arbitrary constant. This condition permits the establishment of the velocity potential ϕ to the approximation of an arbitrary constant. This, however, does not influence the velocity field, which is determined from Eq. (6.113).

Apart from the boundary condition (6.119), the program LABECON can be utilized to solve fluid flow problems, if it is supplemented by a subroutine that evaluates the derivatives of the potential at internal points according to Eqs. (4.15) and (4.16). The modified LABECON is given the name FLUIDCON. The listing of the main program as well as the subroutines DERIV, which computes the derivatives at the internal points, and OUTPUT, which prints the results, are given below. The electronic version of the program FLUIDCON is available on this book's companion website.

```
C=====================================================================
C
      PROGRAM FLUIDCON
C
C
C  This program solves the two dimensional Laplace equation
C  for two-dimensional irrotational inviscid flow of (FLUID)s
C  using the boundary element method with (CON)stant
C  boundary elements
C
C
      IMPLICIT REAL*8 (A-H,O-Z)
      CHARACTER*15 INPUTFILE,OUTPUTFILE
C
C  Set the maximum dimensions
C
      PARAMETER (N=240)
      PARAMETER (IN=50)
```

```
C
C  N= Number of boundary elements equal to number of boundary
C     nodes
C IN= Number of internal points where the function u is calculated
C
      DIMENSION INDEX(N)
      DIMENSION XL(N+1),YL(N+1),XM(N),YM(N),G(N,N),H(N,N)
      DIMENSION UB(N),A(N,N),UNB(N),XIN(IN),YIN(IN),UIN(IN)
      DIMENSION UXIN(IN),UYIN(IN)
C
C  Read the names and open the input and output files
C
      WRITE (*,'(A)')' Name of the INPUTFILE (max.15 characters)'
      READ  (*,'(A)') INPUTFILE
      WRITE (*,'(A)')' Name of the OUTPUTFILE (max.15 characters)'
      READ  (*,'(A)') OUTPUTFILE
      OPEN (UNIT=1, FILE=INPUTFILE)
      OPEN (UNIT=2, FILE=OUTPUTFILE)
C
C  Read date from INPUTFILE
C
      CALL INPUT (XL,YL,XIN,YIN,INDEX,UB,N,IN)
C
C  Compute the G matrix
C
      CALL GMATR (XL,YL,XM,YM,G,N)
C
C  Compute the H matrix
C
      CALL HMATR (XL,YL,XM,YM,H,N)
C
C  Form the system of equations AX=B
C
      CALL ABMATR (G,H,A,UNB,UB,INDEX,N)
C
C  Solve the system of equations
C
      CALL SOLVEQ(A,UNB,N,LSING)
C
C  Form the vectors U and UN of all the boundary values
C
      CALL REORDER (UB,UNB,INDEX,N)
C

C  Compute the values of U at the internal points
C
      CALL UINTER (XL,YL,XIN,YIN,UB,UNB,UIN,N,IN)
C
C  Compute the values of Ux and Uy at the internal points
C
      CALL DERIV(XL,YL,XIN,YIN,UB,UNB,UXIN,UYIN,N,IN)
C
C  Print the results in the OUTPUTFILE
C
      CALL OUTPUT (XM,YM,UB,UNB,XIN,YIN,UIN,UXIN,UYIN,N,IN)
C
C  Close input and output files
C
      CLOSE(1)
      CLOSE(2)
      STOP
      END
C
```

```
C=================================================================
C
      SUBROUTINE DERIV(XL,YL,XIN,YIN,UB,UNB,UXIN,UYIN,N,IN)
C
C
C  This subroutine computes the values of the derivatives
C  Ux and Uy at the internal points
C
      IMPLICIT REAL*8 (A-H,O-Z)
      DIMENSION XL(N+1),YL(N+1),UB(N),UNB(N)
      DIMENSION XIN(IN),YIN(IN),UXIN(IN),UYIN(IN)
      DIMENSION XC(4),YC(4),XI(4),WG(4)
C  N         =Number of boundary elements
C  XL,YL     =Coordinates of the extreme points of the
C              boundary elements
C  XIN,YIN   =Coordinates of the internal points
C  UB,UNB    =Boundary values of U and Un
C  UXIN,UYIN=Derivatives Ux and Uy at the internal points
C  WG        =Weights of the Gauss integration
C  XI        =Coordinates of the Gauss integration points
C              in the interval [-1,1]
C  XC,YC     =Global coordinates of the Gauss integration points
C
      DATA XI/-0.86113631,-0.33998104,0.33998104,0.86113631/
      DATA WG/0.34785485,0.65214515,0.65214515,0.34785485/
      PI=ACOS(-1.)
      XL(N+1)=XL(1)
      YL(N+1)=YL(1)
      PI=ACOS(-1.)
      DO 20 I=1,IN
      UXIN(I)=0.
      UYIN(I)=0.
      DO 10 J=1,N
      AX=(XL(J+1)-XL(J))/2.
      AY=(YL(J+1)-YL(J))/2.
      BX=(XL(J+1)+XL(J))/2.
      BY=(YL(J+1)+YL(J))/2.
      SL=SQRT(AX**2+AY**2)
      ANGLE=ATAN2(AY,AX)-PI/2.
      ENX=COS(ANGLE)
      ENY=SIN(ANGLE)
      GX=0.
      GY=0.
      HX=0.
      HY=0.
        DO 40 K=1,4
      XC(K)=AX*XI(K)+BX
      YC(K)=AY*XI(K)+BY
      DX=XC(K)-XIN(I)
      DY=YC(K)-YIN(I)
      R=SQRT(DX**2+DY**2)
      RX=-DX/R
      RY=-DY/R
      RN=-(RX*ENX+RY*ENY)
      RT=-(-RX*ENY+RY*ENX)
      UX=RX/(2.*PI*R)
      UY=RY/(2.*PI*R)
      UNX=-(RX*RN-RY*RT)/(2.*PI*R**2)
      UNY=-(RX*RT+RY*RN)/(2.*PI*R**2)
      GX=GX+UX*WG(K)*SL
      GY=GY+UY*WG(K)*SL
      HX=HX+UNX*WG(K)*SL
      HY=HY+UNY*WG(K)*SL
 40   CONTINUE
    UXIN(I)=UXIN(I)+HX*UB(J)-GX*UNB(J)
    UYIN(I)=UYIN(I)+HY*UB(J)-GY*UNB(J)
 10 CONTINUE
 20 CONTINUE
    RETURN
    END
C
```

```
      UXIN(I)=UXIN(I)+HX*UB(J)-GX*UNB(J)
      UYIN(I)=UYIN(I)+HY*UB(J)-GY*UNB(J)
   10 CONTINUE
   20 CONTINUE
      RETURN
      END
C
C=====================================================================
C
      SUBROUTINE OUTPUT(XM,YM,UB,UNB,XIN,YIN,UIN,UXIN,UYIN,N,IN)
C
C
C  This subroutine prints the results in the outputfile
C
      IMPLICIT REAL*8 (A-H,O-Z)
      DIMENSION XM(N),YM(N),UB(N),UNB(N)
      DIMENSION XIN(IN),YIN(IN),UIN(IN),UXIN(IN),UYIN(IN)
C
      WRITE(2,100)
  100 FORMAT('',79('*')//1X,'RESULTS'//2X,'BOUNDARY NODES'//
     1         11X,'X',15X,'Y',15X,'U',15X,'Un'/)
C
      DO 10 I=1,N
   10 WRITE(2,200) XM(I),YM(I),UB(I),UNB(I)
  200 FORMAT(4(2X,E14.5))
C
      WRITE (2,300)
  300 FORMAT(//,2X,'INTERNAL POINTS'//10X,'X',15X,'Y',14X,
     1 'U',14X,'UX',16X,'UY'/)
      DO 20 K=1,IN
C
   20 WRITE(2,400)  XIN(K),YIN(K),UIN(K),UXIN(K),UYIN(K)
  400 FORMAT(5(2X,E14.5))
      WRITE(2,500)
  500 FORMAT(' ',79('*'))
      RETURN
      END
C
C=====================================================================
```

EXAMPLE 6.6

Consider the steady laminar flow of an incompressible and nonviscous fluid through the tube of Fig. 6.16. Determine the velocity potential and the velocity

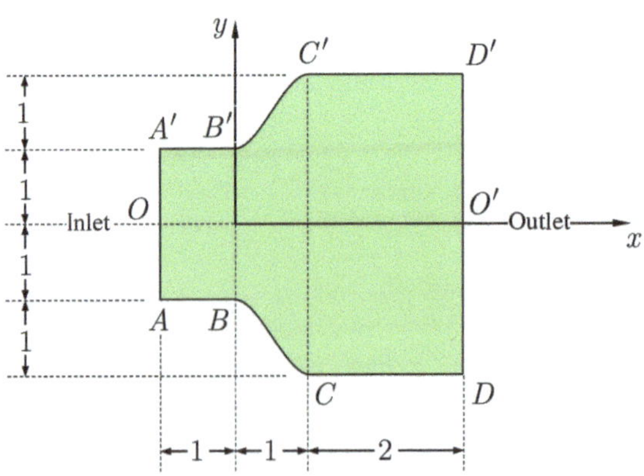

FIGURE 6.16 The tube segment in Example 6.6.

field in the tube. The shape of the curvilinear segments are determined by the equations:

$$\text{Segment } BC\text{: } y = 2x^3 - 3x^2 - 1$$

$$\text{Segment } B'C'\text{: } y = -2x^3 + 3x^2 + 1$$

The assumed boundary conditions are:

$$\text{at the inlet cross-section } AA'\text{: } \frac{\partial \phi}{\partial n} = v_n = -1$$

$$\text{at the outlet cross-section } DD'\text{: } \phi = 0$$

$$\text{at the rigid walls } ABCD \text{ and } A'B'C'D'\text{: } \frac{\partial \phi}{\partial n} = v_n = 0$$

The problem is solved using the program FLUIDCON. The boundary is divided into constant boundary elements in the following way: N1 elements on each of the segments OA and AB; 2N1 elements on each of the segments BC, CD, and DO'. The same discretization is employed for the corresponding symmetric segments above the x-axis. Thus, the total number of elements is N = 16N1.

Figure 6.17 depicts the distribution of the normal velocity component v_n at the outlet cross-section DD' for various values of N. We notice that the flow becomes sufficiently smooth at this cross-section, since it reaches the anticipated mean value $v_n = 0.5$, which is half of the velocity at the inlet cross-section AA'. Moreover, the distributions of the velocity components v_x and v_y at characteristic cross-sections of the tube are shown in

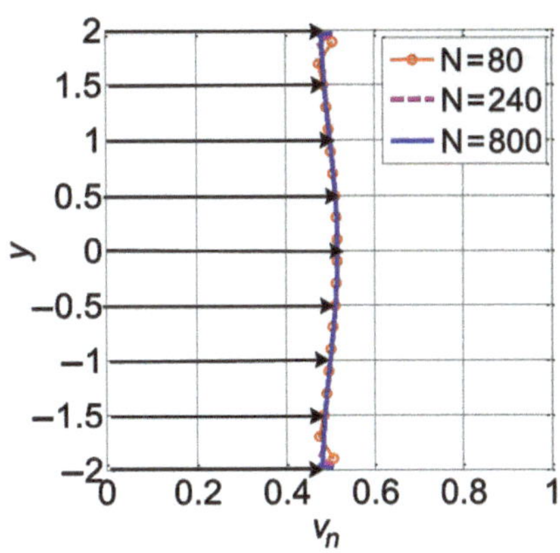

FIGURE 6.17 Distribution of the normal velocity component v_n at the outlet cross-section for various values of N in Example 6.6.

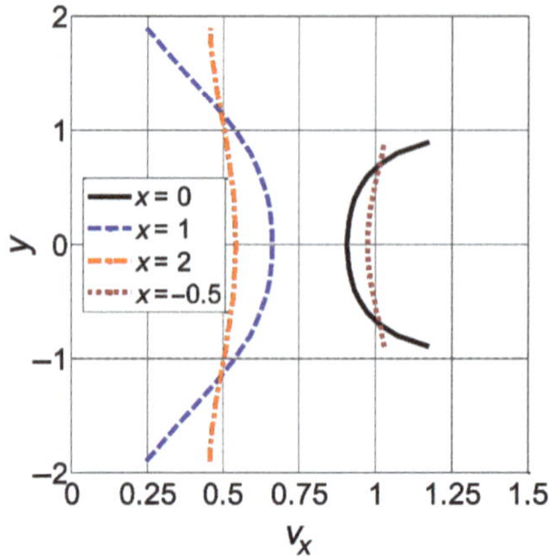

FIGURE 6.18 Distribution of the velocity component v_x at characteristic cross-sections of the tube in Example 6.6.

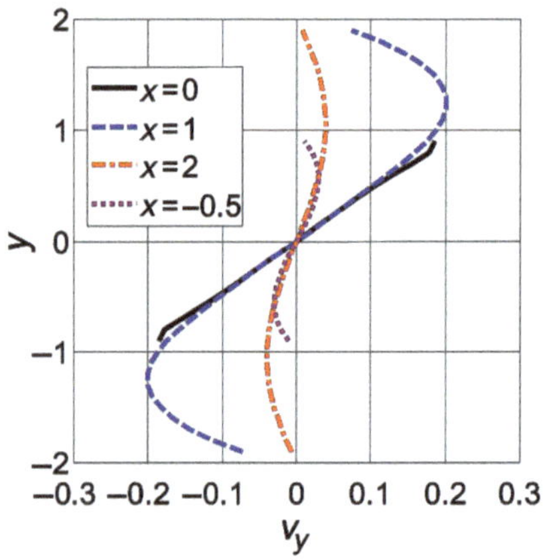

FIGURE 6.19 Distribution of the velocity component v_y at characteristic cross-sections of the tube in Example 6.6.

Figs. 6.18 and 6.19, while the contour plots of the potential are shown in Fig. 6.20.

6.7 CONCLUSIONS

In this chapter, we studied a variety of important engineering problems, which are formulated as boundary value problems for the Laplace or Poisson equation. Other problems described by these equations are, for example, the fluid flow through porous media (Darcy's law), diffusion of ions (Fick's law), electric potential in a body (Ohm's law), etc. For all these problems, the flux **q** is expressed by a law quite analogous to that of Fourier

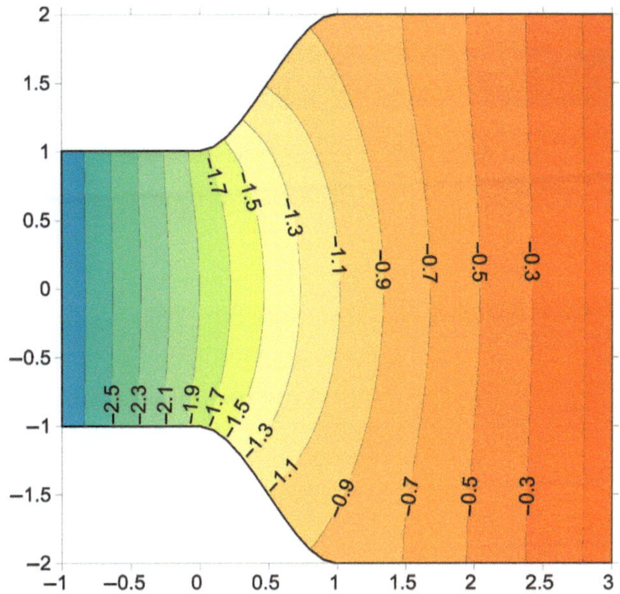

FIGURE 6.20 Contour plots of the velocity potential ϕ in Example 6.6.

TABLE 6.5 Examples of Physical Problems Described by the Laplace or Poisson Equation

Differential Equation	Physical Problem	Physical Quantities	Constitutive Relations
$\nabla \cdot (\nabla \phi) = 0$	Saint-Venant torsion of elastic bars	$\phi = \phi(x, y)$ = warping function	Hooke
			$\begin{Bmatrix} \tau_{xz} \\ \tau_{yz} \end{Bmatrix} = [D] \begin{Bmatrix} \gamma_{xz} \\ \gamma_{yz} \end{Bmatrix}$
			$\mathbf{D} = \mathbf{a}^{-1}$
$\nabla \cdot (S \nabla u) + f = 0$	Small deflections of membranes	$u = u(x, y)$ = deflection surface	$\mathbf{D} = S\,\mathbf{I}$
		$f = f(x, y)$ = transverse load	$\mathbf{I}$ = unit matrix
		S = constant tension force per unit length	
$\nabla \cdot (\mathbf{D} \cdot \nabla T) + f = 0$	Heat flow	$T = T(x, y)$ = temperature	Fourier
		$\mathbf{D}$ = constitutive matrix for thermal conductivity	$\mathbf{q} = -\mathbf{D} \cdot \nabla T$
		$f = f(x, y)$ = internal heat source density	$\mathbf{q}$ = heat flux vector

Continued

TABLE 6.5 continued

Differential Equation	Physical Problem	Physical Quantities	Constitutive Relations
$\nabla\cdot(\nabla\phi)+f=0$	Irrotational, incompressible and inviscid fluid flow	$\phi=\phi(x,y)=$ potential function of the velocity field	$\mathbf{v}=\nabla\phi$
		$f=$ internal heat source density	$\mathbf{v}=$ ***velocity***
			There is no constitutive law
$\nabla\cdot(\mathbf{D}\cdot\nabla\phi)+f=0$	Fluid flow through porous media	$f=$ internal source density	Darcy
		$\phi=$ piezometric head	$\mathbf{q}=-\mathbf{D}\cdot\nabla\phi$
		$\mathbf{D}=$ constitutive matrix for permeability coefficients	$\mathbf{q}=$ volume flux vector
$\nabla\cdot(\mathbf{D}\cdot\nabla V)+f=0$	Electric potential in bodies	$V=$ potential	Ohm
		$\mathbf{D}=$ constitutive matrix for electric conductivity	$\mathbf{q}=-\mathbf{D}\cdot\nabla V$
		$f=$ internal source density of electric charge	$\mathbf{q}=$ electric charge flux vector

(see Section 6.5). Of course, the field function u and the constitutive matrix $\mathbf{D}$ have a different physical meaning in each problem. Table 6.5 presents some of the problems mentioned above by giving the physical meaning of the involved quantities.

A general conclusion that can be drawn from the numerical examples presented in this chapter, is that the BEM with constant elements provides a very good numerical solution for problems described by the Laplace or the Poisson equations. The obtained accuracy is very good (see Examples 6.1, 6.2, 6.4). Moreover, the preparation of the data is very simple, since the discretization is limited only to the boundary.

6.8 REFERENCES

There is voluminous literature on the topics of this chapter. Therefore, we can only make reference to some well-known and commonly used books. The Saint-Venant torsion problem of noncircular prismatic bars is clearly treated in the books of Timoshenko and Goodier [1] and Muskhelishvili [2]. The reader, however, may supplement his knowledge on this subject by studying the books of Kollbrunner and Basler [15], Friemann [16], and Novozhilov [17]. For heat flow problems the interested reader is referred to the books of Carslaw and Jaeger [12] and Uzisik [13]. Finally, we mention the book of Hirsch for fluid flow problems [14].

[1] Timoshenko S, Goodier JN. Theory of elasticity. 2nd ed. New York: McGraw-Hill; 1951.
[2] Muskhelishvili NI. Some basic problems of the mathematical theory of elasticity. 4th ed. Groningen-The Netherlands: P. Noordhoff Ltd.; 1963.

[3] Sauer E. Schub und torsion bei elastischen prismatischen balken. Berlin: Verlag von Wilhelm Erst & Sohn; 1980.
[4] Lekhnitskii SG. Theory of elasticity of an anisotropic elastic body. San Francisco: Holden-Day; 1963.
[5] Timoshenko S, Woinowsky-Krieger S. Theory of plates and shells. New York: McGraw-Hill; 1959.
[6] Katsikadelis JT. *The boundary element method for plate analysis*. Oxford, UK: Academic Press, Elsevier; 2014.
[7] Katsikadelis JT. The analysis of plates on elastic foundation by the integral equation method. N.Y: Polytechnic University of New York; 1982. Ph.D. Dissertation.
[8] Marcus H. 1932. Die Theorie Elastischer Gewebe und ihre Anwendung auf die Berechnung biegsamer Platten. Julius Springer, Berlin.
[9] Paris F, De Leon S. Simply supported plates by the integral equation method. Int J Numer Methods Eng 1987;25:225–33.
[10] Katsikadelis JT. Special methods for plate analysis. In: Beskos D, editor. Boundary element analysis of plates and shells. Berlin: Springer-Verlag; 1991. p. 220–311.
[11] Bialecki R. Solving non-linear heat transfer problems using the boundary element method. In: Wrobel LC, Brebbia CA, editors. Boundary element methods in heat transfer. Southampton: Computational mechanics publications; 1992. p. 87–113. Chapter 4.
[12] Carslaw HS, Jaeger JC. Conduction of heat in solids. London: Oxford, University Press; 1959.
[13] Uzisik MN. Heat conduction. New York: John Wiley & Sons; 1980.
[14] Hirsch C. Numerical computation of internal and external flows, Vol. 1 & 2. Chichester, U.K: John Wiley & Sons; 1988.
[15] Kollbrunner C, Basler K. Torsion. Berlin: Springer-Verlag; 1966.
[16] Friemann H. Schub und torsion in geraden staeben. 2te Auflage Dusseldorf: Werner-Verlag; 1993.
[17] Novozhilov VV. Theory of elasticity. Oxford: Pergamon Student Editions; 1961.
[18] Roark RJ, Young WC. *Formulas for* $\overline{\tau} = \tau / G\theta r$ *stress and strain*. 5th ed. Kogakusha, Tokyo: McGraw-Hill, International Student Edition; 1975.

PROBLEMS

6.1. Compute the dimensionless torsional constant I_t/h^4 and distribution of stress $\tau_{tz}/G\theta h$ on the boundary of a bar having the cross-sections shown below ($\rho = h/4$). Also, compare the obtained results to those of analytical or approximate solutions [18].

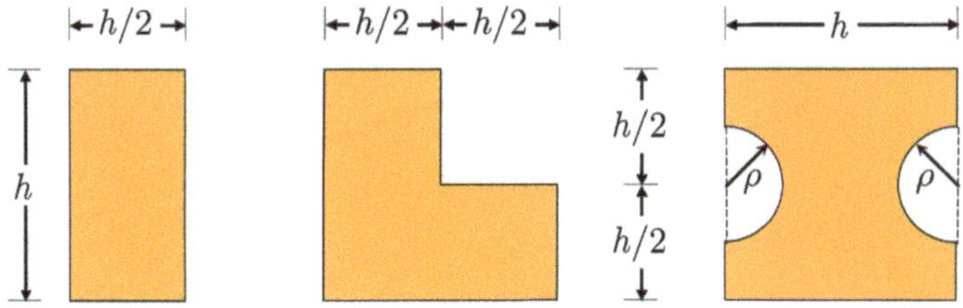

FIGURE P6.1

6.2. Determine the deflection surface of a rectangular membrane ($0 \le x \le a$, $0 \le y \le b$), which is subjected to hydrostatic pressure $f = q_o x/a$. It is given $a = 3.00$ m, $b = 2.00$ m, $q_o = 1$ kN/m^2, $S = 10$ kN/m.

6.3. Solve the problem of Example 6.4 using the whole domain of Fig. 6.9 (two boundaries) instead of using only the upper-right quadrant of Fig. 6.10 (one boundary).

6.4. Determine the temperature distribution in the cross-section of the T-beam insulated as shown in Fig. P6.4.

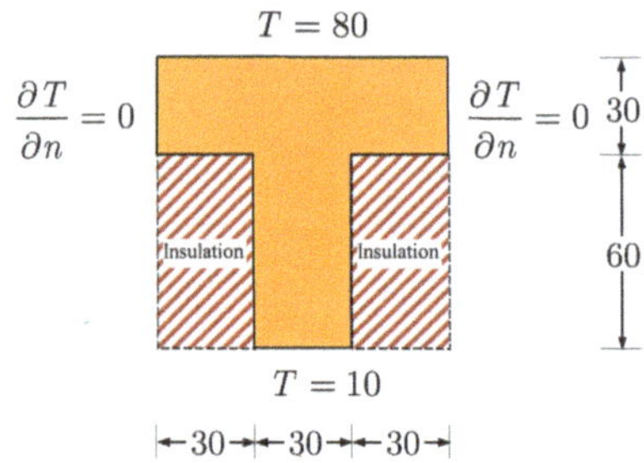

FIGURE P6.4

6.5. Solve the torsion problem for the composite bar of Fig. P6.5, if $G_2 = 2G_1$, where G_1 and G_2 are the shear moduli in the subregions Ω_1 and Ω_2, respectively. Compare the results with those of a homogeneous bar having the same dimensions and $G = 1.5G_1$.

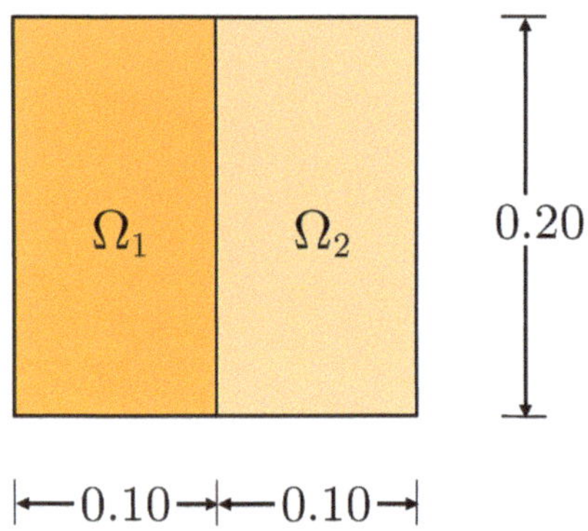

FIGURE P6.5

6.6. Solve the torsion problem for a bar consisting of an orthotropic material with $G_{yz} = 2G_{xz}$ and having square cross-section of side $a = 0.20$.

6.7. Determine the velocity distribution at the outlet cross-section of the tube in Fig. P6.7, if the constant velocity at the inlet cross-section is $v_n = -1$.

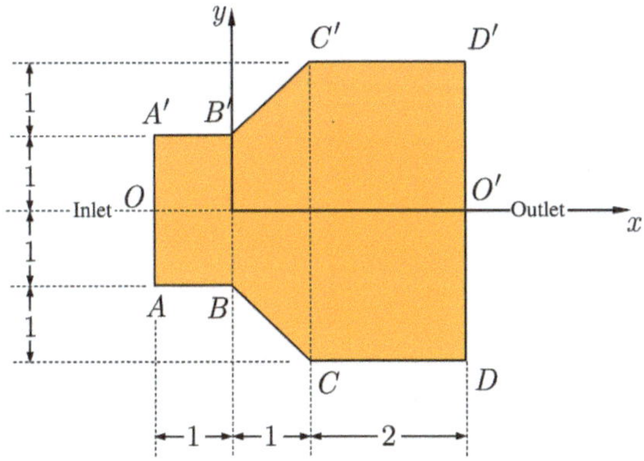

FIGURE P6.7

Chapter | Seven

The BEM for Two-Dimensional Elastostatic Problems

CHAPTER OUTLINE

The Boundary Element Method for Engineers and Scientists.

7.1 INTRODUCTION

This chapter presents the boundary element method (BEM) for the solution of linear elastostatic problems in two dimensions. The development of the BEM for the plane elastostatic problems is analogous to that for the plane potential problems discussed in previous chapters. There is, however, an essential difference. Here the problem is formulated in terms of two basic unknowns which are the two displacement components. Therefore, the resulting boundary integral equations are two and coupled, in contrast to the potential problems for which only one integral equation has to be solved. The consequence of this is that the establishment of the fundamental solution is much more complicated. Both problems of plane elasticity, namely the plane strain and the plane stress, are studied in this chapter. Several example problems are presented, which demonstrate that the BEM offers an efficient computational tool for solving difficult problems in engineering and mathematical physics.

7.2 EQUATIONS OF PLANE ELASTICITY

7.2.1 Plane strain

Plane strain in linear elasticity is considered the case for which:

a. One of the three displacement components, say w along the z-axis, is constant.

b. The other two displacements, u and v along the x- and y-axes, respectively, are functions only of the two variables, x and y.

This state of deformation appears in infinitely long (practically very long) prismatic or cylindrical bodies, whose axis coincides with the z-axis, and the

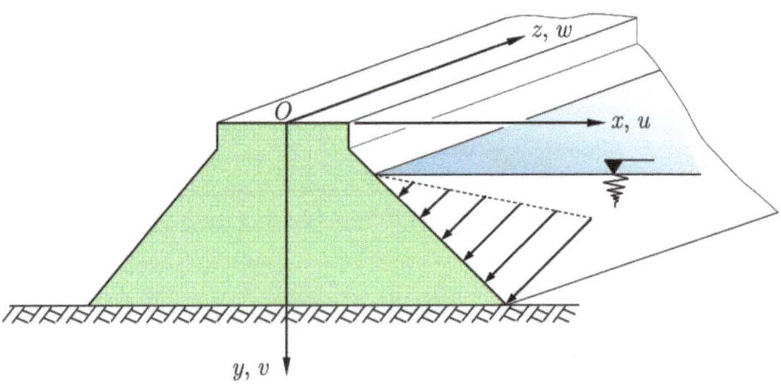

FIGURE 7.1 Cross-section of a long dam under plane strain.

loading is normal to this axis and independent of the z variable (Fig. 7.1). Another example of plane strain is the deformation that occurs in a plane through the axis of a circular cylinder (diametric plane) when the loading is axisymmetric and does not vary in the axial direction.

KINEMATIC RELATIONS

The previously mentioned conditions are stated mathematically as

$$w = C, u = u(x, y), v = v(x, y) \tag{7.1}$$

where C is an arbitrary constant.

Thus, the components of the strain tensor are [1,2]

$$\left.\begin{aligned} \varepsilon_x &= \frac{\partial u}{\partial x} \\ \varepsilon_y &= \frac{\partial v}{\partial y} \\ \gamma_{xy} &= \frac{\partial u}{\partial y} + \frac{\partial v}{\partial x} \\ \varepsilon_z &= 0, \quad \gamma_{xz} = 0, \quad \gamma_{yz} = 0 \end{aligned}\right\} \tag{7.2}$$

CONSTITUTIVE RELATIONS

Assuming linearly elastic and isotropic material, the constitutive relations for plane strain are

$$\left.\begin{aligned} \sigma_x &= \lambda(\varepsilon_x + \varepsilon_y) + 2\mu\varepsilon_x \\ \sigma_y &= \lambda(\varepsilon_x + \varepsilon_y) + 2\mu\varepsilon_y \\ \sigma_z &= \lambda(\varepsilon_x + \varepsilon_y) \\ \tau_{xy} &= \mu\gamma_{xy} \\ \tau_{xz} &= 0 \\ \tau_{yz} &= 0 \end{aligned}\right\} \tag{7.3}$$

where λ and μ are the Lamé constants, which are related to the elastic constants E, G, and ν through the following expressions

$$\mu = G = \frac{E}{2(1+\nu)}, \quad \lambda = \frac{\nu E}{(1+\nu)(1-2\nu)} \tag{7.4}$$

We readily conclude by combining Eqs. (7.2) and (7.3) that the nonvanishing components of the stress tensor are functions only of the variables x and y.

Equations (7.3) are solved for the strain components to yield

$$\varepsilon_x = \frac{1}{E}\left[\sigma_x - \nu(\sigma_y + \sigma_z)\right] \tag{7.5a}$$

$$\varepsilon_y = \frac{1}{E}\left[\sigma_y - \nu(\sigma_x + \sigma_z)\right] \tag{7.5b}$$

$$\sigma_z - \nu(\sigma_x + \sigma_y) = 0 \tag{7.5c}$$

$$\gamma_{xy} = \frac{1}{G}\tau_{xy} \tag{7.5d}$$

$$\gamma_{xz} = 0 \tag{7.5e}$$

$$\gamma_{yz} = 0 \tag{7.5f}$$

Equation (7.5c) gives

$$\sigma_z = \nu(\sigma_x + \sigma_y)$$

Substituting this expression for σ_z into Eqs. (7.5a,b), we obtain the following two equations for the normal strains

$$\varepsilon_x = \frac{1-\nu^2}{E}\left(\sigma_x - \frac{\nu}{1-\nu}\sigma_y\right)$$

$$\varepsilon_y = \frac{1-\nu^2}{E}\left(\sigma_y - \frac{\nu}{1-\nu}\sigma_x\right)$$

Setting

$$\overline{\nu} = \frac{\nu}{1-\nu} \quad \text{and} \quad \overline{E} = \frac{E}{1-\nu^2} \tag{7.6}$$

we find that

$$\mu = G = \frac{E}{2(1+\nu)} = \frac{\overline{E}}{2(1+\overline{\nu})} \quad \text{and} \quad \lambda = \frac{\overline{\nu}\,\overline{E}}{1-\overline{\nu}^2} \tag{7.7}$$

and the strain components may then be written as

$$\varepsilon_x = \frac{1}{\overline{E}}(\sigma_x - \overline{\nu}\sigma_y) \tag{7.8a}$$

$$\varepsilon_y = \frac{1}{\overline{E}}(\sigma_y - \overline{\nu}\sigma_x) \tag{7.8b}$$

$$\gamma_{xy} = \frac{2(1+\overline{\nu})}{\overline{E}}\tau_{xy} \tag{7.8c}$$

The elastic constants $\overline{E}$ and $\overline{\nu}$ are referred to as *effective elastic constants* for plain strain. As we will see later, the effective elastic constants allow to use equations of the same form for both plane strain and plane stress problems.

Equations (7.8a,b,c) are combined in a matrix equation as

$$\{\varepsilon\} = [S]\{\sigma\} \tag{7.9}$$

where $\{\varepsilon\}$ and $\{\sigma\}$ are referred to as the strain and stress vectors, respectively, and are defined as

$$\{\varepsilon\} = \left\{ \begin{array}{c} \varepsilon_x \\ \varepsilon_y \\ \gamma_{xy} \end{array} \right\} \tag{7.10}$$

and

$$\{\sigma\} = \left\{ \begin{array}{c} \sigma_x \\ \sigma_y \\ \tau_{xy} \end{array} \right\} \tag{7.11}$$

The matrix $[S]$ is known as the *compliance* or *flexibility matrix* and has the form

$$[S] = \frac{1}{\overline{E}} \begin{bmatrix} 1 & -\overline{\nu} & 0 \\ -\overline{\nu} & 1 & 0 \\ 0 & 0 & 2(1+\overline{\nu}) \end{bmatrix} \tag{7.12}$$

Solving Eq. (7.9) for the stress vector $\{\sigma\}$, we obtain

$$\{\sigma\} = [C]\{\varepsilon\} \tag{7.13}$$

where

$$[C] = [S]^{-1} = \frac{\overline{E}}{1-\overline{\nu}^2} \begin{bmatrix} 1 & \overline{\nu} & 0 \\ \overline{\nu} & 1 & 0 \\ 0 & 0 & \frac{1}{2}(1-\overline{\nu}) \end{bmatrix} \tag{7.14}$$

The matrix $[C]$ is known as the *stiffness matrix*.

The component form of Eq. (7.13) is

$$\left.\begin{aligned} \sigma_x &= \frac{\overline{E}}{1-\overline{\nu}^2}(\varepsilon_x + \overline{\nu}\varepsilon_y) \\ \sigma_y &= \frac{\overline{E}}{1-\overline{\nu}^2}(\varepsilon_y + \overline{\nu}\varepsilon_x) \\ \tau_{xy} &= \frac{\overline{E}}{2(1+\overline{\nu})}\gamma_{xy} \end{aligned}\right\} \tag{7.15}$$

EQUILIBRIUM EQUATIONS

The equilibrium equations for a three-dimensional body are [2]

$$\frac{\partial \sigma_x}{\partial x} + \frac{\partial \tau_{xy}}{\partial y} + \frac{\partial \tau_{xz}}{\partial z} + b_x = 0 \tag{7.16a}$$

$$\frac{\partial \tau_{xy}}{\partial x} + \frac{\partial \sigma_y}{\partial y} + \frac{\partial \tau_{yz}}{\partial z} + b_y = 0 \tag{7.16b}$$

$$\frac{\partial \tau_{xz}}{\partial x} + \frac{\partial \tau_{yz}}{\partial y} + \frac{\partial \sigma_z}{\partial z} + b_z = 0 \tag{7.16c}$$

where b_x, b_y, and b_z are the components of the body force per unit volume. For the plane problem, it is

$$b_x = b_x(x, y), \quad b_y = b_y(x, y), \quad b_z = 0 \tag{7.17}$$

Taking into account that the stress components are independent of z and using Eq. (7.3), we readily conclude that the Eq. (7.16c) is identically satisfied, while the first two become

$$\left.\begin{aligned} \frac{\partial \sigma_x}{\partial x} + \frac{\partial \tau_{xy}}{\partial y} + b_x = 0 \\ \frac{\partial \tau_{xy}}{\partial x} + \frac{\partial \sigma_y}{\partial y} + b_y = 0 \end{aligned}\right\} \tag{7.18}$$

Substituting Eqs. (7.15) into Eqs. (7.18) and using Eqs. (7.2), we derive the equilibrium equations in terms of the displacement components as

$$\left.\begin{aligned} \nabla^2 u + \frac{1+\overline{\nu}}{1-\overline{\nu}}\left(\frac{\partial^2 u}{\partial x^2} + \frac{\partial^2 v}{\partial x \partial y}\right) + \frac{1}{G} b_x = 0 \\ \nabla^2 v + \frac{1+\overline{\nu}}{1-\overline{\nu}}\left(\frac{\partial^2 u}{\partial x \partial y} + \frac{\partial^2 v}{\partial y^2}\right) + \frac{1}{G} b_y = 0 \end{aligned}\right\} \text{ in } \Omega \tag{7.19}$$

or substituting $\overline{\nu}$ from Eqs. (7.6), we obtain the governing equations for the plane strain problem in the form

$$\left.\begin{aligned} \nabla^2 u + \frac{1}{1-2\nu}\left(\frac{\partial^2 u}{\partial x^2} + \frac{\partial^2 v}{\partial x \partial y}\right) + \frac{1}{G} b_x = 0 \\ \nabla^2 v + \frac{1}{1-2\nu}\left(\frac{\partial^2 u}{\partial x \partial y} + \frac{\partial^2 v}{\partial y^2}\right) + \frac{1}{G} b_y = 0 \end{aligned}\right\} \text{ in } \Omega \tag{7.20}$$

Equations (7.20) are known as the *Navier equations of equilibrium* for the plane elastostatic problem in a body occupying the two-dimensional domain Ω.

BOUNDARY CONDITIONS

The solution of Eqs. (7.19) must satisfy prescribed boundary conditions on the boundary Γ of the body, which are imposed either on the displacements u and v, or on the boundary tractions t_x and t_y. The boundary conditions can be classified into the following four types:

$$\left.\begin{array}{llll}\text{(i)} & u=\bar{u}, & v=\bar{v} & \text{on} \quad \Gamma_1 \\ \text{(ii)} & u=\bar{u}, & t_y=\bar{t}_y & \text{on} \quad \Gamma_2 \\ \text{(iii)} & t_x=\bar{t}_x, & v=\bar{v} & \text{on} \quad \Gamma_3 \\ \text{(iv)} & t_x=\bar{t}_x, & t_y=\bar{t}_y & \text{on} \quad \Gamma_4\end{array}\right\} \tag{7.21}$$

where $\Gamma=\Gamma_1 \cup \Gamma_2 \cup \Gamma_3 \cup \Gamma_4$. The prescribed quantities are designated by an over-bar. Of course, any of the boundary parts Γ_1, Γ_2, Γ_3, Γ_4 may be identical to the whole boundary Γ, that is, the boundary conditions may be only of one type. The boundary conditions are mixed, if different boundary conditions are prescribed over two or more parts of the boundary. Attention should be paid when $\Gamma_4=\Gamma$. In this case, the boundary tractions $\bar{t}_x$ and $\bar{t}_y$ cannot be prescribed arbitrarily, but they must ensure the overall equilibrium of the body, namely

$$\int_\Omega b_x \, d\Omega + \int_\Gamma t_x \, ds = 0$$

$$\int_\Omega b_y \, d\Omega + \int_\Gamma t_y \, ds = 0$$

$$\int_\Omega (xb_y - yb_x) d\Omega + \int_\Gamma (xt_y - yt_x) ds = 0$$

For this type of boundary conditions, the solution of the Navier equations is not uniquely determined as it contains an arbitrary rigid body motion. An example of boundary conditions for an infinitely long body of rectangular cross-section is shown in Fig. 7.2.

The relations between the boundary tractions t_x, t_y and the stress components σ_x, σ_y, and τ_{xy} may be derived from the equilibrium of an infinitesimal body element on the boundary (*Cauchy tetrahedron* [2]). In two dimensions these relations read

$$\left.\begin{array}{l} t_x = \sigma_x n_x + \tau_{xy} n_y \\ t_y = \tau_{xy} n_x + \sigma_y n_y \end{array}\right\} \tag{7.22}$$

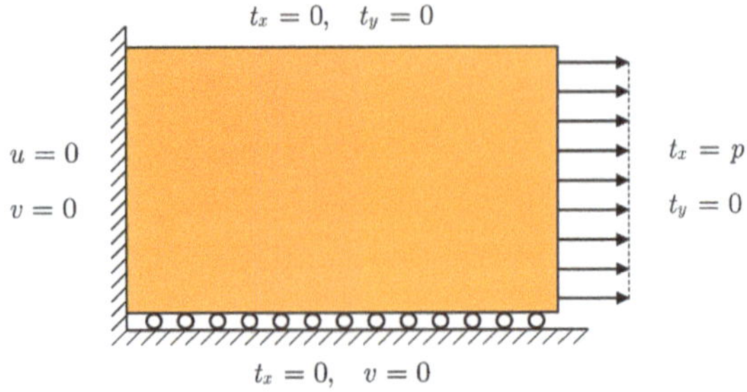

FIGURE 7.2 Support and loading conditions on the boundary of a rectangular plane body.

where n_x and n_y are the direction cosines of the unit vector normal to the boundary. If the stress components in Eqs. (7.22) are replaced by their expressions in Eqs. (7.3) and then the kinematic relations (7.2) are employed, we arrive at the following displacement-based expressions for the boundary tractions

$$\left.\begin{aligned} t_x &= \lambda\left(\frac{\partial u}{\partial x} + \frac{\partial v}{\partial y}\right)n_x + \mu\left(\frac{\partial u}{\partial x}n_x + \frac{\partial v}{\partial x}n_y\right) + \mu\frac{\partial u}{\partial n} \\ t_y &= \lambda\left(\frac{\partial u}{\partial x} + \frac{\partial v}{\partial y}\right)n_y + \mu\left(\frac{\partial u}{\partial y}n_x + \frac{\partial v}{\partial y}n_y\right) + \mu\frac{\partial v}{\partial n} \end{aligned}\right\} \tag{7.23}$$

INITIAL STRESSES AND STRAINS

In many problems initial state of stress or strain may be present, which are due to temperature variations or other causes. For instance, we consider an initial state of strain, whose components are denoted by

$$\{\varepsilon_0\} = \begin{Bmatrix} \varepsilon_x^0 \\ \varepsilon_y^0 \\ \gamma_{xy}^0 \end{Bmatrix} \tag{7.24}$$

Denoting by $\{\varepsilon_t\}$ the total strain, the elastic strain $\{\varepsilon_e\}$ is obtained by subtracting the initial strain from the total one, that is,

$$\{\varepsilon_e\} = \{\varepsilon_t\} - \{\varepsilon_0\} \tag{7.25}$$

By means of Eq. (7.13), we obtain

$$\{\sigma_e\} = [C]\{\varepsilon_e\} = [C](\{\varepsilon_t\} - \{\varepsilon_0\})$$

or

$$\{\sigma_e\} = [C]\{\varepsilon_t\} - [C]\{\varepsilon_0\} = \{\sigma_t\} - \{\sigma_0\} \tag{7.26}$$

where the stresses $\{\sigma_0\} = [C]\{\varepsilon_0\}$ are the initial stresses.

When the initial strain is due to a temperature variation, we have

$$\{\varepsilon_0\} = \alpha \Delta T \begin{Bmatrix} 1 \\ 1 \\ 0 \end{Bmatrix} \tag{7.27}$$

where α is the coefficient of thermal dilatation and ΔT the temperature change. The resulting initial stresses are

$$\{\sigma_0\} = [C]\{\varepsilon_0\} = \frac{\overline{E}\alpha\Delta T}{1-\overline{\nu}} \begin{Bmatrix} 1 \\ 1 \\ 0 \end{Bmatrix} \tag{7.28}$$

7.2.2 Plane stress

The theory of plane stress is another application of the plane theory of elasticity of great practical significance. It describes the state of stress, which is produced in thin plane bodies subjected to in-plane loads. Such a state of stress occurs in *shear walls*, *thin plates* under combined transverse and in-plane loads, *large deflections* of thin plates, etc. [3]. The state of plane stress is described as follows.

We consider a thin elastic body, namely a body whose thickness h is very small compared to the other two dimensions (Fig. 7.3).

The loading is due to the body forces b_x, b_y and the boundary tractions t_x, t_y. The tractions are assumed to be symmetrically distributed with respect

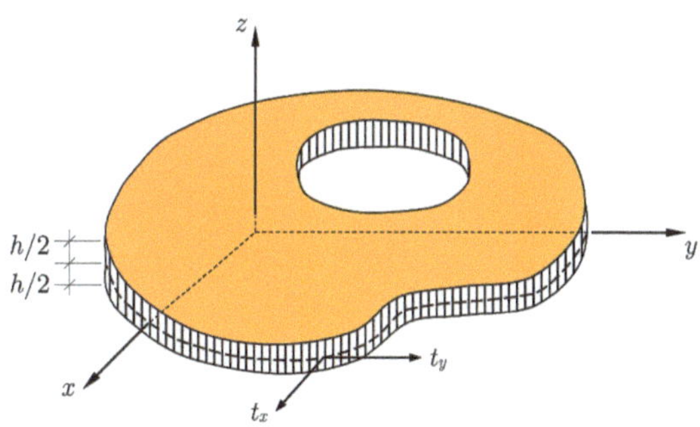

FIGURE 7.3 Thin elastic body.

to the mid-plane of the body, although more often the variation along the thickness h is considered to be constant. In this case the resulting state of stress does not depend on z, but, if the thickness h is very small, it is quite accurate to assume [2] that

$$\sigma_z = 0, \quad \tau_{xz} = 0, \quad \tau_{yz} = 0$$

through the thickness h, whereas the remaining stress components do not depend on the variable z, that is,

$$\sigma_x = \sigma_x(x, y), \quad \sigma_y = \sigma_y(x, y), \quad \tau_{xy} = \tau_{xy}(x, y)$$

Thus, the constitutive equations become

$$\left.\begin{aligned} \varepsilon_x &= \frac{1}{E}(\sigma_x - \nu\sigma_y) \\ \varepsilon_y &= \frac{1}{E}(\sigma_y - \nu\sigma_x) \\ \gamma_{xy} &= \frac{2(1+\nu)}{E}\tau_{xy} \end{aligned}\right\} \tag{7.29}$$

and the equilibrium equations are reduced to two as in the case of plane strain,

$$\left.\begin{aligned} \frac{\partial\sigma_x}{\partial x} + \frac{\partial\tau_{xy}}{\partial y} + b_x = 0 \\ \frac{\partial\tau_{xy}}{\partial x} + \frac{\partial\sigma_y}{\partial y} + b_y = 0 \end{aligned}\right\} \tag{7.30}$$

We notice that Eqs. (7.29) and (7.30) are identical in form to Eqs. (7.8a, b,c) and (7.18), respectively. Therefore, all equations for plane stress can be obtained by the respective equations of plane strain, if the effective elastic constants $\overline{\nu}$ and $\overline{E}$ are replaced with the actual ones ν and E. Thus, we have:

Elastic constants

$$\mu = G = \frac{E}{2(1+\nu)} \tag{7.31a}$$

$$\lambda^* = \frac{\nu E}{1-\nu^2} \tag{7.31b}$$

where λ^* plays the role of the Lamé constant.

Constitutive equations

$$\{\varepsilon\} = [S]\{\sigma\} \tag{7.32}$$

$$\{\sigma\} = [C]\{\varepsilon\} \tag{7.33}$$

$$[S] = \frac{1}{E}\begin{bmatrix} 1 & -\nu & 0 \\ -\nu & 1 & 0 \\ 0 & 0 & 2(1+\nu) \end{bmatrix} \tag{7.34}$$

$$[C] = \frac{E}{1-\nu^2}\begin{bmatrix} 1 & \nu & 0 \\ \nu & 1 & 0 \\ 0 & 0 & \frac{1}{2}(1-\nu) \end{bmatrix} \tag{7.35}$$

Navier's equations of equilibrium

$$\left.\begin{aligned} \nabla^2 u + \frac{1+\nu}{1-\nu}\left(\frac{\partial^2 u}{\partial x^2} + \frac{\partial^2 v}{\partial x \partial y}\right) + \frac{1}{G}b_x = 0 \\ \nabla^2 v + \frac{1+\nu}{1-\nu}\left(\frac{\partial^2 u}{\partial x \partial y} + \frac{\partial^2 v}{\partial y^2}\right) + \frac{1}{G}b_y = 0 \end{aligned}\right\} \tag{7.36}$$

Boundary tractions

$$\left.\begin{aligned} t_x = \lambda^*\left(\frac{\partial u}{\partial x} + \frac{\partial v}{\partial y}\right)n_x + \mu\left(\frac{\partial u}{\partial x}n_x + \frac{\partial v}{\partial x}n_y\right) + \mu\frac{\partial u}{\partial n} \\ t_y = \lambda^*\left(\frac{\partial u}{\partial x} + \frac{\partial v}{\partial y}\right)n_y + \mu\left(\frac{\partial u}{\partial y}n_x + \frac{\partial v}{\partial y}n_y\right) + \mu\frac{\partial v}{\partial n} \end{aligned}\right\} \tag{7.37}$$

Initial stresses due to temperature variation

$$\{\sigma_0\} = \frac{E\alpha\Delta T}{1-\nu}\begin{Bmatrix} 1 \\ 1 \\ 0 \end{Bmatrix} \tag{7.38}$$

7.3 BETTI'S RECIPROCAL IDENTITY

The derivation of the integral representation of the solution for two-dimensional elasticity problems requires the establishment of a reciprocal identity for the Navier operator, Eq. (7.19), similar to that of Green's for the Laplace operator, Eq. (2.16). Betti's reciprocal identity plays this role. It can be readily derived from the known Betti's theorem for the reciprocity of works, which is valid for the linear theory of elasticity. For this purpose, we consider an elastic body occupying the volume V of the three-dimensional space and bounded by the surface S. Moreover, we consider two states of stress due to two different distributions of body forces and boundary quantities (displacements and tractions). We designate the displacements, body forces and tractions for the two states of stress as follows:

State I:

$$\mathbf{u} = \begin{Bmatrix} u \\ v \\ w \end{Bmatrix}, \mathbf{b} = \begin{Bmatrix} b_x \\ b_y \\ b_z \end{Bmatrix}, \mathbf{t} = \begin{Bmatrix} t_x \\ t_y \\ t_z \end{Bmatrix} \tag{7.39}$$

State II:

$$\mathbf{u}^* = \begin{Bmatrix} u^* \\ v^* \\ w^* \end{Bmatrix}, \mathbf{b}^* = \begin{Bmatrix} b_x^* \\ b_y^* \\ b_z^* \end{Bmatrix}, \mathbf{t}^* = \begin{Bmatrix} t_x^* \\ t_y^* \\ t_z^* \end{Bmatrix} \tag{7.40}$$

According to Betti's theorem, the work produced by the displacements of state (I) and the forces of state (II) is equal to the work produced by the displacements of state (II) and the forces of state (I). This may be expressed as

$$\int_V \mathbf{u}\cdot\mathbf{b}^* dV + \int_S \mathbf{u}\cdot\mathbf{t}^* dS = \int_V \mathbf{u}^*\cdot\mathbf{b}\, dV + \int_S \mathbf{u}^*\cdot\mathbf{t}\, dS \tag{7.41}$$

or using Eqs. (7.39) and (7.40)

$$\begin{aligned} &\int_V \left(ub_x^* + vb_y^* + wb_z^*\right) dV + \int_S \left(ut_x^* + vt_y^* + wt_z^*\right) dS \\ &= \int_V \left(u^* b_x + v^* b_y + w^* b_z\right) dV + \int_S \left(u^* t_x + v^* t_y + w^* t_z\right) dS \end{aligned} \tag{7.42}$$

For the plane problem we distinguish the following two cases:

Plain strain. We consider the part of the cylindrical body cut by the two planes z and $z+1$, that is, a slice of unit thickness. In this case, it is $b_z = 0$, $b_z^* = 0$ inside the body, $t_z = 0$, $t_z^* = 0$ on the cylindrical surface and by virtue of Eqs. (7.22) or (7.23) it is $t_x = t_y = 0$, $t_x^* = t_y^* = 0$ on the plane sections. Moreover, taking into account that $w = c$, $w^* = c^*$ on the plane sections and, thus, the works produced on them are of equal magnitude but opposite sign, Eq. (7.42) becomes

$$\begin{aligned}&\int_\Omega \left(ub_x^* + vb_y^*\right) d\Omega + \int_\Gamma \left(ut_x^* + vt_y^*\right) ds\\&= \int_\Omega \left(u^* b_x + v^* b_y\right) d\Omega + \int_\Gamma \left(u^* t_x + v^* t_y\right) ds\end{aligned} \tag{7.43}$$

Plane stress. We consider a thin plate of constant thickness h. In this case, it is $b_z = 0$, $b_z^* = 0$ inside the body and $t_z = 0$, $t_z^* = 0$ on the whole surface S and by virtue of Eqs. (7.37) it is $t_x = t_y = 0$, $t_x^* = t_y^* = 0$ on the plane sections. Hence, also Eq. (7.42) takes the form of Eq. (7.43) when applied to this body.

Furthermore, if the body forces in Eq. (7.43) are replaced by their expressions from Eqs. (7.19), we obtain the reciprocal identity for the Navier operator

$$\begin{aligned}&\int_\Omega \left\{\left[uN_x(u^*, v^*) + vN_y(u^*, v^*)\right] - \left[u^* N_x(u, v) + v^* N_y(u, v)\right]\right\} d\Omega\\&= -\int_\Gamma \left[\left(ut_x^* + vt_y^*\right) - \left(u^* t_x + v^* t_y\right)\right] ds\end{aligned} \tag{7.44}$$

where the operators $N_x(.,.)$ and $N_y(.,.)$ are defined on the basis of Eqs. (7.19) as:

$$\left.\begin{aligned}N_x(u, v) &= -G\left[\nabla^2 u + \frac{1+\bar{\nu}}{1-\bar{\nu}}\left(\frac{\partial^2 u}{\partial x^2} + \frac{\partial^2 v}{\partial x \partial y}\right)\right]\\N_y(u, v) &= -G\left[\nabla^2 v + \frac{1+\bar{\nu}}{1-\bar{\nu}}\left(\frac{\partial^2 u}{\partial x \partial y} + \frac{\partial^2 v}{\partial y^2}\right)\right]\end{aligned}\right\} \tag{7.45}$$

Equations (7.45) are valid for plane strain. We recall that for plane stress $\bar{\nu}$ should be replaced with ν.

7.4 FUNDAMENTAL SOLUTION

In order to derive the boundary integral equations pertaining to the plane elastostatic problem, it is necessary to establish first the fundamental solution of the Navier equations, Eqs. (7.19). From the physical point of view, the fundamental solution expresses the displacements produced in an infinite plane body by a concentrated unit body force. This solution is attributed to Kelvin and this is why it is known in the literature as *Kelvin's solution.* It can be established using the procedure described below.

Consider the concentrated force $\mathbf{F}(F_\xi, F_\eta)$, $|\mathbf{F}| = 1$, applied at point $Q(\xi, \eta)$ of the plane (see Fig. 7.4). It is apparent that the components F_ξ and F_η of the force $\mathbf{F}$ are the direction cosines of the unit vector representing this

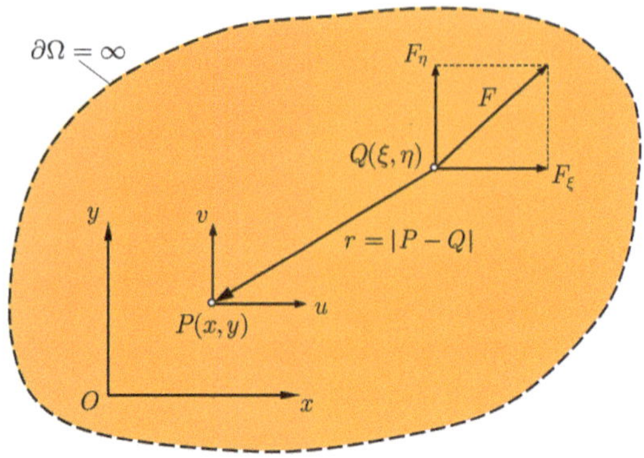

FIGURE 7.4 Concentrated unit force $\mathbf{F}(F_\xi, F_\eta)$ applied at point $Q(\xi, \eta)$ of the plane.

force. The density of the body forces produced by the force **F** at a point $P(x, y)$ can be represented using the delta function. Thus, we have:

$$\mathbf{b} = \delta(P - Q)\mathbf{F} \tag{7.46}$$

or

$$\left.\begin{aligned} b_x &= \delta(P - Q)F_\xi \\ b_y &= \delta(P - Q)F_\eta \end{aligned}\right\} \tag{7.47}$$

In this case, Eqs. (7.19) are written as

$$\left.\begin{aligned} \nabla^2 u + \frac{1+\overline{\nu}}{1-\overline{\nu}}\left(\frac{\partial^2 u}{\partial x^2} + \frac{\partial^2 v}{\partial x \partial y}\right) + \frac{1}{G}\delta(P - Q)F_\xi = 0 \\ \nabla^2 v + \frac{1+\overline{\nu}}{1-\overline{\nu}}\left(\frac{\partial^2 u}{\partial x \partial y} + \frac{\partial^2 v}{\partial y^2}\right) + \frac{1}{G}\delta(P - Q)F_\eta = 0 \end{aligned}\right\} \tag{7.48}$$

The fundamental solution for the Navier operator is a singular solution of Eqs. (7.48), which can be established by expressing the displacement components in terms of the *Galerkin functions*. Thus, we set

$$\left.\begin{aligned} 2Gu &= \frac{2}{1+\overline{\nu}}\nabla^2\phi - \frac{\partial}{\partial x}\left(\frac{\partial \phi}{\partial x} + \frac{\partial \psi}{\partial y}\right) \\ 2Gv &= \frac{2}{1+\overline{\nu}}\nabla^2\psi - \frac{\partial}{\partial y}\left(\frac{\partial \phi}{\partial x} + \frac{\partial \psi}{\partial y}\right) \end{aligned}\right\} \tag{7.49}$$

where $\phi = \phi(x, y)$ and $\psi = \psi(x, y)$ are the Galerkin functions. They represent the components of a vector, the so-called *Galerkin vector* [4].

Introducing Eqs. (7.49) into the equilibrium Eq. (7.48), we arrive at

$$\left.\begin{aligned}\nabla^4\phi &= -(1+\overline{\nu})\,\delta(P-Q)F_\xi\\ \nabla^4\psi &= -(1+\overline{\nu})\,\delta(P-Q)F_\eta\end{aligned}\right\} \tag{7.50}$$

where

$$\nabla^4 = \nabla^2\nabla^2 = \frac{\partial^4}{\partial x^4} + 2\frac{\partial^2}{\partial x^2}\frac{\partial^2}{\partial y^2} + \frac{\partial^4}{\partial y^4}$$

is the biharmonic operator.

Hence, Eqs. (7.49) are a solution to Eqs. (7.48), if the functions ϕ and ψ represent singular particular solutions of Eqs. (7.50). These solutions can be established by working as follows:

The first of Eqs. (7.50) is written as

$$\nabla^2\Phi = -(1+\overline{\nu})\,\delta(P-Q)F_\xi \tag{7.51}$$

where it was set

$$\nabla^2\phi = \Phi \tag{7.52}$$

Equation (7.51) has the form of Eq. (3.8). Thus, a singular particular solution of this equation is

$$\Phi = -\frac{(1+\overline{\nu})}{2\pi}(\ln r + B)F_\xi$$

where $r = |P-Q|$ and B is an arbitrary constant. Consequently, Eq. (7.52) becomes

$$\nabla^2\phi = -\frac{(1+\overline{\nu})}{2\pi}(\ln r + B)F_\xi \tag{7.53}$$

Since the solution is independent of the polar angle at the source point Q, one can write for $r \neq 0$

$$\frac{1}{r}\frac{d}{dr}\left(r\frac{d\phi}{dr}\right) = -\frac{(1+\overline{\nu})}{2\pi}(\ln r + B)F_\xi$$

Two consecutive integrations result in

$$\phi = -\frac{(1+\overline{\nu})}{2\pi}F_\xi\left[\frac{r^2}{4}\ln r + \frac{1}{4}(B-1)r^2 + C\ln r + D\right] \tag{7.54}$$

where C and D are also arbitrary constants. Since we are interested for a particular solution, we can set $B=1$ and $C=D=0$ so that the solution has the simplest form. Thus, we have

$$\phi=-\frac{(1+\overline{\nu})}{8\pi}F_\xi\, r^2\ln r \tag{7.55}$$

Similarly, we obtain

$$\psi=-\frac{(1+\overline{\nu})}{8\pi}F_\eta\, r^2\ln r \tag{7.56}$$

Substitution of Eqs. (7.55) and (7.56) into expressions (7.49) yields the fundamental solution for the Navier equations. As we will see in the next section, the fundamental solution for (1) $F_\xi=1$, $F_\eta=0$ and (2) $F_\xi=0$, $F_\eta=1$ are required for deriving the boundary integral equations. Therefore, the derivation of the expressions for the fundamental solution is limited to these two cases.

(i) $F_\xi=1,\ F_\eta=0$

We readily have

$$\left.\begin{aligned}
&\phi=-\frac{1+\overline{\nu}}{8\pi}r^2\ln r\\
&\psi=0\\
&\nabla^2\phi=-\frac{1+\overline{\nu}}{2\pi}(\ln r+1)\\
&\frac{\partial^2\phi}{\partial x^2}=-\frac{1+\overline{\nu}}{8\pi}\left(2\ln r+2r^2{}_{,x}+1\right)\\
&\frac{\partial^2\phi}{\partial x\partial y}=-\frac{1+\overline{\nu}}{8\pi}2r_{,x}\,r_{,y}
\end{aligned}\right\} \tag{7.57}$$

In the previous expressions and in what follows $r_{,x}$ and $r_{,y}$ express the derivatives of the distance r with respect to x and y, respectively, and are given as (see Appendix A)

$$r_{,x}=-\frac{\xi-x}{r},\qquad r_{,y}=-\frac{\eta-y}{r}$$

which obviously satisfy the relation $r^2{}_{,x}+r^2{}_{,y}=1$.

Equations (7.57) are substituted into expressions (7.49) to yield

$$\left.\begin{aligned}
&U_{x\xi}=-\frac{1}{8\pi G}\left[(3-\overline{\nu})\ln r-(1+\overline{\nu})r^2{}_{,x}+\frac{7-\overline{\nu}}{2}\right]\\
&U_{y\xi}=\frac{1}{8\pi G}(1+\overline{\nu})r_{,x}r_{,y}
\end{aligned}\right\} \tag{7.58}$$

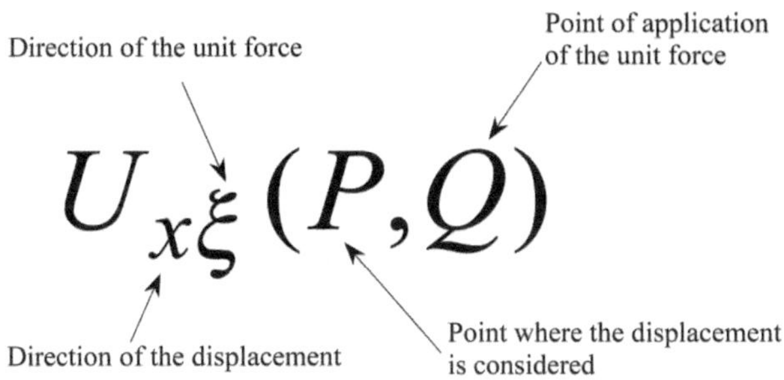

FIGURE 7.5 Component of the two-dimensional elasticity fundamental solution.

where $U_{x\xi}$ and $U_{y\xi}$ represent the displacements u and v, respectively. The first subscript in U denotes the direction of the displacement, whereas the second one the direction of the unit force (see Fig. 7.5).

(ii) $F_\xi = 0$, $F_\eta = 1$

Following the same procedure as in case (1), we obtain

$$\left.\begin{aligned} &\phi = 0 \\ &\psi = -\frac{1+\bar{\nu}}{8\pi} r^2 \ln r \\ &\nabla^2\psi = -\frac{1+\bar{\nu}}{2\pi}(\ln r + 1) \\ &\frac{\partial^2\psi}{\partial x \partial y} = -\frac{1+\bar{\nu}}{8\pi} 2 r_{,x} r_{,y} \\ &\frac{\partial^2\psi}{\partial y^2} = -\frac{1+\bar{\nu}}{8\pi}\left(2\ln r + 2 r^2_{,y} + 1\right) \end{aligned}\right\} \tag{7.59}$$

and substituting Eqs. (7.59) into the Eqs. (7.49), we find

$$\left.\begin{aligned} U_{x\eta} &= \frac{1}{8\pi G}(1+\bar{\nu}) r_{,x} r_{,y} \\ U_{y\eta} &= -\frac{1}{8\pi G}\left[(3-\bar{\nu})\ln r - (1+\bar{\nu}) r^2_{,y} + \frac{7-\bar{\nu}}{2}\right] \end{aligned}\right\} \tag{7.60}$$

Employing indicial notation for the coordinates of points P and Q, that is, x_1, x_2 and ξ_1, ξ_2, respectively, Eqs. (7.58) and (7.60) may be written as

$$U_{ij} = -\frac{1}{8\pi G}\left[C_1 \delta_{ij} \ln r - C_2 r_{,i} r_{,j} + \delta_{ij} C_3\right] \tag{7.61}$$

where

$$C_1 = 3 - \bar{\nu}, \quad C_2 = 1 + \bar{\nu}, \quad C_3 = \frac{7-\bar{\nu}}{2} \tag{7.62}$$

The constant C_3 produces only rigid body displacements, which do not influence the stresses and the strains.

Equations (7.61) may also be written in matrix form as

$$\mathbf{U}(P,Q)=\begin{bmatrix} U_{x\xi} & U_{x\eta} \\ U_{y\xi} & U_{y\eta} \end{bmatrix}=\begin{bmatrix} U_{11} & U_{12} \\ U_{21} & U_{22} \end{bmatrix} \tag{7.63}$$

The components of the fundamental solution represent the components of a *two-point* second order tensor [4,5] known as *Green's tensor* [6].

Since it is (see Appendix A)

$$r=\sqrt{(x-\xi)^2+(y-\eta)^2}, \quad r_{,x}=-r_{,\xi}, \quad r_{,y}=-r_{,\eta}$$

one can easily find out that the tensor defined above is symmetric with respect to points P and Q. This means that its components do not change, if the role of the two points is interchanged, namely, P becomes the point where the force is applied (source point) and Q the point where the displacement is evaluated (field point). Thus, it is

$$\mathbf{U}(P,Q)=\mathbf{U}(Q,P)$$

or

$$\begin{bmatrix} U_{x\xi} & U_{x\eta} \\ U_{y\xi} & U_{y\eta} \end{bmatrix}=\begin{bmatrix} U_{\xi x} & U_{\xi y} \\ U_{\eta x} & U_{\eta y} \end{bmatrix} \tag{7.64}$$

This symmetry expresses the Betti–Maxwell law for the *reciprocity of the displacements*.

7.5 STRESSES DUE TO A UNIT CONCENTRATED FORCE

The expressions for the stress components due to the unit force can be evaluated using Eqs. (7.3). We distinguish again two cases:

(i) $F_\xi=1,\ F_\eta=0$

$$\left.\begin{aligned} \sigma_{x\xi}&=\lambda(U_{x\xi,x}+U_{y\xi,y})+2\mu U_{x\xi,x} \\ \sigma_{y\xi}&=\lambda(U_{x\xi,x}+U_{y\xi,y})+2\mu U_{y\xi,y} \\ \tau_{xy\xi}&=\mu(U_{x\xi,y}+U_{y\xi,x}) \end{aligned}\right\} \tag{7.65}$$

Substituting Eqs. (7.58) into the above equations, yields

$$\left.\begin{aligned}\sigma_{x\xi} &= \frac{A_1}{r}(A_2 r_{,x} + 2r^3{}_{,x})\\ \sigma_{y\xi} &= \frac{A_1}{r}(-A_2 r_{,x} + 2r_{,x}r^2{}_{,y})\\ \tau_{xy\xi} &= \frac{A_1}{r}(A_2 r_{,y} + 2r^2{}_{,x}r_{,y})\end{aligned}\right\} \tag{7.66}$$

where

$$A_1 = -\frac{1+\overline{\nu}}{4\pi}, \quad A_2 = \frac{1-\overline{\nu}}{1+\overline{\nu}} \tag{7.67}$$

(ii) $F_\xi = 0,\ F_\eta = 1$

$$\left.\begin{aligned}\sigma_{x\eta} &= \lambda(U_{x\eta,x} + U_{y\eta,y}) + 2\mu U_{x\eta,x}\\ \sigma_{y\eta} &= \lambda(U_{x\eta,x} + U_{y\eta,y}) + 2\mu U_{y\eta,y}\\ \tau_{xy\eta} &= \mu(U_{x\eta,y} + U_{y\eta,x})\end{aligned}\right\} \tag{7.68}$$

Substituting Eqs. (7.60) into the above equations, results in

$$\left.\begin{aligned}\sigma_{x\eta} &= \frac{A_1}{r}(-A_2 r_{,y} + 2r^2{}_{,x}r_{,y})\\ \sigma_{y\eta} &= \frac{A_1}{r}(A_2 r_{,y} + 2r^3{}_{,y})\\ \tau_{xy\eta} &= \frac{A_1}{r}(A_2 r_{,x} + 2r_{,x}r^2{}_{,y})\end{aligned}\right\} \tag{7.69}$$

If indicial notation is employed for the coordinates, that is, x_1, x_2, then Eqs. (7.66) and (7.69) are reduced to [6]

$$\sigma_{ijk} = \frac{A_1}{r}\left[A_2(\delta_{ik}r_{,j} + \delta_{jk}r_{,i} - \delta_{ij}r_{,k}) + 2r_{,i}r_{,j}r_{,k}\right](i,j,k=1,2) \tag{7.70}$$

The subscript $k = 1, 2$ pertains to the directions ξ and η of the unit force, respectively. It should be noted that $\sigma_{11} = \sigma_x$, $\sigma_{22} = \sigma_y$, $\sigma_{12} = \tau_{xy}$.

7.6 BOUNDARY TRACTIONS DUE TO A UNIT CONCENTRATED FORCE

Expressions for the tractions on the boundary due to the concentrated force can be derived using Eq. (7.22). We distinguish two cases:

(i) $F_\xi = 1,\ F_\eta = 0$

$$T_{x\xi} = \sigma_{x\xi}\, n_x + \tau_{xy\xi}\, n_y$$

$$T_{y\xi} = \tau_{xy\xi}\, n_x + \sigma_{y\xi}\, n_y$$

or using Eqs. (7.66)

$$\left.\begin{aligned} T_{x\xi} &= \frac{A_1}{r}(A_2 + 2r^2{,}_x)r{,}_n \\ T_{y\xi} &= \frac{A_1}{r}(2r{,}_x r{,}_y r{,}_n + A_2 r{,}_t) \end{aligned}\right\} \tag{7.71}$$

(ii) $F_\xi = 0,\ F_\eta = 1$

$$T_{x\eta} = \sigma_{x\eta}\, n_x + \tau_{xy\eta}\, n_y$$

$$T_{y\eta} = \tau_{xy\eta}\, n_x + \sigma_{y\eta}\, n_y$$

or using Eqs. (7.69)

$$\left.\begin{aligned} T_{x\eta} &= \frac{A_1}{r}(2r{,}_x r{,}_y r{,}_n - A_2 r{,}_t) \\ T_{y\eta} &= \frac{A_1}{r}(A_2 + 2r^2{,}_y)r{,}_n \end{aligned}\right\} \tag{7.72}$$

where $r{,}_n = r{,}_x\, n_x + r{,}_y\, n_y$ expresses the derivative of r in the direction of the outward normal to the curve (boundary) passing through the point (x, y), and $r{,}_t = -\, r{,}_x\, n_y + r{,}_y\, n_x$ the derivative along the tangent to this curve. The vectors $\mathbf{n}$ and $\mathbf{t}$ define a right handed system of axes.

Using indicial notation Eqs. (7.71) and (7.72) are reduced to [7]

$$T_{ik} = \frac{A_1}{r}[(A_2\delta_{ik} + 2r{,}_i r{,}_k)r{,}_n + A_2(r{,}_i n_k - r{,}_k\, n_i)] \tag{7.73}$$

$$(i, k = 1, 2)$$

7.7 INTEGRAL REPRESENTATION OF THE SOLUTION

The integral representation of the solution for the two-dimensional elastostatic problem is derived from the reciprocal identity (7.43) by considering as state (II) the state of stress produced by a unit body force at point Q. The unit force is applied first in the ξ-direction and then in the η-direction.

(i) $F_\xi = 1,\ F_\eta = 0$

For this case, the state (II) inside the domain Ω is defined as

$$b_x^* = \delta(P - Q),\quad b_y^* = 0$$

$$u^* = U_{x\xi}(P, Q),\quad v^* = U_{y\xi}(P, Q)$$

while on the boundary Γ it is

$$t_x^* = T_{x\xi}(p, Q),\quad t_y^* = T_{y\xi}(p, Q)$$

where $P \in \Omega$ and $p \in \Gamma$.

Introducing the above expressions into Eq. (7.43) and taking into account that

$$\int_\Omega u b_x^* d\Omega = \int_\Omega u(P)\delta(P - Q)d\Omega_P = u(Q)$$

we obtain the integral representation of the solution for the displacement in the x-direction at points Q inside the domain Ω in the following form

$$\begin{aligned} u(Q) &= \int_\Omega \left[U_{x\xi}(P, Q)b_x(P) + U_{y\xi}(P, Q)b_y(P)\right] d\Omega_P \\ &+ \int_\Gamma \left[U_{x\xi}(p, Q)t_x(p) + U_{y\xi}(p, Q)t_y(p)\right] ds_p \\ &- \int_\Gamma \left[T_{x\xi}(p, Q)u(p) + T_{y\xi}(p, Q)v(p)\right] ds_p \end{aligned} \tag{7.74}$$

(ii) $F_\xi = 0,\ F_\eta = 1$

For this case, inside the domain Ω is

$$b_x^* = 0,\quad b_y^* = \delta(P - Q)$$

$$u^* = U_{x\eta}(P, Q),\quad v^* = U_{y\eta}(P, Q)$$

and on the boundary Γ

$$t_x^* = T_{x\eta}(p, Q),\quad t_y^* = T_{y\eta}(p, Q)$$

Introducing the above expressions into Eq. (7.43) and taking into account that

$$\int_\Omega v b_y^* d\Omega = \int_\Omega v(P)\ \delta(P - Q)d\Omega_P = v(Q)$$

we obtain

$$\begin{aligned} v(Q) &= \int_{\Omega}\left[U_{x\eta}(P,Q)\,b_x(P)+U_{y\eta}(P,Q)\,b_y(P)\right]d\Omega_P \\ &+\int_{\Gamma}\left[U_{x\eta}(p,Q)\,t_x(p)+U_{y\eta}(p,Q)\,t_y(p)\right]ds_p \\ &-\int_{\Gamma}\left[T_{x\eta}(p,Q)\,u(p)+T_{y\eta}(p,Q)\,v(p)\right]ds_p \end{aligned} \tag{7.75}$$

Equations (7.74) and (7.75) represent the integral representation of the solution to Navier equations for plane elasticity. We notice that the role of points $P(x,y)$ and $Q(\xi,\eta)$ in these equations has been interchanged due to reciprocity. Thus, point $Q(\xi,\eta)$ is now the field point and $P(x,y)$ the source point. The vector $\mathbf{n}$ in the expressions (7.71) and (7.72) of $T_{x\xi}$, $T_{y\xi}$, $T_{x\eta}$, and $T_{y\eta}$ is normal to the boundary at the point $p\in\Gamma$ where the load is applied. For reasons of consistency, the initial notation is restored. Hence, $P\in\Omega$ or $p\in\Gamma$ will designate field points, while $Q\in\Omega$ or $q\in\Gamma$ will designate source points, that is, points where unit forces are applied. Thus, Eqs. (7.74) and (7.75) may be rewritten under this notation and at the same time may be combined in a single matrix equation as

$$\begin{aligned} \begin{Bmatrix} u(P) \\ v(P) \end{Bmatrix} &= \int_{\Omega}\begin{bmatrix} U_{\xi x}(Q,P) & U_{\eta x}(Q,P) \\ U_{\xi y}(Q,P) & U_{\eta y}(Q,P) \end{bmatrix}\begin{Bmatrix} b_\xi(Q) \\ b_\eta(Q) \end{Bmatrix} d\Omega_Q \\ &+\int_{\Gamma}\begin{bmatrix} U_{\xi x}(q,P) & U_{\eta x}(q,P) \\ U_{\xi y}(q,P) & U_{\eta y}(q,P) \end{bmatrix}\begin{Bmatrix} t_\xi(q) \\ t_\eta(q) \end{Bmatrix} ds_q \\ &-\int_{\Gamma}\begin{bmatrix} T_{\xi x}(q,P) & T_{\eta x}(q,P) \\ T_{\xi y}(q,P) & T_{\eta y}(q,P) \end{bmatrix}\begin{Bmatrix} u(q) \\ v(q) \end{Bmatrix} ds_q \end{aligned} \tag{7.76}$$

This change in notation necessitates the replacement of derivatives $r_{,x}$ and $r_{,y}$ in Eqs. (7.58), (7.60), (7.71), and (7.72) with $r_{,\xi}$ and $r_{,\eta}$, respectively, without however changing their sign. Thus, we have

$$\left.\begin{aligned} U_{\xi x} &= -\frac{1}{8\pi G}\left[(3-\overline{\nu})\ln r-(1+\overline{\nu})r^2{}_{,\xi}+\frac{7-\overline{\nu}}{2}\right] \\ U_{\eta x} &= U_{\xi y}=\frac{1}{8\pi G}(1+\overline{\nu})r_{,\xi}r_{,\eta} \\ U_{\eta y} &= -\frac{1}{8\pi G}\left[(3-\overline{\nu})\ln r-(1+\overline{\nu})r^2{}_{,\eta}+\frac{7-\overline{\nu}}{2}\right] \\ T_{\xi x} &= \frac{A_1}{r}(A_2+2r^2{}_{,\xi})r_{,n} \\ T_{\eta x} &= \frac{A_1}{r}(2r_{,\xi}\,r_{,\eta}\,r_{,n}+A_2r_{,t}) \\ T_{\xi y} &= \frac{A_1}{r}(2r_{,\xi}\,r_{,\eta}\,r_{,n}-A_2r_{,t}) \\ T_{\eta y} &= \frac{A_1}{r}(A_2+2r^2{}_{,\eta})r_{,n} \end{aligned}\right\} \tag{7.77}$$

where $r_{,n}=r_{,\xi}\,n_x+r_{,\eta}\,n_y$ and $r_{,t}=-\,r_{,\xi}\,n_y+r_{,\eta}\,n_x$.

Employing indicial notation for the two directions, that is, x_1, x_2, Eq. (7.76) is reduced to [7]

$$u_i(P) = \int_{\Omega} U_{ji} b_j d\Omega + \int_{\Gamma} \left(U_{ji} t_j - T_{ji} u_j \right) ds, \quad (i, j = 1, 2) \tag{7.78}$$

7.8 BOUNDARY INTEGRAL EQUATIONS

As in the case of the Laplace equation discussed in Chapter 3, the boundary integral equations for the problem at hand are derived by letting point $P \in \Omega$ in Eq. (7.76) move to a point $p \in \Gamma$. We notice from Eqs. (7.58), (7.60), (7.71), and (7.72) that the kernels in the boundary integrals of Eq. (7.76) are singular, that is, they exhibit a singular behavior when $P \in \Omega \to p \in \Gamma$, because $r \to 0$ when $p \to q$. Therefore, it is necessary to examine the behavior of the boundary integrals in Eq. (7.76), when we let point $P \in \Omega$ coincide with a point $p \in \Gamma$.

We examine the general case of a nonsmooth boundary, and we assume that $P \equiv p$ is a corner point (see Fig. 7.6). Next we consider the domain Ω^*, which results from Ω by deleting a small circular sector with center P, radius ε, and confined by the arcs PA and PB. We denote the circular arc AB by Γ_ε and the sum of the arcs AP and PB by ℓ. The outward normal to Γ_ε coincides with the radius. Moreover, α is the angle between the two tangents to the boundary at point P. Obviously, it is

$$\lim_{\varepsilon \to 0} \Gamma_\varepsilon = 0$$

$$\lim_{\varepsilon \to 0} (\Gamma - \ell) = \Gamma$$

while the chords PA and PB become tangents to the boundary at point P for $\varepsilon \to 0$.

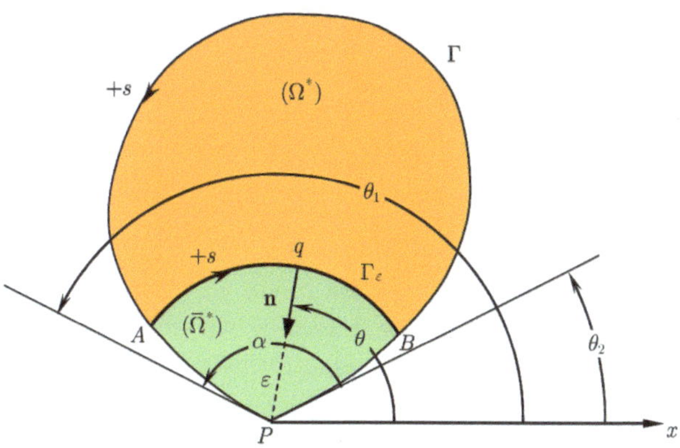

FIGURE 7.6 Geometric definitions related to a corner point P of a nonsmooth boundary.

Using the aforementioned notation, we apply the reciprocity relation (7.43) for the domain Ω^* with (a) $b_x^* = \delta(Q-P)$, $b_y^* = 0$ and (b) $b_x^* = 0$, $b_y^* = \delta(Q-P)$, where $Q \in \Omega^*$, $P \in \overline{\Omega}^* \equiv \Omega - \Omega^*$. In both cases the unit load is applied at point $P(x,y)$. Since point P lies outside the domain, the domain integrals in the left-hand side of Eq. (7.43) vanish. By utilizing indicial notation, we can write

$$\begin{aligned}&\int_{\Gamma-\ell} T_{ji}(P,q)u_j(q)\,ds_q + \int_{\Gamma_\varepsilon} T_{ji}(P,q)u_j(q)\,ds_q\\ &= \int_{\Omega^*} U_{ji}(P,Q)b_j\,d\Omega_Q + \int_{\Gamma-\ell} U_{ji}(P,q)t_j(q)\,ds_q\\ &+ \int_{\Gamma_\varepsilon} U_{ji}(P,q)t_j(q)\,ds_q\end{aligned} \tag{7.79}$$

We will examine the behavior of the integrals in Eq. (7.79) for $\varepsilon \to 0$. It is apparent that, in the limit, line integrals on $\Gamma - \ell$ become integrals on Γ, while domain integrals over Ω^* become integrals over Ω. Therefore, it remains to examine the behavior of integrals on Γ_ε. Namely,

$$\lim_{\varepsilon\to 0}\int_{\Gamma_\varepsilon} U_{ji}\, t_j\, ds \tag{7.80}$$

and

$$\lim_{\varepsilon\to 0}\int_{\Gamma_\varepsilon} T_{ji}\, u_j\, ds \tag{7.81}$$

We examine first the integrals (7.80) whose kernels are the fundamental solution U_{ji}.

Using the mean value theorem of integral calculus, the integrals (7.80) can be written in expanded form as

$$\left.\begin{aligned}&t_\xi(q^*)\lim_{\varepsilon\to 0}\int_{\Gamma_\varepsilon} U_{\xi x}\,ds + t_\eta(q^*)\lim_{\varepsilon\to 0}\int_{\Gamma_\varepsilon} U_{\eta x}\,ds\\ &t_\xi(q^*)\lim_{\varepsilon\to 0}\int_{\Gamma_\varepsilon} U_{\xi y}\,ds + t_\eta(q^*)\lim_{\varepsilon\to 0}\int_{\Gamma_\varepsilon} U_{\eta y}\,ds\end{aligned}\right\} \tag{7.82}$$

where q^* is a point on Γ_ε, which is generally different for each of the four terms. Obviously it is $q^* \to p \equiv P$, when $\varepsilon \to 0$.

On the basis of Eqs. (7.58) and (7.60), we note that the integrals in Eqs. (7.82) involve terms of the following forms:

$$I_1 = \int_{\Gamma_\varepsilon} \ln r\,ds, \quad I_2 = \int_{\Gamma_\varepsilon} r^2{}_{,\xi}\,ds, \quad I_3 = \int_{\Gamma_\varepsilon} r_{,\xi}\,r_{,\eta}\,ds, \quad I_4 = \int_{\Gamma_\varepsilon} r^2{}_{,\eta}\,ds$$

As it was shown in Section 3.3, it is

$$I_1 = \int_{\Gamma_\varepsilon} \ln r \, ds = \int_{\theta_1}^{\theta_2} \varepsilon \ln\varepsilon \, d(-\theta) = \varepsilon \ln\varepsilon(\theta_1 - \theta_2)$$

which in the limit yields

$$\lim_{\varepsilon \to 0} \int_{\Gamma_\varepsilon} \ln r \, ds = (\theta_1 - \theta_2) \lim_{\varepsilon \to 0} (\varepsilon \ln\varepsilon) = 0$$

and consequently

$$I_1 = 0$$

It remains to find the limits of the other three integrals. We can write (see Appendix A)

$$r_{,\xi} = \cos\theta \text{ and } r_{,\eta} = \sin\theta$$

Hence,

$$I_2 = \int_{\Gamma_\varepsilon} r^2{}_{,\xi} \, ds = \int_{\theta_1}^{\theta_2} \cos^2\theta \, \varepsilon \, d(-\theta) = -\varepsilon \left[\frac{\theta}{2} + \frac{\sin 2\theta}{4} \right]_{\theta_1}^{\theta_2}$$

$$I_3 = \int_{\Gamma_\varepsilon} r_{,\xi} r_{,\eta} \, ds = \int_{\theta_1}^{\theta_2} \cos\theta \sin\theta \, \varepsilon \, d(-\theta) = \varepsilon \left[\frac{\cos 2\theta}{4} \right]_{\theta_1}^{\theta_2}$$

$$I_4 = \int_{\Gamma_\varepsilon} r^2{}_{,\eta} \, ds = \int_{\theta_1}^{\theta_2} \sin^2\theta \, \varepsilon \, d(-\theta) = -\varepsilon \left[\frac{\theta}{2} - \frac{\sin 2\theta}{4} \right]_{\theta_1}^{\theta_2}$$

and consequently

$$\lim_{\varepsilon \to 0} I_2 = 0, \quad \lim_{\varepsilon \to 0} I_3 = 0, \quad \lim_{\varepsilon \to 0} I_4 = 0$$

Thus, the last integral of Eqs. (7.80) vanishes for $\varepsilon \to 0$ and the remaining three integrals in the right-hand side of Eq. (7.79) vary continuously as point $P \in \Omega$ approaches point $p \in \Gamma$.

Using again the mean value theorem of integral calculus, the integrals (7.81) with kernels T_{ji} can be written in expanded form as

$$\left. \begin{aligned} & u(q^*) \lim_{\varepsilon \to 0} \int_{\Gamma_\varepsilon} T_{\xi x} \, ds + v(q^*) \lim_{\varepsilon \to 0} \int_{\Gamma_\varepsilon} T_{\eta x} \, ds \\ & u(q^*) \lim_{\varepsilon \to 0} \int_{\Gamma_\varepsilon} T_{\xi y} \, ds + v(q^*) \lim_{\varepsilon \to 0} \int_{\Gamma_\varepsilon} T_{\eta y} \, ds \end{aligned} \right\} \tag{7.83}$$

where q^* is a point on Γ_ε, generally different in each of the four terms.

Referring to Fig. 7.6, we note that $\phi = \text{angle}(\mathbf{r}, \mathbf{n}) = \pi$, where $r \equiv \varepsilon$. Therefore,

$$r,_n = \cos\phi = -1 \quad \text{and} \quad r,_t = \sin\phi = 0$$

Consequently, using Eqs. (7.71), (7.72), and (7.67) and taking into account that $ds = r\, d(-\theta) = -\varepsilon\, d\theta$ (see Fig. 7.6 and Section 3.3), we obtain

$$\begin{aligned} \varepsilon_{\xi x} &= \lim_{\varepsilon\to 0}\int_{\Gamma_\varepsilon} T_{\xi x}\, ds = \lim_{\varepsilon\to 0}\int_{\Gamma_\varepsilon} \frac{A_1}{r}(A_2 + 2r^2,_\xi)r,_n\, ds \\ &= \frac{1+\overline{\nu}}{8\pi}\left(\frac{4}{1+\overline{\nu}}[\theta]_{\theta_2}^{\theta_1} + [\sin 2\theta]_{\theta_2}^{\theta_1}\right) \end{aligned} \tag{7.84}$$

$$\begin{aligned} \varepsilon_{\eta x} &= \lim_{\varepsilon\to 0}\int_{\Gamma_\varepsilon} T_{\eta x}\, ds = \lim_{\varepsilon\to 0}\int_{\Gamma_\varepsilon} \frac{A_1}{r}(2r,_\xi r,_\eta r,_n + A_2 r,_t)\, ds \\ &= -\frac{1+\overline{\nu}}{8\pi}[\cos 2\theta]_{\theta_2}^{\theta_1} \end{aligned} \tag{7.85}$$

$$\begin{aligned} \varepsilon_{\xi y} &= \lim_{\varepsilon\to 0}\int_{\Gamma_\varepsilon} T_{\xi y}\, ds = \lim_{\varepsilon\to 0}\int_{\Gamma_\varepsilon} \frac{A_1}{r}(2r,_\xi r,_\eta r,_n - A_2 r,_t)\, ds \\ &= -\frac{1+\overline{\nu}}{8\pi}[\cos 2\theta]_{\theta_2}^{\theta_1} \end{aligned} \tag{7.86}$$

$$\begin{aligned} \varepsilon_{\eta y} &= \lim_{\varepsilon\to 0}\int_{\Gamma_\varepsilon} T_{\eta y}\, ds = \lim_{\varepsilon\to 0}\int_{\Gamma_\varepsilon} \frac{A_1}{r}(A_2 + 2r^2,_\eta)r,_n\, ds \\ &= \frac{1+\overline{\nu}}{8\pi}\left(\frac{4}{1+\overline{\nu}}[\theta]_{\theta_2}^{\theta_1} - [\sin 2\theta]_{\theta_2}^{\theta_1}\right) \end{aligned} \tag{7.87}$$

Obviously, at points p where the boundary is smooth it is

$$[\theta]_{\theta_2}^{\theta_1} = \theta_1 - \theta_2 = \pi$$

$$[\cos 2\theta]_{\theta_2}^{\theta_1} = \cos 2\theta_1 - \cos 2\theta_2 = \cos 2(\theta_2 + \pi) - \cos 2\theta_2 = 0$$

$$[\sin 2\theta]_{\theta_2}^{\theta_1} = \sin 2\theta_1 - \sin 2\theta_2 = \sin 2(\theta_2 + \pi) - \sin 2\theta_2 = 0$$

and Eqs. (7.84) through (7.87) become

$$\left.\begin{aligned} \varepsilon_{\xi x} &= \frac{1}{2}, \quad \varepsilon_{\eta x} = 0 \\ \varepsilon_{\xi y} &= 0, \quad \varepsilon_{\eta y} = \frac{1}{2} \end{aligned}\right\} \tag{7.88}$$

On the basis of the previous discussion, the boundary integral Eq. (7.79) may be written now in matrix form as

$$\begin{bmatrix} \varepsilon_{\xi x} & \varepsilon_{\eta x} \\ \varepsilon_{\xi y} & \varepsilon_{\eta y} \end{bmatrix} \begin{Bmatrix} u \\ v \end{Bmatrix} = \int_\Omega \begin{bmatrix} U_{\xi x} & U_{\eta x} \\ U_{\xi y} & U_{\eta y} \end{bmatrix} \begin{Bmatrix} b_\xi \\ b_\eta \end{Bmatrix} d\Omega + \int_\Gamma \begin{bmatrix} U_{\xi x} & U_{\eta x} \\ U_{\xi y} & U_{\eta y} \end{bmatrix} \begin{Bmatrix} t_\xi \\ t_\eta \end{Bmatrix} ds - \int_\Gamma \begin{bmatrix} T_{\xi x} & T_{\eta x} \\ T_{\xi y} & T_{\eta y} \end{bmatrix} \begin{Bmatrix} u \\ v \end{Bmatrix} ds \tag{7.89}$$

or using indicial notation

$$\varepsilon_{ji} u_j = \int_\Omega U_{ji} b_j \, d\Omega + \int_\Gamma (U_{ji} t_j - T_{ji} u_j) \, ds \tag{7.90}$$

Apparently, it is $\varepsilon_{ij} = \frac{1}{2}\delta_{ij}$ at points p where the boundary is smooth.

7.9 INTEGRAL REPRESENTATION OF THE STRESSES

The stress components σ_x, σ_y, and τ_{xy} at a point $P(x, y)$ inside the domain Ω are derived from Eq. (7.3). Thus, using the expressions Eq. (7.76) for the displacement components u and v we obtain

$$\begin{aligned} \sigma_x = & \int_\Omega \left\{ \left[\lambda(U_{\xi x,x} + U_{\xi y,y}) + 2\mu U_{\xi x,x} \right] b_\xi + \left[\lambda(U_{\eta x,x} + U_{\eta y,y}) + 2\mu U_{\eta x,x} \right] b_\eta \right\} d\Omega \\ & + \int_\Gamma \left\{ \left[\lambda(U_{\xi x,x} + U_{\xi y,y}) + 2\mu U_{\xi x,x} \right] t_\xi + \left[\lambda(U_{\eta x,x} + U_{\eta y,y}) + 2\mu U_{\eta x,x} \right] t_\eta \right\} ds \\ & - \int_\Gamma \left\{ \left[\lambda(T_{\xi x,x} + T_{\xi y,y}) + 2\mu T_{\xi x,x} \right] u + \left[\lambda(T_{\eta x,x} + T_{\eta y,y}) + 2\mu T_{\eta x,x} \right] v \right\} ds \end{aligned} \tag{7.91}$$

or setting

$$\left. \begin{aligned} \sigma_{x\xi} &= \lambda(U_{\xi x,x} + U_{\xi y,y}) + 2\mu U_{\xi x,x} \\ \sigma_{x\eta} &= \lambda(U_{\eta x,x} + U_{\eta y,y}) + 2\mu U_{\eta x,x} \end{aligned} \right\} \tag{7.92}$$

$$\left. \begin{aligned} \overline{\sigma}_{x\xi} &= \lambda(T_{\xi x,x} + T_{\xi y,y}) + 2\mu T_{\xi x,x} \\ \overline{\sigma}_{x\eta} &= \lambda(T_{\eta x,x} + T_{\eta y,y}) + 2\mu T_{\eta x,x} \end{aligned} \right\} \tag{7.93}$$

Equation (7.91) becomes

$$\begin{aligned} \sigma_x = & \int_\Omega (\sigma_{x\xi} b_\xi + \sigma_{x\eta} b_\eta) d\Omega + \int_\Gamma (\sigma_{x\xi} t_\xi + \sigma_{x\eta} t_\eta) ds \\ & - \int_\Gamma (\overline{\sigma}_{x\xi} u + \overline{\sigma}_{x\eta} v) ds \end{aligned} \tag{7.94}$$

The quantities $\sigma_{x\xi}$ and $\sigma_{x\eta}$ express the stresses σ_x at point $(x, y)\in\Omega$ for unit loads applied at point (ξ,η) in the x- and y-direction, respectively (see Eqs. (7.65) and (7.68)). Similarly, the quantities $\overline{\sigma}_{x\xi}$ and $\overline{\sigma}_{x\eta}$ express the stresses σ_x at point $(x, y)\in\Omega$, due to unit displacements applied at point (ξ,η) in the x- and y-direction, respectively.

Similarly, we obtain

$$\sigma_y = \int_\Omega (\sigma_{y\xi} b_\xi + \sigma_{y\eta} b_\eta) d\Omega + \int_\Gamma (\sigma_{y\xi} t_\xi + \sigma_{y\eta} t_\eta) ds - \int_\Gamma (\overline{\sigma}_{y\xi} u + \overline{\sigma}_{y\eta} v) ds \tag{7.95}$$

$$\tau_{xy} = \int_\Omega (\tau_{xy\xi} b_\xi + \tau_{xy\eta} b_\eta) d\Omega + \int_\Gamma (\tau_{xy\xi} t_\xi + \tau_{xy\eta} t_\eta) ds - \int_\Gamma (\overline{\tau}_{xy\xi} u + \overline{\tau}_{xy\eta} v) ds \tag{7.96}$$

where

$$\left.\begin{aligned} \sigma_{y\xi} &= \lambda(U_{\xi x,x} + U_{\xi y,y}) + 2\mu U_{\xi y,y} \\ \sigma_{y\eta} &= \lambda(U_{\eta x,x} + U_{\eta y,y}) + 2\mu U_{\eta y,y} \end{aligned}\right\} \tag{7.97}$$

$$\left.\begin{aligned} \overline{\sigma}_{y\xi} &= \lambda(T_{\xi x,x} + T_{\xi y,y}) + 2\mu T_{\xi y,y} \\ \overline{\sigma}_{y\eta} &= \lambda(T_{\eta x,x} + T_{\eta y,y}) + 2\mu T_{\eta y,y} \end{aligned}\right\} \tag{7.98}$$

$$\left.\begin{aligned} \tau_{xy\xi} &= \mu(U_{\xi x,y} + U_{\xi y,x}) \\ \tau_{xy\eta} &= \mu(U_{\eta x,y} + U_{\eta y,x}) \end{aligned}\right\} \tag{7.99}$$

$$\left.\begin{aligned} \overline{\tau}_{xy\xi} &= \mu(T_{\xi x,y} + T_{\xi y,x}) \\ \overline{\tau}_{xy\eta} &= \mu(T_{\eta x,y} + T_{\eta y,x}) \end{aligned}\right\} \tag{7.100}$$

The stresses $\sigma_{x\xi}$, $\sigma_{x\eta}$, $\sigma_{y\xi}$, $\sigma_{y\eta}$, $\tau_{xy\xi}$, and $\tau_{xy\eta}$ are given by Eqs. (7.66) and (7.69), whereas $\overline{\sigma}_{x\xi}$, $\overline{\sigma}_{x\eta}$, $\overline{\sigma}_{y\xi}$, $\overline{\sigma}_{y\eta}$, $\overline{\tau}_{xy\xi}$, and $\overline{\tau}_{xy\eta}$ are derived by introducing Eqs. (7.71) and (7.72) in Eqs. (7.93), (7.98), and (7.100), and by performing the involved differentiations. Thus, we obtain

$$\left.\begin{aligned} \overline{\sigma}_{x\xi} &= \frac{A_3}{r^2}\left[2r_{,x} r_{,n}(1 - 4r^2_{,x}) + (2r^2_{,x} + 1)n_x\right] \\ \overline{\sigma}_{x\eta} &= \frac{A_3}{r^2}(-8r^2_{,x} r_{,y} r_{,n} + 2r_{,x} r_{,y} n_x + n_y) \\ \overline{\sigma}_{y\xi} &= \frac{A_3}{r^2}(-8r^2_{,y} r_{,x} r_{,n} + 2r_{,x} r_{,y} n_y + n_x) \\ \overline{\sigma}_{y\eta} &= \frac{A_3}{r^2}\left[2r_{,y} r_{,n}(1 - 4r^2_{,y}) + (2r^2_{,y} + 1)n_y\right] \\ \overline{\tau}_{xy\xi} &= \overline{\sigma}_{x\eta} \\ \overline{\tau}_{xy\eta} &= \overline{\sigma}_{y\xi} \end{aligned}\right\} \tag{7.101}$$

where $A_3 = -2\mu A_1$.

7.10 NUMERICAL SOLUTION OF THE BOUNDARY INTEGRAL EQUATIONS

7.10.1 Evaluation of the unknown boundary quantities

In this Section, the boundary integral equations are solved using the BEM with constant boundary elements. The boundary is divided into N constant elements. Thus the distribution of the displacements and tractions are assumed constant on each element and equal to their value at the nodal point, which lies at the midpoint of the element.

Denoting by $\{u\}^i = \{u^i\ v^i\}^T$ and $\{t\}^i = \{t_x^i\ t_y^i\}^T$ the displacements and tractions at the i-th node and taking into account that the boundary is smooth at the nodal point of the constant element, Eq. (7.89) yield

$$\frac{1}{2}\{u\}^i + \sum_{j=1}^{N} [\hat{H}]^{ij}\{u\}^j = \sum_{j=1}^{N} [G]^{ij}\{t\}^j + \{F\}^i \tag{7.102}$$

where

$$[G]^{ij} = \begin{bmatrix} \int_{\Gamma_j} U_{\xi x}(q,p_i)ds_q & \int_{\Gamma_j} U_{\eta x}(q,p_i)ds_q \\ \int_{\Gamma_j} U_{\xi y}(q,p_i)ds_q & \int_{\Gamma_j} U_{\eta y}(q,p_i)ds_q \end{bmatrix} \tag{7.103}$$

$$[\hat{H}]^{ij} = \begin{bmatrix} \int_{\Gamma_j} T_{\xi x}(q,p_i)ds_q & \int_{\Gamma_j} T_{\eta x}(q,p_i)ds_q \\ \int_{\Gamma_j} T_{\xi y}(q,p_i)ds_q & \int_{\Gamma_j} T_{\eta y}(q,p_i)ds_q \end{bmatrix} \tag{7.104}$$

and

$$\{F\}^i = \left\{ \begin{array}{l} \int_{\Omega} \left[U_{\xi x}(Q,p_i)b_\xi(Q) + U_{\eta x}(Q,p_i)b_\eta(Q) \right] d\Omega_Q \\ \int_{\Omega} \left[U_{\xi y}(Q,p_i)b_\xi(Q) + U_{\eta y}(Q,p_i)b_\eta(Q) \right] d\Omega_Q \end{array} \right\} \tag{7.105}$$

with $p_i, q \in \Gamma$ and $Q \in \Omega$.

Equation (7.102) relates the displacements of the i-th node to the displacements and tractions of all the nodes including the i-th node.

Applying Eq. (7.102) to all the boundary nodal points yields $2N$ equations, which can be written in matrix form as

$$[H]\{u\} = [G]\{t\} + \{F\} \tag{7.106}$$

where

$$[H] = \left[\hat{H}\right] + \frac{1}{2}[I] \tag{7.107}$$

The dimensions of the matrices $[\hat{H}]$ and $[G]$ are $2N \times 2N$, and those of the vectors $\{u\}$, $\{t\}$, and $\{F\}$ are $2N$. They are defined as

$$[G] = \begin{bmatrix} [G]^{11} & [G]^{12} & \cdots & [G]^{1N} \\ [G]^{21} & [G]^{22} & \cdots & [G]^{2N} \\ \vdots & \vdots & \ddots & \vdots \\ [G]^{N1} & [G]^{N2} & \cdots & [G]^{NN} \end{bmatrix} \tag{7.108}$$

$$[\hat{H}] = \begin{bmatrix} [\hat{H}]^{11} & [\hat{H}]^{12} & \cdots & [\hat{H}]^{1N} \\ [\hat{H}]^{21} & [\hat{H}]^{22} & \cdots & [\hat{H}]^{2N} \\ \vdots & \vdots & \ddots & \vdots \\ [\hat{H}]^{N1} & [\hat{H}]^{N2} & \cdots & [\hat{H}]^{NN} \end{bmatrix} \tag{7.109}$$

and

$$\{u\} = \begin{Bmatrix} \{u\}^1 \\ \{u\}^2 \\ \vdots \\ \{u\}^N \end{Bmatrix}, \quad \{t\} = \begin{Bmatrix} \{t\}^1 \\ \{t\}^2 \\ \vdots \\ \{t\}^N \end{Bmatrix}, \quad \{F\} = \begin{Bmatrix} \{F\}^1 \\ \{F\}^2 \\ \vdots \\ \{F\}^N \end{Bmatrix} \tag{7.110}$$

The $2N$ equations (7.106) contain $4N$ boundary values, that is, $2N$ values of displacements and $2N$ values of tractions. However, a total of $2N$ values are known from the boundary conditions. Consequently, Eq. (7.106) can be used to determine the $2N$ unknown boundary values. It should be noted that rearrangement of the unknowns is necessary for mixed boundary conditions. After doing so the following system of $2N$ linear equations is obtained

$$[A]\{X\} = \{R\} + \{F\} \tag{7.111}$$

where $[A]$ is a square coefficient matrix having dimensions $2N \times 2N$, $\{X\}$ is the vector including the $2N$ unknown boundary values and $\{R\}$ is a vector resulting as the sum of the columns of the matrices $[G]$ and $[H]$ multiplied by the respective known boundary values. Columns originating from matrices $[H]$ and $[G]$ should have their sign switched, when they are moved to the other side of the equation. Special attention should be paid in the case where only boundary tractions are prescribed, that is when the vector $\{t\}$ in Eq. (7.106) is known. For boundary conditions of this kind, (case (iv) of Eq. (7.21)), the displacements are not determined uniquely, because they also include a rigid body motion. This is reflected in the matrix $[H]$ whose rank is $2N - 3$ and therefore cannot be inverted. To overcome this problem we restrain the rigid body motion. For this purpose the body is supported by setting three elements of the vector $\{u\}$ equal to zero. Attention should be paid in selecting the elements in order to exclude any infinitesimal kinematic indeterminacy of the body, which would lead to an ill-conditioned matrix $[A]$.

7.10.2 Evaluation of displacements in the interior of the body

The system of Eq. (7.111) is solved for the unknown boundary values of the displacements and tractions. Thereafter all the boundary quantities are known and, consequently, the displacements at any point $P_i(x_i, y_i)$ inside the domain Ω can be evaluated using Eq. (7.76), which after discretization becomes

$$\{u\}^i = \sum_{j=1}^{N} [G]^{ij}\{t\}^j - \sum_{j=1}^{N} [\hat{H}]^{ij}\{u\}^j + \{F\}^i \tag{7.112}$$

The matrices $[G]^{ij}$ and $[\hat{H}]^{ij}$ as well as the vector $\{F\}^i$ are evaluated according to Eqs. (7.103)–(7.105), respectively, with $P \in \Omega$ in place of $p_i \in \Gamma$. The superscript i is pertaining now to point $P_i(x_i, y_i)$ inside Ω and not to that of the boundary nodes.

7.10.3 Evaluation of stresses in the interior of the body

The stresses at any point $P_i(x_i, y_i)$ inside the domain Ω are evaluated using Eqs. (7.94)–(7.96), which after discretization are written as

$$\begin{Bmatrix} \sigma_x \\ \sigma_y \\ \tau_{xy} \end{Bmatrix}^i = \sum_{j=1}^{N} [\sigma]^{ij}\{t\}^j - \sum_{j=1}^{N} [\overline{\sigma}]^{ij}\{u\}^j + \{S\}^i \tag{7.113}$$

The matrices $[\sigma]^{ij}$ and $[\overline{\sigma}]^{ij}$ as well as the vector $\{S\}^i$ are computed from the following relations

$$[\sigma]^{ij} = \begin{bmatrix} \int_{\Gamma_j} \sigma_{x\xi}(q, P_i)ds_q & \int_{\Gamma_j} \sigma_{x\eta}(q, P_i)ds_q \\ \int_{\Gamma_j} \sigma_{y\xi}(q, P_i)ds_q & \int_{\Gamma_j} \sigma_{y\eta}(q, P_i)ds_q \\ \int_{\Gamma_j} \tau_{xy\xi}(q, P_i)ds_q & \int_{\Gamma_j} \tau_{xy\eta}(q, P_i)ds_q \end{bmatrix} \tag{7.114}$$

$$[\overline{\sigma}]^{ij} = \begin{bmatrix} \int_{\Gamma_j} \overline{\sigma}_{x\xi}(q, P_i)ds_q & \int_{\Gamma_j} \overline{\sigma}_{x\eta}(q, P_i)ds_q \\ \int_{\Gamma_j} \overline{\sigma}_{y\xi}(q, P_i)ds_q & \int_{\Gamma_j} \overline{\sigma}_{y\eta}(q, P_i)ds_q \\ \int_{\Gamma_j} \overline{\tau}_{xy\xi}(q, P_i)ds_q & \int_{\Gamma_j} \overline{\tau}_{xy\eta}(q, P_i)ds_q \end{bmatrix} \tag{7.115}$$

$$\{S\}^i = \begin{Bmatrix} \int_\Omega \left[\sigma_{x\xi}(Q, P_i)b_\xi(Q) + \sigma_{y\eta}(Q, P_i)b_\eta(Q)\right] d\Omega_Q \\ \int_\Omega \left[\sigma_{x\xi}(Q, P_i)b_\xi(Q) + \sigma_{y\eta}(Q, P_i)b_\eta(Q)\right] d\Omega_Q \\ \int_\Omega \left[\tau_{xy\xi}(Q, P_i)b_\xi(Q) + \tau_{yx\eta}(Q, P_i)b_\eta(Q)\right] d\Omega_Q \end{Bmatrix} \tag{7.116}$$

7.10.4 Evaluation of stresses on the boundary

The stresses σ_x, σ_y, and τ_{xy} on the boundary can be evaluated from Eqs. (7.94)–(7.96) by letting point $P \in \Omega$ approach point $p \in \Gamma$ and following a limiting process similar to that presented in Section7.8. This procedure, however, is not recommended due to the difficulties arising from the behavior of the line integrals as $P \in \Omega$ moves to $p \in \Gamma$, and on the other hand from the need to treat singular and hypersingular line integrals. For this reason, the technique that follows is preferred as being much simpler and straightforward.

The derivatives of the displacements u and v along the tangent to the boundary are given as

$$\left.\begin{aligned} \frac{\partial u}{\partial t} &= -\frac{\partial u}{\partial x} n_y + \frac{\partial u}{\partial y} n_x \\ \frac{\partial v}{\partial t} &= -\frac{\partial v}{\partial x} n_y + \frac{\partial v}{\partial y} n_x \end{aligned}\right\} \tag{7.117}$$

These derivatives can be computed from the boundary values of u and v by numerical differentiation (see Section 6.2.2).

Combining Eqs. (7.23) and (7.117), we can write

$$\begin{bmatrix} (\lambda + 2\mu) n_x & \mu\, n_y & \mu\, n_y & \lambda\, n_x \\ \lambda\, n_y & \mu\, n_x & \mu\, n_x & (\lambda + 2\mu) n_y \\ -n_y & n_x & 0 & 0 \\ 0 & 0 & -n_y & n_x \end{bmatrix} \begin{Bmatrix} u_{,x} \\ u_{,y} \\ v_{,x} \\ v_{,y} \end{Bmatrix} = \begin{Bmatrix} t_x \\ t_y \\ u_{,t} \\ v_{,t} \end{Bmatrix} \tag{7.118}$$

Taking into account that $n_x^2 + n_y^2 = 1$, it can be readily shown that the determinant of the coefficient matrix in Eq. (7.118) is

$$D = -\mu(\lambda + 2\mu) \neq 0 \tag{7.119}$$

Hence the derivatives of u and v with respect to x and y can always be evaluated and the stresses can be computed from the expressions

$$\left.\begin{aligned} \sigma_x &= \lambda(u_{,x} + v_{,y}) + 2\mu u_{,x} \\ \sigma_y &= \lambda(u_{,x} + v_{,y}) + 2\mu u_{,y} \\ \tau_{xy} &= \mu(u_{,y} + v_{,x}) \end{aligned}\right\} \tag{7.120}$$

7.11 BODY FORCES

It becomes apparent from Eqs. (7.105) and (7.116) that the inclusion of body forces requires the evaluation of domain integrals. These integrals can be treated using one of the following ways.

7.11.1 Direct numerical evaluation

The domain Ω is discretized into cells and the integration is performed using the procedure developed for Poisson's equation in Section 4.4. This method, however, has two drawbacks. The first one is that it diminishes the elegance and computational efficiency of the BEM as it involves domain discretization, which spoils the pure boundary character of the method. Of course, the problem does not appear as complex as in the finite element method (FEM), where the continuous body is approximated by the discretized one. In the BEM, the domain discretization is employed only to approximate the domain integral. Even the assumption of constant value of the integrand on each cell (discontinuous finite element) results in a good approximation for the integral.

The second drawback of the domain integration originates from Eqs. (7.58), (7.60), (7.66), (7.69), and (7.101) whose integrands behave like $\ln r$, $1/r$, $1/r^2$, or $1/r^3$, where $r = |P - Q|$. These integrals become singular or hypersingular, when point P lies on the cell over which the integration is performed, because point Q also lies on the same cell and thus r will take among others the value $r = 0$. Several methods have been developed for the evaluation of these integrals [8]. A relatively simple but effective and accurate method is that developed by Katsikadelis [9], which converts the singular or hypersingular domain integrals to regular line integrals on the boundary of the cell. This method is presented in Appendix B. Another, efficient method for the evaluation of singular domain integrals is presented in [3].

7.11.2 Evaluation using a particular solution

In this case the solution of Navier's equations, Eq. (7.19), is obtained as a sum of two solutions

$$u = u_0 + u_1, \quad v = v_0 + v_1 \tag{7.121}$$

where u_0, v_0 is the solution of the homogeneous equation and u_1, v_1 a particular one of the nonhomogeneous equation. Using the notation of Eqs. (7.45) for the differential operators, the Navier equations are written as

$$\left.\begin{aligned} N_x(u, v) &= b_x \\ N_y(u, v) &= b_y \end{aligned}\right\} \quad \text{in } \Omega \tag{7.122}$$

or using Eqs. (7.121) and taking into account that the operators N_x and N_y are linear, we can obtain the solution u_0, v_0 from the following boundary value problem

$$\left.\begin{aligned} N_x(u_0, v_0) &= 0 \\ N_y(u_0, v_0) &= 0 \end{aligned}\right\} \quad \text{in } \Omega \tag{7.123}$$

and

i.	$u_0 = \overline{u} - u_1,$	$v_0 = \overline{v} - v_1$	on	Γ_1	(7.124a)
ii.	$u_0 = \overline{u} - u_1,$	$(t_y)_0 = \overline{t}_y - t_y^1$	on	Γ_2	(7.124b)
iii.	$(t_x)_0 = \overline{t}_x - t_x^1,$	$v_0 = \overline{v} - v_1$	on	Γ_3	(7.124c)
iv.	$(t_x)_0 = \overline{t}_x - t_x^1,$	$(t_y)_0 = \overline{t}_y - t_y^1$	on	Γ_4	(7.124d)

The above types of boundary conditions result from Eqs. (7.21) when these are expressed in terms of u_0 and v_0. The traction components t_x^1 and t_y^1 are obtained from Eqs. (7.23), if u, v are replaced by u_1, v_1.

The particular solution is obtained from the equations

$$\left.\begin{aligned} N_x(u_1, v_1) &= b_x \\ N_y(u_1, v_1) &= b_y \end{aligned}\right\}$$

It is apparent that the determination of the particular solution should precede the solution of the homogeneous equations. The particular solution can be obtained from Eqs. (7.49)

$$\left.\begin{aligned} 2Gu_1 &= \frac{2}{1+\overline{\nu}}\nabla^2\phi_1 - \frac{\partial}{\partial x}\left(\frac{\partial\phi_1}{\partial x} + \frac{\partial\psi_1}{\partial y}\right) \\ 2Gv_1 &= \frac{2}{1+\overline{\nu}}\nabla^2\psi_1 - \frac{\partial}{\partial y}\left(\frac{\partial\phi_1}{\partial x} + \frac{\partial\psi_1}{\partial y}\right) \end{aligned}\right\} \tag{7.125}$$

The functions ϕ_1, ψ_1 are the components of the *Galerkin vector* and are determined as a particular solution of the following equations

$$\nabla^4\phi_1 = -(1+\overline{\nu})b_x \tag{7.126a}$$

$$\nabla^4\psi_1 = -(1+\overline{\nu})b_y \tag{7.126b}$$

Equations (7.126a,b) represent the bending equation of a thin plate subjected to transverse loads $-(1+\overline{\nu})b_x$ and $-(1+\overline{\nu})b_y$, respectively. A particular solution of these equations can be obtained using the method presented in [10]. This method, which is discussed below, is an extension of that presented in Section 3.4.2 for the harmonic equation to the case of the biharmonic equation (see also Ref. [3]).

By introducing the variables

$$z = x + iy \text{ and } \overline{z} = x - iy \tag{7.127}$$

Equation (7.126a) is transformed to

$$16\frac{\partial^4\overline{\phi}_1}{\partial z^2\partial\overline{z}^2} = b_x(z,\overline{z}) \tag{7.128}$$

where it has been set

$$\overline{\phi}_1 = -\frac{\phi_1}{(1+\overline{\nu})} \tag{7.129}$$

A particular solution $\overline{\phi}_1(z,\overline{z})$ is readily obtained by integrating Eq. (7.128) successively four times. The resulting arbitrary integration functions can be neglected, since we are looking for a particular solution.

Subsequently, back substitution of z and $\overline{z}$ by virtue of Eq. (7.127) yields $\overline{\phi}_1(x, y)$. Similarly, we find the particular solution $\overline{\psi}_1(x, y)$ of Eq. (7.126b).

EXAMPLE 7.1

Determine the particular solution, if the body force $b_x = \rho g\cos\theta$, $b_y = \rho g\sin\theta$ is the weight of the body, which acts in the direction of the vector $(\cos\theta,\ \sin\theta)$. We denote by ρ the mass density per unit area and by g the acceleration of gravity.

First we establish the components of the Galerkin vector. Equation (7.128) is written

$$16\frac{\partial^4\overline{\phi}_1}{\partial z^2\partial\overline{z}^2} = \rho g\cos\theta$$

Successive integrations yield

$$\overline{\phi}_1 = \frac{\rho g\cos\theta}{64}z^2\overline{z}^2 = \frac{\rho g}{64}r^4\cos\theta$$

or

$$\phi_1 = kr^4\cos\theta$$

where it was set

$$k = -\frac{\rho g(1+\overline{\nu})}{64} \quad \text{and} \quad r^2 = x^2 + y^2$$

Similarly, we obtain

$$\psi_1 = kr^4\sin\theta$$

Substituting the above expressions for ϕ_1 and ψ_1 into Eqs. (7.125) yields

$$u_1 = -\frac{\rho g}{4G}\left\{r^2\cos\theta - \frac{1+\overline{\nu}}{8}\left[(3x^2+y^2)\cos\theta + 2xy\sin\theta\right]\right\}$$

and

$$v_1 = -\frac{\rho g}{4G}\left\{r^2\sin\theta - \frac{1+\overline{\nu}}{8}\left[2xy\cos\theta + (x^2+3y^2)\sin\theta\right]\right\}$$

If the y-axis is taken in the vertical direction, it will be $\theta = -\pi/2$, hence $\cos\theta = 0$, $\sin\theta = -1$, and the particular solution is simplified as

$$u_1 = -\frac{\rho g(1+\overline{\nu})}{16G}xy$$

$$v_1 = -\frac{\rho g}{16G}\left[-4(x^2+y^2) + \frac{1}{2}(1+\overline{\nu})(x^2+3y^2)\right]$$

7.11.3 Transformation of the domain integrals to boundary integrals

This method is closer to the BEM's logic, since domain integration is avoided and the pure boundary character of the method is maintained. The transformation of the domain integrals to boundary line integrals may be accomplished by different techniques. Two methods are going to be presented in the sequel. One of them is general and applies to body forces having an arbitrary distribution, while the other is used for body forces derived from a potential function.

a. *Body forces having arbitrary distribution*

This method is analogous to that presented for Poisson's equation in Section 3.5 (ii).

First, using the procedure presented in the previous section, we determine a particular solution u_1, v_1 for Navier's equations,

$$N_x(u_1, v_1) = b_x$$

$$N_y(u_1, v_1) = b_y$$

Next the reciprocal relation (7.43) is employed consecutively for

i. $u = u_1, \quad v = v_1 \quad \text{and} \quad u^* = U_{\xi x}, \quad v^* = U_{\eta x}$
ii. $u = u_1, \quad v = v_1 \quad \text{and} \quad u^* = U_{\xi y}, \quad v^* = U_{\eta y}$

and having in mind that the fundamental solution $U_{\xi x}$, $U_{\eta x}$, $U_{\xi y}$, $U_{\eta y}$ satisfies the equations

$$\left.\begin{aligned} N_x(U_{\xi x}, U_{\eta x}) &= \delta(Q-P) \\ N_y(U_{\xi x}, U_{\eta x}) &= 0 \end{aligned}\right\} \tag{7.130a}$$

and

$$\left.\begin{aligned} N_x(U_{\xi y}, U_{\eta y}) &= 0 \\ N_y(U_{\xi y}, U_{\eta y}) &= \delta(Q-P) \end{aligned}\right\} \tag{7.130b}$$

we obtain the components F_x and F_y of the vector (7.105) as

$$\begin{aligned} F_x(P) &= \varepsilon_{11} u_1(P) + \varepsilon_{21} v_1(P) \\ &\quad - \int_\Gamma \left[U_{\xi x}(q,P) t_\xi^1(q) + U_{\eta x}(q,P) t_\eta^1(q) \right] ds_q \\ &\quad + \int_\Gamma \left[T_{\xi x}(q,P) u_1(q) + T_{\eta x}(q,P) v_1(q) \right] ds_q \end{aligned} \tag{7.131}$$

$$\begin{aligned} F_y(P) &= \varepsilon_{12} u_1(P) + \varepsilon_{22} v_1(P) \\ &\quad - \int_\Gamma \left[U_{\xi y}(q,P) t_\xi^1(q) + U_{\eta y}(q,P) t_\eta^1(q) \right] ds_q \\ &\quad + \int_\Gamma \left[T_{\xi y}(q,P) u_1(q) + T_{\eta y}(q,P) v_1(q) \right] ds_q \end{aligned} \tag{7.132}$$

where the coefficients ε_{ij} take the value $\varepsilon_{ij} = \delta_{ij}$ for $P \in \Omega$, while for $P \in \Gamma$ they are evaluated from Eqs. (7.84)–(7.87). If the boundary is smooth at point $P \in \Gamma$, it is $\varepsilon_{ij} = \frac{1}{2}\delta_{ij}$.

It should be noted that Eqs. (7.131) and (7.132) hold for body forces b_ξ and b_η which are distributed over the whole domain Ω. If b_ξ or b_η act only on a subregion $\Omega^* \subset \Omega$, then for points P outside Ω^* it is $\varepsilon_{ij} = 0$ and thus terms outside the integrals vanish. Therefore, Eqs. (7.131) and (7.132) can only be employed for points $P \in (\Omega^* \cup \Gamma^*)$.

b. *Body forces derived from a potential function*

The previous case, where b_ξ and b_η are arbitrary functions, is rather theoretical. On the other hand, the case where the body forces are derived from a potential function is important and of special practical interest, for example, gravitational forces. The conversion of the domain integrals to boundary integrals can be accomplished as follows.

If $V = V(\xi, \eta)$ is the function that represents the potential, then the components of the body force are obtained as

$$b_\xi = \frac{\partial V}{\partial \xi}, \quad b_\eta = \frac{\partial V}{\partial \eta} \tag{7.133}$$

Moreover, the potential function satisfies the equation

$$\nabla^2 V = 0 \tag{7.134}$$

The integral F_x defined by Eq. (7.105) is written as

$$\begin{aligned} F_x &= \int_\Omega (U_{\xi x} V_{,\xi} + U_{\eta x} V_{,\eta})\, d\Omega \\ &= -\int_\Omega (U_{\xi x,\xi} + U_{\eta x,\eta}) V\, d\Omega + \int_\Gamma (U_{\xi x} n_x + U_{\eta x} n_y) V\, ds \end{aligned} \tag{7.135}$$

where n_x and n_y are the direction cosines of the unit vector $\mathbf{n}$ which is normal to the boundary at point (ξ, η). The above transformation of the domain integral is achieved by employing Eqs. (2.7) and (2.8).

Incorporating Eqs. (7.58) and (7.60), the domain integral of the right-hand side of Eq. (7.135) becomes

$$\int_\Omega (U_{\xi x,\xi} + U_{\eta x,\eta}) V\, d\Omega = -\frac{2(1-\overline{\nu})}{8\pi G} \int_\Omega (\ln r)_{,\xi} V\, d\Omega \tag{7.136}$$

Further, setting

$$\phi = \frac{1}{4} r^2 \ln r \tag{7.137}$$

we find that

$$\nabla^2 \phi_{,\xi} = (\ln r)_{,\xi}$$

and then applying Green's identity (2.16) for $v = V$ and $u = \phi_{,\xi}$, we obtain

$$\int_\Omega V \nabla^2 \phi_{,\xi}\, d\Omega = \int_\Gamma \left(V \frac{\partial \phi_{,\xi}}{\partial n} - \phi_{,\xi} \frac{\partial V}{\partial n} \right) ds \tag{7.138}$$

Hence, Eq. (7.135) can be finally written as

$$F_x = \frac{(1-\overline{\nu})}{4\pi G} \int_\Gamma \left(V \frac{\partial \phi_{,\xi}}{\partial n} - \phi_{,\xi} \frac{\partial V}{\partial n} \right) ds + \int_\Gamma (U_{\xi x} n_x + U_{\eta x} n_y)\, ds \tag{7.139}$$

Similarly, the contribution of the body forces in the y-direction may be obtained in terms of boundary-only integrals as

$$F_y = \frac{(1-\overline{\nu})}{4\pi G} \int_\Gamma \left(V \frac{\partial \phi_{,\eta}}{\partial n} - \phi_{,\eta} \frac{\partial V}{\partial n} \right) ds + \int_\Gamma (U_{\xi y} n_x + U_{\eta y} n_y)\, ds \tag{7.140}$$

7.12 PROGRAM ELBECON FOR SOLVING THE PLANE ELASTOSTATIC PROBLEM WITH CONSTANT BOUNDARY ELEMENTS

On the basis of the analysis presented in the previous sections a computer program has been written in the FORTRAN language. It solves the two plane elasticity problems, namely the plane strain and the plane stress for plane bodies which may have holes. For the simplicity of the program, the body forces are not included. The program employs constant elements for the discretization of the boundary integral equations. The structure of program ELBECON is shown in the macro flow chart of Fig. 7.7.

7.12.1 Main program

The main program defines the parameters N, NB, and IN, which specify the number of boundary elements, the number of boundaries, and the number of internal points where the solution will be computed, respectively. It opens two files, the file INPUTFILE, which contains the data and the OUTPUTFILE file, in which the results are rendered. Subsequently it calls the following ten subroutines:

INPUTEL	Reads the data from INPUTFILE
GMATREL	Forms the matrix $[G]$ defined by Eq. (7.108)
HMATREL	Forms the matrix $[H]$ defined by Eqs. (7.107) and (7.109)
ABMATREL	Rearranges the matrices $[H]$ and $[G]$ on the basis of the given boundary conditions and forms the matrix $[A]$ and the vector $\{B\}=\{R\}$ of Eq. (7.111)
SOLVEQ	Solves the system of linear equations $[A]\{X\}=\{R\}$ using Gauss elimination
REORDEREL	Rearranges the boundary values and forms the matrices $\{u\}$, $\{v\}$, $\{t_x\}$, and $\{t_y\}$
UVINTER	Computes the displacements u and v at the internal points using Eq. (7.112)
STRESSB	Computes the stresses σ_x, σ_y, and τ_{xy} at the boundary nodal points using Eqs. (7.120)
STRESSIN	Computes the stresses σ_x, σ_y and τ_{xy} at the internal points using Eq. (7.113)
OUTPUTEL	Writes the results in OUTPUTFILE

The variables and the arrays used in the program together with their meaning are given below:

N	Total number of boundary elements and hence of boundary nodes for constant boundary elements
IN	Number of internal points, where the displacements and the stresses are computed

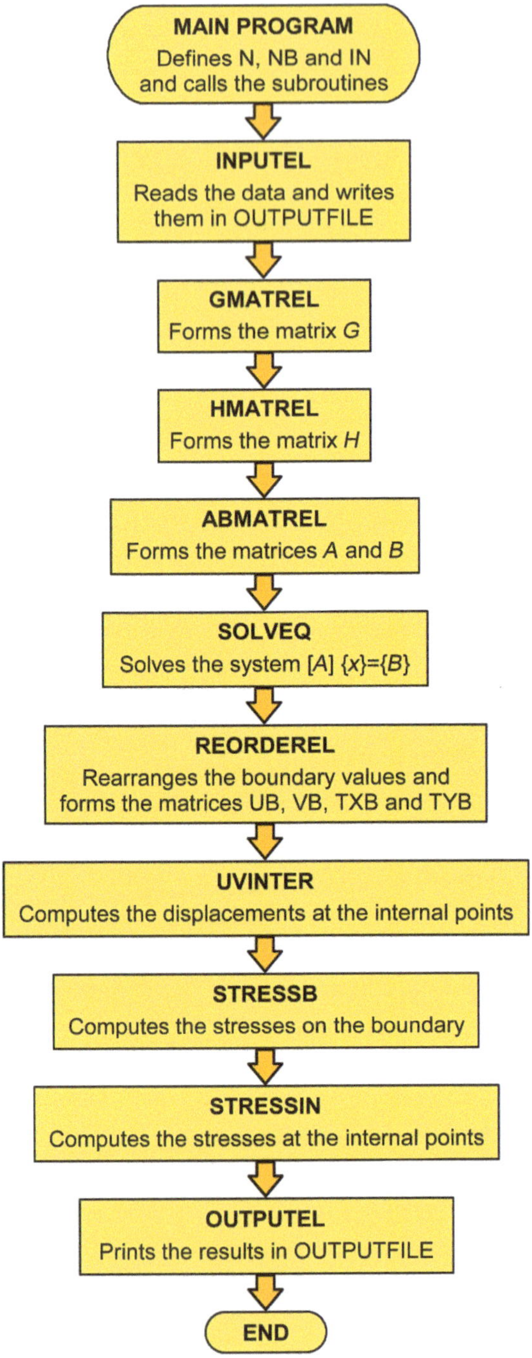

FIGURE 7.7 Macro flow chart of program ELBECON.

NB	Number of boundaries
NL	One-dimensional array of dimension NB containing the serial number of the last element of each boundary
IPLANE	Integer specifying the type of elasticity problem. IPLANE = 0 for plane strain; IPLANE = 1 for plane stress
EL	Young's modulus of elasticity, E
GL	Shear modulus, G
PN	Poisson's ratio, ν
KCODE	One-dimensional array specifying the type of boundary conditions at node I ($I = 1,2,\ldots,N$) taking the values: KCODE(I) = 1 when u and v are prescribed KCODE(I) = 2 when t_x and t_y are prescribed KCODE(I) = 3 when u and t_y are prescribed KCODE(I) = 4 when t_x and v are prescribed
XL, YL	One-dimensional arrays containing the x and y coordinates of the extreme points of the elements
XM, YM	One-dimensional arrays containing the x and y coordinates of the boundary nodes
XIN, YIN	One-dimensional arrays containing the x and y coordinates of the internal points, where the displacements and stresses are computed
R	One-dimensional array defined in Eq. (7.111)
UB, VB	One-dimensional arrays. At input they contain the prescribed boundary values, that is, displacements and/or tractions. At output they contain the boundary nodal values of u and v
TXB, TYB	One-dimensional arrays containing the values of the boundary tractions t_x and t_y
UIN, VIN	One-dimensional arrays containing the computed values of the displacements u and v at the internal points
SXIN, SYIN, SXYIN	One-dimensional arrays containing the computed values of the stress components σ_x, σ_y, and τ_{xy} at the IN internal points
SXB, SYB, SXYB	One-dimensional arrays containing the computed values of the stress components σ_x, σ_y, and τ_{xy} at the N boundary points

7.12.2 INPUTEL subroutine

The INPUTEL subroutine reads all the data required by the program using free FORMAT. The data have been written in INPUTFILE, to which the user has given a specific name as required by the main program. This file contains the following data:

1. *User's name*. One line containing the name of the user.
2. *Title*. One line containing the title of the program.

3. *The code number of the problem type.* IPLANE = 0 for plane strain; IPLANE = 1 for plane stress.
4. *Elastic constants.* Modulus of elasticity EL and Poisson's ratio PN.
5. The serial number of the last element of each boundary. NB integers forming the entries of the array NL.
6. *The extreme points of the boundary elements.* N couples of values consisting of the coordinates XL, YL of the extreme points of the boundary elements. They are read in the positive sense, that is, counter-clockwise on the external boundary, and clockwise on the internal ones.
7. *Boundary conditions.* N triples of numbers consisting of the values of KCODE, u or t_x and v or t_y. More specifically:

   KCODE = 1: u, v

   KCODE = 2: t_x, t_y

   KCODE = 3: u, t_y

   KCODE = 4: t_x, v
8. *Coordinates of the internal points.* IN couples of values consisting of the coordinates XIN and YIN of the internal points, where the values of the displacements u,v and stresses σ_x, σ_y, τ_{xy} will be computed.

Finally, the subroutine INPUTEL writes the data in OUTPUTFILE, to which the user gives a specific name.

7.12.3 GMATREL subroutine

The GMATREL subroutine forms the matrix $[G]$ defined by Eq. (7.108). First, the submatrices $[G]^{ij}(i,j=1,2,\ldots,N)$, given by Eq. (7.103) are evaluated and subsequently placed in matrix $[G]$. The elements of the matrix $[G]^{ij}$ are line integrals of the fundamental solution along the constant boundary element. We distinguish two cases for the position of the elements with respect to the diagonal of the matrix:

i. *Off-diagonal elements, $i \neq j$*

In this case the reference point $P_i(x_i, y_i)$ lies outside the j-th element, over which the integration is performed, and thus the distance $r = |q - P_i|$ never vanishes. Consequently, the line integrals in Eq. (7.103) are regular and they are evaluated using a four-point Gauss integration by calling the RLINTG subroutine (see program LABECON).

ii. *Diagonal elements, $i = j$*

In this case the reference point $P_i(x_i, y_i)$ lies on the j-th element, over which the integration is performed. The distance $r = |q - P_i|$ vanishes for $P_i \equiv q$ and consequently the line integrals in Eq. (7.103) become singular. For this reason, their evaluation is carried out analytically in the interval $[\varepsilon, \ell_i/2]$ and then, we take the limit $\varepsilon \to 0$. It should be mentioned that the integral in the interval $[\varepsilon, \ell_i/2]$ is equal to that in $[-\ell_i/2, -\varepsilon]$.

Noting that

$$r_{,x} = \cos\alpha = \frac{x_{i+1} - x_i}{\ell_i}, \quad r_{,y} = \sin\alpha = \frac{y_{i+1} - y_i}{\ell_i} \tag{7.141}$$

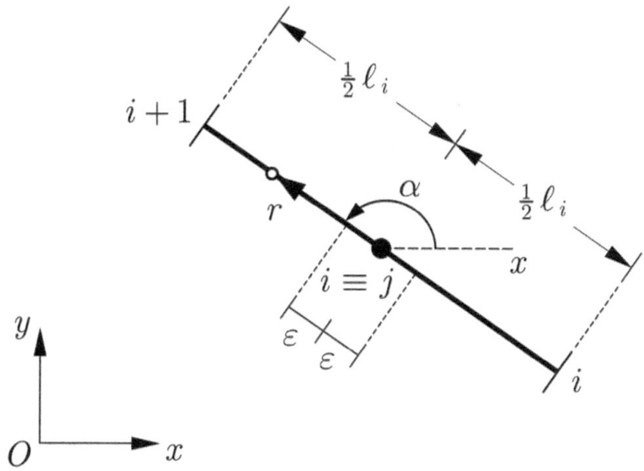

FIGURE 7.8 Element subdivision for the analytical evaluation of the singular integrals of submatrices $[G]^{ii}$.

where $\alpha = \text{angle}\,(\mathbf{x}, \mathbf{r}) = \text{constant}$ for each element (see Fig. 7.8), Eqs. (7.58), (7.60), and (7.103) yield

$$\begin{aligned} G_{11}^{ii} &= \int_{\Gamma_i} U_{\xi x}\, ds = \int_{\Gamma_i} U_{x\xi}\, ds \\ &= -\frac{1}{8\pi G}\lim_{\varepsilon\to 0}\left\{ 2(3-\overline{\nu})\int_{\varepsilon}^{\ell_i/2} \ln r\, dr + \left[-2(1+\overline{\nu})\cos^2\alpha + \frac{7-\overline{\nu}}{2}\right]\int_{\varepsilon}^{\ell_i/2} dr \right\} \\ &= -\frac{1}{8\pi G}\lim_{\varepsilon\to 0}\left\{ 2(3-\overline{\nu})\left[r\ln r - r\right]_{\varepsilon}^{\ell_i/2} + \left[-2(1+\overline{\nu})\cos^2\alpha + \frac{7-\overline{\nu}}{2}\right]\left[r\right]_{\varepsilon}^{\ell_i/2} \right\} \\ &= -\frac{\ell_i}{8\pi G}\left[(3-\overline{\nu})\left(\ln\frac{\ell_i}{2} - 1\right) - (1+\overline{\nu})\cos^2\alpha + \frac{7-\overline{\nu}}{2}\right] \end{aligned} \tag{7.142a}$$

$$\begin{aligned} G_{12}^{ii} &= G_{21}^{ii} = \int_{\Gamma_i} U_{\xi y}\, ds = \int_{\Gamma_i} U_{y\xi}\, ds \\ &= \frac{1}{8\pi G}\lim_{\varepsilon\to 0}\left\{ (1+\overline{\nu})\, 2\int_{\varepsilon}^{\ell_i/2} \cos\alpha \sin\alpha\, dr \right\} \\ &= \frac{1}{8\pi G}(1+\overline{\nu})\ 2\cos\alpha\ \sin\alpha \int_{0}^{\ell_i/2} dr \\ &= \frac{\ell_i}{8\pi G}(1+\overline{\nu})\cos\alpha\ \sin\alpha \end{aligned} \tag{7.142b}$$

$$
\begin{aligned}
G_{22}^{ii} &= \int_{\Gamma_i} U_{\eta y}\, ds = \int_{\Gamma_i} U_{y\eta}\, ds \\
&= -\frac{1}{8\pi G}\lim_{\varepsilon\to 0}\left\{ 2(3-\overline{\nu})\int_{\varepsilon}^{\ell_i/2} \ln r\, dr + \left[-2(1+\overline{\nu})\sin^2\alpha + \frac{7-\overline{\nu}}{2}\right]\int_{\varepsilon}^{\ell_i/2} dr \right\} \\
&= -\frac{1}{8\pi G}\lim_{\varepsilon\to 0}\left\{ 2(3-\overline{\nu})\big[r\ln r - r\big]_{\varepsilon}^{\ell_i/2} + \left[-2(1+\overline{\nu})\sin^2\alpha + \frac{7-\overline{\nu}}{2}\right]\big[r\big]_{\varepsilon}^{\ell_i/2} \right\} \\
&= -\frac{\ell_i}{8\pi G}\left[(3-\overline{\nu})\left(\ln\frac{\ell_i}{2}-1\right) - (1+\overline{\nu})\sin^2\alpha + \frac{7-\overline{\nu}}{2}\right]
\end{aligned}
\tag{7.142c}
$$

7.12.4 HMATREL subroutine

The HMATREL subroutine forms the matrix $[H]$ defined by Eq. (7.107). First, it evaluates the submatrices $[\hat{H}]^{ij}$ $(i, j = 1, 2, \ldots, N)$ using Eq. (7.104), which are then placed in matrix $[\hat{H}]$. We distinguish again two cases for the position of the elements with respect to the diagonal of the matrix:

i. *Off-diagonal elements,* $i \neq j$

As it was explained in the case of matrices $[G]^{ij}$, the distance $r = |q - P_i|$ does not vanish. Therefore, the line integrals are always regular and can be evaluated using a four-point Gauss integration by calling subroutine RLINTH (see program LABECON).

ii. *Diagonal elements,* $i = j$

In this case the reference point $P_i(x_i, y_i)$ lies on the j-th element over which the integration is performed. The distance $r = |q - P_i|$ vanishes for $P_i \equiv q$ and consequently the line integrals of Eq. (7.104) become singular.

The integration of the kernels given by Eqs. (7.71) and (7.72) requires evaluation of singular integrals having the form

$$
\int_{\Gamma_i} \frac{r_{,n}}{r}\, ds \quad \text{and} \quad \int_{\Gamma_i} \frac{r_{,t}}{r}\, ds
$$

which can also be written as (see Section 4.3)

$$
\int_{\Gamma_i} \frac{\cos\phi}{r}\, ds \quad \text{and} \quad \int_{\Gamma_i} \frac{\sin\phi}{r}\, ds
$$

where $\phi = \text{angle}(\mathbf{r}, \mathbf{n})$ (see Fig. 4.4).

The first integral equals to zero as it was proved in Eq. (4.28b). Moreover, by virtue of Eqs. (4.21) and (4.27) the second integral is written

$$
\begin{aligned}
\int_{\Gamma_i} \frac{\sin\phi}{r}\, ds &= \int_{-1}^{1} \frac{\sin\phi}{|\xi|}\, d\xi \\
&= -\int_{0}^{1} \frac{d|\xi|}{|\xi|} + \int_{0}^{-1} \frac{d|\xi|}{|\xi|} = 0
\end{aligned}
$$

since $\phi = 3\pi/2$ when $0 \le \xi \le 1$, and $\phi = \pi/2$ when $-1 \le \xi \le 0$. Hence

$$[\hat{H}]^{ii} = [0]$$

and

$$[H]^{ii} = [\hat{H}]^{ii} + \frac{1}{2}[I] = \begin{bmatrix} 1/2 & 0 \\ 0 & 1/2 \end{bmatrix} \tag{7.143}$$

7.12.5 ABMATREL subroutine

The ABMATREL subroutine rearranges the columns of matrices $[G]$ and $[H]$, and creates the matrix $[A]$ and the vector $\{R\}$ of Eq. (7.111). The columns of matrix $[A]$ consist of all the columns of matrices $[G]$ and $[H]$ that correspond to the unknown boundary values of u, v, t_x, and t_y. The vector $\{R\}$ results as the sum of those columns of $[G]$ and $[H]$ which correspond to the known values of u, v, t_x, and t_y after they have been multiplied by the respective values. It should be noted that a change of sign occurs, when the columns of $[G]$ or $[H]$ are transferred to the other side of Eq. (7.106).

7.12.6 SOLVEQ subroutine

The SOLVEQ subroutine solves the system of equations $AX = R$ calling the subroutine LEQS. The solution is obtained by Gauss elimination and stored in the vector R. The output parameter LSING takes the value LSING = 0, when the matrix A is regular, or LSING = 1, when the matrix A is singular (see program LABECON).

7.12.7 REORDEREL subroutine

The REORDEREL subroutine rearranges the vector $\{R\}$ on the basis of the given boundary conditions and forms the vectors $\{u\}$, $\{v\}$, $\{t_x\}$ and $\{t_y\}$ of all boundary displacements and tractions.

7.12.8 UVINTER subroutine

The UVINTER subroutine computes the displacements u and v at internal points utilizing Eqs. (7.112). The matrices $[G]^{ij}$ and $[\hat{H}]^{ij}$ are evaluated by subroutines RLINTG and RLINTH. This is possible because $r = |P - q| \neq 0$ for all points $P \in \Omega$ and thus the involved line integrals are always regular.

For the evaluation of G_{11}^{ij} and G_{22}^{ij} the constant term in the expressions of $U_{\xi x}$ and $U_{\eta y}$ was neglected, since it does not influence the displacements. Indeed, the contribution of a constant C to Eq. (7.76) is

$$\left.\begin{aligned} C\left(\int_\Omega b_\xi \, d\Omega + \int_\Gamma t_\xi \, ds\right) &= 0 \quad \text{in } u(p) \\ C\left(\int_\Omega b_\eta \, d\Omega + \int_\Gamma t_\eta \, ds\right) &= 0 \quad \text{in } v(p) \end{aligned}\right\} \tag{7.144}$$

according to the equilibrium equations of the external forces acting on the body in the x- and y-directions, respectively.

7.12.9 STRESSB subroutine

The STRESSB subroutine computes the stresses σ_x, σ_y and τ_{xy} at the boundary nodal points. First, it evaluates the derivatives $u_{,t}$ and $v_{,t}$ in the direction of the tangent to the boundary via finite differences (see Section 6.2.2), and then computes the stresses using Eqs. (7.118) and (7.120).

7.12.10 STRESSIN subroutine

The STRESSIN subroutine computes the stresses σ_x, σ_y and τ_{xy} at the internal points on the basis of Eq. (7.113).

7.12.11 OUTPUTEL subroutine

The OUTPUTEL subroutine writes all the results in the output file.

The listing of program ELBECON is given below. The electronic version of the code is available on this book's companion website.

```
C=====================================================================
C

      PROGRAM ELBECON
C
C
C   This program solves the two dimensional (EL)asticiy problem
C   using the (B)oundary (E)lement method with (CON)stant
C   boundary elements for domain with multiple boundaries
C
      USE MSIMSL
      IMPLICIT REAL*8 (A-H,O-Z)
      CHARACTER*15 INPUTFILE,OUTPUTFILE
C
C   Set the maximum dimensions
C
      PARAMETER (N=44)
      PARAMETER (IN=3)
      PARAMETER (NB=1)
C

C   N= Number of boundary elements equal to number of boundary
C      nodes
C
C  IN= Number of internal points where the function u is calculated
C
C  NB= Number of boundaries of the multiple boundary domain
C
      COMMON EL,GL,PN,ALAMDA
      DIMENSION NL(NB),KCODE(N)
      DIMENSION XL(N+1),YL(N+1),XM(N),YM(N),G(2*N,2*N),H(2*N,2*N)
      DIMENSION UB(N),VB(N),TXB(N),TYB(N),A(2*N,2*N),R(2*N)
      DIMENSION XIN(IN),YIN(IN),UIN(IN),VIN(IN)
      DIMENSION SXIN(IN),SYIN(IN),SXYIN(IN),SXB(N),SYB(N),SXYB(N)
C
C   Read the names and open the input and output files
C
      WRITE (*,'(A)')' Name of the INPUTFILE (max.15 characters)'
      READ  (*,'(A)') INPUTFILE
      WRITE (*,'(A)')' Name of the OUTPUTFILE (max.15 characters)'
      READ  (*,'(A)') OUTPUTFILE
      OPEN (1, FILE=INPUTFILE)
      OPEN (2, FILE=OUTPUTFILE)
```

```
C
C  Read data from INPUTFILE
C
      CALL INPUTEL(XL,YL,XIN,YIN,KCODE,IPLANE,NL,UB,VB,N,IN,NB)
C
C  Compute the G matrix
C
      CALL GMATREL(XL,YL,XM,YM,G,N,NL,NB)
C
C  Compute the H matrix
C
      CALL HMATREL(XL,YL,XM,YM,H,N,NL,NB)
C
C  Form the system of equations AX=B
C
      CALL ABMATREL(G,H,A,R,UB,VB,KCODE,N)
C
C  Solve the system of equations
C
      CALL SOLVEQ(A,R,N,LSING)
C
C  Form the vectors U and UN of all the boundary values
C
      CALL REORDEREL(R,UB,VB,TXB,TYB,KCODE,N)
C
C  Compute the values of the displacements at the internal points
C
      CALL UVINTER(XL,YL,XIN,YIN,UB,VB,TXB,TYB,UIN,VIN,N,IN,NL,NB)
C
C  Compute the values of the stresses at the internal points
C
      CALL STRESSIN(XL,YL,XIN,YIN,UB,VB,TXB,TYB,
     1                    SXIN,SYIN,SXYIN,N,IN,NL,NB)
C
C  Compute the values of the boundary stresses
C
      CALL STRESSB(XL,YL,UB,VB,TXB,TYB,SXB,SYB,SXYB,N,NL,NB)
C
C  Print the results in the OUTPUTFILE
C
      CALL OUTPUTEL(XM,YM,UB,VB,TXB,TYB,XIN,YIN,UIN,VIN,
     1               SXB,SYB,SXYB,SXIN,SYIN,SXYIN,N,IN)
C
C  Close input and output files
C
      CLOSE(1)
      CLOSE(2)
      STOP
      END
C
C
C======================================================================
C
      SUBROUTINE INPUTEL(XL,YL,XIN,YIN,KCODE,IPLANE,NL,UB,VB,N,IN,NB)
C
C
C  This subroutine reads the input data from the input file
C  and writes them in the output file
C
      IMPLICIT REAL*8 (A-H,O-Z)
      COMMON EL,GL,PN,ALAMDA
      CHARACTER*80 NAME,TITLE
      DIMENSION NL(NB),KCODE(N)
```

```
      DIMENSION UB(N),VB(N)
      DIMENSION XL(N+1),YL(N+1),XIN(IN),YIN(IN)
C
      WRITE(2,100)
  100 FORMAT('',76('*'))
      READ(1,'(A)') NAME
      WRITE(2,'(A)') NAME
C
C  Read the title of the program
C
      READ(1,'(A)')TITLE
      WRITE(2,'(A)')TITLE
C
      WRITE(2,200)N,IN,NB
  200 FORMAT(//'DATA'//2X,'NUMBER OF BOUNDARY ELEMENTS ='
     1      ,I3/2X,'NUMBER OF INTERNAL POINTS   =',I3/2X,
     1       'NUMBER OF BOUNDARIES         =',I3)
C
C  Read the type of the problem: IPLANE=0 for plane strain,
C                                IPLANE=1 for plane stress
C
      READ(1,*)IPLANE
      IF(IPLANE.EQ.0)THEN
      WRITE(2,700)
  700 FORMAT(/'  PLANE STRAIN PROBLEM'/)
      ELSE
      WRITE(2,450)
  450 FORMAT(/'  PLANE STRESS PROBLEM'/)
      ENDIF
C
C  Read the elastic constants
C
      READ(1,*) EL,PN
C
      WRITE(2,150) EL,PN
  150 FORMAT(2X,'ELASTIC CONSTANTS:','   ELASTIC MODULUS =',E10.4,
     1    '  POISSON RATIO =',F5.2/)
C
      IF(IPLANE.EQ.0)THEN
      EL=EL/(1.-PN**2)
      PN=PN/(1-PN)
      GL=EL/(2.*(1+PN))
      ALAMDA=PN*EL/(1-PN**2)
      ELSEIF(IPLANE.NE.0)THEN
      GL=EL/(2.*(1+PN))
      ALAMDA=PN*EL/(1-PN**2)
      ENDIF
C
C  Read the of the last element at each boundary
C
      READ(1,*)(NL(I), I=1,NB)
C
C  Read the coordinates of the extreme points of the boundary
C  elements XL,YL
C
      READ(1,*)  (XL(I),YL(I),I=1,N)
C
C  Write the coordinates in the output file
C
      WRITE(2,300)
  300 FORMAT(//'  COORDINATES OF THE EXTREME POINTS OF THE BOUNDARY',
     1          ' ELEMENTS',//1X,'NODE',9X,'  XL',13X,'  YL'/)
      DO 20 I=1,N
   20 WRITE(2,400) I,XL(I),YL(I)
```

```
  400 FORMAT(2X,I3,2(3X,E14.5))
C
C  Read the boundary conditions and store in UB(I) and VB(I)
C
C    KCODE(I)=1 when u  and v  are prescribed
C    KCODE(I)=2 when tx and ty are prescribed
C    KCODE(I)=3 when u  and ty are prescribed
C    KCODE(I)=4 when tx and v  are prescribed
C
      READ(1,*) (KCODE(I),UB(I),VB(I),I=1,N)
C
C  Write the boundary conditions in the output file
C
      WRITE(2,500)
  500 FORMAT(//2X,'BOUNDARY CONDITIONS'//2X,'NODE',6X,'  KCODE',
     1 20X,'PRESCRIBED VALUES'/)
      DO 30 I=1,N
   30 WRITE(2,600) I,KCODE(I),UB(I), VB(I)
  600 FORMAT (2X,I3,9X,I3,8X,E14.5,8X,E14.5)
C
C  Read the coordinates of the internal points
C
      READ(1,*) (XIN(I),YIN(I),I=1,IN)
      RETURN
      END
C
C=====================================================================
C
      SUBROUTINE GMATREL (XL,YL,XM,YM,G,N,NL,NB)
C
C  This subroutine computes the elements of the G matrix
C
      IMPLICIT REAL*8 (A-H,O-Z)
      COMMON EL,GL,PN,ALAMDA
      DIMENSION XL(N+1),YL(N+1),XM(N),YM(N)
      DIMENSION NL(NB),G(2*N,2*N)
C
C  Compute the nodal coordinates and store them in the arrays XM
C  and YM
C
      PI=ACOS(-1.)
      XL(N+1)=XL(1)
      YL(N+1)=YL(1)
      DO 10  I=1,N
      XM(I)=(XL(I)+XL(I+1))/2.
   10 YM(I)=(YL(I)+YL(I+1))/2.
      IF(NB.LE.1) GOTO 40
      XM(NL(1))=(XL(NL(1))+XL(1))/2.
      YM(NL(1))=(YL(NL(1))+YL(1))/2.
      DO 15 K=2,NB
      XM(NL(K))=(XL(NL(K))+XL(NL(K-1)+1))/2.
   15 YM(NL(K))=(YL(NL(K))+YL(NL(K-1)+1))/2.
C
C  Compute the elements of matrix G
C
   40 DO 20 I=1,N
      X0=XM(I)
      Y0=YM(I)
      DO 20 J=1,N
      X1=XL(J)
      Y1=YL(J)
      X2=XL(J+1)
      Y2=YL(J+1)
      IF(NB.LE.1) GOTO 60
```

```
      IF(J.NE.NL(1))GOTO 50
            X2=XL(1)
            Y2=YL(1)
            GOTO 60
   50     DO 30 K=2,NB
          IF(J.NE.NL(K))GOTO 30
            X2=XL(NL(K-1)+1)
            Y2=YL(NL(K-1)+1)
            GOTO 60
   30       CONTINUE
   60  IF(I.NE.J)THEN
C
C       IFLAG=1: for the boundary nodes:  IFLAG=2: for the internal points
C
      IFLAG=1
      CALL RLINTG(X0,Y0,X1,Y1,X2,Y2,RES11,RES12,RES22,IFLAG)
      G(2*I-1,2*J-1)=RES11
      G(2*I-1,2*J)=RES12
      G(2*I,2*J-1)=RES12
      G(2*I,2*J)=RES22
      ELSEIF(I.EQ.J)THEN
      DX=X2-X1
      DY=Y2-Y1
      SL=DSQRT(DX**2+DY**2)
      COSA=DX/SL
      SINA=DY/SL
      SPIG=SL/(8.*PI*GL)
C
       C3=(7.-PN)/2.
C
      RES11=-SPIG*((3.-PN)*(DLOG(SL/2.)-1.)-(1.+PN)*COSA**2+C3)
      RES12=SPIG*(1.+PN)*COSA*SINA
      RES22=-SPIG*((3.-PN)*(DLOG(SL/2.)-1.)-(1.+PN)*SINA**2+C3)
      G(2*I-1,2*J-1)=RES11
      G(2*I-1,2*J)=RES12
      G(2*I,2*J-1)=RES12
      G(2*I,2*J)=RES22
      ENDIF
   20 CONTINUE
      RETURN
      END
C
C==================================================================
C
      SUBROUTINE RLINTG(X0,Y0,X1,Y1,X2,Y2,RES11,RES12,RES22,IFLAG)
C
C  This subroutine computes the off-diagonal elements of the
C  matrix G
C
C  RA= The distance of the point O to the Gauss integration point
C      on the boundary element
C
C  WG= The weights of the Gauss integration
C
C  XI= The coordinates of the Gauss integration points in the
C      interval [-1,1]
C
C  XC,YC=The global coordinates of the Gauss integration points
C
      IMPLICIT REAL*8 (A-H,O-Z)
      COMMON EL,GL,PN,ALAMDA
      DIMENSION XC(4),YC(4),XI(4),WG(4)
      DATA XI/-0.86113631,-0.33998104,0.33998104,0.86113631/
      DATA WG/0.34785485,0.65214515,0.65214515,0.34785485/
```

```
      PI=ACOS(-1.)
      AX=(X2-X1)/2.
      AY=(Y2-Y1)/2.
      BX=(X2+X1)/2.
      BY=(Y2+Y1)/2.
C
C  Compute the Jacobian
C
      SL=DSQRT(AX**2+AY**2)
C
C  Compute the line integral
C
      RES11=0.
      RES12=0.
      RES22=0.
C
      PIG=1./(8.*PI*GL)
      DO 30 I=1,4
      XC(I)=AX*XI(I)+BX
      YC(I)=AY*XI(I)+BY
      DX=XC(I)-X0
      DY=YC(I)-Y0
      RA=DSQRT(DX**2+DY**2)
      COSA=DX/RA
      SINA=DY/RA
C
        IF (IFLAG.EQ.1) THEN
        C3=(7.-PN)/2.
        ELSEIF (IFLAG.EQ.2) THEN
        C3=0.
        ENDIF
C
      RES11=RES11-PIG*((3.-PN)*DLOG(RA)-(1+PN)*COSA**2+C3)*WG(I)*SL
      RES12=RES12+PIG*(1.+PN)*COSA*SINA*WG(I)*SL
   30 RES22=RES22-PIG*((3.-PN)*DLOG(RA)-(1+PN)*SINA**2+C3)*WG(I)*SL
      RETURN
      END
C
C=================================================================
C
      SUBROUTINE HMATREL(XL,YL,XM,YM,H,N,NL,NB)
C
C  This subroutine computes the elements of the H matrix
C
      IMPLICIT REAL*8 (A-H,O-Z)
      COMMON EL,GL,PN,ALAMDA
      DIMENSION XL(N+1),YL(N+1),XM(N),YM(N),NL(NB)
      DIMENSION H(2*N,2*N)
C
C  Compute the nodal coordinates and store them in the arrays XM
C  and YM
C
      PI=ACOS(-1.)
      XL(N+1)=XL(1)
      YL(N+1)=YL(1)
      DO 10  I=1,N
      XM(I)=(XL(I)+XL(I+1))/2.
   10 YM(I)=(YL(I)+YL(I+1))/2.
      IF(NB.LE.1) GOTO 40
      XM(NL(1))=(XL(NL(1))+XL(1))/2.
      YM(NL(1))=(YL(NL(1))+YL(1))/2.
      DO 15 K=2,NB
      XM(NL(K))=(XL(NL(K))+XL(NL(K-1)+1))/2.
   15 YM(NL(K))=(YL(NL(K))+YL(NL(K-1)+1))/2.
```

```
C
C  Compute the elements of matrix H
C
   40 DO 20 I=1,N
      X0=XM(I)
      Y0=YM(I)
      DO 20 J=1,N
      X1=XL(J)
      Y1=YL(J)
      X2=XL(J+1)
      Y2=YL(J+1)
      IF(NB.LE.1) GOTO 60
      IF(J.NE.NL(1))GOTO 50
            X2=XL(1)
            Y2=YL(1)
            GOTO 60
   50      DO 30 K=2,NB
          IF(J.NE.NL(K))GOTO 30
            X2=XL(NL(K-1)+1)
            Y2=YL(NL(K-1)+1)
            GOTO 60
   30       CONTINUE
   60 IF(I.NE.J)THEN
      CALL RLINTH(X0,Y0,X1,Y1,X2,Y2,RES11,RES12,RES21,RES22)
      H(2*I-1,2*J-1)=RES11
      H(2*I-1,2*J)=RES21
      H(2*I,2*J-1)=RES12
      H(2*I,2*J)=RES22
      ELSEIF (I.EQ.J) THEN
      H(2*I-1,2*J-1)=.5
      H(2*I-1,2*J)=0.
      H(2*I,2*J-1)=0.
      H(2*I,2*J)=.5
      ENDIF
   20 CONTINUE
      RETURN
      END
C
C==================================================================
C
      SUBROUTINE RLINTH(X0,Y0,X1,Y1,X2,Y2,RES11,RES12,RES21,RES22)
C
C
C  This subroutine computes the off-diagonal elements of the
C  matrix H
C
C  RA= The distance of the point O to the Gauss integration point
C      on the boundary element
C  WG= The weights of the Gauss integration
C  XI= The coordinates of the Gauss integration points in the
C      interval [-1,1]
C  XC,YC=The global coordinates of the Gauss integration points
C
      IMPLICIT REAL*8 (A-H,O-Z)
      COMMON EL,GL,PN,ALAMDA
      DIMENSION XC(4),YC(4),XI(4),WG(4)
      DATA XI/-0.86113631,-0.33998104,0.33998104,0.86113631/
      DATA WG/0.34785485,0.65214515,0.65214515,0.34785485/
      PI=ACOS(-1.)
      AX=(X2-X1)/2.
      AY=(Y2-Y1)/2.
      BX=(X2+X1)/2.
      BY=(Y2+Y1)/2.
```

```
      SL=DSQRT(AX**2+AY**2)
      ENX=AY/SL
      ENY=-AX/SL
C
C  Compute the line integral
C
      RES11=0.
      RES12=0.
      RES21=0.
      RES22=0.
C
      A1=-(1.+PN)/(4.*PI)
      A2=(1.-PN)/(1.+PN)
      DO 30  I=1,4
      XC(I)=AX*XI(I)+BX
      YC(I)=AY*XI(I)+BY
      DX=XC(I)-X0
      DY=YC(I)-Y0
      RA=DSQRT(DX**2+DY**2)
      COSA=DX/RA
      SINA=DY/RA
      RN=COSA*ENX+SINA*ENY
      RT=SINA*ENX-COSA*ENY
      RES11=RES11+(A1/RA)*(A2+2.*COSA**2)*RN*WG(I)*SL
      RES21=RES21+(A1/RA)*(2.*COSA*SINA*RN+A2*RT)*WG(I)*SL
      RES12=RES12+(A1/RA)*(2.*COSA*SINA*RN-A2*RT)*WG(I)*SL
   30 RES22=RES22+(A1/RA)*(A2+2.*SINA**2)*RN*WG(I)*SL
      RETURN
      END
C
C=====================================================================
C
      SUBROUTINE ABMATREL(G,H,A,R,UB,VB,KCODE,N)
C
C
C  This subroutine rearranges the matrices G and H and produces the
C  matrices A and R
C
      IMPLICIT REAL*8 (A-H,O-Z)
      DIMENSION G(2*N,2*N),H(2*N,2*N),A(2*N,2*N)
      DIMENSION UB(N),VB(N),KCODE(N),R(2*N)
C
C  Reorder the columns the system of equations and store them in A
C
      N2=2*N
      DO 20 I=1,N2
      R(I)=0.
      DO 20 J=1,N2
   20 A(I,J)=0.
      DO 70 J=1,N
      GOTO(30,40,50,60)KCODE(J)
   30 DO 35 I=1,N2
      A(I,2*J-1)=-G(I,2*J-1)
      A(I,2*J)=-G(I,2*J)
   35 R(I)=R(I)-H(I,2*J-1)*UB(J)-H(I,2*J)*VB(J)
      GOTO 70
   40 DO 45 I=1,N2
      A(I,2*J-1)=H(I,2*J-1)
      A(I,2*J)=H(I,2*J)
   45 R(I)=R(I)+G(I,2*J-1)*UB(J)+G(I,2*J)*VB(J)
      GOTO 70
   50 DO 55 I=1,N2
      A(I,2*J-1)=-G(I,2*J-1)
      A(I,2*J)=H(I,2*J)
```

```
   55 R(I)=R(I)-H(I,2*J-1)*UB(J)+G(I,2*J)*VB(J)
      GOTO 70
   60 DO 65 I=1,N2
      A(I,2*J-1)=H(I,2*J-1)
      A(I,2*J)=-G(I,2*J)
   65 R(I)=R(I)+G(I,2*J-1)*UB(J)-H(I,2*J)*VB(J)
   70 CONTINUE
      RETURN
      END
C
C=====================================================================
C
      SUBROUTINE SOLVEQ(A,R,N,LSING)
C
C
      IMPLICIT REAL*8 (A-H,O-Z)
      DIMENSION A(2*N,2*N),R(2*N)
      NN=2*N
      CALL LEQS(A,R,NN,LSING)
      IF(LSING.EQ.0)THEN
      WRITE(2,150)
  150 FORMAT(/,'',76('*')//2X'The system has been solved regularly'/)
      ELSEIF(LSING.EQ.1)THEN
      WRITE(2,170)
  170 FORMAT(/,'',76('*')//2X'The system is singular'/)
      ENDIF
      RETURN
      END
C
C=====================================================================
C
      SUBROUTINE LEQS(A,B,N,LSING)
C
C    This subroutine uses Gauss elimination to solve
C    a system of linear equations, [A]{X}={B}, where
C    A : One-dimensional array which contains the occasional row of
C        the two-dimensional array of the coefficients of the unknowns
C    B : One-dimensional array which contains the known coefficients
C    N : Integer denoting the number of the unknowns
C    LSING: Integer taking the values:
C           LSING = 0, if the system has been solved regularly
C           LSING = 1, if the system is singular
C
      IMPLICIT REAL*8 (A-H,O-Z)
      DIMENSION A(1),B(1)
C
      LSING=0
      JJ=-N
      DO 10 J=1,N
        JY=J+1
        JJ=JJ+N+1
        AMAX=0.0
        IHELP=JJ-J
      DO 20 I=J,N
        IJ=IHELP+I
        IF(ABS(AMAX)-ABS(A(IJ))) 30,20,20
   30      AMAX=A(IJ)
           IMAX=I
   20 CONTINUE
         IF(ABS(AMAX).EQ.0.) THEN
            LSING=1
            RETURN
         END IF
      I1=J+N*(J-2)
```

```
         IHELP=IMAX-J
         DO 40 K=J,N
           I1=I1+N
           I2=I1+IHELP
           ATEMP=A(I1)
           A(I1)=A(I2)
           A(I2)=ATEMP
      40 A(I1)=A(I1)/AMAX
         ATEMP=B(IMAX)
         B(IMAX)=B(J)
         B(J)=ATEMP/AMAX
            IF(J-N) 50,70,50
      50          IQS=N*(J-1)
                  DO 10 IX=JY,N
                    IXJ=IQS+IX
                    IHELP=J-IX
                    DO 60 JX=JY,N
                      IJREF=N*(JX-1)+IX
                      JJX=IJREF+IHELP
      60              A(IJREF)=A(IJREF)-(A(IXJ)*A(JJX))
      10 B(IX)=B(IX)-B(J)*A(IXJ)
      70    NY=N-1
         NN=N*N
         DO 80 J=1,NY
           I1=NN-J
           I2=N-J
           I3=N
         DO 80 K=1,J
         B(I2)=B(I2)-A(I1)*B(I3)
         I1=I1-N
   80    I3=I3-1
       RETURN
       END
C
C
C====================================================================
C
       SUBROUTINE REORDEREL(R,UB,VB,TXB,TYB,KCODE,N)
C
C
C  This subroutine rearranges the arrays UB and VB
C  in such a way that all boundary values of u and v
C  are stored in UB and VB, while all  boundary
C  values of tx and ty in TXB and TYB
C
       IMPLICIT REAL*8 (A-H,O-Z)
       DIMENSION UB(N),VB(N),R(2*N),TXB(N),TYB(N),KCODE(N)
C
       DO 50  I=1,N
       GOTO(10,20,30,40) KCODE(I)
    10 TXB(I)=R(2*I-1)
       TYB(I)=R(2*I)
       GOTO 50
    20 TXB(I)=UB(I)
       TYB(I)=VB(I)
       UB(I)=R(2*I-1)
       VB(I)=R(2*I)
       GOTO 50
    30 TYB(I)=VB(I)
       TXB(I)=R(2*I-1)
       VB(I)=R(2*I)
       GOTO 50
    40 TXB(I)=UB(I)
       UB(I)=R(2*I-1)
       TYB(I)=R(2*I)
    50 CONTINUE
```

```
C
      RETURN
      END
C
C=====================================================================
C
      SUBROUTINE UVINTER(XL,YL,XIN,YIN,UB,VB,TXB,TYB,UIN,VIN,N,IN,NL,NB)
C
C  This subroutine computes the values of the displacements u and v
C  at the internal points
C
      IMPLICIT REAL*8 (A-H,O-Z)
      DIMENSION XL(N+1),YL(N+1),XIN(IN),YIN(IN),UB(N),VB(N)
      DIMENSION TXB(N),TYB(N),UIN(IN),VIN(IN)
      DIMENSION NL(NB)
C
C  Compute the values of u and v at the internal points
C
      IF(NB.GT.1) GOTO 5
      XL(N+1)=XL(1)
      YL(N+1)=YL(1)
C
    5 DO 10  KK=1,IN
      UIN(KK)=0.
      VIN(KK)=0.
      DO 20  J=1,N
      X0=XIN(KK)
      Y0=YIN(KK)
      X1=XL(J)
      Y1=YL(J)
      X2=XL(J+1)
      Y2=YL(J+1)
      IF(NB.LE.1)GOTO 60
        IF(J.NE.NL(1))GOTO 50
            X2=XL(1)
            Y2=YL(1)
            GOTO 60
   50      DO 30 K=2,NB
          IF(J.NE.NL(K))GOTO 30
            X2=XL(NL(K-1)+1)
            Y2=YL(NL(K-1)+1)
            GOTO 60
   30       CONTINUE
C
   60 IFLAG=2
C
      CALL RLINTG(X0,Y0,X1,Y1,X2,Y2,G11,G12,G22,IFLAG)
      CALL RLINTH(X0,Y0,X1,Y1,X2,Y2,H11,H12,H21,H22)
      UIN(KK)=UIN(KK)-H11*UB(J)-H21*VB(J)+G11*TXB(J)+G12*TYB(J)
   20 VIN(KK)=VIN(KK)-H12*UB(J)-H22*VB(J)+G12*TXB(J)+G22*TYB(J)
   10 CONTINUE
C
      RETURN
      END
C
C=====================================================================
C

      SUBROUTINE STRESSIN(XL,YL,XIN,YIN,UB,VB,TXB,TYB,
     1                    SXIN,SYIN,SXYIN,N,IN,NL,NB)
C
C
C  This subroutine computes the stresses at the internal points
C
      IMPLICIT REAL*8 (A-H,O-Z)
      DIMENSION XL(N+1),YL(N+1),XIN(IN),YIN(IN),UB(N),VB(N)
      DIMENSION TXB(N),TYB(N)
```

```
      DIMENSION SXIN(IN),SYIN(IN),SXYIN(IN)
      DIMENSION NL(NB)
C
      IF(NB.GT.1) GOTO 5
      XL(N+1)=XL(1)
      YL(N+1)=YL(1)
C
    5 DO 10  KK=1,IN
      SXIN(KK)=0.
      SYIN(KK)=0.
      SXYIN(KK)=0.
      DO 20  J=1,N
      X0=XIN(KK)
      Y0=YIN(KK)
      X1=XL(J)
      Y1=YL(J)
      X2=XL(J+1)
      Y2=YL(J+1)
      IF(NB.LE.1)GOTO 60
        IF(J.NE.NL(1))GOTO 50
            X2=XL(1)
            Y2=YL(1)
            GOTO 60
   50      DO 30 K=2,NB
          IF(J.NE.NL(K))GOTO 30
            X2=XL(NL(K-1)+1)
            Y2=YL(NL(K-1)+1)
            GOTO 60
   30       CONTINUE
   60 CALL SLINTH(X0,Y0,X1,Y1,X2,Y2,RES11,RES21,RES31,RES12,RES22,
     1          RES32,RESB11,RESB21,RESB31,RESB12,RESB22,RESB32)
C
      SXIN(KK)=SXIN(KK)+RES11*TXB(J)+RES12*TYB(J)
     1                 -RESB11*UB(J)-RESB12*VB(J)
      SYIN(KK)=SYIN(KK)+RES21*TXB(J)+RES22*TYB(J)
     1                 -RESB21*UB(J)-RESB22*VB(J)
   20 SXYIN(KK)=SXYIN(KK)+RES31*TXB(J)+RES32*TYB(J)
     1                   -RESB31*UB(J)-RESB32*VB(J)
   10 CONTINUE
      RETURN
      END
C
C=====================================================================
C
      SUBROUTINE SLINTH (X0,Y0,X1,Y1,X2,Y2,RES11,RES21,RES31,RES12,
     1       RES22,RES32,RESB11,RESB21,RESB31,RESB12,RESB22,RESB32)
C
C
C  This subroutine computes the line integrals of the kernels in the
C  integral representations of the stresses
C
C  RA= The distance of the point O to the Gauss integration point
C      on the boundary element
C
C  WG= The weights of the Gauss integration
C
C  XI= The coordinates of the Gauss integration points in the
C      interval [-1,1]
C
C  XC,YC=The global coordinates of the Gauss integration points
C
C
      IMPLICIT REAL*8 (A-H,O-Z)
      COMMON EL,GL,PN,ALAMDA
      DIMENSION XC(4),YC(4),XI(4),WG(4)
      DATA XI/-0.86113631,-0.33998104,0.33998104,0.86113631/
      DATA WG/0.34785485,0.65214515,0.65214515,0.34785485/
```

```
      PI=ACOS(-1.)
      AX=(X2-X1)/2.
      AY=(Y2-Y1)/2.
      BX=(X2+X1)/2.
      BY=(Y2+Y1)/2.
      SL=DSQRT(AX**2+AY**2)
      ENX=AY/SL
      ENY=-AX/SL
C
C  Compute the line integral
C
      RES11=0.
      RES21=0.
      RES31=0.
      RES12=0.
      RES22=0.
      RES32=0.
      RESB11=0.
      RESB21=0.
      RESB31=0.
      RESB12=0.
      RESB22=0.
      RESB32=0.
C
      A1=-(1.+PN)/(4.*PI)
      A2=(1.-PN)/(1.+PN)
      A3=-2.*GL*A1
      DO 30  I=1,4
      XC(I)=AX*XI(I)+BX
      YC(I)=AY*XI(I)+BY
      DX=XC(I)-X0
      DY=YC(I)-Y0
      RA=DSQRT(DX**2+DY**2)
      COSA=-DX/RA
      SINA=-DY/RA
      RN=COSA*ENX+SINA*ENY
C
      RES11=RES11+(A1/RA)*(A2*COSA+2.*COSA**3)*WG(I)*SL
      RES21=RES21+(A1/RA)*(-A2*COSA+2.*COSA*SINA**2)*WG(I)*SL
      RES31=RES31+(A1/RA)*(A2*SINA+2.*COSA**2*SINA)*WG(I)*SL
      RES12=RES12+(A1/RA)*(-A2*SINA+2.*COSA**2*SINA)*WG(I)*SL
      RES22=RES22+(A1/RA)*(A2*SINA+2.*SINA**3)*WG(I)*SL
      RES32=RES32+(A1/RA)*(A2*COSA+2.*COSA*SINA**2)*WG(I)*SL
C
      RESB11=RESB11+(A3/RA**2)*(2.*COSA*RN*(1.-4.*COSA**2)+
     1                          (2.*COSA**2+1.)*ENX)*WG(I)*SL

      RESB21=RESB21+(A3/RA**2)*(-8.*SINA**2*COSA*RN+2.*COSA*SINA*ENY+
     1                                                ENX)*WG(I)*SL

      RESB12=RESB12+(A3/RA**2)*(-8.*COSA**2*SINA*RN+2.*COSA*SINA*ENX+
     1                                                ENY)*WG(I)*SL

      RESB22=RESB22+(A3/RA**2)*(2.*SINA*RN*(1.-4.*SINA**2)+
     1                          (2.*SINA**2+1.)*ENY)*WG(I)*SL
   30 CONTINUE
      RESB31=RESB12
      RESB32=RESB21
      RETURN
      END
C
C
C=====================================================================
C
      SUBROUTINE STRESSB(XL,YL,UB,VB,TXB,TYB,SXB,SYB,SXYB,N,NL,NB)
```

```
C
C
C   This subroutine computes the stress at the boundary
C   nodal points
C
      IMPLICIT REAL*8 (A-H, O-Z)
      COMMON EL,GL,PN,ALAMDA
      DIMENSION XL(N+1),YL(N+1),UB(N),VB(N),TXB(N),TYB(N)
      DIMENSION SXB(N),SYB(N),SXYB(N),SL(N),ENX(N),ENY(N),A(4,4),B(4)
      DIMENSION NL(NB)
C
      IF(NB.GT.1) GOTO 5
      XL(N+1)=XL(1)
      YL(N+1)=YL(1)
C
C   Computation of the element half lengths and direction
C   cosines of the normal vector
C
    5 DO 10 K=1,NB
      IF(K.EQ.1)THEN
      NF=1
      ELSE
      NF=NL(K-1)+1
      ENDIF
      DO 10 I=NF,NL(K)
      IF(I.EQ.NL(K))THEN
      AX=(XL(NF)-XL(NL(K)))/2.
      AY=(YL(NF)-YL(NL(K)))/2.
      SL(I)=DSQRT(AX**2+AY**2)
      ENX(I)=AY/SL(I)
      ENY(I)=-AX/SL(I)
      ELSE
      AX=(XL(I+1)-XL(I))/2.
      AY=(YL(I+1)-YL(I))/2.
      SL(I)=DSQRT(AX**2+AY**2)
      ENX(I)=AY/SL(I)
      ENY(I)=-AX/SL(I)
      ENDIF
   10 CONTINUE
C
C   Computation of the tangential derivatives UT and VT
C
      DO 20 K=1,NB
      IF(K.EQ.1)THEN
      NF=1
      ELSE
      NF=NL(K-1)+1
      ENDIF
      DO 20 I=NF,NL(K)
      IF (I.EQ.NF) THEN
      S1=SL(NL(K))+SL(NF)
      S2=SL(NF)+SL(NF+1)
      U1=UB(NL(K))
      U2=UB(NF)
      U3=UB(NF+1)
      V1=VB(NL(K))
      V2=VB(NF)
      V3=VB(NF+1)
      ELSE IF (I.EQ.NL(K)) THEN
      S1=SL(NL(K)-1)+SL(NL(K))
      S2=SL(NL(K))+SL(NF)
      U1=UB(NL(K)-1)
      U2=UB(NL(K))
      U3=UB(NF)
```

```
      V1=VB(NL(K)-1)
      V2=VB(NL(K))
      V3=VB(NF)
      ELSE
      S1=SL(I-1)+SL(I)
      S2=SL(I)+SL(I+1)
      U1=UB(I-1)
      U2=UB(I)
      U3=UB(I+1)
      V1=VB(I-1)
      V2=VB(I)
      V3=VB(I+1)
      ENDIF
C
C  Computation of the boundary stresses
C
      UT=(S1**2*U3-S2**2*U1+(S2**2-S1**2)*U2)
     1       /(S1*S2*(S1+S2))
      VT=(S1**2*V3-S2**2*V1+(S2**2-S1**2)*V2)
     1       /(S1*S2*(S1+S2))
      A(1,1)=(ALAMDA+2.*GL)*ENX(I)
      A(1,2)=GL*ENY(I)
      A(1,3)=GL*ENY(I)
      A(1,4)=ALAMDA*ENX(I)
      A(2,1)=ALAMDA*ENY(I)
      A(2,2)=GL*ENX(I)
      A(2,3)=GL*ENX(I)
      A(2,4)=(ALAMDA+2.*GL)*ENY(I)
      A(3,1)=-ENY(I)
      A(3,2)=ENX(I)
      A(3,3)=0.
      A(3,4)=0.
      A(4,1)=0.
      A(4,2)=0.
      A(4,3)=-ENY(I)
      A(4,4)=ENX(I)
      B(1)=TXB(I)
      B(2)=TYB(I)
      B(3)=UT
      B(4)=VT
      CALL LEQS(A,B,4,LSING)
      SXB(I)=ALAMDA*(B(1)+B(4))+2.*GL*B(1)
      SYB(I)=ALAMDA*(B(1)+B(4))+2.*GL*B(4)
      SXYB(I)=GL*(B(2)+B(3))
   20 CONTINUE
      RETURN
      END
C
C
C===================================================================
C
      SUBROUTINE OUTPUTEL(XM,YM,UB,VB,TXB,TYB,XIN,YIN,UIN,VIN,
     1                     SXB,SYB,SXYB,SXIN,SYIN,SXYIN,N,IN)
C
C
C  This subroutine prints the results in the outputfile.
C
      IMPLICIT REAL*8 (A-H,O-Z)
      DIMENSION XM(N),YM(N),UB(N),VB(N),TXB(N),TYB(N)
      DIMENSION XIN(N),YIN(N),UIN(IN),VIN(IN)
      DIMENSION SXB(N),SYB(N),SXYB(N)
      DIMENSION SXIN(IN),SYIN(IN),SXYIN(IN)
C
      WRITE(2,100)
```

```
  100 FORMAT('',76('*')///'RESULTS'///2X,'BOUNDARY NODES'//'NODE',
     1            5X,'X',11X,'Y',12X,'U',11X,'V',10X,'TXB',9X,'TYB'/)
      DO 10 I=1,N
   10 WRITE(2,200)I,XM(I),YM(I),UB(I),VB(I),TXB(I),TYB(I)
  200 FORMAT(I3,2(1X,E11.5),1X,4(1X,E11.5))
C
      WRITE (2,300)
  300 FORMAT(//,2X,'DISPLACEMENTS AT INTERNAL POINTS'//'  POINT',9X,
     1           'X',14X,'Y',15X,'U',15X,'V'/)
      DO 20 K=1,IN
   20 WRITE(2,400)K,XIN(K),YIN(K),UIN(K),VIN(K)
  400 FORMAT(I5,4(2X,E14.5))
      WRITE(2,600)
  600 FORMAT(//,2X,'STRESSES AT THE BOUNDARY NODAL POINTS'/)
      WRITE(2,700)
  700 FORMAT('NODE',7X,'X',14X,'Y',11X,'SXB',12X,'SYB',10X,'SXYB'/)
      DO 30 I=1,N
   30 WRITE(2,800)I,XM(I),YM(I),SXB(I),SYB(I),SXYB(I)
  800 FORMAT(I3,5(2X,E12.5))
      WRITE (2,900)
  900 FORMAT(//,2X,'STRESSES AT THE INTERNAL POINTS'/)
      WRITE (2,1000)
 1000 FORMAT('NODE',7X,'X',14X,'Y',11X,'SXIN',10X,'SYIN',10X,'SXYIN'/)
      DO 40 I=1,IN
   40 WRITE(2,800)I,XIN(I),YIN(I),SXIN(I),SYIN(I),SXYIN(I)
      WRITE(2,500)
  500 FORMAT(/,'',76('*'))
      RETURN
      END
C
C
C==================================================================
```

Example 7.2

The scope of this example is to illustrate the use of the program ELBECON by solving a simple plane stress problem. The body under consideration is a deep beam (i.e., the length of the beam is not large in comparison with its depth) clamped at its two ends. It has a thickness $h = 0.1$ m and material constants $E = 2 \times 10^5$ kN/m^2, $\nu = 0.20$. All the other data are shown in Fig. 7.9.

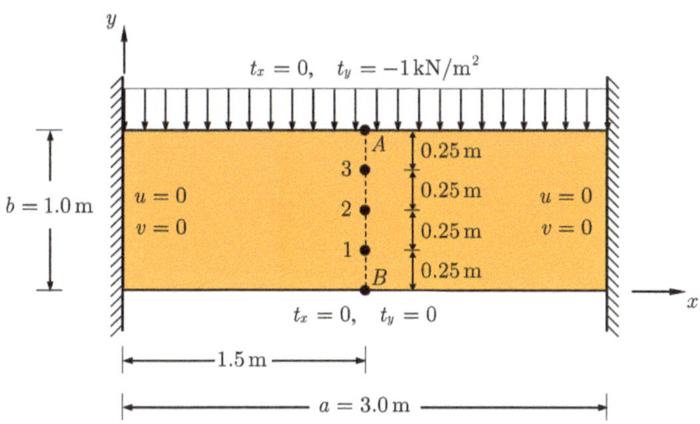

FIGURE 7.9 Deep beam clamped at both ends.

The results are obtained using a total of N = 44 boundary elements. The horizontal boundaries (upper and lower) are divided each into NX = 15 boundary elements, whereas the vertical ones (left and right) into NY = 7 elements each. The data file for elasticity problems of rectangular domains can be constructed automatically by program RECT-4.FOR which has been written for this purpose. This program is listed below for the values N = 44, NX = 15, NY = 7, IN = 3, NB = 1, IPLANE = 1, IX = 1, JY = 3 of the Example 7.2. The electronic version of the program is available on this book's companion website.

```
C=====================================================================
C
      PROGRAM RECT4
C
C  This program creates the INPUTFILE for ELBECON when the
C  domain is a rectangle AA*BB
C
      IMPLICIT REAL*8 (A-H,O-Z)
      CHARACTER*15 INPUTFILE
      CHARACTER*80 NAME,TITLE
      PARAMETER (N=44)
      PARAMETER (IN=3)
      PARAMETER (NB=1)
      DIMENSION XL(N+1),YL(N+1),XIN(IN),YIN(IN)
      DIMENSION KCODE(N),UB(N),VB(N),NL(NB)
C
      WRITE (*,'(A)')' Name of the INPUTFILE (max.15 characters)'
      READ  (*,'(A)') INPUTFILE
      OPEN (1, FILE=INPUTFILE)
      WRITE (*,'(A)')' User NAME (max.15 characters)'
      READ(*,'(A)')NAME
      WRITE(1,'(A)')NAME
      WRITE (*,'(A)')' Program TITLE (max.15 characters)'
      READ(*,'(A)')TITLE
      WRITE(1,'(A)')TITLE
      WRITE(1,100)
  100 FORMAT(' ',79(' '))
C
      IPLANE=1
      WRITE(1,150)IPLANE
  150 FORMAT(I4)
      EL=200000.
      PN=.2
      WRITE(1,200)EL,PN
  200 FORMAT(F10.0,F6.3)
      WRITE(1,100)
C
      NL(1)=N
      WRITE(1,250) (NL(I),I=1,NB)
  250 FORMAT(I4)
C
C  Boundary coordinates
C
      AA=3.
      BB=1.
      NX=15
      NY=7
      DA=AA/NX
      DB=BB/NY
```

```
      DO 1 I=1,NX
      XL(I)=(I-1)*DA
      YL(I)=0.
    1 WRITE(1,300)XL(I),YL(I)
  300 FORMAT(2(F14.8,2X))
C
      DO 2 I=1,NY
      XL(I+NX)=AA
      YL(I+NX)=(I-1)*DB
    2 WRITE(1,300)XL(I+NX),YL(I+NX)

C
      DO 3 I=1,NX
      XL(I+NX+NY)=AA-(I-1)*DA
      YL(I+NX+NY)=BB
    3 WRITE(1,300)XL(I+NX+NY),YL(I+NX+NY)
C
      DO 4 I=1,NY
      XL(I+NX+NY+NX)=0.
      YL(I+NX+NY+NX)=BB-(I-1)*DB
    4 WRITE(1,300)XL(I+NX+NY+NX),YL(I+NX+NY+NX)
      WRITE(1,100)
C
C  Boundary conditions
C
      XL(N+1)=XL(1)
      YL(N+1)=YL(1)
      DO 10 I=1,NX
      KCODE(I)=2
      XM=(XL(I)+XL(I+1))/2.
      YM=(YL(I)+YL(I+1))/2.
      UB(I)=0.
   10 VB(I)=0.
      DO 20 I=1,NY
      II=I+NX
      KCODE(II)=1
      XM=(XL(II)+XL(II+1))/2.
      YM=(YL(II)+YL(II+1))/2.
      UB(II)=0.
   20 VB(II)=0.
      DO 30 I=1,NX
      II=I+NX+NY
      KCODE(II)=2
      XM=(XL(II)+XL(II+1))/2.
      YM=(YL(II)+YL(II+1))/2.
      UB(II)=0.
   30 VB(II)=-1.
      DO 40 I=1,NY
      II=I+NX+NY+NX
      KCODE(II)=1
      XM=(XL(II)+XL(II+1))/2.
      YM=(YL(II)+YL(II+1))/2.
      UB(II)=0.
   40 VB(II)=0.
      DO 50 I=1,N
   50 WRITE(1,400)KCODE(I),UB(I),VB(I)
  400 FORMAT(I4,3X,2(F18.8))
```

```
      C
             WRITE(1,100)
      C
      C  Coordinates of the internal points
      C
             IX=1
             JY=3
             DX=AA/IX
             DY=BB/(JY+1)
             WRITE(1,100)
      C
             DO 5 J=1,JY
             DO 5 I=1,IX
             K=(J-1)*JY+I
             XIN(K)=DX/2.+(I-1)*DX
             YIN(K)=DY+(J-1)*DY
             WRITE(1,300)XIN(K),YIN(K)
           5 CONTINUE
      C


      WRITE(1,100)
C
      STOP
      END
C
C==================================================================
```

EXAMPLE 7.2 (DATA)

```
J.T.KATSIKADELIS
EXAMPLE 7.2

   1
   200000.   .200

  44
     .00000000       .00000000
     .20000000       .00000000
     .40000000       .00000000
     .60000000       .00000000
     .80000000       .00000000
    1.00000000       .00000000
    1.20000000       .00000000
    1.40000000       .00000000
    1.60000000       .00000000
    1.80000000       .00000000
    2.00000000       .00000000
    2.20000000       .00000000
    2.40000000       .00000000
    2.60000000       .00000000
    2.80000000       .00000000
    3.00000000       .00000000
    3.00000000       .14285714
    3.00000000       .28571429
    3.00000000       .42857143
    3.00000000       .57142857
    3.00000000       .71428571
    3.00000000       .85714286
    3.00000000      1.00000000
    2.80000000      1.00000000
```

```
  2.60000000      1.00000000
  2.40000000      1.00000000
  2.20000000      1.00000000
  2.00000000      1.00000000
  1.80000000      1.00000000
  1.60000000      1.00000000
  1.40000000      1.00000000
  1.20000000      1.00000000
  1.00000000      1.00000000
   .80000000      1.00000000
   .60000000      1.00000000
   .40000000      1.00000000
   .20000000      1.00000000
   .00000000      1.00000000
   .00000000       .85714286
   .00000000       .71428571
   .00000000       .57142857
   .00000000       .42857143
   .00000000       .28571429
   .00000000       .14285714

 2           .00000000          .00000000
 2           .00000000          .00000000
 2           .00000000          .00000000
 2           .00000000          .00000000
 2           .00000000          .00000000
 2           .00000000          .00000000
 2           .00000000          .00000000
 2           .00000000          .00000000
 2           .00000000          .00000000
 2           .00000000          .00000000
 2           .00000000          .00000000
 2           .00000000          .00000000
 2           .00000000          .00000000
 2           .00000000          .00000000
 2           .00000000          .00000000
 1           .00000000          .00000000
 1           .00000000          .00000000
 1           .00000000          .00000000
 1           .00000000          .00000000
 1           .00000000          .00000000
 1           .00000000          .00000000
 1           .00000000          .00000000
 2           .00000000        -1.00000000
 2           .00000000        -1.00000000
 2           .00000000        -1.00000000
 2           .00000000        -1.00000000
 2           .00000000        -1.00000000
 2           .00000000        -1.00000000
 2           .00000000        -1.00000000
 2           .00000000        -1.00000000
 2           .00000000        -1.00000000
 2           .00000000        -1.00000000
 2           .00000000        -1.00000000
 2           .00000000        -1.00000000
 2           .00000000        -1.00000000
 2           .00000000        -1.00000000
 2           .00000000        -1.00000000
```

```
1            .00000000            .00000000
1            .00000000            .00000000
1            .00000000            .00000000
1            .00000000            .00000000
1            .00000000            .00000000
1            .00000000            .00000000
1            .00000000            .00000000


 1.50000000        .25000000
 1.50000000        .50000000
 1.50000000        .75000000
```

EXAMPLE 7.2 (RESULTS)

```
************************************************************************
 J.T.KATSIKADELIS
 EXAMPLE 7.2


DATA

  NUMBER OF BOUNDARY ELEMENTS = 44
  NUMBER OF INTERNAL POINTS   =  3
  NUMBER OF BOUNDARIES        =  1

  PLANE STRESS PROBLEM

  ELASTIC CONSTANTS:   ELASTIC MODULUS = .2000E+06  POISSON RATIO =  .20


  COORDINATES OF THE EXTREME POINTS OF THE BOUNDARY ELEMENTS

 NODE           XL                YL

    1       .00000E+00       .00000E+00
    2       .20000E+00       .00000E+00
    3       .40000E+00       .00000E+00
    4       .60000E+00       .00000E+00
    5       .80000E+00       .00000E+00
    6       .10000E+01       .00000E+00
    7       .12000E+01       .00000E+00
    8       .14000E+01       .00000E+00
    9       .16000E+01       .00000E+00
   10       .18000E+01       .00000E+00
   11       .20000E+01       .00000E+00
   12       .22000E+01       .00000E+00
   13       .24000E+01       .00000E+00
   14       .26000E+01       .00000E+00
   15       .28000E+01       .00000E+00
   16       .30000E+01       .00000E+00
   17       .30000E+01       .14290E+00
   18       .30000E+01       .28570E+00
   19       .30000E+01       .42860E+00
   20       .30000E+01       .57140E+00
   21       .30000E+01       .71430E+00
   22       .30000E+01       .85710E+00
   23       .30000E+01       .10000E+01
   24       .28000E+01       .10000E+01
   25       .26000E+01       .10000E+01
```

26	.24000E+01	.10000E+01
27	.22000E+01	.10000E+01
28	.20000E+01	.10000E+01
29	.18000E+01	.10000E+01
30	.16000E+01	.10000E+01
31	.14000E+01	.10000E+01
32	.12000E+01	.10000E+01
33	.10000E+01	.10000E+01
34	.80000E+00	.10000E+01
35	.60000E+00	.10000E+01
36	.40000E+00	.10000E+01
37	.20000E+00	.10000E+01
38	.00000E+00	.10000E+01
39	.00000E+00	.85710E+00
40	.00000E+00	.71430E+00
41	.00000E+00	.57140E+00
42	.00000E+00	.42860E+00
43	.00000E+00	.28570E+00
44	.00000E+00	.14290E+00

BOUNDARY CONDITIONS

NODE	KCODE	PRESCRIBED VALUES	
1	2	.00000E+00	.00000E+00
2	2	.00000E+00	.00000E+00
3	2	.00000E+00	.00000E+00
4	2	.00000E+00	.00000E+00
5	2	.00000E+00	.00000E+00
6	2	.00000E+00	.00000E+00
7	2	.00000E+00	.00000E+00
8	2	.00000E+00	.00000E+00
9	2	.00000E+00	.00000E+00
10	2	.00000E+00	.00000E+00
11	2	.00000E+00	.00000E+00
12	2	.00000E+00	.00000E+00
13	2	.00000E+00	.00000E+00
14	2	.00000E+00	.00000E+00
15	2	.00000E+00	.00000E+00
16	1	.00000E+00	.00000E+00
17	1	.00000E+00	.00000E+00
18	1	.00000E+00	.00000E+00
19	1	.00000E+00	.00000E+00
20	1	.00000E+00	.00000E+00
21	1	.00000E+00	.00000E+00
22	1	.00000E+00	.00000E+00
23	2	.00000E+00	-.10000E+01
24	2	.00000E+00	-.10000E+01
25	2	.00000E+00	-.10000E+01
26	2	.00000E+00	-.10000E+01
27	2	.00000E+00	-.10000E+01
28	2	.00000E+00	-.10000E+01
29	2	.00000E+00	-.10000E+01
30	2	.00000E+00	-.10000E+01
31	2	.00000E+00	-.10000E+01
32	2	.00000E+00	-.10000E+01
33	2	.00000E+00	-.10000E+01
34	2	.00000E+00	-.10000E+01
35	2	.00000E+00	-.10000E+01
36	2	.00000E+00	-.10000E+01
37	2	.00000E+00	-.10000E+01
38	1	.00000E+00	.00000E+00
39	1	.00000E+00	.00000E+00
40	1	.00000E+00	.00000E+00
41	1	.00000E+00	.00000E+00
42	1	.00000E+00	.00000E+00

43	1	.00000E+00	.00000E+00
44	1	.00000E+00	.00000E+00

****

The system has been solved regularly

****

RESULTS

BOUNDARY NODES

NODE	X	Y	U	V	TXB	TYB
1	.10000E+00	.00000E+00	-.35454E-05	-.27273E-05	.00000E+00	.00000E+00
2	.30000E+00	.00000E+00	-.59237E-05	-.73057E-05	.00000E+00	.00000E+00
3	.50000E+00	.00000E+00	-.68889E-05	-.12152E-04	.00000E+00	.00000E+00
4	.70000E+00	.00000E+00	-.67681E-05	-.16769E-04	.00000E+00	.00000E+00
5	.90000E+00	.00000E+00	-.58006E-05	-.20756E-04	.00000E+00	.00000E+00
6	.11000E+01	.00000E+00	-.42088E-05	-.23812E-04	.00000E+00	.00000E+00
7	.13000E+01	.00000E+00	-.22062E-05	-.25727E-04	.00000E+00	.00000E+00
8	.15000E+01	.00000E+00	.46889E-20	-.26379E-04	.00000E+00	.00000E+00
9	.17000E+01	.00000E+00	.22062E-05	-.25727E-04	.00000E+00	.00000E+00
10	.19000E+01	.00000E+00	.42088E-05	-.23812E-04	.00000E+00	.00000E+00
11	.21000E+01	.00000E+00	.58006E-05	-.20756E-04	.00000E+00	.00000E+00
12	.23000E+01	.00000E+00	.67681E-05	-.16769E-04	.00000E+00	.00000E+00
13	.25000E+01	.00000E+00	.68889E-05	-.12152E-04	.00000E+00	.00000E+00
14	.27000E+01	.00000E+00	.59237E-05	-.73057E-05	.00000E+00	.00000E+00
15	.29000E+01	.00000E+00	.35454E-05	-.27273E-05	.00000E+00	.00000E+00
16	.30000E+01	.71450E-01	.00000E+00	.00000E+00	-.33787E+01	.11540E+01
17	.30000E+01	.21430E+00	.00000E+00	.00000E+00	-.16686E+01	.12294E+01
18	.30000E+01	.35715E+00	.00000E+00	.00000E+00	-.82195E+00	.13507E+01
19	.30000E+01	.50000E+00	.00000E+00	.00000E+00	-.17941E+00	.14455E+01
20	.30000E+01	.64285E+00	.00000E+00	.00000E+00	.48140E+00	.15057E+01
21	.30000E+01	.78570E+00	.00000E+00	.00000E+00	.14103E+01	.15686E+01
22	.30000E+01	.92855E+00	.00000E+00	.00000E+00	.35425E+01	.18746E+01
23	.29000E+01	.10000E+01	-.40575E-05	-.44337E-05	.00000E+00	-.10000E+01
24	.27000E+01	.10000E+01	-.62476E-05	-.98013E-05	.00000E+00	-.10000E+01
25	.25000E+01	.10000E+01	-.70260E-05	-.14902E-04	.00000E+00	-.10000E+01
26	.23000E+01	.10000E+01	-.68021E-05	-.19577E-04	.00000E+00	-.10000E+01
27	.21000E+01	.10000E+01	-.57929E-05	-.23561E-04	.00000E+00	-.10000E+01
28	.19000E+01	.10000E+01	-.41924E-05	-.26602E-04	.00000E+00	-.10000E+01
29	.17000E+01	.10000E+01	-.21956E-05	-.28506E-04	.00000E+00	-.10000E+01
30	.15000E+01	.10000E+01	.84703E-21	-.29154E-04	.00000E+00	-.10000E+01
31	.13000E+01	.10000E+01	.21956E-05	-.28506E-04	.00000E+00	-.10000E+01
32	.11000E+01	.10000E+01	.41924E-05	-.26602E-04	.00000E+00	-.10000E+01
33	.90000E+00	.10000E+01	.57929E-05	-.23561E-04	.00000E+00	-.10000E+01
34	.70000E+00	.10000E+01	.68021E-05	-.19577E-04	.00000E+00	-.10000E+01
35	.50000E+00	.10000E+01	.70260E-05	-.14902E-04	.00000E+00	-.10000E+01
36	.30000E+00	.10000E+01	.62476E-05	-.98013E-05	.00000E+00	-.10000E+01
37	.10000E+00	.10000E+01	.40575E-05	-.44337E-05	.00000E+00	-.10000E+01
38	.00000E+00	.92855E+00	.00000E+00	.00000E+00	-.35425E+01	.18746E+01
39	.00000E+00	.78570E+00	.00000E+00	.00000E+00	-.14103E+01	.15686E+01
40	.00000E+00	.64285E+00	.00000E+00	.00000E+00	-.48140E+00	.15057E+01
41	.00000E+00	.50000E+00	.00000E+00	.00000E+00	.17941E+00	.14455E+01
42	.00000E+00	.35715E+00	.00000E+00	.00000E+00	.82195E+00	.13507E+01
43	.00000E+00	.21430E+00	.00000E+00	.00000E+00	.16686E+01	.12294E+01
44	.00000E+00	.71450E-01	.00000E+00	.00000E+00	.33787E+01	.11540E+01

DISPLACEMENTS AT INTERNAL POINTS

POINT	X	Y	U	V
1	.15000E+01	.25000E+00	.15419E-20	-.27150E-04
2	.15000E+01	.50000E+00	.62204E-21	-.27688E-04
3	.15000E+01	.75000E+00	.59557E-22	-.28377E-04

STRESSES AT THE BOUNDARY NODAL POINTS

NODE	X	Y	SXB	SYB	SXYB
1	.10000E+00	.00000E+00	-.33246E+01	.00000E+00	.00000E+00
2	.30000E+00	.00000E+00	-.16718E+01	.00000E+00	.00000E+00
3	.50000E+00	.00000E+00	-.42220E+00	.00000E+00	.00000E+00
4	.70000E+00	.00000E+00	.54417E+00	.00000E+00	.00000E+00
5	.90000E+00	.00000E+00	.12797E+01	.00000E+00	.00000E+00
6	.11000E+01	.00000E+00	.17972E+01	.00000E+00	.00000E+00
7	.13000E+01	.00000E+00	.21044E+01	.00000E+00	.00000E+00
8	.15000E+01	.00000E+00	.22062E+01	.00000E+00	.00000E+00
9	.17000E+01	.00000E+00	.21044E+01	.00000E+00	.00000E+00
10	.19000E+01	.00000E+00	.17972E+01	.00000E+00	.00000E+00
11	.21000E+01	.00000E+00	.12797E+01	.00000E+00	.00000E+00
12	.23000E+01	.00000E+00	.54417E+00	.00000E+00	.00000E+00
13	.25000E+01	.00000E+00	-.42220E+00	.00000E+00	.00000E+00
14	.27000E+01	.00000E+00	-.16718E+01	.00000E+00	.00000E+00
15	.29000E+01	.00000E+00	-.33246E+01	.00000E+00	.00000E+00
16	.30000E+01	.71450E-01	-.33787E+01	.77025E+00	.11540E+01
17	.30000E+01	.21430E+00	-.16686E+01	-.33372E+00	.12294E+01
18	.30000E+01	.35715E+00	-.82195E+00	-.16439E+00	.13507E+01
19	.30000E+01	.50000E+00	-.17941E+00	-.35883E-01	.14455E+01
20	.30000E+01	.64285E+00	.48140E+00	.96279E-01	.15057E+01
21	.30000E+01	.78570E+00	.14103E+01	.28205E+00	.15686E+01
22	.30000E+01	.92855E+00	.35425E+01	-.16422E+01	.18746E+01
23	.29000E+01	.10000E+01	.33594E+01	-.10000E+01	.00000E+00
24	.27000E+01	.10000E+01	.12843E+01	-.10000E+01	.00000E+00
25	.25000E+01	.10000E+01	.77255E-01	-.10000E+01	.00000E+00
26	.23000E+01	.10000E+01	-.81653E+00	-.10000E+01	.00000E+00
27	.21000E+01	.10000E+01	-.15048E+01	-.10000E+01	.00000E+00
28	.19000E+01	.10000E+01	-.19987E+01	-.10000E+01	.00000E+00
29	.17000E+01	.10000E+01	-.22962E+01	-.10000E+01	.00000E+00
30	.15000E+01	.10000E+01	-.23956E+01	-.10000E+01	.00000E+00
31	.13000E+01	.10000E+01	-.22962E+01	-.10000E+01	.00000E+00
32	.11000E+01	.10000E+01	-.19987E+01	-.10000E+01	.00000E+00
33	.90000E+00	.10000E+01	-.15048E+01	-.10000E+01	.00000E+00
34	.70000E+00	.10000E+01	-.81653E+00	-.10000E+01	.00000E+00
35	.50000E+00	.10000E+01	.77255E-01	-.10000E+01	.00000E+00
36	.30000E+00	.10000E+01	.12843E+01	-.10000E+01	.00000E+00
37	.10000E+00	.10000E+01	.33594E+01	-.10000E+01	.00000E+00
38	.00000E+00	.92855E+00	.35425E+01	-.16422E+01	-.18746E+01
39	.00000E+00	.78570E+00	.14103E+01	.28205E+00	-.15686E+01
40	.00000E+00	.64285E+00	.48140E+00	.96279E-01	-.15057E+01
41	.00000E+00	.50000E+00	-.17941E+00	-.35883E-01	-.14455E+01
42	.00000E+00	.35715E+00	-.82195E+00	-.16439E+00	-.13507E+01
43	.00000E+00	.21430E+00	-.16686E+01	-.33372E+00	-.12294E+01
44	.00000E+00	.71450E-01	-.33787E+01	.77025E+00	-.11540E+01

STRESSES AT THE INTERNAL POINTS

NODE	X	Y	SXIN	SYIN	SXYIN
1	.15000E+01	.25000E+00	.93386E+00	-.20348E+00	.49613E-15
2	.15000E+01	.50000E+00	-.95648E-01	-.51066E+00	-.79103E-15
3	.15000E+01	.75000E+00	-.11270E+01	-.81510E+00	-.95063E-15

****

TABLE 7.1 Computed Values at Boundary and Internal Points of Example 7.2 for Various Values of N

Point	BEM						FEM	Beam Theory
	Number of Boundary Elements, N							
	24	44	120	200	300	400		
Displacement v × 10^4								
A	−0.2977	−0.2915	−0.3001	−0.2987	−0.2979	−0.2979	−0.2924	
2	−0.2825	−0.2768	−0.2859	−0.2845	−0.2837	−0.2832	−0.2783	−0.2885
B	−0.2675	−0.2638	−0.2748	−0.2739	−0.2733	−0.2730	−0.2682	
Stress σ_x								
A	−2.285	−2.395	−2.609	−2.626	−2.633	−2.635	−2.630	−2.250
3	−1.146	−1.127	−1.183	−1.183	−1.182	−1.181	−1.170	−1.125
2	−0.100	−0.095	−0.092	−0.091	−0.091	−0.091	−0.089	0.000
1	0.942	0.934	0.997	0.998	0.998	0.998	0.986	1.125
B	2.087	2.206	2.425	2.443	2.451	2.452	2.450	2.250

Table 7.1 presents the computed values of the displacement v and the stress σ_x at the cross-section $x = 1.5$ for various values of N. They are compared with those obtained using the FEM and the Timoshenko beam theory (i.e., including shear deformation). The deflection at point 2 is given by the expression

$$v_2 = \frac{q\alpha^4}{384EI}\left[1 + 8\kappa(1+\nu)\left(\frac{b}{a}\right)^2\right]$$

where κ is the shape coefficient of the cross-section ($\kappa = 1.2$ for a rectangular cross-section) and q is the load per unit length ($q = -h\ t_y$).

For the beam theory, the stress σ_x is computed from the expression

$$\sigma_x = \frac{M}{I}\ (0.5 - y)$$

where $M = qa^2/24$ is the bending moment at the mid-section.

We observe that a number of 24 constant boundary elements is not sufficient for an accurate solution. The convergence of the BEM was achieved with N = 300 (NX = 113, NY = 37), whereas of the FEM with 432 rectangular hybrid elements. Finally, the distribution of σ_x at the cross-section $x = 3.0$ is shown in Fig. 7.10 as it was obtained using BEM, FEM and the Timoshenko beam theory.

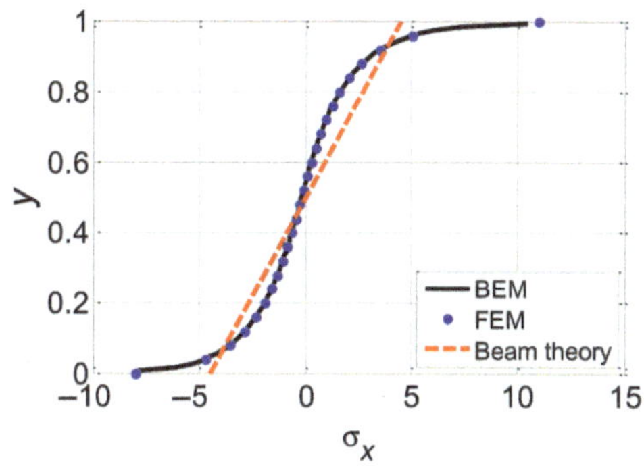

FIGURE 7.10 Distribution of σ_x at the cross-section $x = 3.0$ in Example 7.2.

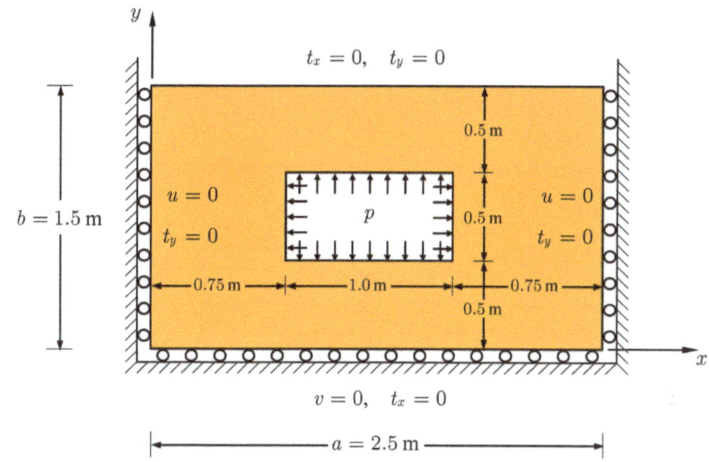

FIGURE 7.11 Pipe under uniform pressure.

EXAMPLE 7.3

Determine the deformation and state of stress for the pipe of Fig. 7.11, which is subjected to internal pressure $p = 1$ MPa. It has a uniform cross-section. Since it is very long in the z-direction, the produced state of stress is plane strain. The material constants are $E = 2 \times 10^5$ kN/m^2 and $\nu = 0.20$. All the other data are given in Fig. 7.11.

The results have been obtained using program ELBECON with INPLANE = 0, N = 348, NB = 2, NL(1) = 284, and NL(2) = 348. The external boundary is divided into N = 142 (NX = 91, NY = 51) constant boundary elements, while the inner boundary is discretized into N = 32 (NX = 21, NY = 11) elements. The data file is created using program RECTEL-MU.FOR. Its electronic version is available on this book's companion website. Due to space considerations, the obtained results are given only at selected

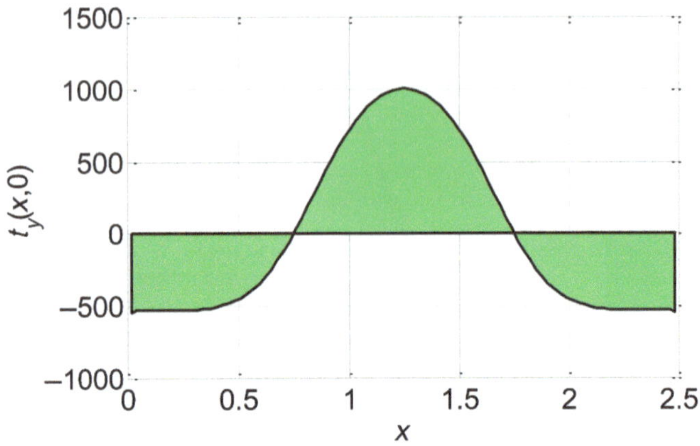

FIGURE 7.12 Distribution of t_y along the boundary $y = 0$ in Example 7.3.

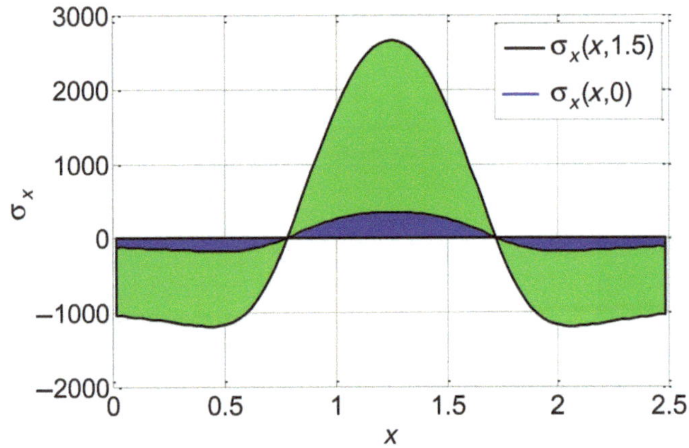

FIGURE 7.13 Distribution of σ_x along the boundaries $y = 0$ and $y = 1.5$ in Example 7.3.

points. The distribution of t_y along the boundary $y = 0$ is shown in Fig. 7.12, whereas the distribution of σ_x along the boundaries $y = 0$ and $y = 1.5$ are shown in Fig. 7.13. Fig. 7.14 depicts the deformed shape of the pipe's cross-section. Finally, the stress contours σ_x, σ_y, and τ_{xy} on the cross-section of the pipe are presented in Figs. 7.15–7.17, respectively.

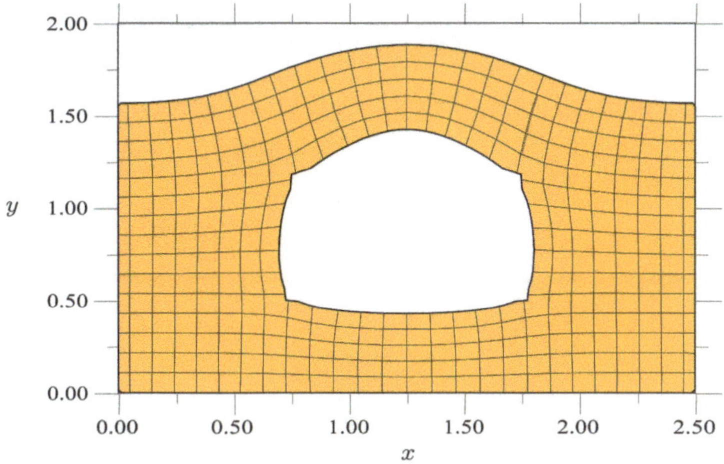

FIGURE 7.14 Deformed pipe cross-section in Example 7.3.

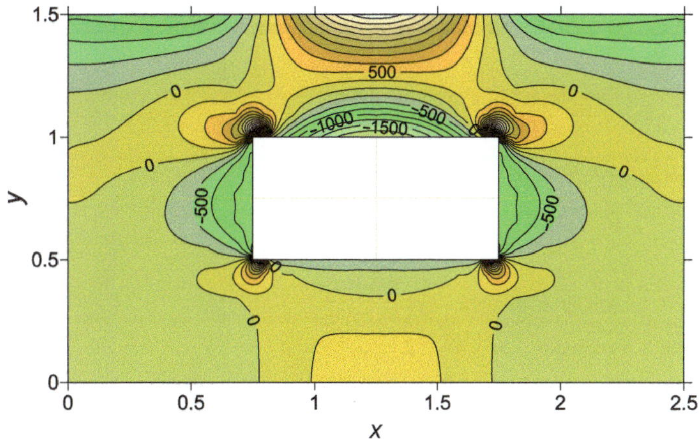

FIGURE 7.15 Contours of σ_x at the pipe cross-section in Example 7.3.

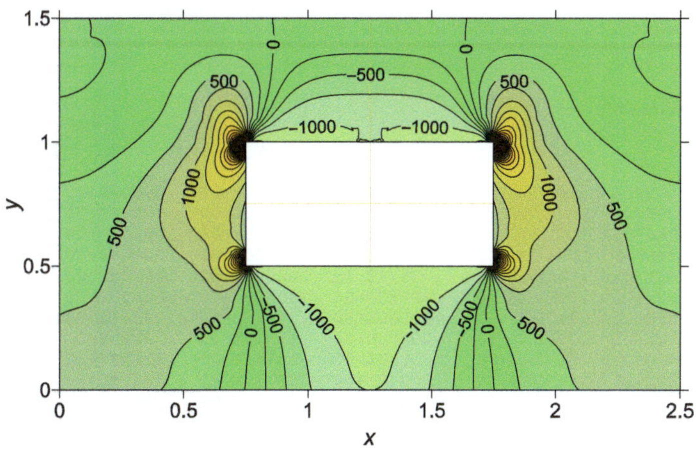

FIGURE 7.16 Contours of σ_y at the pipe cross-section in Example 7.3.

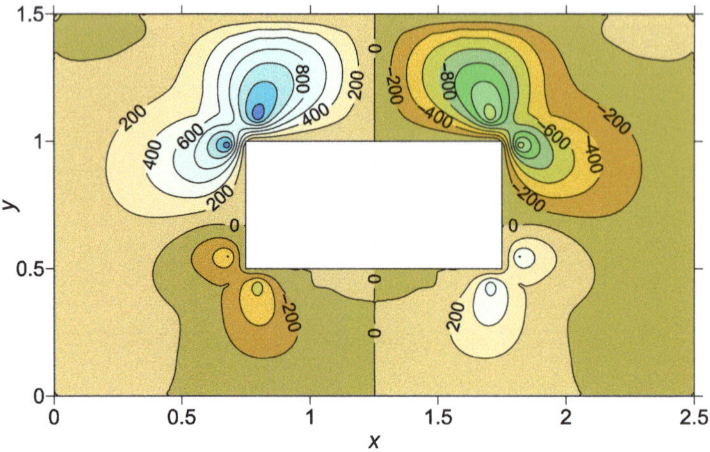

FIGURE 7.17 Contours of τ_{xy} at the pipe cross-section in Example 7.3.

EXAMPLE 7.3 (RESULTS)

```
********************************************************************************
J.T. KATSIKADELIS
EXAMPLE 7.3


DATA

  NUMBER OF BOUNDARY ELEMENTS =348
  NUMBER OF INTERNAL POINTS   = 28
  NUMBER OF BOUNDARIES        =  2

  PLANE STRAIN PROBLEM

  ELASTIC CONSTANTS:   ELASTIC MODULUS = .2000E+06   POISSON RATIO =   .20


********************************************************************************


RESULTS


  BOUNDARY NODES

NODE      X            Y             U            V            TXB          TYB

   1  .13750E-01  .00000E+00  -.40863E-04  .00000E+00  .00000E+00 -.54084E+03
  21  .56320E+00  .00000E+00  -.77574E-03  .00000E+00  .00000E+00 -.38160E+03
  41  .11126E+01  .00000E+00  -.38778E-03  .00000E+00  .00000E+00  .91767E+03
  51  .13874E+01  .00000E+00   .38778E-03  .00000E+00  .00000E+00  .91767E+03
  71  .19368E+01  .00000E+00   .77574E-03  .00000E+00  .00000E+00 -.38160E+03
  91  .24863E+01  .00000E+00   .40863E-04  .00000E+00  .00000E+00 -.54084E+03
  92  .25000E+01  .14700E-01   .00000E+00  .77291E-04 -.15566E+03  .00000E+00
 117  .25000E+01  .75000E+00   .00000E+00  .17836E-02  .33229E+01  .00000E+00
 122  .25000E+01  .89710E+00   .00000E+00  .19644E-02  .12538E+02  .00000E+00
 142  .25000E+01  .14853E+01   .00000E+00  .22776E-02 -.10209E+04  .00000E+00
 143  .24863E+01  .15000E+01   .14349E-03  .23207E-02  .00000E+00  .00000E+00
 163  .19368E+01  .15000E+01   .31103E-02  .50269E-02  .00000E+00  .00000E+00
 188  .12500E+01  .15000E+01   .18543E-16  .13954E-01  .00000E+00  .00000E+00
 193  .11126E+01  .15000E+01  -.16963E-02  .13351E-01  .00000E+00  .00000E+00
 213  .56320E+00  .15000E+01  -.31103E-02  .50269E-02  .00000E+00  .00000E+00
```

233	.13750E-01	.15000E+01	-.14349E-03	.23207E-02	.00000E+00	.00000E+00
234	.00000E+00	.14853E+01	.00000E+00	.22776E-02	.10209E+04	.00000E+00
254	.00000E+00	.89710E+00	.00000E+00	.19644E-02	-.12538E+02	.00000E+00
259	.00000E+00	.75000E+00	.00000E+00	.17836E-02	-.33229E+01	.00000E+00
264	.00000E+00	.60290E+00	.00000E+00	.15190E-02	.28356E+02	.00000E+00
284	.00000E+00	.14700E-01	.00000E+00	.77291E-04	.15566E+03	.00000E+00
285	.75000E+00	.52275E+00	-.12769E-02	.17301E-02	-.10000E+04	.00000E+00
295	.75000E+00	.97725E+00	-.25935E-03	.42063E-02	-.10000E+04	.00000E+00
296	.77380E+00	.10000E+01	.22635E-02	.78607E-02	.00000E+00	.10000E+04
305	.12024E+01	.10000E+01	.40034E-03	.15508E-01	.00000E+00	.10000E+04
306	.12500E+01	.10000E+01	.24666E-16	.15582E-01	.00000E+00	.10000E+04
333	.14881E+01	.50000E+00	-.13884E-03	-.23804E-02	.00000E+00	-.10000E+04
334	.14405E+01	.50000E+00	-.10429E-03	-.25202E-02	.00000E+00	-.10000E+04
335	.13929E+01	.50000E+00	-.73750E-04	-.26211E-02	.00000E+00	-.10000E+04
336	.13452E+01	.50000E+00	-.46907E-04	-.26884E-02	.00000E+00	-.10000E+04
337	.12976E+01	.50000E+00	-.22767E-04	-.27274E-02	.00000E+00	-.10000E+04
338	.12500E+01	.50000E+00	.24666E-17	-.27401E-02	.00000E+00	-.10000E+04
339	.12024E+01	.50000E+00	.22767E-04	-.27274E-02	.00000E+00	-.10000E+04
340	.11548E+01	.50000E+00	.46907E-04	-.26884E-02	.00000E+00	-.10000E+04
341	.11071E+01	.50000E+00	.73750E-04	-.26211E-02	.00000E+00	-.10000E+04
342	.10595E+01	.50000E+00	.10429E-03	-.25202E-02	.00000E+00	-.10000E+04
343	.10119E+01	.50000E+00	.13884E-03	-.23804E-02	.00000E+00	-.10000E+04
344	.96430E+00	.50000E+00	.17738E-03	-.21916E-02	.00000E+00	-.10000E+04
345	.91670E+00	.50000E+00	.21854E-03	-.19392E-02	.00000E+00	-.10000E+04
346	.86905E+00	.50000E+00	.25904E-03	-.15988E-02	.00000E+00	-.10000E+04
347	.82140E+00	.50000E+00	.29284E-03	-.11218E-02	.00000E+00	-.10000E+04
348	.77380E+00	.50000E+00	.31598E-03	-.36762E-03	.00000E+00	-.10000E+04

DISPLACEMENTS AT INTERNAL POINTS

POINT	X	Y	U	V
1	.50000E-01	.75000E+00	-.20003E-04	.17918E-02
5	.18000E+01	.75000E+00	.16284E-02	.27481E-02
8	.22500E+01	.75000E+00	.13165E-03	.20022E-02
9	.24500E+01	.75000E+00	.20003E-04	.17918E-02
10	.12500E+01	.50000E-01	-.51161E-18	-.31340E-03
11	.12500E+01	.25000E+00	-.86228E-18	-.13846E-02
14	.12500E+01	.12500E+01	.57259E-18	.14663E-01
16	.11360E+00	.10940E+00	-.15068E-03	.32212E-03
18	.34090E+00	.10940E+00	-.44789E-03	.32159E-03
19	.45450E+00	.10940E+00	-.60576E-03	.30355E-03
21	.68180E+00	.10940E+00	-.87774E-03	.11796E-03
22	.79550E+00	.10940E+00	-.90670E-03	-.85431E-04
24	.10227E+01	.10940E+00	-.59246E-03	-.49048E-03
28	.14773E+01	.10940E+00	.59246E-03	-.49048E-03

STRESSES AT THE BOUNDARY NODAL POINTS

NODE	X	Y	SXB	SYB	SXYB
1	.13750E-01	.00000E+00	-.12473E+03	.54084E+03	.00000E+00
21	.56320E+00	.00000E+00	-.16853E+03	.38160E+03	.00000E+00
51	.13874E+01	.00000E+00	.32799E+03	-.91767E+03	.00000E+00
71	.19368E+01	.00000E+00	-.16853E+03	.38160E+03	.00000E+00
91	.24863E+01	.00000E+00	-.12473E+03	.54084E+03	.00000E+00
92	.25000E+01	.14700E-01	-.15566E+03	.51405E+03	.00000E+00
112	.25000E+01	.60290E+00	-.28356E+02	.41419E+03	.00000E+00
117	.25000E+01	.75000E+00	.33229E+01	.32136E+03	.00000E+00
122	.25000E+01	.89710E+00	.12538E+02	.19337E+03	.00000E+00
142	.25000E+01	.14853E+01	-.10209E+04	-.30121E+02	.00000E+00
143	.24863E+01	.15000E+01	-.10410E+04	-.28422E-13	.00000E+00
163	.19368E+01	.15000E+01	-.10806E+04	.00000E+00	.00000E+00
183	.13874E+01	.15000E+01	.23879E+04	.00000E+00	.00000E+00

189	.12226E+01	.15000E+01	.26490E+04	.11369E-12	.00000E+00
190	.11951E+01	.15000E+01	.26161E+04	.11369E-12	.00000E+00
193	.11126E+01	.15000E+01	.23879E+04	.00000E+00	.00000E+00
213	.56320E+00	.15000E+01	-.10806E+04	.00000E+00	.00000E+00
233	.13750E-01	.15000E+01	-.10410E+04	-.28422E-13	.00000E+00
234	.00000E+00	.14853E+01	-.10209E+04	-.30121E+02	.00000E+00
254	.00000E+00	.89710E+00	.12538E+02	.19337E+03	.00000E+00
259	.00000E+00	.75000E+00	.33229E+01	.32136E+03	.00000E+00
264	.00000E+00	.60290E+00	-.28356E+02	.41419E+03	.00000E+00
284	.00000E+00	.14700E-01	-.15566E+03	.51405E+03	.00000E+00
285	.75000E+00	.52275E+00	-.10000E+04	.48286E+04	.00000E+00
290	.75000E+00	.75000E+00	-.10000E+04	.71607E+03	.00000E+00
295	.75000E+00	.97725E+00	-.10000E+04	.87511E+04	.00000E+00
296	.77380E+00	.10000E+01	.56641E+04	-.10000E+04	.00000E+00
304	.11548E+01	.10000E+01	-.19111E+04	-.10000E+04	.00000E+00
306	.12500E+01	.10000E+01	-.20022E+04	-.10000E+04	.00000E+00
308	.13452E+01	.10000E+01	-.19111E+04	-.10000E+04	.00000E+00
316	.17262E+01	.10000E+01	.56641E+04	-.10000E+04	.00000E+00
317	.17500E+01	.97725E+00	-.10000E+04	.87511E+04	.00000E+00
322	.17500E+01	.75000E+00	-.10000E+04	.71607E+03	.00000E+00
323	.17500E+01	.70455E+00	-.10000E+04	.64376E+03	.00000E+00
327	.17500E+01	.52275E+00	-.10000E+04	.48286E+04	.00000E+00
328	.17262E+01	.50000E+00	.33042E+04	-.10000E+04	.00000E+00
336	.13452E+01	.50000E+00	-.36151E+03	-.10000E+04	.00000E+00
338	.12500E+01	.50000E+00	-.34965E+03	-.10000E+04	.00000E+00
340	.11548E+01	.50000E+00	-.36151E+03	-.10000E+04	.00000E+00
348	.77380E+00	.50000E+00	.33042E+04	-.10000E+04	.00000E+00

STRESSES AT THE INTERNAL POINTS

NODE	X	Y	SXIN	SYIN	SXYIN
1	.50000E-01	.75000E+00	.32732E+01	.32797E+03	.39129E+02
3	.45000E+00	.75000E+00	-.33895E+03	.85178E+03	.30682E+03
5	.18000E+01	.75000E+00	-.10504E+04	.10058E+04	-.10739E+03
8	.22500E+01	.75000E+00	-.63993E+02	.46921E+03	-.19058E+03
9	.24500E+01	.75000E+00	.32732E+01	.32797E+03	-.39129E+02
10	.12500E+01	.50000E-01	.34120E+03	-.10214E+04	-.25867E-11
11	.12500E+01	.25000E+00	.18654E+03	-.10732E+04	-.24088E-11
14	.12500E+01	.12500E+01	.39998E+03	-.57256E+03	.29360E-11
16	.11360E+00	.10940E+00	-.12971E+03	.53151E+03	.85015E+01
18	.34090E+00	.10940E+00	-.15320E+03	.52814E+03	.16767E+02
19	.45450E+00	.10940E+00	-.16599E+03	.49999E+03	.51534E+01
21	.68180E+00	.10940E+00	-.10149E+03	.20709E+03	-.98670E+02
23	.90910E+00	.10940E+00	.17098E+03	-.51830E+03	-.13063E+03
24	.10227E+01	.10940E+00	.26594E+03	-.81266E+03	-.77126E+02
26	.12500E+01	.10940E+00	.31853E+03	-.10328E+04	-.10625E-11
28	.14773E+01	.10940E+00	.26594E+03	-.81266E+03	.77126E+02

****

7.13 REFERENCES

The boundary value problems of the theory of elasticity are well known and the interested reader will find them in many relevant books, some of which are listed below. References [1,2,4,11–13] are addressed mainly to engineers, while [14–16] are more theoretical. Regarding the application of the BEM to two- or three-dimensional elastostatic problems, the reader is advised to consult references [7] and [17–27]. The BEM has also been employed to analyze composite plane bodies with perfect [28] or unilateral [29] bonding at the interfaces. The two-dimensional elastostatic problem has been also treated by the analog equation method [30], which converts Navier's equations into two uncoupled Poisson's equations, thus, simplifying significantly the solution (see Section 8.5).

[1] Ugural AC, Fenster SK. Advanced strength and applied elasticity. 3rd ed. New Jersey: Prentice Hall PTR; 1995.

[2] Timoshenko S, Goodier JN. Theory of elasticity. 2nd ed. New York: McGraw-Hill; 1951.

[3] Katsikadelis JT. The boundary element method for plate analysis. Elsevier: Academic Press; 2014.

[4] Malvern LE. Introduction to the mechanics of a continuous medium. Englewood Cliffs, New Jersey: Prentice Hall; 1969.

[5] Lass H. Vector and tensor analysis. New York: McGraw-Hill; 1950.

[6] Kupradze VD. Potential methods in the theory of elasticity. Jerusalem: Israel Program for Scientific Translations; 1965.

[7] Banerjee PK, Butterfield R. Boundary elements in engineering science. New York: McGraw-Hill; 1981.

[8] Sladek V, Sladek J, editors. Singular integrals in boundary element methods. Southampton: Computational Mechanics Publications; 1998.

[9] Katsikadelis JT. The analog equation method. A powerful BEM-based solution technique for linear and nonlinear engineering problems. In: Brebbia CA, editor. Boundary element XVI. Southampton: Computational Mechanics Publications; 1994. p. 167–83.

[10] Katsikadelis JT, Armenakas AE. A new boundary equation solution to the plate problem. ASME J Appl Mech, 56. 1989. p. 364–74.

[11] Boversi AP. Elasticity in engineering mechanics. Englewood Cliffs, New Jersey: Prentice Hall; 1965.

[12] Fung YC. Foundations in solids mechanics. Englewood Cliffs, New Jersey: Prentice Hall; 1965.

[13] Wang CT. Applied elasticity. New York: McGraw-Hill; 1953.

[14] Novozhilov VV. Theory of elasticity. Oxford: Pergamon Student editions; 1961.

[15] Muskhelishvili NI. Some basic problems of the mathematical theory of elasticity. 4th ed. Groningen-The Netherlands: P. Noordhoff Ltd.; 1963.

[16] Sokolnikoff I. Mathematical theory of elasticity. 2nd ed. New York: McGraw-Hill; 1950.

[17] Rizzo FJ. An integral equation approach to boundary value problems of classical elastostatics. Quart Appl Math 1967;25:83–95.

[18] Brebbia CA. The boundary element method for engineers. London: Pentech Press; 1978.

[19] Brebbia AA, Telles JCF, Wrobel LC. Boundary element techniques. Southampton: Computational Mechanics Publications; 1984.

[20] Brebbia CA, Dominguez J. Boundary elements: an introductory course. 2nd ed. Southampton: Computational Mechanics Publications; 2001.

[21] Becker AA. The boundary element method in engineering. New York: McGraw-Hill; 1992.

[22] Beer G, Watson JO. Introduction to finite and boundary element methods for engineers. New York: John Wiley and Sons; 1992.

[23] Chen G, Zhou J. Boundary element methods. London: Academic Press; 1992.

[24] Trevelyan J. Boundary elements for engineers. Theory and applications. Southampton: Computational Mechanics Publications; 1994.

[25] Kane JH. Boundary element analysis in engineering continuum mechanics. Englewood Cliffs, New Jersey: Prentice Hall; 1994.

[26] Gaul L, Kögl M, Wagner M. Boundary elements for engineers and scientists. Berlin: Springer-Verlag; 2003.

[27] Aliabadi MH. The boundary element method, vol. 2. England: John Wiley & Sons Ltd; 2003.

[28] Katsikadelis JT, Kokkinos FT. Static and dynamic analysis of composite shear walls by the boundary element method. Acta Mechanica 1987;68:231–50.

[29] Katsikadelis JT, Kokkinos FT. Analysis of composite shear walls with interface separation, friction and slip using BEM. Int J Solids Struct 1993;30:1825–48.
[30] Katsikadelis JT, Kandilas CV. Solving the plane elastostatic problems by the analog equation method. Comput Struct 1997;64:305–12.

PROBLEMS

7.1. Derive the boundary integral equations for the plain strain problem taking into account initial strains.

7.2. Determine the deformation and the stresses in the pipe of Fig. 7.11, which are produced by a temperature change of $\Delta T = 80°C$. The coefficient of thermal dilatation is $\alpha = 10^{-5}$, while the elastic constants are those of Example 7.3.

7.3. Determine the deformation and the stresses in the pipe of Fig. 7.11, when the temperature of the fluid in the pipe is $T_{in} = 300°C$, the ambient temperature is $T_{out} = 25°C$, the reference temperature (temperature of manufacturing for the pipe) is $T_0 = 10°C$, and the internal pressure is $p = 1$ MPa. The coefficient of thermal dilatation is $\alpha = 10^{-5}$.

7.4. Determine the deformation and the stresses in the deep beam of Fig. 7.9 considering also the weight of the body, $\rho_w = 25\ \text{kN/m}^3$.

7.5. The two-column frame is subjected to uniformly distributed horizontal loads $p = 750$ kN/m as shown in Fig. P7.5. The supporting wall is fixed ($u = 0$ and $v = 0$) along its sides AB, BC, CD, DE, and EF. Determine the deformed shape of the frame and evaluate its lateral stiffness K_{el} using the theory of elasticity. The thickness of the structure (beam and supporting wall) is 0.25 m and the elastic constants are $E = 2.1 \times 10^7$ kN/m^2 and $\nu = 0.20$.

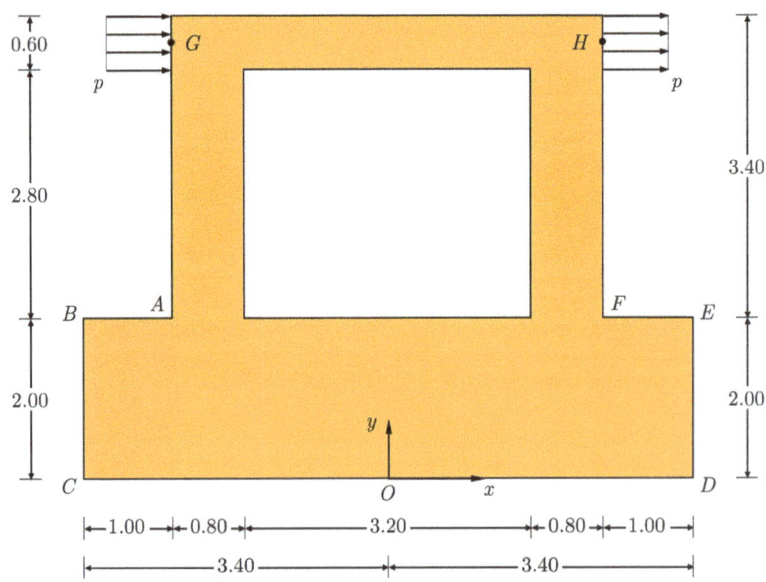

FIGURE P7.5

7.6. The shear wall of Fig. P7.6 is subjected to distributed horizontal loads $p = 50\ \text{kN/m}$. Determine the deformation of the plane body and find the stress distribution along the cross-sections $x = 0.00, 0.90, 1.70\ \text{m}$ and $y = 0.40, 1.60, 4.00\ \text{m}$. The thickness is $0.25\ \text{m}$ and the elastic constants are $E = 2.1 \times 10^7\ \text{kN/m}^2$ and $\nu = 0.15$. The support conditions of the structure are $u(x, 0) = 0$ and $v(x, 0) = 0$.

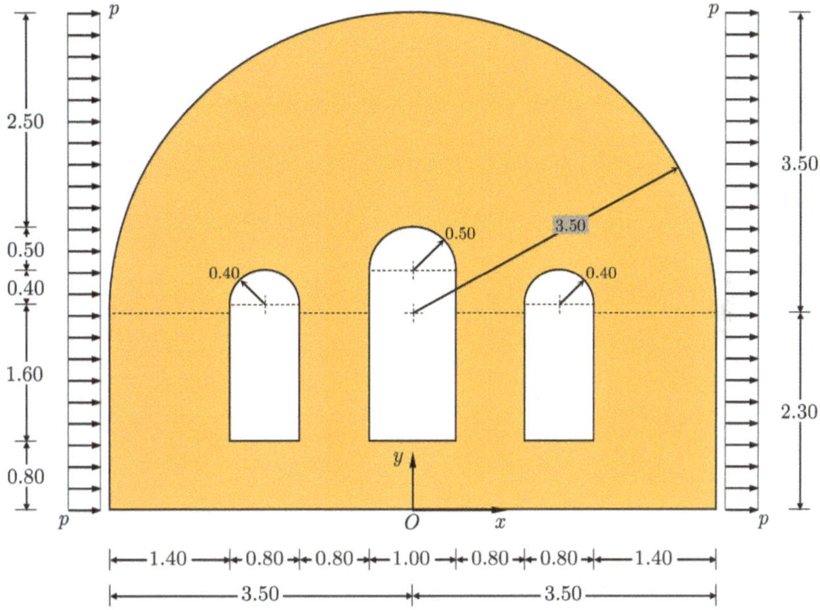

FIGURE P7.6

Chapter | Eight

The BEM for Potential Problems in Inhomogeneous Anisotropic Bodies

CHAPTER OUTLINE

8.1 INTRODUCTION

In this chapter, the boundary element method (BEM) is developed for solving problems described by the general second order elliptic partial differential equation with variable coefficients. Many potential problems in physics and engineering science are described by this equation. As we know, the BEM can be applied only when the integral representation of the solution is available. This requires the establishment of a reciprocal identity for the governing operator and its fundamental solution. While for the problem at hand the reciprocal identity is known (see Section 2.5), unfortunately, it is not possible

The Boundary Element Method for Engineers and Scientists.

to establish the fundamental solution for the general second order elliptic partial differential equation, except only for some special forms of the equation governing particular problems, for example, Helmholtz equation, heat conduction equation in homogeneous bodies, potential problems in anisotropic bodies with constant material properties (see Sections 3.6 and 6.2.4). Even when the fundamental solution can be established, it is different for different problems and it is expressed by special mathematical functions, whose manipulation to derive the boundary integral equation is a difficult and tedious task. The whole procedure demands special care for the evaluation of the derivatives, the derivation of boundary integral equation, as well as their numerical solution.

These problems discouraged the researchers to develop and use the BEM as a computational method for solving boundary value problems described by such equations. Thus, very early on the investigators looked for BEM formulations that would solve the general potential problems using the known simple fundamental solution of the Laplace equation at the cost of introducing domain integrals, which evidently annul the pure boundary character of the BEM and may reduce to certain degree its advantages over the domain type methods, for example, finite different method (FDM) and finite element method (FEM). Several such methods are reported in the literature. Among them we distinguish the dual reciprocity method (DRM) and the analog equation method (AEM) as the most efficient methods. Both methods maintain the boundary-only character in the sense that the discretization is restricted to the boundary. The DRM introduced by Nardini and Brebbia [1] appeared as the most promising method to overcome the lack of a known fundamental solution. Many problems have been successfully solved by this method [2,3,4]. However, this method is subject to a major limitation. It works only if for a given nonstandard differential operator a dominant operator with known fundamental solution can be extracted. In general this is not feasible, for example, in differential equations with variable coefficients, coupled differential equations, or nonlinear ones [5,6]. On the other hand the AEM not only reduces a given problem to one with a simple known fundamental solution but it is alleviated from any restriction inherent in DRM [7]. Thus, an integral representation of the solution can always be established and the standard BEM can be applied. It can solve efficiently not only linear but also nonlinear problems, static as well as dynamic, and it is problem independent for systems described by a differential equation of the same order. In this chapter, both methods will be presented for the solution of the potential problems described by the complete second order elliptic partial differential equation.

8.2 THE GENERAL SECOND ORDER ELLIPTIC PARTIAL DIFFERENTIAL EQUATION

The boundary value problem for the complete second order elliptic partial differential equation in two dimensions is stated as

$$A\,u_{,xx} + 2Bu_{,xy} + Cu_{,yy} + Du_{,x} + Eu_{,y} + Fu = f(\mathbf{x}), \quad \mathbf{x}\,(x, y) \in \Omega \tag{8.1}$$

subject to the boundary conditions

$$u = \alpha(\mathbf{x}), \quad \mathbf{x} \in \Gamma_1 \tag{8.2a}$$

$$\nabla u \cdot \mathbf{m} = \gamma(\mathbf{x}), \quad \mathbf{x} \in \Gamma_2 \tag{8.2b}$$

where $\Gamma_1 \cup \Gamma_2 = \Gamma$; $u = u(\mathbf{x})$ represents the unknown field function; $A(\mathbf{x}), B(\mathbf{x}), \ldots, F(\mathbf{x})$ are position dependent coefficients satisfying the ellipticity condition $B^2 - AC < 0$ [8]. The quantity $\nabla u \cdot \mathbf{m}$ represents the physical boundary quantity, that is, the flux in the direction of the *conormal* to the boundary $\mathbf{m} = (An_x + Bn_y)\mathbf{i} + (Bn_x + Cn_y)\mathbf{j}$. Apparently, the direction of the conormal coincides with the normal to the boundary if $A = C$ and $B = 0$. Finally, $\alpha(\mathbf{x})$ and $\gamma(\mathbf{x})$ are functions specified on boundary. The boundary Γ may be multiply-connected. The coefficients satisfy the self-adjointness conditions

$$A_{,x} + B_{,y} = D, \quad B_{,x} + C_{,y} = E \tag{8.3a, b}$$

which ensure that the operator

$$L = A\frac{\partial^2}{\partial x^2} + 2B\frac{\partial^2}{\partial x \partial y} + C\frac{\partial^2}{\partial y^2} + D\frac{\partial}{\partial x} + E\frac{\partial}{\partial y} + F \tag{8.4}$$

is self-adjoint, that is, $L^* = L$ (see Section 2.5). Note that condition (8.2a) represents the Dirichlet boundary condition, while (8.2b) the Neumann boundary condition (see Fig. 8.1).

We consider the functional

$$J(u) = \int_\Omega \left[\frac{1}{2}\left(Au_{,x}^2 + 2Bu_{,x}u_{,y} + Cu_{,y}^2 - Fu^2\right) + fu\right] d\Omega - \int_\Gamma \gamma u \, ds \tag{8.5}$$

The condition $\delta J(u) = 0$ produces the boundary value problem described by Eqs. (8.1) and (8.2a,b) as the *Euler-Lagrange* equation with the associated

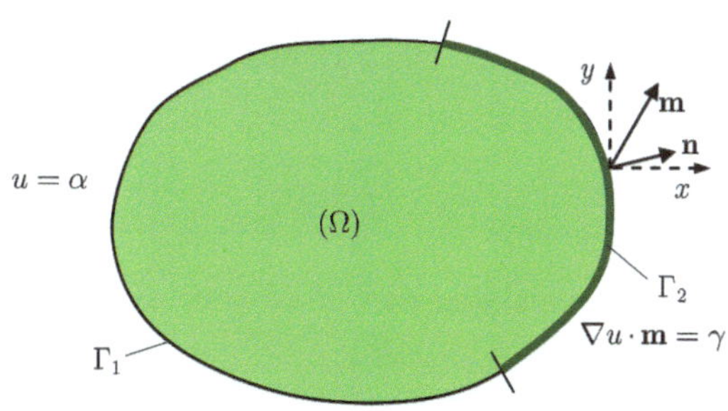

FIGURE 8.1 Domain Ω with mixed boundary conditions.

boundary conditions. The derivation is achieved via the calculus of variations (see Section 2.7.3). This derivation is proposed as an exercise.

The boundary value problem (8.1), (8.2a,b) under the conditions (8.3a,b) for suitable meaning of its coefficients occurs in many problems in engineering and mathematical physics such as heat transfer, electrostatic, seepage problems, membranes on elastic subgrade, etc., where the involved media may exhibit heterogeneous anisotropic properties.

8.3 THE DUAL RECIPROCITY METHOD

8.3.1 The fundamentals of the DRM

From the historical point of view, the DRM was introduced by Nardini and Brebbia [1] in their effort to avoid the use of the dynamic fundamental solution for the wave equation. The dynamic fundamental solution complicates the derivation of the boundary integral equations as it requires the dynamic reciprocal theorem, and special care for the subsequent numerical solution [9]. They developed a BEM formulation using the simple static fundamental solution, namely that of the Laplace equation. This formulation, however, introduced domain integrals of the unknown field function. Thus, the resulting integral equations are boundary-domain integral equations. In order to maintain the pure boundary character of their method, they developed a technique to convert the domain integrals to boundary line integrals, which allowed the problem to be solved using the standard BEM. The method was further developed and applied to solve linear problems for which the fundamental solution either could not be determined or, even available, it was too difficult to derive the boundary integral equations and solve them numerically. The method was also extended to nonlinear problems, where no reciprocal identity and fundamental solution exist.

The fundamental concept of the DRM is described as follows:

Let us consider the differential equation

$$N(u) = f \tag{8.6}$$

where N is a generic in general nonlinear operator with variable coefficients and f a generic source term. Let us suppose now that a linear dominant operator L, that is an operator of the same order as N, can be extracted from N, and whose adjoint operator L^* has a known fundamental solution. We can then split the operator N as

$$N = L + \hat{N} \tag{8.7}$$

where $\hat{N}$ represents the remaining part of N after extracting L. This allows us to write Eq. (8.6) as

$$L(u) = f - \hat{N}(u) \tag{8.8}$$

or

$$L(u) = b \tag{8.9}$$

The source $b = f - \hat{N}(u)$ is treated as unknown generalized source term. Now proceeding as the standard boundary element formulation (see Section 3.3) we obtain the integral representation formula

$$u(x) = \int_{\Omega} v\, b\, d\Omega - \int_{\Gamma} [\boldsymbol{P}^*(v)\, \boldsymbol{Q}(u) - \boldsymbol{P}(u)\, \boldsymbol{Q}^*(v)] ds \tag{8.10}$$

for the solution of the differential equation (8.9), in which v is the fundamental solution of the adjoint operator L^*, that is, a singular particular solution of the equation

$$L^*(u) = \delta(\mathbf{x}, \boldsymbol{\xi}) \tag{8.11}$$

and $\boldsymbol{P}$, $\boldsymbol{Q}$ are linear operators related to u and $\boldsymbol{P}^*$, $\boldsymbol{Q}^*$ their adjoint operators related to v. Their order is at least by one less than that of L. Eq. (8.10) can be used to evaluate the solution if first the following two problems are solved: (1) The boundary quantities $\boldsymbol{P}(u)$, $\boldsymbol{Q}(u)$ are expressed in terms of the specified boundary conditions; (2) The domain integral in Eq. (8.10) is evaluated.

To make concrete these concepts we illustrate the DRM starting with the Poisson equation. The procedure adheres to the steps presented in [2] with a slight difference of the notation to comply with that employed in this book.

8.3.2 The DRM for the Poisson equation

The DRM is explained with reference to the Poisson equation

$$\nabla^2 u = b \quad \text{in } \Omega \tag{8.12}$$

Here, the source term $b = b(x, y)$, is a known function of position only.

The integral representation of the solution of Eq. (8.12) is given as (cf. Section 3.4)

$$\varepsilon u(\mathbf{x}) = \int_{\Omega} v\, b\, d\Omega - \int_{\Gamma} \left(v \frac{\partial u}{\partial n} - u \frac{\partial v}{\partial n} \right) ds, \qquad \mathbf{x}\,(x, y) \in (\Omega \cup \Gamma) \tag{8.13}$$

To maintain the pure boundary character of the BEM, hence its advantages over the domain type methods, the domain integral in Eq. (8.13), namely

$$I(\mathbf{x}) = \int_{\Omega} v(\mathbf{x}, \mathbf{y})\, b(\mathbf{y})\, d\Omega_{\mathbf{y}}\,, \quad \mathbf{x}\,(x, y) \in (\Omega \cup \Gamma)\,, \quad \mathbf{y}(x, y) \in \Omega \tag{8.14}$$

must be converted to boundary line integral.

The technique presented in Section 3.5 (ii) could be used. This technique, though effective, has an inherent drawback, which is the determination of a particular solution of the equation

$$\nabla^2 \hat{u} = b \tag{8.15}$$

for a given source $b = b(x, y)$. Apparently, this procedure cannot be embedded in a computer code, since the user is responsible for providing the function $\hat{u}$ and its normal derivative on the boundary according to Eq. (3.53). This drawback can be overcome by employing the DRM, which proposes the use of a series of particular solutions $\hat{u}_j$ instead of a single function $\hat{u}$. The number of $\hat{u}_j$ used is taken equal to the total number of nodes in the problem, which may consist of N boundary nodes and L internal nodes, Fig. 8.2.

In the DRM the source $b(x, y)$ is approximated by the series

$$b \simeq \sum_{j=1}^{N+L} a_j \phi_j \tag{8.16}$$

where a_j are initially unknown coefficients and $\phi_j = \phi_j(x, y)$ spatial approximation functions. The particular solutions $\hat{u}_j$ are obtained from the solution of the equation

$$\nabla^2 \hat{u}_j = \phi_j \tag{8.17}$$

Evidently, Eq. (8.16) is exact at the nodal points.

Next application of Green's reciprocal identity (2.16) for $u = \nabla^2 \hat{u}_j$ yields

$$\int_\Omega \left(v\nabla^2 \hat{u}_j - \hat{u}_j \nabla^2 v \right) d\Omega = \int_\Gamma \left(v \frac{\partial \hat{u}_j}{\partial n} - \hat{u}_j \frac{\partial v}{\partial n} \right) ds \tag{8.18}$$

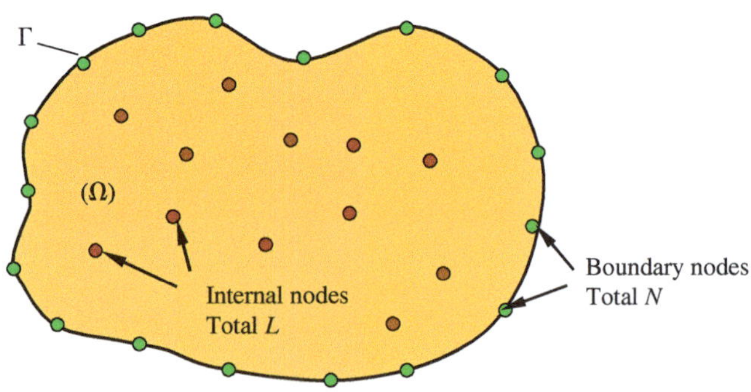

FIGURE 8.2 Boundary and internal nodes in DRM.

or taking into account Eq. (8.17) and $v = \ln r/2\pi$ we obtain

$$\int_{\Omega} v(\mathbf{x},\mathbf{y})\phi_j(r_{j\mathbf{y}})\, d\Omega_{\mathbf{y}} = \varepsilon \hat{u}_j(r_{j\mathbf{x}}) + \int_{\Gamma}\left(v(\mathbf{x},\boldsymbol{\xi})\frac{\partial \hat{u}_j(r_{j\boldsymbol{\xi}})}{\partial n_{\boldsymbol{\xi}}} - \hat{u}_j(r_{j\boldsymbol{\xi}})\frac{\partial v(\mathbf{x},\boldsymbol{\xi})}{\partial n_{\boldsymbol{\xi}}}\right) ds_{\boldsymbol{\xi}} \tag{8.19}$$

Hence, on the bases of Eqs. (8.14) and (8.16) we can write

$$\int_{\Omega} vb\, d\Omega = \sum_{j=1}^{N+L} a_j\left(\varepsilon \hat{u}_j + \int_{\Gamma} v\frac{\partial \hat{u}_j}{\partial n}\, ds - \int_{\Gamma} \hat{u}_j\frac{\partial v}{\partial n}\, ds\right) \tag{8.20}$$

Thus, Eq. (8.13) when written for the node i reads

$$\varepsilon_i u_i - \int_{\Gamma} u\frac{\partial v}{\partial n}\, ds + \int_{\Gamma} v\frac{\partial u}{\partial n}\, ds = \sum_{j=1}^{N+L} a_j\left(\varepsilon_i \hat{u}_{ij} + \int_{\Gamma} v\frac{\partial \hat{u}_j}{\partial n}\, ds - \int_{\Gamma} \hat{u}_j\frac{\partial v}{\partial n}\, ds\right) \tag{8.21}$$

We observe that Eq. (8.21) was derived by applying Green's reciprocal identity, i.e., the reciprocity principle, twice to take all terms to the boundary, hence the name DRM.

The next step is to write Eq. (8.21) in discretized form, where the integrals are replaced with summations over the elements.

$$\varepsilon_i u_i + \sum_{k=1}^{N}\int_k v^* q\, ds - \sum_{k=1}^{N}\int_k uq^*\, ds = \sum_{j=1}^{N+L} a_j\left(\varepsilon_i \hat{u}_{ij} + \sum_{k=1}^{N}\int_k v^*\hat{q}_j\, ds - \sum_{k=1}^{N}\int_k \hat{u}_j q^*\, ds\right) \tag{8.22}$$

where it has been set for convenience $\partial u/\partial n = q$, $v = v^*$, $\partial v/\partial n = q^*$, $\partial \hat{u}_j/\partial n = \hat{q}_j$; $\hat{u}_{ij}$ is the value of the function $\hat{u}_j$ at point i. Since $\hat{u}_j$ and $\hat{q}_j$ are known functions, there is no need to approximate them within the elements. However, it is convenient to do so and use the matrices **H** and **G** defined in Section 4.2. This procedure introduces an approximation in the evaluation of the terms in the right-hand side of Eq. (8.22). However, it is shown that this error is small, while the efficiency of the method is increased. Thus, after using the boundary element technique for the evaluation of the integrals over the elements, the above equation is written in terms of the nodal values as

$$\varepsilon_i u_i + \sum_{k=1}^{N} G_{ik} q_k - \sum_{k=1}^{N} \hat{H}_{ik} u_k = \sum_{j=1}^{N+L} a_j \left(\varepsilon_i \hat{u}_{ij} + \sum_{k=1}^{N} G_{ik} \hat{q}_k - \sum_{k=1}^{N} \hat{H}_{ik} \hat{u}_k \right) \tag{8.23}$$

After application of the above equation to all boundary nodes, we can write Eq. (8.23) in matrix form

$$\mathbf{Hu} - \mathbf{Gq} = \sum_{j=1}^{N+L} a_j (\mathbf{Hu}_j - \mathbf{Gq}_j) \tag{8.24}$$

The free term coefficient ε_i has been incorporated in matrix **H**. For constant elements the matrices **H** and **G** are given by Eqs. (4.3) and (4.5).

If each of the vectors $\mathbf{u}_j$, $\mathbf{q}_j$ is considered to represent a column of the matrices $\hat{\mathbf{U}}$ and $\hat{\mathbf{Q}}$, respectively, we may define

$$\hat{\mathbf{U}} = [\hat{\mathbf{u}}_1 \quad \hat{\mathbf{u}}_2 \quad \ldots \quad \hat{\mathbf{u}}_{N+L}] \tag{8.25a}$$

$$\hat{\mathbf{Q}} = [\hat{\mathbf{q}}_1 \quad \hat{\mathbf{q}}_2 \quad \ldots \quad \hat{\mathbf{q}}_{N+L}] \tag{8.25b}$$

$$\boldsymbol{\alpha} = \{\alpha_1 \quad \alpha_2 \quad \ldots \quad \alpha_{N+L}\}^T \tag{8.25c}$$

Then the summations in Eq. (8.24) can be dropped and written

$$\mathbf{Hu} - \mathbf{Gq} = (\mathbf{H}\hat{\mathbf{U}} - \mathbf{G}\hat{\mathbf{Q}})\boldsymbol{\alpha} \tag{8.26}$$

Equation (8.26) is the basis for the development of the DRM. Apparently, it is a boundary-only method in the sense that only boundary discretization is performed. Domain nodes may be included, when it is desirable to evaluate the solution at these points. The use of internal nodes depends also on the type of the approximation functions ϕ_j. Anyhow, the use of domain nodes makes the method more robust. For a better inspection, Eq. (8.26) can be represented schematically as shown in Fig. 8.3.

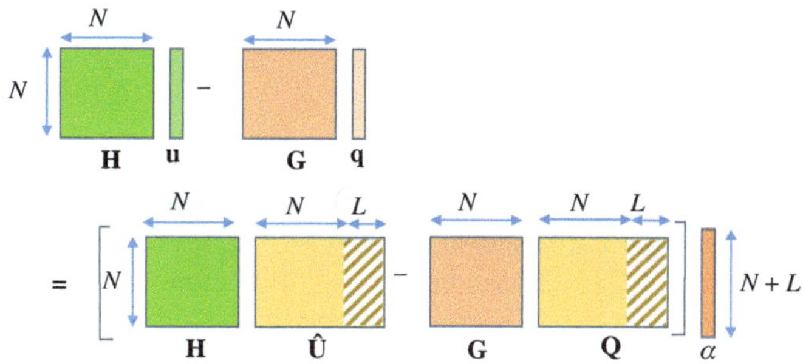

FIGURE 8.3 Schematic representation of the matrices in DRM (Eq. 8.26).

THE VECTOR $\boldsymbol{\alpha}$

The source term $b(x, y)$ is known for Poisson's equation. This allows the evaluation of the coefficients a_j, which is achieved as follows:

By taking the value of b at the $(N + L)$ nodal points, a set of equations of the form

$$b_i = \sum_{j=1}^{N+L} \alpha_j \phi_{ji} \tag{8.27}$$

is obtained; this may be expressed in matrix form as

$$\mathbf{b} = \boldsymbol{\Phi}\boldsymbol{\alpha} \tag{8.28}$$

where each column of $\boldsymbol{\Phi}$ consists of a vector $\boldsymbol{\phi}_j$ containing the values of the function ϕ_j at the $(N + L)$ DRM collocation points, that is, $\phi_j(x_i, y_i)$, $i = 1, 2, \ldots, (N + L)$. Thus Eq. (8.28) may be inverted to give

$$\boldsymbol{\alpha} = \boldsymbol{\Phi}^{-1}\mathbf{b} \tag{8.29}$$

The right-hand side of Eq. (8.26) is thus a known vector. Eq. (8.26) may be further written as

$$\mathbf{Hu} - \mathbf{Gq} = \mathbf{F} \tag{8.30}$$

where

$$\mathbf{F} = (\mathbf{H}\hat{\mathbf{U}} - \mathbf{G}\hat{\mathbf{Q}})\boldsymbol{\Phi}^{-1}\mathbf{b} \tag{8.31}$$

We see that the vector **F** can be obtained directly by multiplying known matrices and vectors. Eq. (8.30) may be compared with Eq. (4.32), which gives the same result by integrating over internal cells.

Applying the boundary conditions (3.35) as explained in Chapter 3, this equation reduces to the form (4.9)

$$\mathbf{A}\mathbf{x} = \mathbf{y} \tag{8.32}$$

where the vector $\mathbf{x}$ contains N unknown boundary values of u or q.

Noting that for the Laplace operator it is $\mathbf{m} = \mathbf{n}$, hence $\nabla u \cdot \mathbf{m} = u_{,n} = q$, the boundary conditions (8.2) may be written in the more convenient form

$$c_1 u + c_2 q = c_3 \tag{8.33}$$

where $c_i(\mathbf{x})$ $(i = 1, 2, 3)$ are quantities specified on the boundary Γ. Apparently, all types of boundary conditions are obtained from Eq. (8.33) for appropriate values of c_i.

Application of the boundary conditions (8.33) to all boundary points yields the system of N equations

$$\mathbf{c}_1 \mathbf{u} + \mathbf{c}_2 \mathbf{q} = \mathbf{c}_3 \tag{8.34}$$

where $\mathbf{c}_1, \mathbf{c}_2$ are diagonal matrices and $\mathbf{c}_3$ a vector containing the N values of c_i.

Equations (8.30) and (8.34) can be solved simultaneously to give the boundary values of u and q. This procedure, though it increases the number of unknowns by N, it simplifies the programming of the solution procedure as it treats all types of boundary conditions in the same way.

INTERIOR SOLUTION

Once the nodal boundary values of u and q are known, the values at any internal node can be computed from Eq. (8.23) with $\varepsilon_i = 1$

$$\begin{aligned} u_i = &- \sum_{k=1}^{N} G_{ik} q_k + \sum_{k=1}^{N} \hat{H}_{ik} u_k \\ &+ \sum_{j=1}^{N+L} a_j \left(\hat{u}_{ij} + \sum_{k=1}^{N} G_{ik} \hat{q}_k - \sum_{k=1}^{N} \hat{H}_{ik} \hat{u}_k \right) \end{aligned} \tag{8.35}$$

Note that the row matrices $[\hat{H}_{ik}]$ and $[G_{ik}]$ are evaluated for each interior point, where the solution is desired.

After application of Eq. (8.35) to all internal nodes we may write it in matrix form

$$\mathbf{H}^* \mathbf{u} - \mathbf{I}\mathbf{u}^* - \mathbf{G}^* \mathbf{q} = (\mathbf{H}^* \hat{\mathbf{U}} - \mathbf{G}^* \hat{\mathbf{Q}})\boldsymbol{\alpha} - \hat{\mathbf{U}}^* \boldsymbol{\alpha} \tag{8.36}$$

The star designates quantities referred to the interior points. Thus, the matrices have the dimensions: $\mathbf{H}^*, \mathbf{G}^*$ $L \times N$; $\hat{\mathbf{U}}, \hat{\mathbf{Q}}$ $N \times (N + L)$; $\hat{\mathbf{U}}^*$ $L \times (N + L)$; $\mathbf{u}^*$ $L \times 1$; $\mathbf{I}$ is the $L \times L$ unit matrix.

Equations (8.26) and (8.36) can be combined as

$$\begin{bmatrix} \mathbf{H} & \mathbf{0} \\ \mathbf{H}^* & -\mathbf{I} \end{bmatrix} \begin{Bmatrix} \mathbf{u} \\ \mathbf{u}^* \end{Bmatrix} - \begin{bmatrix} \mathbf{G} & \mathbf{0} \\ \mathbf{G}^* & \mathbf{0} \end{bmatrix} \begin{Bmatrix} \mathbf{q} \\ \mathbf{0} \end{Bmatrix} = \left(\begin{bmatrix} \mathbf{H} & \mathbf{0} \\ \mathbf{H}^* & -\mathbf{I} \end{bmatrix} \begin{bmatrix} \hat{\mathbf{U}} \\ \hat{\mathbf{U}}^* \end{bmatrix} - \begin{bmatrix} \mathbf{G} & \mathbf{0} \\ \mathbf{G}^* & \mathbf{0} \end{bmatrix} \begin{bmatrix} \hat{\mathbf{Q}} \\ \mathbf{0} \end{bmatrix} \right) \boldsymbol{\alpha} \tag{8.37}$$

For better inspection the above equation is represented schematically in Fig. 8.4. We observe that Eq. (8.37) has the form of Eq. (8.26). Thus, we may write it as

$$\mathbf{Hu} - \mathbf{Gq} = (\mathbf{H}\hat{\mathbf{U}} - \mathbf{G}\hat{\mathbf{Q}})\boldsymbol{\alpha} \tag{8.38}$$

provided that the matrices and vectors are redefined according to Eq. (8.37).

This representation with all matrices in $(N + L) \times (N + L)$ form simplifies the understanding of the method and will be used in the next sections for problems for which the source b is an unknown function.

TYPE OF INTERPOLATION FUNCTIONS

The success of DRM depends on the choice of the interpolation functions $\phi_j(x, y)$ which are used to approximate the source $b(x, y)$. To obtain accurate and reliable results with the DRM, the approximation functions must be appropriately chosen and satisfy the following requirements [10].

1. For $N + L \to \infty$, the series (8.16) must converge to b.
2. The rate of convergence must be high, so that the source term b is represented accurately even when few interpolation points are used.
3. The accuracy of the computations should not be influenced by the geometry of the domain Ω.
4. The particular solutions corresponding to the interpolation functions $\phi_j(x, y)$ should be available in closed form. This ensures fast computations and simple implementation of the computer program.

The interpolation functions $\phi_j(x, y)$ used in DRM analysis can be classified into *local* and *global* functions. The latter interpolate over the entire domain, while the former interpolate only in the neighborhood of a particular

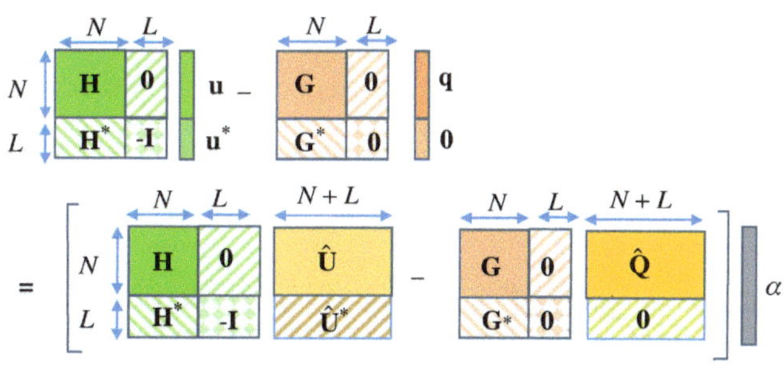

FIGURE 8.4 Schematic representation of the matrices in DRM (Eq. 8.37).

TABLE 8.1 Particular Solutions for Various RBFs

RBF $\phi(r)$	Particular Solution $\hat{u}(r)$
$\phi = 1 + r$	$\hat{u} = \frac{1}{4}r^2 + \frac{1}{9}r^3$
$\phi = 1 + r + r^2 + \dots + r^k$	$\hat{u} = \frac{1}{4}r^2 + \frac{1}{9}r^3 + \frac{1}{16}r^4 + \dots + \frac{1}{(2+k)^2}r^{2+k}$
$\phi = \sqrt{c^2 + r^2}$	$\hat{u} = \frac{1}{9}\phi^3 + \frac{1}{3}\phi c^2 - \frac{c^3}{3}\ln(c + \phi)$

point. The most widely used global functions are the so-called *radial basis functions* (RBFs) $\phi_j = \phi_j(r)$, which are functions of the Euclidean distance

$$r = |\mathbf{x} - \mathbf{x}_j| = \sqrt{(x - x_j)^2 + (y - y_j)^2} \tag{8.39}$$

between the field point $\mathbf{x}(x, y)$ and the collocation point $\mathbf{x}_j(x_j, y_j)$. Many types of RBFs have been reported in the literature. The most commonly used in DRM analysis are parts of the polynomial $\phi_j = 1 + r + r^2 + \dots$, multiquadrics (MQs) $\phi_j = (r^2 + c^2)^{1/2}$ with c being an arbitrary constant, the so-called *shape parameter*, which must be appropriately chosen to obtain optimum accuracy, thin plate spines, compact support RBFs, etc. Best results are obtained by combining RBFs augmented with elements of a polynomial series. An overview of the developments on RBFs can be found in the articles [11–15] and more recently in [16]. However, the use of such special functions requires a lot of experience with the problem at hand and is therefore not recommended for a general analysis. Instead, the use of the very simple function

$$\phi(r) = 1 + r \tag{8.40}$$

has been shown to yield accurate results for a wide variety of problems [1,2,10,17]. Also, the corresponding particular solutions are available in closed form and are therefore easy to implement. Table 8.1 gives the particular solutions for some commonly used RBFs.

8.3.3 The DRM for equations of the type $\nabla^2 u = b(x, y, u, u_{,x}, u_{,y})$

Equations of this type result, if Eq. (8.1) can be written in its canonical form [8], that is,

$$\nabla^2 u + Du_{,x} + Eu_{,y} + Fu = f(x, y) \tag{8.41}$$

Keeping the dominant part of the operator in the left-hand side and shifting the remaining part in the right-hand side, Eq. (8.41) is written as

$$\nabla^2 u = b(x, y, u, u_{,x}, u_{,y}) \tag{8.42}$$

where

$$b(x, y, u, u_{,x}, u_{,y}) = f - (Du_{,x} + Eu_{,y} + Fu) \tag{8.43}$$

THE EQUATION $\nabla^2 u + u = 0$

In order to derive the necessary basic relationships, the DRM will be applied first to the simplest equation of this type

$$\nabla^2 u + u = 0 \tag{8.44}$$

Apparently, in this case it is $b(x, y, u, u_{,x}, u_{,y}) = -u$. Thus, from Eq. (8.29) we have

$$\boldsymbol{\alpha} = -\boldsymbol{\Phi}^{-1}\mathbf{u} \tag{8.45}$$

and Eq. (8.38) becomes

$$\mathbf{Hu} - \mathbf{Gq} = -(\mathbf{H}\hat{\mathbf{U}} - \mathbf{G}\hat{\mathbf{Q}})\boldsymbol{\Phi}^{-1}\mathbf{u} \tag{8.46}$$

or

$$\mathbf{Hu} - \mathbf{Gq} = -\mathbf{Su} \tag{8.47}$$

where

$$\mathbf{S} = (\mathbf{H}\hat{\mathbf{U}} - \mathbf{G}\hat{\mathbf{Q}})\boldsymbol{\Phi}^{-1} \tag{8.48}$$

is a known matrix.

After rearranging, Eq. (8.47) is written as

$$(\mathbf{H} + \mathbf{S})\mathbf{u} = \mathbf{Gq} \tag{8.49}$$

Equation (8.49) represents a system of $(N + L)$ equations for $(2N + L)$ unknowns, that is, $2N$ boundary values of u and q, and L internal values of u. This equation can be combined with Eq. (8.34) and solved to yield all the unknown quantities.

Equations (8.46)–(8.49) form the basis of application of the DRM to equations of the form (8.42).

THE EQUATION $\nabla^2 u + u_{,x} = 0$

We consider now the equation

$$\nabla^2 u + u_{,x} = 0 \tag{8.50}$$

In this case it is $b(x, y, u, u_{,x}, u_{,y}) = -u_{,x}$ and Eq. (8.29) becomes

$$\boldsymbol{\alpha} = -\boldsymbol{\Phi}^{-1}\mathbf{u}_{,x} \tag{8.51}$$

which is inserted in Eq. (8.38) to give

$$\mathbf{Hu} - \mathbf{Gq} = -\mathbf{Su}_{,x} \tag{8.52}$$

A mechanism must now be established to relate the nodal values of u to the nodal values of the derivative $u_{,x}$. This is achieved using the approximation

$$\mathbf{u} = \boldsymbol{\Phi}\boldsymbol{\beta} \tag{8.53}$$

where $\boldsymbol{\beta} \neq \boldsymbol{\alpha}$. Differentiating Eq. (8.53) yields

$$\mathbf{u}_{,x} = \boldsymbol{\Phi}_{,x}\,\boldsymbol{\beta} \tag{8.54}$$

Equation (8.53) gives $\boldsymbol{\beta} = \boldsymbol{\Phi}^{-1}\mathbf{u}$, which is substituted in Eq. (8.54) to give

$$\mathbf{u}_{,x} = \boldsymbol{\Phi}_{,x}\,\boldsymbol{\Phi}^{-1}\mathbf{u} \tag{8.55}$$

Substituting now into Eq. (8.52) gives

$$\mathbf{Hu} - \mathbf{Gq} = -\mathbf{S}\boldsymbol{\Phi}_{,x}\,\boldsymbol{\Phi}^{-1}\mathbf{u} \tag{8.56}$$

where $\mathbf{S}$ is defined by Eq. (8.48).

Setting

$$\mathbf{R} = \mathbf{S}\boldsymbol{\Phi}_{,x}\,\boldsymbol{\Phi}^{-1} \tag{8.57}$$

and rearranging yields the system of equations

$$(\mathbf{H} + \mathbf{R})\mathbf{u} = \mathbf{Gq} \tag{8.58}$$

which is treated exactly as Eq. (8.49) to obtain the solution.

VARIABLE COEFFICIENTS

The DRM can be extended to equations with variable coefficients. Consider the Helmholtz equation

$$\nabla^2 u + F(x, y)u = 0 \tag{8.59}$$

The DRM source is now $b = -Fu$, which is applied to all nodes to give

$$\mathbf{b} = -\mathbf{Fu} \tag{8.60}$$

where $\mathbf{F}$ is an $(N+L) \times (N+L)$ diagonal matrix containing the nodal values of the function $F(x, y)$, that is,

$$\mathbf{F} = \begin{bmatrix} F(x_1, y_1) & 0 & \cdots & 0 \\ 0 & F(x_2, y_2) & \cdots & 0 \\ \vdots & \vdots & \ddots & \vdots \\ 0 & 0 & 0 & F(x_{N+L}, y_{N+L}) \end{bmatrix} \tag{8.61}$$

Thus, Eq. (8.29) becomes

$$\alpha = -\mathbf{\Phi}^{-1}\mathbf{F}\mathbf{u} \tag{8.62}$$

which is inserted in Eq. (8.38) to produce

$$\mathbf{H}\mathbf{u} - \mathbf{G}\mathbf{q} = -\mathbf{S}\mathbf{F}\mathbf{u} \tag{8.63}$$

where $\mathbf{S}$ is defined by Eq. (8.49). After rearranging, Eq. (8.63) becomes

$$(\mathbf{H} + \mathbf{S}\mathbf{F})\mathbf{u} = \mathbf{G}\mathbf{q} \tag{8.64}$$

which is treated exactly as Eq. (8.49) to obtain the solution.

THE DRM FOR THE COMPLETE ELLIPTIC DIFFERENTIAL EQUATION

The cases studied in the previous Sections "The Equation $\nabla^2 u + u = 0$", "The Equation $\nabla^2 u + u_{,x} = 0$", "Variable Coefficients", and Section 8.3.2 enable the application of the DRM to the complete elliptic differential equation Eq. (8.41). We readily derive the following DRM equation

$$(\mathbf{H} + \mathbf{R})\mathbf{u} = \mathbf{G}\mathbf{q} + \mathbf{S}\mathbf{f} \tag{8.65}$$

where

$$\mathbf{R} = \mathbf{S}\left[(\mathbf{D}\mathbf{\Phi}_{,x} + \mathbf{E}\mathbf{\Phi}_{,y})\mathbf{\Phi}^{-1} + \mathbf{F}\right] \tag{8.66}$$

in which $\mathbf{D}, \mathbf{E}, \mathbf{F}$ represent $(N+L) \times (N+L)$ diagonal matrices containing the nodal values of the position dependent coefficients $D(x, y)$, $E(x, y)$, $F(x, y)$, respectively, and $\mathbf{f}$ is the vector containing the nodal values of the source $f(x, y)$; the matrix $\mathbf{S}$ is defined by Eq. (8.48).

Equations (8.65) and (8.34) are combined to give the system of equations

$$\begin{bmatrix} \mathbf{H} + \mathbf{R} & -\mathbf{G} \\ \mathbf{c}_1 & \mathbf{c}_2 \end{bmatrix} \begin{Bmatrix} \mathbf{u} \\ \mathbf{q} \end{Bmatrix} = \begin{Bmatrix} \mathbf{S}\mathbf{f} \\ \mathbf{c}_3 \end{Bmatrix} \tag{8.67}$$

which is solved to yield the solution.

DERIVATIVES OF THE FIELD FUNCTION AT INTERNAL NODES

The derivatives at internal nodes are obtained by direct differentiation of Eq. (8.13) with $\varepsilon = 1$

$$u_{,x}(\mathbf{x}) = \int_{\Omega} v_{,x}\, b\, d\Omega - \int_{\Gamma} \left(v_{,x} \frac{\partial u}{\partial n} - u \frac{\partial v_{,x}}{\partial n} \right) ds, \quad \mathbf{x}\,(x, y) \in \Omega \tag{8.68a}$$

$$u_{,y}(\mathbf{x}) = \int_{\Omega} v_{,y}\, b\, d\Omega - \int_{\Gamma} \left(v_{,y} \frac{\partial u}{\partial n} - u \frac{\partial v_{,y}}{\partial n} \right) ds, \quad \mathbf{x}\,(x, y) \in \Omega \tag{8.68b}$$

This approach requires the evaluation of new influence matrices and transformation of the domain integrals to the boundary. The use of the DRM provides a simple alternative to Eqs. (8.68), using already computed matrices, independently of the governing equation and the type of the source. This is achieved using the relation (8.55), that is,

$$\mathbf{u}_{,x} = \mathbf{\Phi}_{,x}\, \mathbf{\Phi}^{-1}\mathbf{u} \tag{8.69a}$$

$$\mathbf{u}_{,y} = \mathbf{\Phi}_{,y}\, \mathbf{\Phi}^{-1}\mathbf{u} \tag{8.69b}$$

SOLUTION PROCEDURE STEPS

A computer code can be written for the solution of the boundary value problem (8.1), (8.2) by adhering to the following steps:

1. Compute the $N \times N$ matrices $\mathbf{G}$ and $\mathbf{H}$ for the boundary nodes using Eqs. (4.3) and (4.5).
2. Choose RBFs and compute the $N \times (L+N)$ matrices $\hat{\mathbf{U}}$, $\hat{\mathbf{Q}}$ for the boundary nodes using Eqs. (8.25a,b).
3. Compute the $L \times N$ matrices $\mathbf{G}^*$, $\mathbf{H}^*$ for the internal nodes using Eqs. (4.3).
4. Compute the $L \times (L+N)$ matrices $\hat{\mathbf{U}}^*$, $\hat{\mathbf{Q}}^*$ for the internal nodes using Eqs. (8.25a,b).
5. Formulate the matrices $\mathbf{G}$, $\mathbf{H}$, $\hat{\mathbf{U}}$, $\hat{\mathbf{Q}}$ defined in Eq. (8.37).
6. Compute the matrices $\mathbf{\Phi}$, $\mathbf{S}$, and $\mathbf{R}$ using Eqs. (8.28), (8.48), and (8.66).
7. Solve the system of linear equations (8.67).

EXAMPLES

On the basis of the previous solution procedure steps a FORTRAN code has been written which implements the DRM and example problems are solved.

EXAMPLE 8.1

In this example the DRM is employed for the solution of the boundary value problem

$$\nabla^2 u + u_{,x} + u_{,y} - 2u = \sin(x+y) - 4\sin x \sin y \quad \text{in } \Omega \tag{8.70a}$$

$$u = \sin x \sin y \quad \text{on } \Gamma \tag{8.70b}$$

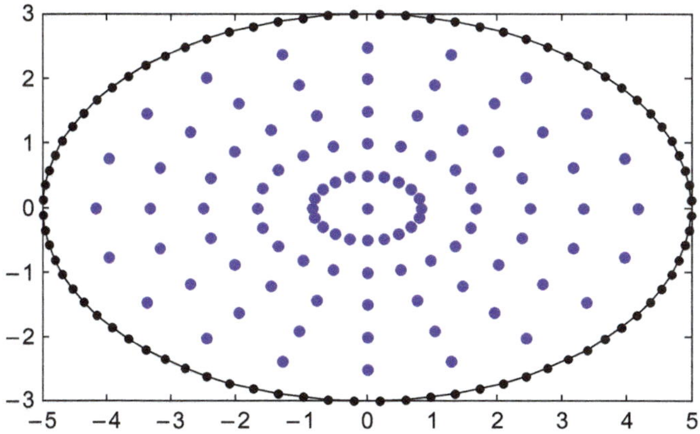

FIGURE 8.5 Boundary and domain nodal points in Example 8.1.

TABLE 8.2 Solution and its Derivatives in Example 8.1. Upper Value: Computed; Lower Value: Exact

x	y	u	u,x	u,y
1.0300	1.9021	0.8102	0.5171	−0.2786
		0.8107	0.4867	−0.2788
1.9593	1.6180	0.9226	−0.3876	−0.0455
		0.9244	−0.3783	−0.0437
3.1702	0.6180	−0.0175	−0.5843	−0.0327
		−0.0165	−0.5792	−0.0233
3.9627	0.7725	−0.5114	−0.4603	−0.5325
		−0.5185	−0.4756	−0.5242
−2.0225	−0.8816	0.6941	0.3337	−0.5752
		0.6944	0.3369	−0.5720
−0.6742	0.2939	−0.1809	0.2254	−0.5970
		−0.1808	0.2263	−0.5975

The domain Ω is the ellipse with semi axes $a=5$, $b=3$. The problem admits an exact solution $u_{exact} = \sin x \sin y$. The employed RBFs are the MQs with shape parameter $c=1$. The numerical results have been obtained using N = 80 constant boundary elements and L = 101 domain nodal points distributed as shown in Fig. 8.5. Table 8.2 shows the computed values of the solution and its first derivatives at certain points as compared with the exact ones. Fig. 8.6 shows the solution u and its derivatives $u_{,x}$, $u_{,y}$ along the line $x=y$. Finally, Fig. 8.7 shows the normal derivative $q = u_{,n}$ along the boundary.

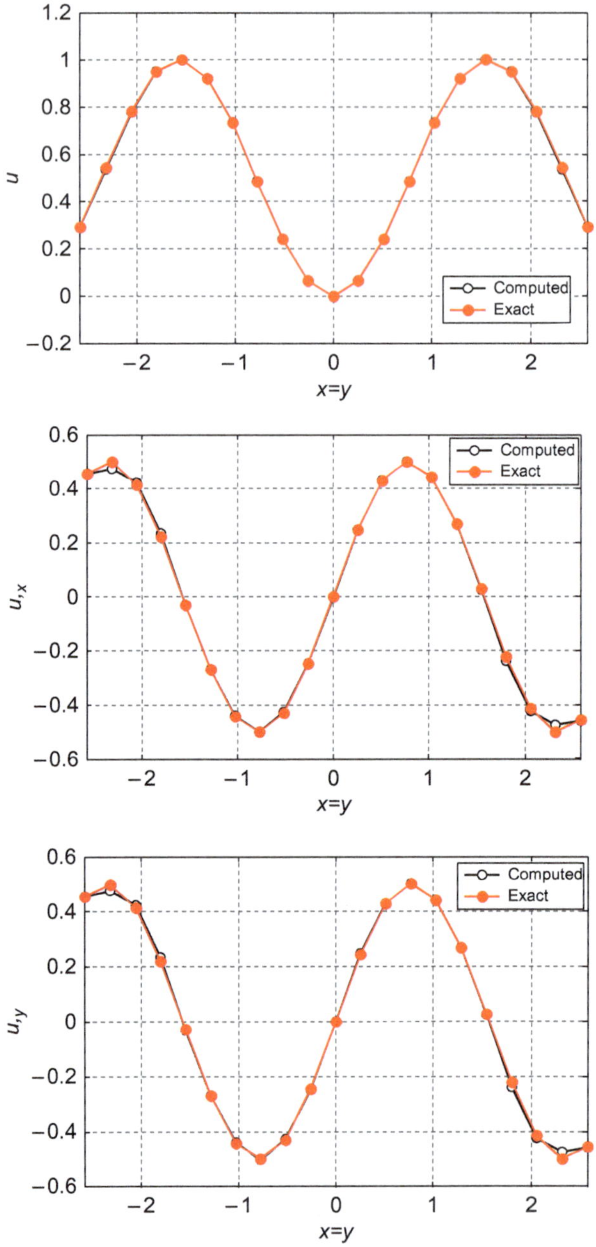

FIGURE 8.6 Solution u and derivatives $u_{,x}$, $u_{,y}$ along the line $x = y$ in Example 8.1.

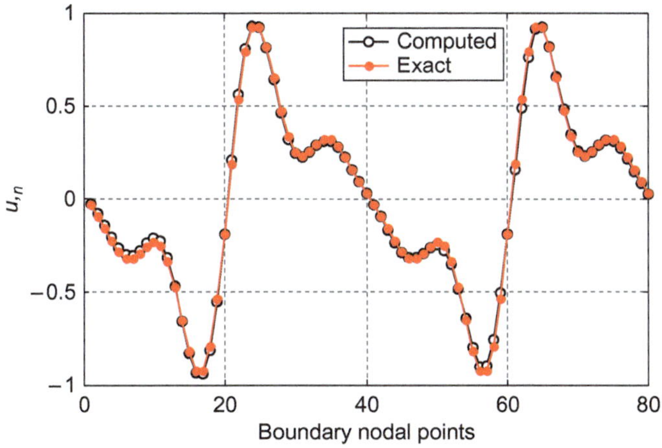

FIGURE 8.7 Normal derivative $q = u_{,n}$ along the boundary in Example 8.1.

8.4 THE ANALOG EQUATION METHOD

In Section 8.3 the DRM BEM was presented for the solution of the second order elliptic equation. It was shown that the application of the method was possible when the equation was in its canonical form, Eq. (8.41). This form allows the use of the Laplace fundamental solution to develop the DRM BEM. However, this is not possible for the general potential equation, Eq. (8.1). The inability to apply DRM may also appear in nonlinear equations. For example the equation of a surface with constant Gaussian curvature K [6]

$$u_{,x}\, u_{,y} - u_{xy}^2 - K(1+u_{,x}^2+u_{,x}^2)^2 = 0 \tag{8.71a}$$

or Plateau's equation for the minimal surface

$$(1+u_{,y}^2)u_{,xx} - 2u_{,x}u_{,y}u_{,xy} + (1+u_{,x}^2)u_{,xx} = 0 \tag{8.71b}$$

cannot be treated by the DRM.

The AEM is alleviated from any restriction and can be employed to solve efficiently any differential equation not only linear but also nonlinear and systems of them as well. The AEM is based on the *principle of the analog equation* (PAE), which reduces all potential problems, linear or nonlinear, to the Poisson equation.

8.4.1 The principle of the analog equation

The PAE was introduced by Katsikadelis in 1994 [18] and was used in conjunction with the BEM to develop the analog equation method or AEM as it is known with its acronym. The PAE can be stated as:

> *Any differential equation, be it linear or nonlinear, can be replaced by another differential equation of the same order under an unknown fictitious source. The substitute equation is termed the analog equation.*

To make this concept more concrete we illustrate the PAE as follows: We consider the boundary value problem

$$N(u) = g(\mathbf{x}), \ \mathbf{x} \in \Omega \tag{8.72a}$$

$$B(u) = \overline{g}(\mathbf{x}), \ \mathbf{x} \in \Gamma \tag{8.72b}$$

where N and B are linear or nonlinear differential operators.

Let $u(\mathbf{x})$ be the solution of the problem (8.72a,b). If $\hat{N}$ is another differential operator (linear or nonlinear) of the same order with N and it is applied to $u(\mathbf{x})$ yields

$$\hat{N}(u) = b(\mathbf{x}), \ \mathbf{x} \in \Omega \tag{8.73}$$

where now $b(\mathbf{x})$ is an unknown function (source, load). Eq. (8.73) together with the boundary condition (8.72b) can give the solution of the original problem, provided that the source $b(\mathbf{x})$ is established. Eq. (8.73) is the *analog equation* of the original problem, which together with the boundary condition (8.72b) constitutes the *equivalent* or the *substitute problem.* The replacement of the actual problem by the substitute one expresses the PAE.

The PAE applies also to a system of coupled differential equations. In this case the analog equations may constitute a set of uncoupled equations. The PAE applies to elliptic, hyperbolic and parabolic differential equations of integer or fractional order derivative [19]. Obviously, the practical significance of the PAE for the BEM is very important as a linear equation with simple known fundamental solution can be chosen as the analog equation. Then the standard BEM is readily employed to solve the substitute problem and the solution of the original problem is obtained from the integral representation of the substitute problem. The fictitious source is first established using a BEM-based procedure. The PAE is depicted in Fig. 8.8. This figure shows the deflection curve $u(x)$ of a beam with variable stiffness $EI(x)$ under the load $g(x)$. According to the PAE, the function $u(x)$ can result as the deflection curve of a beam with constant unit stiffness, say $EI(x) = 1$, under an appropriate load $b(x)$. Obviously, the solution of the substitute beam problem is simple if the fictitious load $b(x)$ is first established. Some examples illustrating the PAE are presented in Table 8.3–8.5.

The Principle of the Analog Equation [18,19]	
Real problem	Substitute problem
$N(u) = g(\mathbf{x}), \quad \mathbf{x} \in \Omega$ $B(u) = \bar{g}(\mathbf{x}), \quad \mathbf{x} \in \Gamma$ $N(\ )$: linear or nonlinear operator	$L(u) = b(\mathbf{x}), \quad \mathbf{x} \in \Omega$ $B(u) = \bar{g}(\mathbf{x}), \quad \mathbf{x} \in \Gamma$ $L(\ )$: linear operator of the order of $N(\ )$
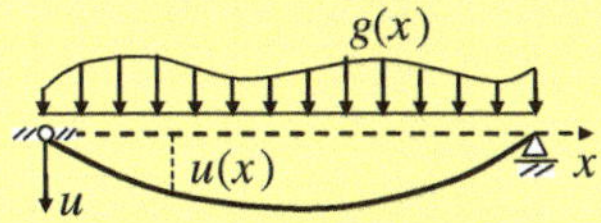 $[EI(x)u(x)_{,xx}]_{,xx} + ku(x)^2 = g(x)$	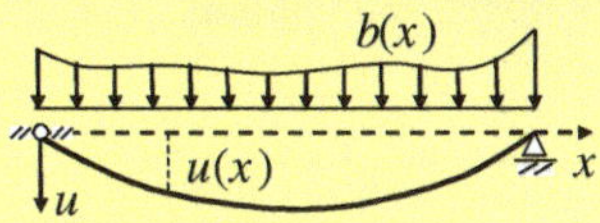 $u(x)_{,xxxx} = b(x)$

FIGURE 8.8 The principle of the analog equation [18,19].

TABLE 8.3 Example 1. The PAE Applied to a Linear Problem

Real Problem	
$(1+y^2)u_{,xx} + 2xyu_{,xy} + (1+2x^2)u_{,yy} + xu_{,x} + yu_{,y} + u = 7x^2 - 5xy + 5y^2 + 4$	in Ω
$u = x^2 - xy + y^2$ on Γ_u, $u_{,n} = 2x - y$ on Γ_n	
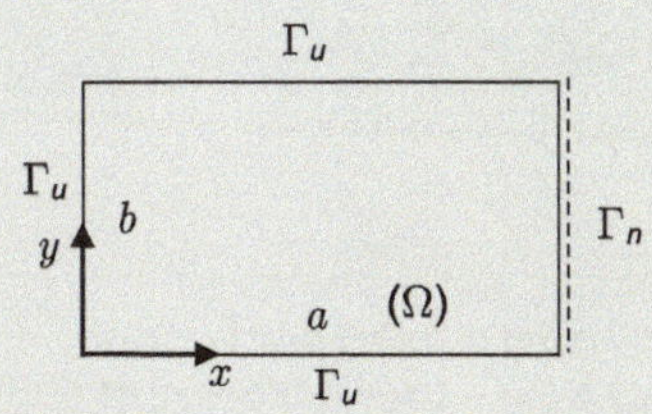	
Substitute Problem	
$\nabla^2 u = b(x, y)$ in Ω $u = x^2 - xy + y^2$ on Γ_u, $u_{,n} = 2x - y$ on Γ_n	
If $b(x, y) = 4$, both problems assume the same solution $u = x^2 - xy + y^2$	

8.4.2 The AEM for the elliptic partial differential equation

Let $u = u(\mathbf{x})$ be the sought solution to the boundary value problem (8.1), (8.2). This function is two times continuously differentiable in Ω. Since Eq. (8.1) is first of the second order, the operator $\hat{N} = \nabla^2$, that is, the Laplace operator, can be employed to produce the analog equation. This yields

$$\nabla^2 u = b(\mathbf{x}), \quad \mathbf{x}(x, y) \in \Omega \tag{8.74}$$

TABLE 8.4 Example 2. The PAE Applied to a Nonlinear Problem

Real Problem
$u_{,xx} + u_{,yy} - (1 + u_{,x}^2 + u_{,y}^2) + 5u = 3(1 + xy) \quad \text{in } \Omega$
$u = x^2 - xy + y^2 \text{ on } \Gamma_u, \quad u_{,n}^2 = 4x^2 - 4xy + y^2 \quad \text{on } \Gamma_n$
The domain Ω is the rectangle of Example 1
Substitute Problem
$\nabla^2 u = b(x, y) \quad \text{in } \Omega$
$u = x^2 - xy + y^2 \quad \text{on } \Gamma_u, \quad u_{,n}^2 = 4x^2 - 4xy + y^2 \quad \text{on } \Gamma_n$
If $b(x, y) = 4$, both problems assume the same solution $u = x^2 - xy + y^2$

TABLE 8.5 Example 3. The PAE Applied to a System of Two Coupled Equations

Real Problem

$$\nabla^2 u + \frac{1+\nu}{1-\nu}(u_{,xx} + v_{,xy}) + \left(\frac{\lambda_{,x}^*}{\mu} + 2\frac{\mu_{,x}}{\mu}\right)u_{,x} + \frac{\mu_{,y}}{\mu}u_{,y} + \frac{\mu_{,y}}{\mu}v_{,x} + \frac{\lambda_{,x}^*}{\mu}v_{,y} = -\frac{f_x}{\mu} \quad \text{in } \Omega$$

$$\nabla^2 v + \frac{1+\nu}{1-\nu}(u_{,xy} + v_{,yy}) + \frac{\lambda_{,y}^*}{\mu}u_{,y} + \frac{\mu_{,x}}{\mu}u_{,y} + \frac{\mu_{,x}}{\mu}v_{,x} + \left(\frac{\lambda_{,y}^*}{\mu} + 2\frac{\mu_{,y}}{\mu}\right)v_{,y} = -\frac{f_y}{\mu} \quad \text{in } \Omega$$

$u = xy + 0.2y^2, \quad v = xy + 0.2x^2 \quad \text{on } \Gamma_u$

$t_x = \lambda^*(x + y) + 2\mu y \quad v = xy + 0.2x^2 \quad \text{on } \Gamma_t$

where

$$\lambda^* = \frac{\nu E}{(1-\nu^2)}, \; \mu = \frac{E}{2(1+\nu)}$$

$f_x = E_0(1 + x)[-0.1(1 + x)(7 + 3\nu) - 2(\nu x + y)]/(1 - \nu^2)$

$f_y = E_0(1 + x)[-0.1(1 + x)(7 + 3\nu) - 1.4(x + y)(1 - \nu)]/(1 - \nu^2)$

$E = E_0(1+x)^2, \quad E_0 = 2 \times 10^5 \text{ kN/m}^2, \quad \nu = 0.2$

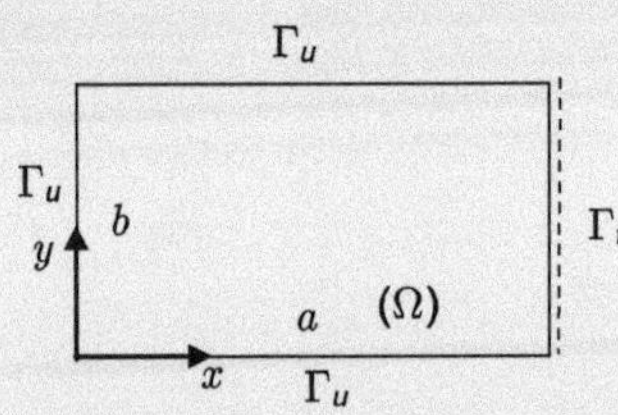

Substitute Problem

$\nabla^2 u = b_x(x, y) \quad \nabla^2 v = b_y(x, y) \quad \text{in } \Omega$

$u = xy + 0.2y^2, \quad v = xy + 0.2x^2 \quad \text{on } \Gamma_u$

$t_x = \lambda^*(x + y) + 2\mu y \quad v = xy + 0.2x^2 \quad \text{on } \Gamma_t$

If $b_x(x, y) = b_y(x, y) = 0.4$, both problems assume the same solution

$u = xy + 0.2y^2, \quad v = xy + 0.2x^2$

Equation (8.74) indicates that the solution of Eq. (8.1) could be established by solving Eq. (8.74) under the boundary conditions (8.2a,b), if the fictitious source $b(\mathbf{x})$ is first established. This is accomplished adhering to the following procedure:

We write the solution of Eq. (8.74) in integral form. Thus, we have (see Section 3.3)

$$\varepsilon u(\mathbf{x}) = \int_{\Omega} u^* b \, d\Omega - \int_{\Gamma} (u^* q - q^* u) ds \quad \mathbf{x} \in \Omega \cup \Gamma \tag{8.75}$$

We recall that $u^* = \ln r / 2\pi$ is the fundamental solution of Eq. (8.74) and $q^* = u^*_{,n}$ its derivative normal to the boundary with $r = |\boldsymbol{\xi} - \mathbf{x}| = [(\xi - x)^2 + (y - \eta)^2]^{1/2}$ being the distance between any two points $\mathbf{x}\{x, y\} \in \Omega \cup \Gamma$ and $\boldsymbol{\xi}(\xi, \eta) \in \Gamma$; ε is the free term coefficient, which takes the values $\varepsilon = 1$ if $\mathbf{x} \in \Omega$, $\varepsilon = \alpha/2\pi$ if $\mathbf{x} \in \Gamma$, and $\varepsilon = 0$ if $\mathbf{x} \notin \Omega \cup \Gamma$; α is the interior angle between the tangents of boundary at point $\mathbf{x}$. Note that it is $\varepsilon = 1/2$ for points where the boundary is smooth.

The next step is to approximate the source in Eq. (8.74) using local or global approximation functions. The first approach has been employed in [15]. Here, the AEM is developed using RBF approximation functions as in DRM. Since $b(\mathbf{x})$ is defined in Ω, only internal collocation points are taken, that is,

$$b \simeq \sum_{j=1}^{L} a_j \phi_j \tag{8.76}$$

where L is the number of the internal nodes.

After converting the domain integral to boundary line integral and discretizing the boundary using N elements, Eq. (8.75) becomes (see Eq. (8.26))

$$\mathbf{Hu} - \mathbf{Gq} = \mathbf{T}\boldsymbol{\alpha} \tag{8.77}$$

where

$$\mathbf{T} = (\mathbf{H}\hat{\mathbf{U}} - \mathbf{G}\hat{\mathbf{Q}}) \tag{8.78}$$

$$\boldsymbol{\alpha} = \boldsymbol{\Phi}^{-1}\mathbf{b}, \ \mathbf{b} = \{ b_1 \quad b_2 \quad \cdots \quad b_L \}^T, \ \boldsymbol{\alpha} = \{ a_1 \quad a_2 \quad \cdots \quad a_L \}^T \tag{8.79}$$

The matrices $\hat{\mathbf{U}}$ and $\hat{\mathbf{Q}}$ have dimensions $N \times L$ and are defined as in Eqs. (8.25a,b), that is,

$$\hat{\mathbf{U}} = [\hat{\mathbf{u}}_1 \quad \hat{\mathbf{u}}_2 \quad \ldots \quad \hat{\mathbf{u}}_L] \tag{8.80a}$$

$$\hat{\mathbf{Q}} = [\hat{\mathbf{q}}_1 \quad \hat{\mathbf{q}}_2 \quad \ldots \quad \hat{\mathbf{q}}_L] \tag{8.80b}$$

Equation (8.77) represents a set of N equations for the boundary quantities u, q. Another set of N equations for these quantities is obtained from the boundary conditions (8.2a,b), which may be combined as

$$c_1 u + c_2 \nabla u \cdot \mathbf{m} = c_3 \tag{8.81}$$

where $c_i(\mathbf{x})$ $(i = 1, 2, 3)$ are quantities specified on the boundary Γ. Apparently, all types of boundary conditions are obtained from Eq. (8.81) for appropriate values of c_i.

Before using Eq. (8.81), the flux $\nabla u \cdot \mathbf{m}$ must be first expressed in terms of $q = u_{,n}$ and u. This is achieved as follows.

It can be shown that(*)

$$\nabla u \cdot \mathbf{m} = (\mathbf{m} \cdot \mathbf{n}) u_{,n} + (\mathbf{m} \cdot \mathbf{t}) u_{,t} \tag{8.82}$$

The tangential derivative $u_{,t}$ can be expressed in terms of u. This can be accomplished either by establishing the integral representation of the tangential derivative (see Problem 3.4) or using finite differences as in Section 5.2.2. The latter technique is accurate and easy to implement. Backward and forward difference schemes are employed at points adjacent to corner points. Thus we obtain

$$\mathbf{u}_{,t} = \mathbf{d}\mathbf{u} \tag{8.83}$$

where

$$\mathbf{d} = \frac{1}{2l}\begin{array}{c} \begin{array}{cccccccccc} & & & & j & j+1 & & & & \end{array} \\ \left[\begin{array}{cccccccccc}
-3 & 4 & -1 & & & & & & & \\
-1 & 0 & 1 & & & & & & & \\
 & -1 & 0 & 1 & & & & & & \\
 & \ddots & \ddots & \ddots & & & & & & \\
 & & -1 & 0 & 1 & & & & & \\
 & & 1 & -4 & 3 & & & & & \\
 & & & & & -3 & 4 & -1 & & \\
 & & & & & -1 & 0 & 1 & & \\
 & & & & & & \ddots & \ddots & \ddots & \\
 & & & & & & & -1 & 0 & 1 \\
 & & & & & & & 1 & -4 & 3
\end{array}\right] \end{array}
\begin{array}{l} \\ \\ \\ \\ \\ j-1 \\ j \\ j+1 \\ j+2 \\ \\ \\ \\ \end{array}$$

(*) $\nabla u \cdot \mathbf{m} = [m_x \quad m_y]\begin{Bmatrix} u_{,x} \\ u_{,y} \end{Bmatrix} = [m_x \quad m_y]\begin{bmatrix} n_x & -n_y \\ n_y & n_x \end{bmatrix}\begin{Bmatrix} u_{,n} \\ u_{,t} \end{Bmatrix} = (\mathbf{m} \cdot \mathbf{n}) u_{,n} + (\mathbf{m} \cdot \mathbf{t}) u_{,t}$

In the above matrix scheme, j and $j+1$ designate nodes adjacent to a corner and l the common length of the boundary elements.

Applying the boundary condition (8.81) at the N boundary nodes and using Eq. (8.82) we obtain

$$\mathbf{c}_1\mathbf{u} + \mathbf{c}_2[(\mathbf{m}\cdot\mathbf{n})\mathbf{q} + (\mathbf{m}\cdot\mathbf{t})\mathbf{u},_t] = \mathbf{c}_3 \tag{8.84}$$

which after the replacement of the tangential derivative becomes

$$\mathbf{C}_1\mathbf{u} + \mathbf{C}_2\mathbf{q} = \mathbf{c}_3 \tag{8.85}$$

where

$$\mathbf{C}_1 = \mathbf{c}_1 + \mathbf{c}_2(\mathbf{m}\cdot\mathbf{t})\mathbf{d},\ \ \mathbf{C}_2 = \mathbf{c}_2(\mathbf{m}\cdot\mathbf{n}) \tag{8.86}$$

with $\mathbf{c}_i$ $(i=1,2)$, $(\mathbf{m}\cdot\mathbf{t})$, and $(\mathbf{m}\cdot\mathbf{n})$ being diagonal matrices and $\mathbf{c}_3$ a vector containing the N values of the respective quantities.

Equations (8.77) and (8.85) can be combined as

$$\begin{bmatrix} \mathbf{H} & -\mathbf{G} \\ \mathbf{C}_1 & \mathbf{C}_2 \end{bmatrix}\begin{bmatrix} \mathbf{u} \\ \mathbf{q} \end{bmatrix} = \begin{bmatrix} \mathbf{T} \\ \mathbf{0} \end{bmatrix}\boldsymbol{\alpha} + \begin{Bmatrix} \mathbf{0} \\ \mathbf{c}_3 \end{Bmatrix} \tag{8.87}$$

which gives

$$\begin{Bmatrix} \mathbf{u} \\ \mathbf{q} \end{Bmatrix} = \mathbf{Z}\boldsymbol{\alpha} + \mathbf{e} \tag{8.88}$$

where

$$\mathbf{Z} = \begin{bmatrix} \mathbf{H} & -\mathbf{G} \\ \mathbf{C}_1 & \mathbf{C}_2 \end{bmatrix}^{-1}\begin{bmatrix} \mathbf{T} \\ \mathbf{0} \end{bmatrix},\ \mathbf{e} = \begin{bmatrix} \mathbf{H} & -\mathbf{G} \\ \mathbf{C}_1 & \mathbf{C}_2 \end{bmatrix}^{-1}\begin{Bmatrix} \mathbf{0} \\ \mathbf{c}_3 \end{Bmatrix} \tag{8.89a, b}$$

The next step is to express the derivatives at the internal points in terms of the boundary quantities and the fictitious source. For this purpose, Eq. (8.75) with $\varepsilon=1$ is differentiated to yield the derivatives involved in Eq. (8.1). Thus, we obtain

$$u,_x(\mathbf{x}) = \int_\Omega u,_x^*\, b\, d\Omega - \int_\Gamma (u,_x^*\, q - q,_x^*\, u)ds \quad \mathbf{x}\in\Omega \tag{8.90a}$$

$$u,_y(\mathbf{x}) = \int_\Omega u,_y^*\, b\, d\Omega - \int_\Gamma (u,_y^*\, q - q,_y^*\, u)ds \quad \mathbf{x}\in\Omega \tag{8.90b}$$

$$u,_{xx}(\mathbf{x}) = \int_\Omega u,_{xx}^*\, b\, d\Omega - \int_\Gamma (u,_{xx}^*\, q - q,_{xx}^*\, u)ds \quad \mathbf{x}\in\Omega \tag{8.90c}$$

$$u,_{xy}(\mathbf{x}) = \int_\Omega u,_{xy}^*\, b\, d\Omega - \int_\Gamma (u,_{xy}^*\, q - q,_{xy}^*\, u)ds \quad \mathbf{x}\in\Omega \tag{8.90d}$$

$$u,_{yy}(\mathbf{x}) = \int_\Omega u,_{yy}^*\, b\, d\Omega - \int_\Gamma (u,_{yy}^*\, q - q,_{yy}^*\, u)ds \quad \mathbf{x}\in\Omega \tag{8.90e}$$

Equations (8.90) can be written in the compact form

$$u_{,kl}(\mathbf{x}) = \int_{\Omega} u_{,kl}^{*}\, b\, d\Omega - \int_{\Gamma} (u_{,kl}^{*}\, q - q_{,kl}^{*}\, u) ds \quad \mathbf{x} \in \Omega, \quad k, l = 0, x, y \tag{8.91}$$

This notation implies $u_{,00} = u$, $u_{,x0} = u_{,x}$, $u_{,0y} = u_{,y}$, etc.

Applying the procedure described in Section 8.3.2 gives

$$\int_{\Omega} u^{*}(\mathbf{x}, \mathbf{y})_{,x}\, \phi_j(r_{j\mathbf{y}}) d\Omega_{\mathbf{y}} = \varepsilon \hat{u}_j(r_{j\mathbf{x}})_{,x} + \int_{\Gamma} \left(u^{*}(\mathbf{x}, \boldsymbol{\xi})_{,x} \frac{\partial \hat{u}_j(r_{j\boldsymbol{\xi}})}{\partial n_{\boldsymbol{\xi}}} - \hat{u}_j(r_{j\boldsymbol{\xi}}) \frac{\partial u^{*}(\mathbf{x}, \boldsymbol{\xi})_{,\mathbf{x}}}{\partial n_{\boldsymbol{\xi}}} \right) ds_{\boldsymbol{\xi}} \tag{8.92}$$

where $\mathbf{x}, \mathbf{y} \in \Omega$, $\boldsymbol{\xi} \in \Gamma$, $r_{j\mathbf{y}} = |\mathbf{x}_j - \mathbf{y}|$, $r_{j\boldsymbol{\xi}} = |\mathbf{x}_j - \boldsymbol{\xi}|$. The subscript in the differentials and normal derivatives indicates the point, with respect to which the integration or differentiation is performed. Fig. 8.9 shows the points and their distances involved in Eq. (8.92).

For convenience, the arguments and subscripts in Eq. (8.92) may be omitted. Thus, this equations reads

$$\int_{\Omega} u^{*}{}_{,x}\, \phi_j\, d\Omega = \varepsilon \hat{u}_{j,x} + \int_{\Gamma} (u_{,x}^{*}\, \hat{q}_j - \hat{u}_j q_{,x}^{*}) ds \tag{8.93}$$

where $q^{*} = \partial u^{*}/\partial n$ and $\hat{q}_j = \partial \hat{q}_j/\partial n$

By virtue of Eqs. (8.93) and (8.76), Eq. (8.90a) becomes

$$u_{,x}(\mathbf{x}) = \sum_{j=1}^{L} \left[\hat{u}_{j,x} + \int_{\Gamma} a_j (u_{,x}^{*}\, q_j - \hat{u}_j q_{,x}^{*}) ds \right] - \int_{\Gamma} (u_{,x}^{*}\, q - q_{,x}^{*}\, u) ds, \quad \mathbf{x} \in \Omega \tag{8.94}$$

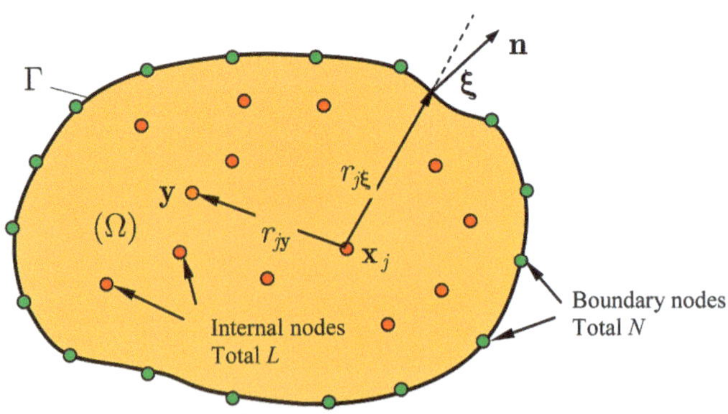

FIGURE 8.9 Boundary and internal collocation points, and radial distances in AEM.

After discretizing the boundary using N elements, Eq. (8.94) is written for the internal node i

$$
\begin{aligned}
u_{i,x} = & -\sum_{k=1}^{N} G_{ik,x}\, q_k + \sum_{k=1}^{N} \hat{H}_{ik,x}\, u_k \\
& + \sum_{j=1}^{N+L} a_j \left(\hat{u}_{ij,x} + \sum_{k=1}^{N} G_{ik,x}\, \hat{q}_k - \sum_{k=1}^{N} \hat{H}_{ik,x}\, \hat{u}_k \right)
\end{aligned}
\tag{8.95}
$$

Next applying Eq. (8.95) to the L internal nodes and using matrix form yields

$$\mathbf{u}_{,x} = -\mathbf{G}_{,x}\mathbf{q} + \hat{\mathbf{H}}_{,x}\mathbf{u} + (\hat{\mathbf{U}}_{,x} + \mathbf{G}_{,x}\hat{\mathbf{Q}} - \hat{\mathbf{H}}_{,x}\hat{\mathbf{U}})\boldsymbol{\alpha} \tag{8.96a}$$

Similarly, we obtain

$$\mathbf{u}_{,y} = -\mathbf{G}_{,y}\mathbf{q} + \hat{\mathbf{H}}_{,y}\mathbf{u} + (\hat{\mathbf{U}}_{,y} + \mathbf{G}_{,y}\hat{\mathbf{Q}} - \hat{\mathbf{H}}_{,y}\hat{\mathbf{U}})\boldsymbol{\alpha} \tag{8.96b}$$

$$\mathbf{u}_{,xx} = -\mathbf{G}_{,xx}\mathbf{q} + \hat{\mathbf{H}}_{,xx}\mathbf{u} + (\hat{\mathbf{U}}_{,xx} + \mathbf{G}_{,xx}\hat{\mathbf{Q}} - \hat{\mathbf{H}}_{,xx}\hat{\mathbf{U}})\boldsymbol{\alpha} \tag{8.96c}$$

$$\mathbf{u}_{,xy} = -\mathbf{G}_{,xy}\mathbf{q} + \hat{\mathbf{H}}_{,xy}\mathbf{u} + (\hat{\mathbf{U}}_{,xy} + \mathbf{G}_{,xy}\hat{\mathbf{Q}} - \hat{\mathbf{H}}_{,xy}\hat{\mathbf{U}})\boldsymbol{\alpha} \tag{8.96d}$$

$$\mathbf{u}_{,yy} = -\mathbf{G}_{,yy}\mathbf{q} + \hat{\mathbf{H}}_{,yy}\mathbf{u} + (\hat{\mathbf{U}}_{,yy} + \mathbf{G}_{,yy}\hat{\mathbf{Q}} - \hat{\mathbf{H}}_{,yy}\hat{\mathbf{U}})\boldsymbol{\alpha} \tag{8.96e}$$

$$\mathbf{u}^* = -\mathbf{G}\mathbf{q} + \hat{\mathbf{H}}\mathbf{u} + (\hat{\mathbf{U}} + \mathbf{G}\hat{\mathbf{Q}} - \hat{\mathbf{H}}\hat{\mathbf{U}})\boldsymbol{\alpha} \tag{8.96f}$$

The matrices $\hat{\mathbf{U}}_{,kl}$, $k, l = 0, x, y$ are derived from $\hat{\mathbf{U}}$ by performing the indicated differentiation. Here, $\mathbf{u}^*$ represents the vector of the values of u at the internal nodes to distinguish it from the vector $\mathbf{u}$ designating the boundary values of u. The superscript star is dropped, if there is no chance of confusion.

Equations (8.96) by virtue of Eq. (8.88) become

$$\mathbf{u}_{,x} = \mathbf{S}_{,x}\boldsymbol{\alpha} + \mathbf{z}_{,x} \tag{8.97a}$$

$$\mathbf{u}_{,y} = \mathbf{S}_{,y}\boldsymbol{\alpha} + \mathbf{z}_{,y} \tag{8.97b}$$

$$\mathbf{u}_{,xx} = \mathbf{S}_{,xx}\boldsymbol{\alpha} + \mathbf{z}_{,xx} \tag{8.97c}$$

$$\mathbf{u}_{,xy} = \mathbf{S}_{,xy}\boldsymbol{\alpha} + \mathbf{z}_{,xy} \tag{8.97d}$$

$$\mathbf{u}_{,yy} = \mathbf{S}_{,yy}\boldsymbol{\alpha} + \mathbf{z}_{,yy} \tag{8.97e}$$

$$\mathbf{u} = \mathbf{S}\boldsymbol{\alpha} + \mathbf{z} \tag{8.97f}$$

where

$$\mathbf{S},_{kl} = [\hat{\mathbf{H}},_{kl} \quad - \mathbf{G},_{kl}]\mathbf{Z} + (\hat{\mathbf{U}},_{kl} + \mathbf{G},_{kl}\hat{\mathbf{Q}} - \hat{\mathbf{H}},_{kl}\hat{\mathbf{U}}), \quad k, l = 0, x, y \tag{8.98a}$$

are known matrices and

$$\mathbf{z},_{kl} = [\hat{\mathbf{H}},_{kl} \quad - \mathbf{G},_{kl}]\mathbf{e}, \quad k, l = 0, x, y \tag{8.98b}$$

are known vectors. Eqs. (8.97) give the vectors containing the values of field function and its derivatives at the L internal nodes.

The final step of the AEM is to collocate Eq. (8.1) at the L internal nodes. This yields

$$\mathbf{A}\ \mathbf{u},_{xx} + 2\mathbf{B}\mathbf{u},_{xy} + \mathbf{C}\mathbf{u},_{yy} + \mathbf{D}\mathbf{u},_{x} + \mathbf{E}\mathbf{u},_{y} + \mathbf{F}\mathbf{u} = \mathbf{f} \tag{8.99}$$

where $\mathbf{A}\,, \mathbf{B}, \ldots, \mathbf{F}$ are $L \times L$ diagonal matrices containing the values of the coefficients $A, B, \ldots, F$ at the internal nodes and $\mathbf{f}$ is a vector containing the values of the actual source at the same points.

Substituting Eqs. (8.97) in Eq. (8.99) gives

$$\overline{\mathbf{K}}\boldsymbol{\alpha} = \mathbf{p} \tag{8.100}$$

where

$$\overline{\mathbf{K}} = \mathbf{A}\ \mathbf{S},_{xx} + 2\mathbf{B}\mathbf{S},_{xy} + \mathbf{C}\mathbf{S},_{yy} + \mathbf{D}\mathbf{S},_{x} + \mathbf{E}\mathbf{S},_{y} + \mathbf{S} \tag{8.101a}$$

$$\mathbf{p} = \mathbf{f} - (\mathbf{A}\ \mathbf{z},_{xx} + 2\mathbf{B}\mathbf{z},_{xy} + \mathbf{C}\mathbf{z},_{yy} + \mathbf{D}\mathbf{z},_{x} + \mathbf{E}\mathbf{z},_{y} + \mathbf{z}) \tag{8.101b}$$

Equation (8.100) constitute a set of L equations, which is solved for the vector $\boldsymbol{\alpha}$.

Finally, the solution and its derivatives at the internal nodes are obtained from Eqs. (8.97). Moreover, the solution and its derivatives at any point $\mathbf{x} \in \Omega$ can be evaluated using also Eq. (8.97), but in this case the involved matrices must be evaluated with respect to the point $\mathbf{x}$.

EVALUATION OF THE INFLUENCE MATRICES

The evaluation of the matrices $\mathbf{G}, \mathbf{H}, \mathbf{G},_{x}, \hat{\mathbf{H}},_{x}, \ldots, \mathbf{G},_{yy}, \hat{\mathbf{H}},_{yy}$ depends on the type of element used for the boundary discretization. For constant elements they are evaluated as follows:

The matrices $\mathbf{H}, \mathbf{G}$, in Eq. (8.77) are evaluated using the relations (cf. Eqs. (4.3) and (4.5))

$$G_{ij} = \int_{\Gamma_j} v(\mathbf{x}_i, \mathbf{y})\, ds_{\mathbf{y}}, \quad H_{ij} = \int_{\Gamma_j} \frac{\partial v(\mathbf{x}_i, \mathbf{y})}{\partial n} ds_{\mathbf{y}} - \frac{1}{2}\delta_{ij}, \tag{8.102}$$

$$(i, j = 1, 2, \ldots, N), \quad \mathbf{x}_i, \mathbf{y} \in \Gamma,$$

Their numerical evaluation is presented in Section 4.2.

The matrices $\mathbf{G}, \hat{\mathbf{H}}, \mathbf{G},_x, \hat{\mathbf{H}},_x, \ldots, \mathbf{G},_{yy}, \hat{\mathbf{H}},_{yy}$ in Eqs. (8.96) are given by the relations (cf. Eq. (4.17))

$$G_{ik} = \int_{\Gamma_k} v(\mathbf{x}_i, \mathbf{y})\, ds_{\mathbf{y}}, \quad \hat{H}_{ik} = \int_{\Gamma_k} q(\mathbf{x}_i, \mathbf{y})\, ds_{\mathbf{y}} \tag{8.103a}$$

$$G_{ik},_x = \int_{\Gamma_k} \frac{\partial v(\mathbf{x}_i, \mathbf{y})}{\partial x}\, ds_{\mathbf{y}}, \quad \hat{H}_{ik},_x = \int_{\Gamma_k} \frac{\partial q(\mathbf{x}_i, \mathbf{y})}{\partial x}\, ds_{\mathbf{y}} \tag{8.103b}$$

$$G_{ik},_y = \int_{\Gamma_k} \frac{\partial v(\mathbf{x}_i, \mathbf{y})}{\partial y}\, ds_{\mathbf{y}}, \quad \hat{H}_{ik},_y = \int_{\Gamma_k} \frac{\partial q(\mathbf{x}_i, \mathbf{y})}{\partial y}\, ds_{\mathbf{y}} \tag{8.103c}$$

$$G_{ik},_{xx} = \int_{\Gamma_k} \frac{\partial^2 v(\mathbf{x}_i, \mathbf{y})}{\partial x^2}\, ds_{\mathbf{y}}, \quad \hat{H}_{ik},_{xx} = \int_{\Gamma_k} \frac{\partial^2 q(\mathbf{x}_i, \mathbf{y})}{\partial x^2}\, ds_{\mathbf{y}} \tag{8.103d}$$

$$G_{ik},_{xy} = \int_{\Gamma_k} \frac{\partial^2 v(\mathbf{x}_i, \mathbf{y})}{\partial x \partial y}\, ds_{\mathbf{y}}, \quad \hat{H}_{ik},_{xy} = \int_{\Gamma_k} \frac{\partial^2 q(\mathbf{x}_i, \mathbf{y})}{\partial x \partial y}\, ds_{\mathbf{y}} \tag{8.103e}$$

$$G_{ik},_{yy} = \int_{\Gamma_k} \frac{\partial^2 v(\mathbf{x}_i, \mathbf{y})}{\partial y^2}\, ds_{\mathbf{y}}, \quad \hat{H}_{ik},_{yy} = \int_{\Gamma_k} \frac{\partial^2 q(\mathbf{x}_i, \mathbf{y})}{\partial y^2}\, ds_{\mathbf{y}} \tag{8.103f}$$

$$(i = 1, 2, \ldots, L), \quad (k = 1, 2, \ldots, N), \quad \mathbf{x}_i \in \Omega, \quad \mathbf{y} \in \Gamma.$$

The boundary integrals in Eqs. (8.103) are regular, because $r \neq 0$ and thus they can be computed using Gauss integration for regular integrals. Attention should be paid to points $\mathbf{x}_i$ near the boundary. At those points, methods for near-singular integrals may be employed to obtain accurate results (see Section 5.6). The expressions for the derivatives of the fundamental solution involved in Eq. (8.103) can be established using the relevant relations given in Appendix A.

EVALUATION OF THE DERIVATIVES OF MATRIX $\hat{\mathbf{U}}$

The matrix $\hat{\mathbf{U}}$ in Eqs. (8.96) contains the values of the particular solution $\hat{u}_j = \hat{u}(r_{j\mathbf{x}})$, $r_{j\mathbf{x}} = |\mathbf{x} - \mathbf{x}_j| = \sqrt{(x - x_j)^2 + (y - y_j)^2}$, $\mathbf{x} \in \Omega$, $\mathbf{x}_j$ at the L internal collocation points. Thus, the elements of matrices $\hat{\mathbf{U}},_x, \hat{\mathbf{U}},_y, \hat{\mathbf{U}},_{xx}, \hat{\mathbf{U}},_{xy}, \hat{\mathbf{U}},_{yy}$ in Eqs. (8.96a)–(8.96e) result by differentiation of the function $\hat{u}_j = \hat{u}(r_{j\mathbf{x}})$. Hence, we have

$$\hat{u}_j,_x = \hat{u}'_j r,_x \tag{8.104a}$$

$$\hat{u}_j,_y = \hat{u}'_j r,_y \tag{8.104b}$$

$$\hat{u}_j,_{xx} = \hat{u}'',_j r,_x^2 + \hat{u}'_j r,_{xx} \tag{8.104c}$$

$$\hat{u}_j,_{xy} = \hat{u}'',_j r,_x r,_y + \hat{u}'_j r,_{xy} \tag{8.104d}$$

$$\hat{u}_j,_{yy} = \hat{u}'',_j r,_y^2 + \hat{u}_j' r,_{yy} \tag{8.104e}$$

where (see Appendix A)

$$r,_x = \frac{x - x_j}{r}, \quad r,_y = \frac{y - y_j}{r} \tag{8.105a, b}$$

$$r,_{xx} = \frac{r,_y^2}{r}, \quad r,_{yy} = \frac{r,_x^2}{r}, \quad r,_{xy} = -\frac{r,_x r,_y}{r} \tag{8.105c, d, e}$$

For the off-diagonal elements $(i \neq j)$ of the matrices $\hat{\mathbf{U}},_x, \hat{\mathbf{U}},_y, \hat{\mathbf{U}},_{xx}, \hat{\mathbf{U}},_{xy}, \hat{\mathbf{U}},_{yy}$, it is $r \neq 0$. Hence, the expressions (8.104) can be used to compute the values of these elements. However, for the diagonal elements $(i = j)$ it is $r = 0$ and their values are computed via a limiting process.

For the RBFs given in Table 8.1, it is shown that

a. $\phi = \sqrt{r^2 + c^2}$

$$\lim_{r \to 0} \hat{u}_j,_x = 0, \quad \lim_{r \to 0} \hat{u}_j,_y = 0 \tag{8.106a, b}$$

$$\lim_{r \to 0} \hat{u}_j,_{xx} = \frac{c}{2}, \quad \lim_{r \to 0} \hat{u}_j,_{yy} = \frac{c}{2}, \quad \lim_{r \to 0} \hat{u}_j,_{xy} = 0 \tag{8.106c, d, e}$$

b. $\phi = 1 + r + r^2 + \ldots + r^k$

$$\lim_{r \to 0} \hat{u}_j,_x = 0, \quad \lim_{r \to 0} \hat{u}_j,_y = 0 \quad \text{for } k \geq 0 \tag{8.107a, b}$$

$$\lim_{r \to 0} \hat{u}_j,_{xx} = \frac{1}{2}, \quad \lim_{r \to 0} \hat{u}_j,_{yy} = \frac{1}{2}, \quad \lim_{r \to 0} \hat{u}_j,_{xy} = 0 \tag{8.107c, d, e}$$

OPTIMAL SOLUTION

The accuracy of the solution depends on the number of internal nodes, their position, the type of the employed RBF, and the shape parameter of the RBF, if it includes such a parameter. A major drawback of the MQs is the uncertainty of the choice of the shape parameter [20]. Extended research has been carried out to obtain an optimum value of c and formulas have been proposed for its approximation [21]. The minimization of the functional $J(u)$, Eq. (8.5), has been employed to obtain optimal values of the shape parameter and the position of the centers of the RBFs [22,23].

The use of the latter method requires the evaluation of the domain integral in Eq. (8.5). This is facilitated, if the domain integral is converted to a boundary line integral using the DRM. Thus denoting the integrand of the domain integral by

$$R(\mathbf{x}) = \frac{1}{2}(Au,_x^2 + 2Bu,_x u,_y + Cu,_y^2 - Fu^2) + fu \tag{8.108}$$

and approximating it with

$$R(\mathbf{x}) \simeq \sum_{j=1}^{L} \beta_j \phi_j(r) \tag{8.109}$$

we obtain

$$\int_{\Omega} R(\mathbf{x}) d\Omega = (\mathbf{H}\hat{\mathbf{U}} - \mathbf{G}\hat{\mathbf{Q}})\mathbf{\Phi}^{-1}\mathbf{R} \tag{8.110}$$

where

$$\mathbf{\Phi} = [\phi(r_{ji}), \quad \mathbf{R} = \{R(x_i)\}, \quad i, j = 1, 2, \ldots, L \tag{8.111}$$

SOLUTION PROCEDURE STEPS

A computer code can be written for the solution of the boundary value problem (8.1), (8.2) using the AEM by adhering to the following steps:

1. Compute the $N \times N$ matrices $\mathbf{G}$ and $\mathbf{H}$ for the boundary nodes using Eqs. (4.3) and (4.5).
2. Compute the $N \times L$ matrices $\hat{\mathbf{U}}$, $\hat{\mathbf{Q}}$ and $\mathbf{T}$ using Eqs. (8.78) and (8.80).
3. Compute the $N \times N$ diagonal matrices $\mathbf{C}_1, \mathbf{C}_2$ and the vector $\mathbf{c}_3$ using Eqs. (8.85) and (8.86).
4. Compute the $2N \times L$ matrix $\mathbf{Z}$ and the vector $\mathbf{e}$ using Eqs. (8.89).
5. Compute the $L \times L$ matrices $\mathbf{G}$, $\mathbf{G}_{,x}$, $\mathbf{G}_{,y}, \ldots, \mathbf{G}_{,yy}$, $\hat{\mathbf{H}}$, $\hat{\mathbf{H}}_{,x}$, $\hat{\mathbf{H}}_{,y}, \ldots, \hat{\mathbf{H}}_{,yy}$, for the internal nodes using Eqs. (8.103).
6. Compute the $L \times L$ matrices $\hat{\mathbf{U}}$, $\hat{\mathbf{U}}_{,x}$, $\hat{\mathbf{U}}_{,y}, \ldots, \hat{\mathbf{U}}_{,yy}$ using Eqs. (8.80) and (8.104).
7. Compute the $L \times L$ matrices $\mathbf{S}_{,kl}$ and the vectors $\mathbf{z}_{,kl}$ using Eqs. (8.98).
8. Compute the $L \times L$ matrix $\overline{\mathbf{K}}$ and the vector $\mathbf{p}$ using Eqs. (8.101) and solve Eq. (8.100) for the vector $\boldsymbol{\alpha}$.
9. Compute the solution using Eqs. (8.97).
10. If desired, optimize the solution by minimizing the functional (8.5).

Remark

In this section, the AEM was presented as a boundary-only method, in the sense that the discretization and integration are restricted only to the boundary as in DRM. Thus the AEM maintains the boundary character and, therefore, the advantages of the BEM. However, the AEM can be efficiently implemented as D/BEM. In this case, the domain integrals in Eqs. (8.75) and (8.91) are approximated using domain discretization. Then after elimination of the boundary quantities by virtue of the boundary conditions (8.85), the values of the field function and its derivatives at the domain nodal points are expressed in terms of the nodal values of the fictitious source $b(\mathbf{x})$. Subsequent substitution in Eq. (8.99) yields the algebraic equation for the evaluation of the fictitious source. In this approach, the domain Ω is divided into a finite number of elements over which the law of variation of $b(\mathbf{x})$ is assumed known, for example, constant, linear, etc. Triangular elements with linear variation of $b(\mathbf{x})$ are very convenient, because, on the one hand, they give good accuracy and, on the other hand, the discretization in triangles is readily performed using the *Delaunay triangulation*. There are ready to use subroutines for this purpose (see the relevant Matlab function). Though the domain discretization spoils the pure boundary character of the method, it may be preferable as it is alleviated from the drawbacks related to the RBFs. The implementation of the AEM with domain discretization is presented in detail in [19].

EXAMPLES

On the basis of the procedure presented in the Section "Solution Procedure Steps", a FORTRAN code has been written for the solution of the boundary value problem (8.1), (8.2). The employed RBFs ϕ_j are multiquadrics (MQs). This computer program uses constant boundary elements and has been given the name AEMLABE. The electronic version of this program is given on this book's companion website.

Certain examples problems are presented which demonstrate the efficiency and accuracy of the AEM.

Example 8.2

As a first example we consider a benchmark problem [24]. This problem is governed by the Poisson equation

$$\nabla^2 u = -\frac{10^6}{52} \text{ in } \Omega, \quad u = 0 \quad \text{on } \Gamma \tag{8.112}$$

where Ω is the rectangular domain $-0.3 \le x \le 0.3$, $-0.2 \le y \le 0.2$. The exact value of u at the center is $u(0,0) = 310.10$. The solution has been obtained using N constant boundary elements and L domain nodal points uniformly distributed on the rectangular domain. The obtained results for various values of N, L and c are shown in Table 8.6. The solution converges for N = 160.

TABLE 8.6 Dependence of the Solution $u(0,0)$ on N, L, and c in Example 8.2

N	c = 1 L = 25	c = 1 L = 49	c = 1.5 L = 25	c = 3 L = 25
20	313.30	313.28	313.29	313.26
40	310.78	310.79	310.79	310.78
80	310.24	310.23	310.24	310.23
160	310.10	310.10	310.10	310.10

Example 8.3

In this example we obtain the solution for the following complete second order elliptic partial differential equation

$$(1+y^2)u_{xx}+2xyu_{xy}+(1+2x^2)u_{yy}+xu_x+yu_y+u = 7x^2-5xy+5y^2+4 \quad \text{in } \Omega \tag{8.113}$$

where the domain Ω is the ellipse with semi axes $a=5$ and $b=3$.

Three types of boundary conditions are studied

i. $u=\alpha(\mathbf{x})$ on Γ (Dirichlet)

ii. $\nabla u\cdot\mathbf{m}=\gamma(\mathbf{x})$ on Γ (Neumann)

iii. $\nabla u\cdot\mathbf{m}=\gamma(\mathbf{x})$ on Γ_m, $u=\alpha(\mathbf{x})$ on Γ_u (mixed)

where $\Gamma_m=\{y=b\sqrt{1-x^2/a^2},\ 0\le x\le a\}$, $\Gamma_u=\Gamma-\Gamma_m$ and

$$\alpha(\mathbf{x})=x^2-xy+y^2 \tag{8.114a}$$

$$\gamma(\mathbf{x})=\frac{\left[(1+y^2)xb^2+xy^2a^2\right](2x-y)+\left[b^2x^2y+a^2(1+2x^2)y\right](-x+2y)}{\sqrt{b^4x^2+a^4y^2}} \tag{8.114b}$$

In all three cases the problem admits an analytical solution

$$u_{exact}=x^2-xy+y^2 \tag{8.115}$$

The solution has been obtained using N = 100 constant boundary elements and L = 93 internal points, located as shown in Fig. 8.10.

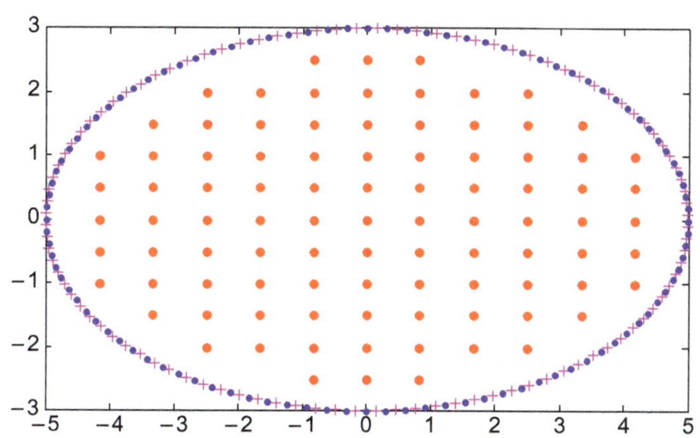

FIGURE 8.10 Elliptic domain and nodal points in Example 8.3.

The computed nodal values of the solution and its derivatives are plotted in Fig. 8.11 as compared with the exact values. In all three cases the computed results are practically identical with the exact ones. Moreover, Figs. 8.12 and 8.13 show the convergence of the mean squared error $\mathrm{MSE} = \frac{1}{L}\sqrt{\sum_{i=1}^{L}[Y(i) - Y_{exact}(i)]^2}$ of the solution and its derivatives with increasing shape parameter and number of the nodes for case (i).

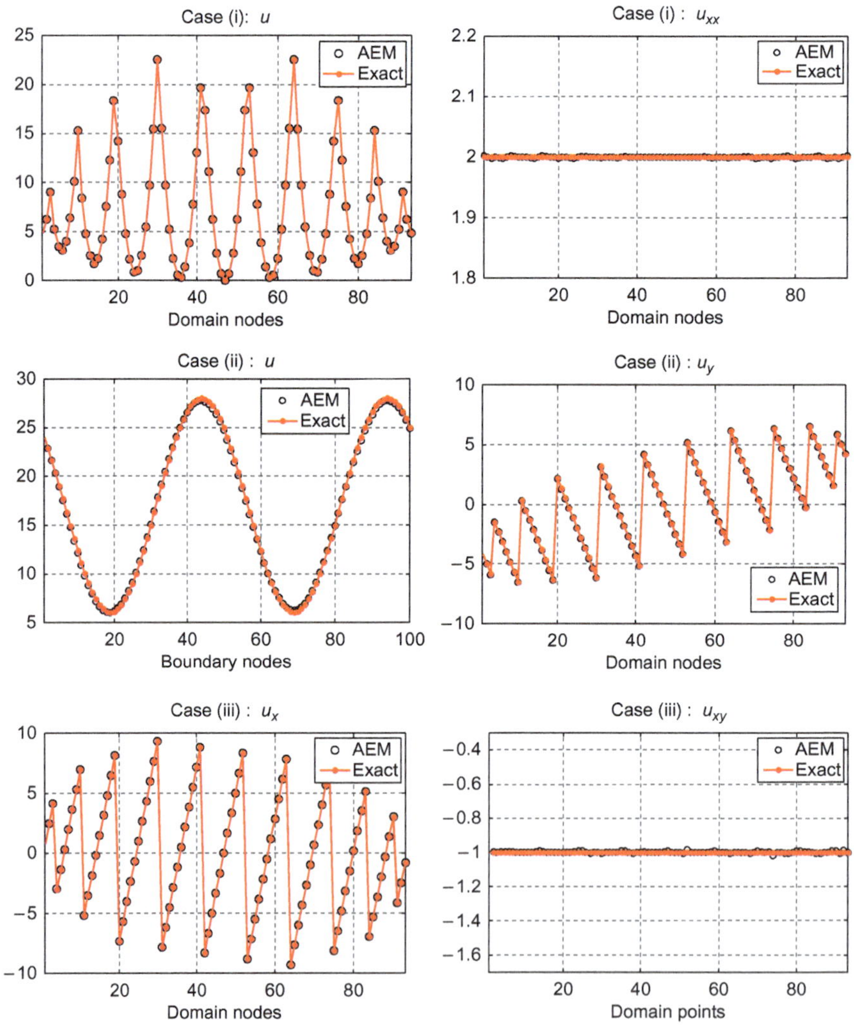

FIGURE 8.11 Nodal values of the solution and its derivatives in Example 8.3.

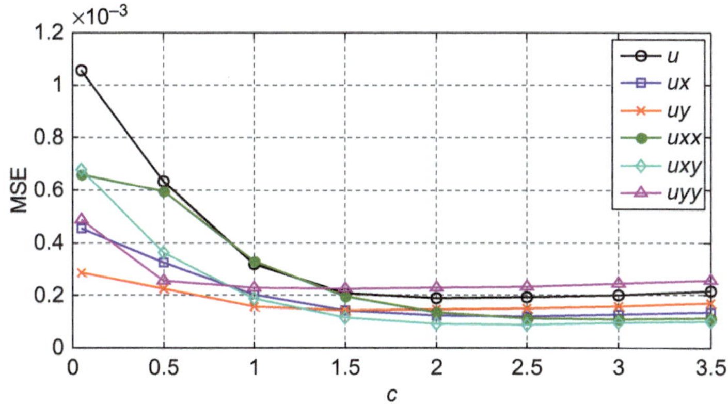

FIGURE 8.12 Dependence of MSE on shape parameter c with $\mathrm{N}=100$, $\mathrm{L}=93$ in Example 8.3, case (i).

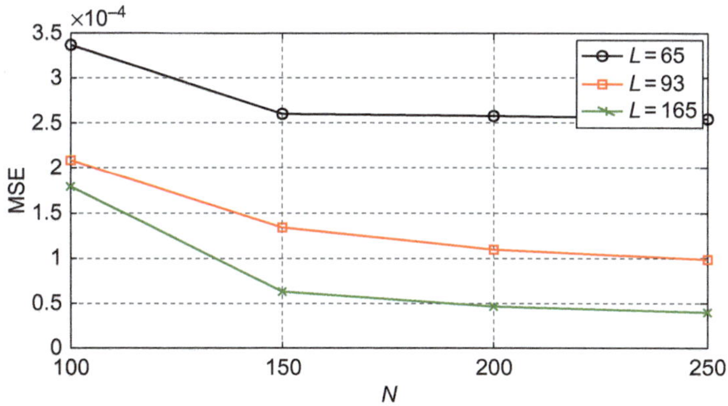

FIGURE 8.13 Dependence of MSE on the number of the boundary and domain nodes N and L for $c=1.5$ in Example 8.3, case (i).

Example 8.4

As a third example we study the temperature distribution in the plane inhomogeneous anisotropic body shown in Fig. 8.14 having a conductivity matrix

$$\mathbf{D}=\begin{bmatrix}(2x+y+2)^2 & (x-y)^2\\ (x-y)^2 & (x+2y+2)^2\end{bmatrix} \tag{8.116}$$

We look for the steady state response under the internal heat source $f=18(x^2+y^2)-16$ and the boundary conditions shown in

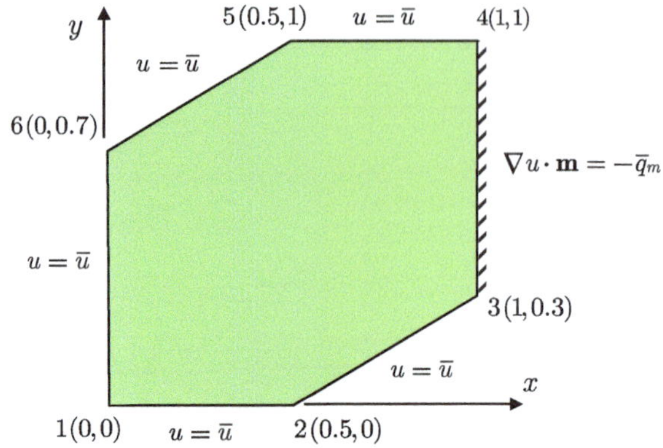

FIGURE 8.14 Geometry of the domain and boundary conditions in Example 8.4.

Fig. 8.14. The temperature distribution is governed by the boundary value problem

$$\nabla \cdot (\mathbf{D}\nabla u) + f = 0 \quad \text{in } \Omega \tag{8.117}$$

$$u = \overline{u} \quad \text{on } \Gamma_u \tag{8.118a}$$

$$\nabla u \cdot \mathbf{m} = -\overline{q}_m \quad \text{on } \Gamma_q \tag{8.118b}$$

where

$$\overline{u} = x^2 + y^2 - 5xy \text{ and } \overline{q}_m = [(2x-5y)(2x+y+2)^2 + (2x-5y)(x-y)]$$

By virtue of Eq. (8.116), Eq. (8.117) is written as

$$\left.\begin{aligned}&(2x+y+2)^2 u_{xx} + 2(x-y)^2 u_{xy} + (x+2y+2)^2 u_{yy}\\ &\quad + (6x+6y+8)u_x + (6x+6y+8)u_y = -18(x^2+y^2) + 16\end{aligned}\right\} \text{ in } \Omega \tag{8.119}$$

The problem has an analytical solution

$$u_{exact} = x^2 + y^2 - 5xy \tag{8.120}$$

The solution was computed using N = 180 constant boundary elements and L = 103 domain nodal points distributed uniformly (Fig. 8.15). Numerical results for the temperature u and the fluxes $q_x = -(k_{xx}u_{,x} + k_{xy}u_{,y})$, $q_y = -(k_{yx}u_{,x} + k_{yy}u_{,y})$ are given in Table 8.7 as compared with the exact ones. Table 8.8 shows the mean squared error $\text{MSE} = \frac{1}{L}\sqrt{\sum_{i=1}^{L}[Y(i) - Y_{exact}(i)]^2}$ for the solution and its derivatives for different values of the shape parameter c. Finally, Fig. 8.16 show the contours of the temperature distribution.

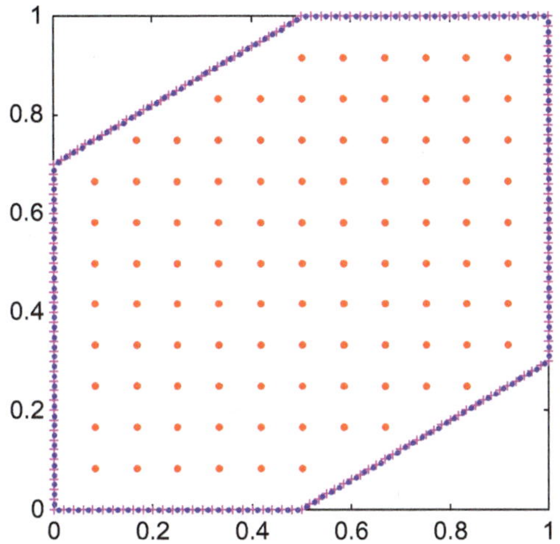

FIGURE 8.15 Nodal points in Example 8.4 ($N = 180, L = 103$).

TABLE 8.7 Temperature and Fluxes Along the Line $x = 0.25$ in Example 8.4. ($N = 180$, $L = 103$, $c = 1$). Upper Value: Computed; Lower Value: Exact

x	y	u	q_x	q_y
0.25	0.167	−0.1180	2.377	6.119
		−0.1181	2.377	6.120
	0.250	−0.1875	5.671	5.671
		−0.1875	5.672	5.672
	0.333	−0.2430	9.368	4.970
		−0.2430	9.369	4.970
	0.417	−0.2847	13.479	4.005
		−0.2847	13.481	4.005
	0.500	−0.3125	18.014	2.766
		−0.3125	18.015	2.765
	0.583	−0.3264	22.983	1.242
		−0.3264	22.984	1.241
	0.667	−0.3264	28.396	−0.576
		−0.3264	28.398	−0.578
	0.750	−0.3125	34.264	−2.700
		−0.3125	34.265	−2.703

TABLE 8.8 Mean Square Error MSE of the Solution and Its Derivatives in Example 8.4

	MSE × 10^3					
c	u	$u_{,x}$	$u_{,y}$	$u_{,xx}$	$u_{,yy}$	$u_{,xy}$
0.05	0.22	0.74	0.82	3.40	5.31	5.10
0.5	0.19	0.64	0.74	3.19	5.00	4.81
1.0	0.18	0.64	0.71	3.22	4.61	4.61
1.5	0.20	0.69	0.78	3.34	5.18	4.92

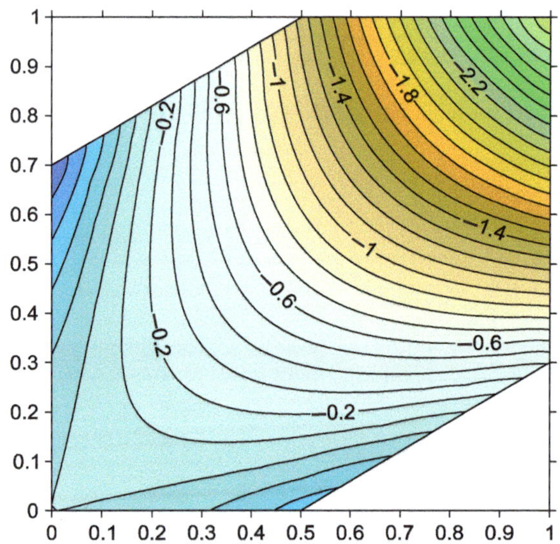

FIGURE 8.16 Contours of temperature distribution in Example 8.4.

8.5 THE BEM FOR COUPLED SECOND ORDER PARTIAL DIFFERENTIAL EQUATIONS

Systems of coupled second order elliptic partial differential equations arise in several fields of engineering and mathematical physics, for example, plane elasticity, large deflections of elastic membranes. These equations can be solved using the standard BEM only if the fundamental solution of the system of equations can be established, for example, for Navier equations (see Chapter 7). However, this is not possible if the equations have variable coefficients as in inhomogeneous bodies. Therefore, recourse to the DRM or AEM is inevitable. The DRM could be employed, but only for certain form of equations, while the AEM can treat the general form of the coupled second order equations, for example, the equations governing the deformation of inhomogeneous anisotropic elastic bodies. Hence, only the AEM is presented in the subsequent sections.

8.5.1 The AEM for the plane elastostatic problem

In this section the AEM is illustrated by applying it to solve the plane elasticity problem, first for homogeneous isotropic bodies (Navier's equations) and then for inhomogeneous anisotropic bodies.

HOMOGENOUS ISOTROPIC ELASTIC BODY

We consider the plane stress problem for a homogeneous isotropic plane body of uniform (unit) thickness made of linear elastic material occupying the two-dimensional, in general multiply connected, domain Ω with boundary $\Gamma = \cup_{k=0}^{K} \Gamma_k$ in xy-plane (Fig. 8.17). Its response is governed by Eqs. (7.36) (Section 7.2), namely

$$\nabla^2 u + \frac{1+\nu}{1-\nu}(u_{,xx} + v_{,xy}) + \frac{1}{G} f_x = 0 \quad \text{in } \Omega \tag{8.121a}$$

$$\nabla^2 v + \frac{1+\nu}{1-\nu}(u_{,xy} + v_{,yy}) + \frac{1}{G} f_y = 0 \quad \text{in } \Omega \tag{8.121b}$$

The pertinent boundary conditions on a part of the boundary may be of the following type:

$$\text{i.} \quad u = \overline{u}, \quad v = \overline{v} \tag{8.122a}$$

$$\text{ii.} \quad u = \overline{u}, \quad t_y = \overline{t}_y \tag{8.122b}$$

$$\text{iii.} \quad t_x = \overline{t}_x, \quad v = \overline{v} \tag{8.122c}$$

$$\text{iv.} \quad t_x = \overline{t}_x, \quad t_y = \overline{t}_y \tag{8.122d}$$

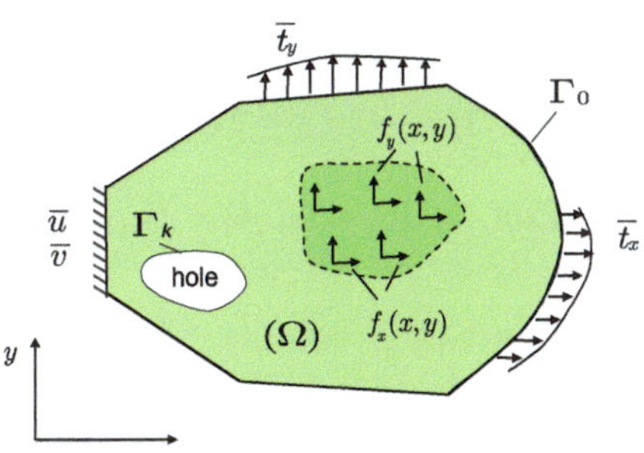

FIGURE 8.17 Domain Ω and boundary $\Gamma = \cup_{k=0}^{K} \Gamma_k$.

where $u = u(x, y)$, $v = v(x.y)$ designate the displacement components, f_x, f_y the body forces, and t_x, t_y the boundary tractions given as

$$t_x = \lambda^*(u_{,x} + v_{,y})n_x + \mu(u_{,x}n_x + v_{,x}n_y) + \mu u_{,n} \tag{8.123a}$$

$$t_y = \lambda^*(u_{,x} + v_{,y})n_y + \mu(u_{,y}n_x + v_{,y}n_y) + \mu v_{,n} \tag{8.123b}$$

and μ, λ^* the effective Lamé constants

$$\mu = G = \frac{E}{2(1+\nu)}, \qquad \lambda^* = \frac{\nu E}{1-\nu^2} \tag{8.124}$$

The overbar in Eqs. (8.122) designates a prescribed quantity. Attention should be paid to boundary condition (iv), Eq. (8.122d). In this case, the boundary tractions cannot be prescribed arbitrarily, but they must ensure overall equilibrium of the body (cf. Section 7.2.1 "Boundary Conditions").

Since Eqs. (8.121) are of the second order, it is convenient to use the analog equations

$$\nabla^2 u = b^{(1)}(\mathbf{x}) \tag{8.125a}$$

$$\nabla^2 v = b^{(2)}(\mathbf{x}) \tag{8.125b}$$

where $b^{(1)}(\mathbf{x})$, $b^{(2)}(\mathbf{x})$ are two unknown fictitious sources.

The solution of Eqs. (8.125) is given in integral form

$$\varepsilon u(\mathbf{x}) = \int_\Omega u^* b^{(1)} d\Omega - \int_\Gamma (u^* u_{,n} - u^*_{,n} u) ds \quad \mathbf{x} \in \Omega \cup \Gamma \tag{8.126a}$$

$$\varepsilon v(\mathbf{x}) = \int_\Omega u^* b^{(2)} d\Omega - \int_\Gamma (u^* v_{,n} - u^*_{,n} v) ds \quad \mathbf{x} \in \Omega \cup \Gamma \tag{8.126b}$$

Approximating the fictitious sources with RBF series

$$b^{(i)} \simeq \sum_{j=1}^{L} a_j^{(i)} \phi_j, \quad i = 1, 2 \tag{8.127}$$

and applying the procedure described in Section 8.4.2 give

$$\mathbf{Hu} - \mathbf{Gu}_{,n} = \mathbf{T}\boldsymbol{\alpha}_1 \tag{8.128a}$$

$$\mathbf{Hv} - \mathbf{Gv}_{,n} = \mathbf{T}\boldsymbol{\alpha}_2 \tag{8.128b}$$

$$\mathbf{u}_{,kl} = -\mathbf{G}_{,kl}\mathbf{u}_{,n} + \hat{\mathbf{H}}_{,kl}\mathbf{u} + (\hat{\mathbf{U}}_{,kl} + \mathbf{G}_{,kl}\hat{\mathbf{Q}} - \hat{\mathbf{H}}_{,kl}\hat{\mathbf{U}})\boldsymbol{\alpha}_1, \quad k, l = 0, x, y \tag{8.129a}$$

$$\mathbf{v}_{,kl} = -\mathbf{G}_{,kl}\mathbf{v}_{,n} + \hat{\mathbf{H}}_{,kl}\mathbf{v} + (\hat{\mathbf{U}}_{,kl} + \mathbf{G}_{,kl}\hat{\mathbf{Q}} - \hat{\mathbf{H}}_{,kl}\hat{\mathbf{U}})\boldsymbol{\alpha}_2, \quad k, l = 0, x, y \tag{8.129b}$$

$$\mathbf{b}_1 = \boldsymbol{\Phi}\boldsymbol{\alpha}_1, \quad \mathbf{b}_2 = \boldsymbol{\Phi}\boldsymbol{\alpha}_2 \tag{8.130a, b}$$

where $\mathbf{u}, \mathbf{u}_{,n}, \mathbf{v}, \mathbf{v}_{,n}$ represent the vectors of the boundary nodal values of $u, u_{,n}, v, v_{,n}$; $\mathbf{u}_{,kl}, \mathbf{v}_{,kl}$ the vectors of the interior nodal values of u, v and their derivatives, $\mathbf{b}_1, \mathbf{b}_2$ the vectors of values of the fictitious sources at the interior nodes, and $\boldsymbol{\alpha}_1, \boldsymbol{\alpha}_2$ the vectors of the RBF series coefficients. All other matrices and vectors have been defined in Section 8.4.2.

Equations (8.128) represent a set of $2N$ equations for the boundary quantities $u, u_{,n}, v, v_{,n}$. Another set of $2N$ equations for these quantities is obtained from the boundary conditions (8.122), which may be combined as

$$\gamma_1 u + \gamma_2 t_x = \gamma_3 \tag{8.131a}$$

$$\delta_1 v + \delta_2 t_y = \delta_3 \tag{8.131b}$$

in which the coefficients γ_1, γ_2 and δ_1, δ_2 take the values 0 or 1, while γ_3 takes the values $\overline{u}$ or $\overline{t}_x$ and δ_3 the values $\overline{v}$ or $\overline{t}_y$.

The tractions t_x, t_y can be expressed in terms of $u_{,n}, u_{,t}$ and $v_{,n}, v_{,t}$ using Eqs. (8.123) (cf. Eq. 7.118).Thus, we may write

$$\begin{aligned} \begin{Bmatrix} t_x \\ t_y \end{Bmatrix} &= C \begin{bmatrix} n_x & \dfrac{1-\nu}{2} n_y \\ \nu n_y & \dfrac{1-\nu}{2} n_x \end{bmatrix} \begin{Bmatrix} u_{,x} \\ u_{,y} \end{Bmatrix} + C \begin{bmatrix} \dfrac{1-\nu}{2} n_y & \nu n_x \\ \dfrac{1-\nu}{2} n_x & n_y \end{bmatrix} \begin{Bmatrix} v_{,x} \\ v_{,y} \end{Bmatrix} \\ &= C \begin{bmatrix} n_x & \dfrac{1-\nu}{2} n_y \\ \nu n_y & \dfrac{1-\nu}{2} n_x \end{bmatrix} \mathbf{R}^T \begin{Bmatrix} u_{,n} \\ u_{,t} \end{Bmatrix} + C \begin{bmatrix} \dfrac{1-\nu}{2} n_y & \nu n_x \\ \dfrac{1-\nu}{2} n_x & n_y \end{bmatrix} \mathbf{R}^T \begin{Bmatrix} v_{,n} \\ v_{,t} \end{Bmatrix} \end{aligned} \tag{8.132}$$

where $C = E/(1-\nu^2)$, and $\mathbf{R}$ the transformation matrix, that is,

$$\mathbf{R} = \begin{bmatrix} n_x & n_y \\ -n_y & n_x \end{bmatrix} = \begin{bmatrix} \cos\alpha & \sin\alpha \\ -\sin\alpha & \cos\alpha \end{bmatrix}, \quad \alpha = \measuredangle\, x, \boldsymbol{n} \tag{8.133}$$

Equation (8.132) may be written as

$$\begin{Bmatrix} t_x \\ t_y \end{Bmatrix} = \begin{bmatrix} T_{11} & T_{12} \\ T_{21} & T_{22} \end{bmatrix} \begin{Bmatrix} u_{,n} \\ u_{,t} \end{Bmatrix} + \begin{bmatrix} \overline{T}_{11} & \overline{T}_{12} \\ \overline{T}_{21} & \overline{T}_{22} \end{bmatrix} \begin{Bmatrix} v_{,n} \\ v_{,t} \end{Bmatrix} \tag{8.134}$$

where

$$\begin{bmatrix} T_{11} & T_{12} \\ T_{21} & T_{22} \end{bmatrix} = C \begin{bmatrix} n_x & \dfrac{1-\nu}{2} n_y \\ \nu n_y & \dfrac{1-\nu}{2} n_x \end{bmatrix} \mathbf{R}^T \tag{8.135a}$$

$$\begin{bmatrix} \overline{T}_{11} & \overline{T}_{12} \\ \overline{T}_{21} & \overline{T}_{22} \end{bmatrix} = C \begin{bmatrix} \dfrac{1-\nu}{2} n_y & \nu n_x \\ \dfrac{1-\nu}{2} n_x & n_y \end{bmatrix} \mathbf{R}^T \tag{8.135b}$$

Applying Eqs. (8.131) at the N boundary nodal points yields

$$\begin{bmatrix} \boldsymbol{\gamma}_1 & \mathbf{0} \\ \mathbf{0} & \boldsymbol{\delta}_1 \end{bmatrix} \begin{Bmatrix} \mathbf{u} \\ \mathbf{v} \end{Bmatrix} + \begin{bmatrix} \boldsymbol{\gamma}_2 & 0 \\ 0 & \boldsymbol{\delta}_2 \end{bmatrix} \begin{Bmatrix} \mathbf{t}_x \\ \mathbf{t}_y \end{Bmatrix} = \begin{Bmatrix} \boldsymbol{\gamma}_3 \\ \boldsymbol{\delta}_3 \end{Bmatrix} \tag{8.136}$$

where $\boldsymbol{\gamma}_1, \boldsymbol{\gamma}_2, \boldsymbol{\delta}_1, \boldsymbol{\delta}_2$ are $N \times N$ diagonal matrices and $\boldsymbol{\gamma}_3, \boldsymbol{\delta}_3$ are $N \times 1$ vectors.

Equation (8.134) is now applied at the boundary nodes and introduced into Eq. (8.136). Then replacing the tangential derivatives $u_{,t}$, $v_{,t}$ with finite differences using the matrix **d**, Eq. (8.83), reduces the boundary conditions in the form

$$\boldsymbol{\beta}_{11}\mathbf{u} + \boldsymbol{\beta}_{12}\mathbf{u}_{,n} + \boldsymbol{\beta}_{13}\mathbf{v} + \boldsymbol{\beta}_{14}\mathbf{v}_{,n} = \boldsymbol{\gamma}_3 \tag{8.137a}$$

$$\boldsymbol{\beta}_{21}\mathbf{u} + \boldsymbol{\beta}_{22}\mathbf{u}_{,n} + \boldsymbol{\beta}_{23}\mathbf{v} + \boldsymbol{\beta}_{24}\mathbf{v}_{,n} = \boldsymbol{\delta}_3 \tag{8.137b}$$

where $\boldsymbol{\beta}_{ij}$ are $N \times N$ known matrices given as

$$\boldsymbol{\beta}_{11} = \boldsymbol{\gamma}_1 + \boldsymbol{\gamma}_2\mathbf{T}_{12}\mathbf{d}, \quad \boldsymbol{\beta}_{12} = \boldsymbol{\gamma}_2\mathbf{T}_{11} \tag{8.138a, b}$$

$$\boldsymbol{\beta}_{13} = \boldsymbol{\gamma}_2\overline{\mathbf{T}}_{12}\mathbf{d}, \quad \boldsymbol{\beta}_{14} = \boldsymbol{\gamma}_2\overline{\mathbf{T}}_{11} \tag{8.138c, d}$$

$$\boldsymbol{\beta}_{21} = \boldsymbol{\delta}_2\mathbf{T}_{22}\mathbf{d}, \quad \boldsymbol{\beta}_{22} = \boldsymbol{\delta}_2\mathbf{T}_{21} \tag{8.138e, f}$$

$$\boldsymbol{\beta}_{23} = \boldsymbol{\delta}_1 + \boldsymbol{\delta}_2\overline{\mathbf{T}}_{22}\mathbf{d}, \quad \boldsymbol{\beta}_{24} = \boldsymbol{\delta}_2\overline{\mathbf{T}}_{21} \tag{8.138g, h}$$

Equations (8.128) and (8.137) are combined and solved for $\mathbf{u}, \mathbf{u}_{,n}, \mathbf{v}, \mathbf{v}_{,n}$. This yields

$$\begin{Bmatrix} \mathbf{u} \\ \mathbf{u}_{,n} \end{Bmatrix} = \mathbf{Z}^{(1)}\boldsymbol{\alpha}_1 + \mathbf{e}^{(1)} \tag{8.139a}$$

$$\begin{Bmatrix} \mathbf{v} \\ \mathbf{v}_{,n} \end{Bmatrix} = \mathbf{Z}^{(2)}\boldsymbol{\alpha}_2 + \mathbf{e}^{(2)} \tag{8.139b}$$

where $\mathbf{Z}^{(1)}$, $\mathbf{Z}^{(2)}$ and $\mathbf{e}^{(1)}$, $\mathbf{e}^{(2)}$ are the upper and lower half of the matrices $\mathbf{Z}$ and $\mathbf{e}$, respectively, that is,

$$\mathbf{Z} = \begin{bmatrix} \mathbf{Z}^{(1)} \\ \mathbf{Z}^{(2)} \end{bmatrix} = \begin{bmatrix} \mathbf{H} & -\mathbf{G} & \mathbf{0} & \mathbf{0} \\ \mathbf{0} & \mathbf{0} & \mathbf{H} & -\mathbf{G} \\ \boldsymbol{\beta}_{11} & \boldsymbol{\beta}_{12} & \boldsymbol{\beta}_{13} & \boldsymbol{\beta}_{14} \\ \boldsymbol{\beta}_{21} & \boldsymbol{\beta}_{22} & \boldsymbol{\beta}_{23} & \boldsymbol{\beta}_{24} \end{bmatrix}^{-1} \begin{bmatrix} \mathbf{T} & \mathbf{0} \\ \mathbf{0} & \mathbf{T} \\ \mathbf{0} & \mathbf{0} \\ \mathbf{0} & \mathbf{0} \end{bmatrix} \tag{8.140a}$$

$$\mathbf{e} = \begin{bmatrix} \mathbf{e}^{(1)} \\ \mathbf{e}^{(2)} \end{bmatrix} = \begin{bmatrix} \mathbf{H} & -\mathbf{G} & \mathbf{0} & \mathbf{0} \\ \mathbf{0} & \mathbf{0} & \mathbf{H} & -\mathbf{G} \\ \boldsymbol{\beta}_{11} & \boldsymbol{\beta}_{12} & \boldsymbol{\beta}_{13} & \boldsymbol{\beta}_{14} \\ \boldsymbol{\beta}_{21} & \boldsymbol{\beta}_{22} & \boldsymbol{\beta}_{23} & \boldsymbol{\beta}_{24} \end{bmatrix}^{-1} \begin{Bmatrix} \mathbf{0} \\ \mathbf{0} \\ \mathbf{F}_1 \\ \mathbf{F}_1 \end{Bmatrix} \tag{8.140b}$$

Equations (8.129) by virtue of Eqs. (8.139) and (8.140) become

$$\mathbf{u}_{,kl} = \mathbf{S}_{,kl}^{(1)} \boldsymbol{\alpha}_1 + \mathbf{z}_{,kl}^{(1)} \tag{8.141a}$$

$$\mathbf{v}_{,kl} = \mathbf{S}_{,kl}^{(2)} \boldsymbol{\alpha}_2 + \mathbf{z}_{,kl}^{(2)} \tag{8.141b}$$

where

$$\mathbf{S}_{,kl}^{(i)} = [\hat{\mathbf{H}}_{,kl} - \mathbf{G}_{,kl}]\mathbf{Z}^{(i)} + (\hat{\mathbf{U}}_{,kl} + \mathbf{G}_{,kl}\hat{\mathbf{Q}} - \hat{\mathbf{H}}_{,kl}\hat{\mathbf{U}}) \tag{8.142a}$$

$$\mathbf{z}_{,kl}^{(i)} = [\hat{\mathbf{H}}_{,kl} - \mathbf{G}_{,kl}]\mathbf{e}^{(i)} \tag{8.142b}$$

are known matrices and vectors, respectively ($i = 1, 2,\ k, l = 0, x, y$).

Next collocating Eqs. (8.121) at the L internal nodes gives

$$\mathbf{b}_1 + \frac{1+\nu}{1-\nu}(\mathbf{u}_{,xx} + \mathbf{v}_{,xy}) + \frac{1}{G}\mathbf{f}_x = 0 \tag{8.143a}$$

$$\mathbf{b}_2 + \frac{1+\nu}{1-\nu}(\mathbf{u}_{,xy} + \mathbf{v}_{,yy}) + \frac{1}{G}\mathbf{f}_y = 0 \tag{8.143b}$$

where $\mathbf{b}_1, \mathbf{b}_2,\ \mathbf{u}_{,xx}, \mathbf{u}_{,xy}, \mathbf{v}_{,xy}, \mathbf{v}_{,yy}\ \mathbf{f}_{,x}, \mathbf{f}_{,y}$ are the vectors containing the values of the respective quantities at the L internal nodes.

Using Eqs. (8.130) and (8.141) to substitute the fictitious sources and the derivatives in Eq. (8.143) gives

$$\mathbf{A}_{11}\boldsymbol{\alpha}_1 + \mathbf{A}_{12}\boldsymbol{\alpha}_2 = \mathbf{p}_1 \tag{8.144a}$$

$$\mathbf{A}_{21}\boldsymbol{\alpha}_1 + \mathbf{A}_{22}\boldsymbol{\alpha}_2 = \mathbf{p}_2 \tag{8.144b}$$

where

$$\mathbf{A}_{11} = \frac{1-\nu}{1+\nu}\mathbf{\Phi} + \mathbf{S},^{(1)}_{xx} \tag{8.145a}$$

$$\mathbf{A}_{12} = \mathbf{S},^{(2)}_{xy} \tag{8.145b}$$

$$\mathbf{A}_{21} = \mathbf{S},^{(1)}_{xy} \tag{8.145c}$$

$$\mathbf{A}_{2} = \frac{1-\nu}{1+\nu}\mathbf{\Phi} + \mathbf{S},^{(2)}_{yy} \tag{8.145d}$$

$$\mathbf{p}_1 = -\frac{1}{G}\frac{1-\nu}{1+\nu}\mathbf{f}_x - \mathbf{z},^{(1)}_{xx} - \mathbf{z},^{(2)}_{xy} \tag{8.145e}$$

$$\mathbf{p}_2 = -\frac{1}{G}\frac{1-\nu}{1+\nu}\mathbf{f}_y - \mathbf{z},^{(1)}_{xy} - \mathbf{z},^{(2)}_{yy} \tag{8.145f}$$

Equations (8.144a,b) constitute a system of $2L$ equations, which can be solved to obtain $2L$ coefficients $\boldsymbol{\alpha}_1, \boldsymbol{\alpha}_2$. The displacements and their derivatives at the internal nodes are obtained using Eqs. (8.141). Moreover, the solution and its derivatives at any point $\mathbf{x} \in \Omega$ can be evaluated using also Eqs. (8.141), but in this case the involved matrices must be evaluated with respect to the point $\mathbf{x}$.

EXAMPLES

On the basis of the procedure presented in the Section "Homogenous Isotropic Elastic Body", a FORTRAN code has been written for the solution of the boundary value problem (8.121)–(8.122). The employed RBFs ϕ_j are MQs. Certain example problems are presented which demonstrate the efficiency and accuracy of the AEM.

Example 8.5

In this example the deformation and the state of stress of the pipe of Example 7.3 is determined using the AEM with RBFs of MQ type as presented in the Section "Homogenous Isotropic Elastic Body". The obtained numerical results are shown in Figs. 8.18 through 8.23 as compared with the BEM solution (see Chapter 7), the AEM with domain discretization for the approximation of the fictitious sources (see [19], Chapter 3) and the FEM. The BEM results have been obtained with N = 660 constant boundary elements, the AEM-MQs results with N = 660 constant boundary elements and L = 488 domain nodal points, the AEM-Domain results with N = 660 constant boundary elements and L = 488 domain nodal points resulting from 786 linear triangular elements, while the FEM results with 8188 quadrilateral elements using the NASTRAN code.

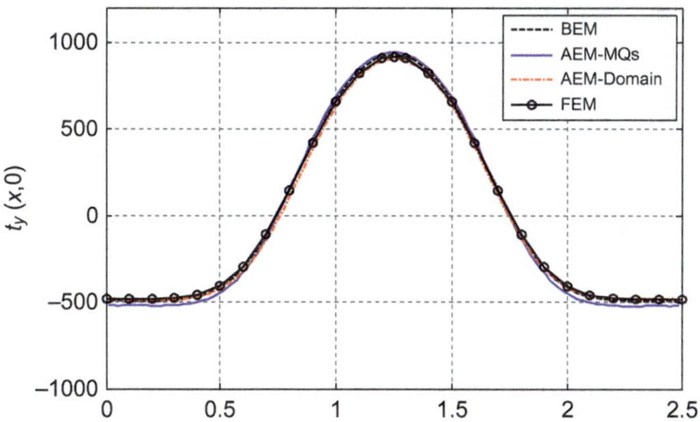

FIGURE 8.18 Distribution of traction t_y along the boundary $y = 0$ in Example 8.5.

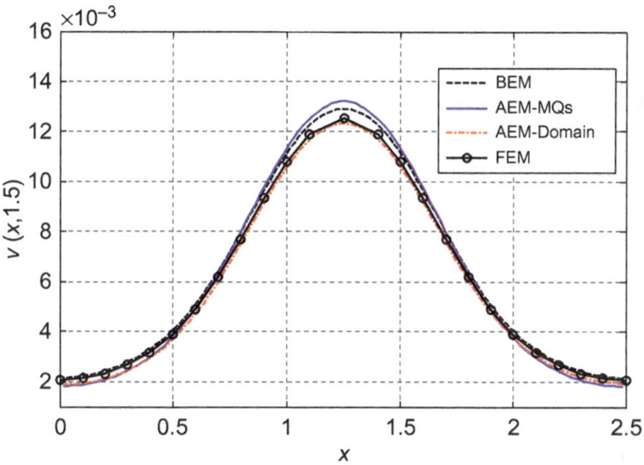

FIGURE 8.19 Distribution of the displacement v along the upper boundary $y = 1.5$ in Example 8.5.

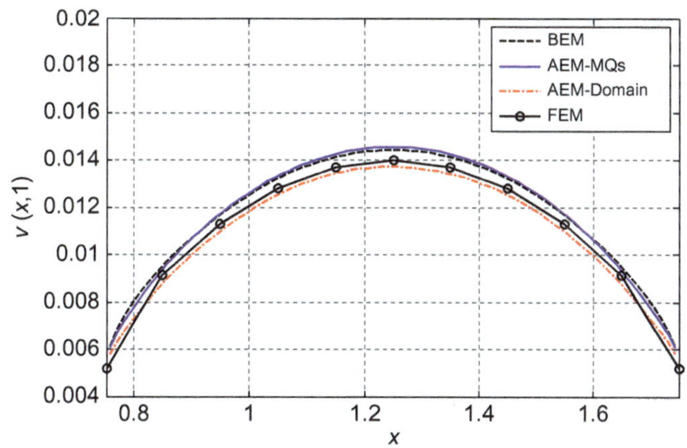

FIGURE 8.20 Distribution of the displacement v along the line $y = 1.0$ in Example 8.5.

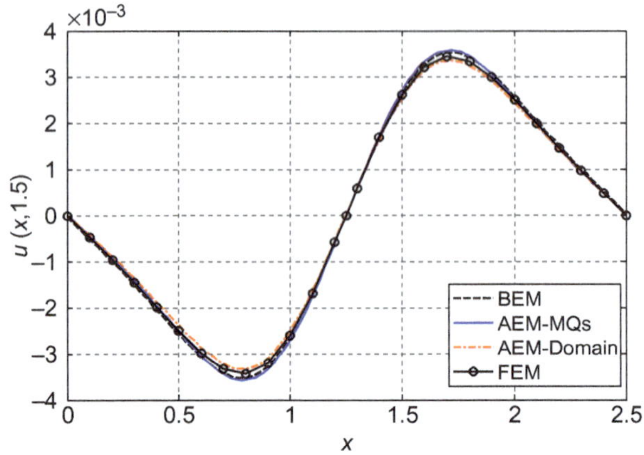

FIGURE 8.21 Distribution of the displacement u along the boundary $y = 1.5$ in Example 8.5.

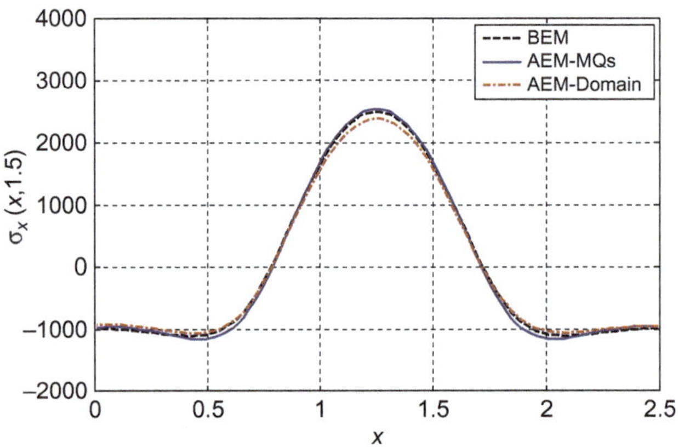

FIGURE 8.22 Distribution of the stress σ_x along the boundary $y = 1.5$ in Example 8.5.

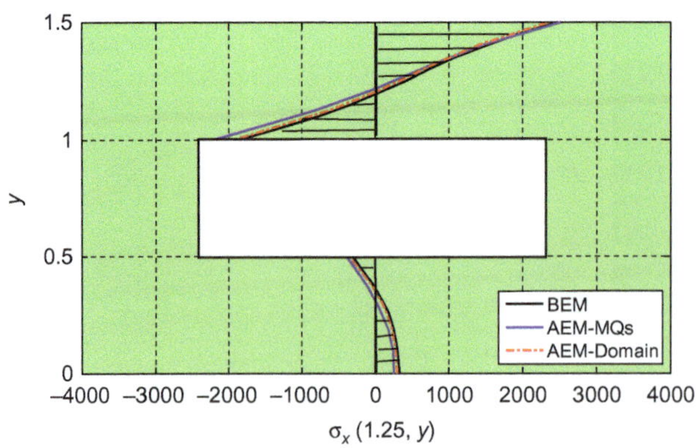

FIGURE 8.23 Distribution of the stress σ_x along the line $x = 1.25$ in Example 8.5.

INHOMOGENOUS ANISOTRPIC ELASTIC BODY

We consider an inhomogeneous anisotropic plane body of uniform (unit) thickness under plane stress ($\sigma_z = \tau_{xz} = \tau_{yz} = 0$) made of linear elastic material occupying the two-dimensional, in general multiply connected, domain Ω with boundary $\Gamma = \cup_{k=0}^{K} \Gamma_k$ in xy-plane (cf. Fig. 8.17). The governing equations result by taking the equilibrium of an element of the plane body.

Thus we have

$$\hat{\nabla}^T \boldsymbol{\sigma} + \mathbf{f} = \mathbf{0} \tag{8.146}$$

where

$$\mathbf{u} = \begin{Bmatrix} u \\ v \end{Bmatrix}, \ \boldsymbol{\sigma} = \begin{Bmatrix} \sigma_x \\ \sigma_y \\ \tau_{xy} \end{Bmatrix}, \ \mathbf{f} = \begin{Bmatrix} f_x \\ f_y \end{Bmatrix} \tag{8.147a, b, c}$$

are the displacement, the stress and the body force vector, respectively, and $\hat{\nabla}$ is the differential operator defined as

$$\hat{\nabla} = \begin{bmatrix} \dfrac{\partial}{\partial x} & 0 \\ 0 & \dfrac{\partial}{\partial y} \\ \dfrac{\partial}{\partial y} & \dfrac{\partial}{\partial x} \end{bmatrix} \tag{8.148}$$

Moreover, the constitutive relations and the kinematic relations (cf. Eqs. (7.32) and (7.33)) are written as

$$\boldsymbol{\sigma} = \mathbf{C}\boldsymbol{\varepsilon} \tag{8.149}$$

$$\boldsymbol{\varepsilon} = \hat{\nabla}\mathbf{u} \tag{8.150}$$

where $\boldsymbol{\varepsilon} = \{ \varepsilon_x \quad \varepsilon_y \quad \gamma_{xy} \}^T$ represents the strain vector and

$$\mathbf{C} = \begin{bmatrix} C_{11} & C_{12} & C_{13} \\ C_{21} & C_{22} & C_{23} \\ C_{31} & C_{32} & C_{33} \end{bmatrix} \tag{8.151}$$

the elastic constitutive matrix, which is position dependent, $C_{ij} = C_{ij}(x, y)$, symmetric $C_{ij} = C_{ji}$ and invertible, $\det\mathbf{C} \neq 0$.

Equation (8.146) by virtue of Eqs. (8.149)–(8.151) is written as

$$\hat{\nabla}^T \mathbf{C}\hat{\nabla}\mathbf{u} + \mathbf{f} = \mathbf{0} \tag{8.152}$$

which in terms of the displacements components become

$$L_{11}(u) + L_{12}(v) + f_x = 0 \tag{8.153a}$$

$$L_{21}(u) + L_{22}(v) + f_y = 0 \tag{8.153b}$$

where

$$\begin{aligned} L_{11} = {} & C_{11}\frac{\partial^2}{\partial x^2} + 2C_{13}\frac{\partial^2}{\partial x \partial y} + C_{33}\frac{\partial^2}{\partial y^2} \\ & + (C_{11},_x + C_{13},_y)\frac{\partial}{\partial x} + (C_{13},_x + C_{33},_y)\frac{\partial}{\partial y} \end{aligned} \tag{8.154a}$$

$$\begin{aligned} L_{12} = {} & C_{13}\frac{\partial^2}{\partial x^2} + (C_{12} + C_{33})\frac{\partial^2}{\partial x \partial y} + C_{23}\frac{\partial^2}{\partial y^2} \\ & + (C_{13},_x + C_{33},_y)\frac{\partial}{\partial x} + (C_{12},_x + C_{23},_y)\frac{\partial}{\partial y} \end{aligned} \tag{8.154b}$$

$$\begin{aligned} L_{21} = {} & C_{13}\frac{\partial^2}{\partial x^2} + (C_{12} + C_{33})\frac{\partial^2}{\partial x \partial y} + C_{23}\frac{\partial^2}{\partial y^2} \\ & + (C_{13},_x + C_{12},_y)\frac{\partial}{\partial x} + (C_{33},_x + C_{23},_y)\frac{\partial}{\partial y} \end{aligned} \tag{8.154c}$$

$$\begin{aligned} L_{22} = {} & C_{33}\frac{\partial^2}{\partial x^2} + 2C_{23}\frac{\partial^2}{\partial x \partial y} + C_{22}\frac{\partial^2}{\partial y^2} \\ & + (C_{33},_x + C_{23},_y)\frac{\partial}{\partial x} + (C_{23},_x + C_{22},_y)\frac{\partial}{\partial y} \end{aligned} \tag{8.154d}$$

The boundary conditions are given by Eqs. (8.122).

The boundary tractions are expressed in terms of the stress components

$$t_x = \sigma_x n_x + \tau_{xy} n_y = \bar{t}_x \tag{8.155a}$$

$$t_y = \tau_{xy} n_x + \sigma_y n_y = \bar{t}_y \tag{8.155b}$$

which by virtue of Eqs. (8.149) and (8.150) are written in terms of the displacements

$$\begin{aligned} & (C_{11}n_x + C_{13}n_y)u,_x + (C_{13}n_x + C_{33}n_y)u,_y \\ & + (C_{13}n_x + C_{33}n_y)v,_x + (C_{12}n_x + C_{23}n_y)v,_y = \bar{t}_x \end{aligned} \tag{8.156a}$$

$$\begin{aligned} & (C_{13}n_x + C_{12}n_y)u,_x + (C_{33}n_x + C_{23}n_y)u,_y \\ & + (C_{33}n_x + C_{23}n_y)v,_x + (C_{23}n_x + C_{22}n_y)v,_y = \bar{t}_y \end{aligned} \tag{8.156b}$$

where $\mathbf{n}(n_x, n_y)$ is the unit vector normal to the boundary.

If the body is orthotropic, the constitutive matrix for plane stress may be written in terms of the elastic moduli and Poisson's ratios as

$$\mathbf{C} = \begin{bmatrix} \dfrac{E_1}{1-\nu_{12}\nu_{21}} & \dfrac{\nu_{21}E_1}{1-\nu_{12}\nu_{21}} & 0 \\ \dfrac{\nu_{12}E_2}{1-\nu_{12}\nu_{21}} & \dfrac{E_2}{1-\nu_{12}\nu_{21}} & 0 \\ 0 & 0 & G_{12} \end{bmatrix} \tag{8.157}$$

subject to the constraint $\nu_{21}E_1 = \nu_{12}E_2$.

For an isotropic body it is $E_1 = E_2 = E$ and $\nu_{21} = \nu_{12} = \nu$ and the constitutive matrix is simplified as

$$\mathbf{C} = \frac{E}{1-\nu^2}\begin{bmatrix} 1 & \nu & 0 \\ \nu & 1 & 0 \\ 0 & 0 & \dfrac{1}{2}(1-\nu) \end{bmatrix} \tag{8.158}$$

The AEM solution adheres to the same steps as in the Section "Homogenous Isotropic Elastic Body". Therefore Eqs. (8.126)–(8.130) apply to this case, too.

The boundary tractions are expressed in terms of the normal and tangential derivatives as

$$\begin{Bmatrix} t_x \\ t_x \end{Bmatrix} = \begin{bmatrix} C_{11}n_x + C_{13}n_y & C_{13}n_x + C_{33}n_y \\ C_{13}n_x + C_{12}n_y & C_{33}n_x + C_{23}n_y \end{bmatrix} \mathbf{R}^T \begin{Bmatrix} u_{,n} \\ u_{,t} \end{Bmatrix} + \begin{bmatrix} C_{13}n_x + C_{33}n_y & C_{12}n_x + C_{23}n_y \\ C_{33}n_x + C_{23}n_y & C_{23}n_x + C_{22}n_y \end{bmatrix} \mathbf{R}^T \begin{Bmatrix} v_{,n} \\ v_{,t} \end{Bmatrix} \tag{8.159}$$

Equation (8.159) is now applied to the boundary nodes and subsequently introduced into Eq. (8.136). Then replacing the tangential derivatives $u_{,t}$, $v_{,t}$ with finite differences using the matrix $\mathbf{d}$, Eq. (8.83), gives the boundary conditions in the form

$$\overline{\boldsymbol{\beta}}_{11}\mathbf{u} + \overline{\boldsymbol{\beta}}_{12}\mathbf{u}_{,n} + \overline{\boldsymbol{\beta}}_{13}\mathbf{v} + \overline{\boldsymbol{\beta}}_{14}\mathbf{v}_{,n} = \boldsymbol{\gamma}_3 \tag{8.160a}$$

$$\overline{\boldsymbol{\beta}}_{21}\mathbf{u} + \overline{\boldsymbol{\beta}}_{22}\mathbf{u}_{,n} + \overline{\boldsymbol{\beta}}_{23}\mathbf{v} + \overline{\boldsymbol{\beta}}_{24}\mathbf{v}_{,n} = \boldsymbol{\delta}_3 \tag{8.160b}$$

where $\overline{\boldsymbol{\beta}}_{ij}$ are $N \times N$ known matrices. They are derived as the matrices $\boldsymbol{\beta}_{ij}$ in Eqs. (8.138). Their expressions are not given here as it is more convenient to derive them in the computer program for the numerical implementation of the solution procedure. Therefore, Eqs. (8.139)–(8.142) are valid, too, provided that the matrices $\boldsymbol{\beta}_{ij}$ in Eq. (8.140) are replaced with $\overline{\boldsymbol{\beta}}_{ij}$.

The last step of the solution procedure is to apply Eqs. (8.153) to the L internal nodes. Thus we obtain

$$L_{11}(\mathbf{u}) + L_{12}(\mathbf{v}) + \mathbf{f}_x = 0 \tag{8.161a}$$

$$L_{21}(\mathbf{u}) + L_{22}(\mathbf{v}) + \mathbf{f}_y = 0 \tag{8.161b}$$

Then using Eqs. (8.141) to substitute the derivatives indicated by the operators L_{ij} produces the system of equations

$$\mathbf{A}_{11}\boldsymbol{\alpha}_1 + \mathbf{A}_{12}\boldsymbol{\alpha}_2 = \mathbf{p}_1 \tag{8.162a}$$

$$\mathbf{A}_{21}\boldsymbol{\alpha}_1 + \mathbf{A}_{22}\boldsymbol{\alpha}_2 = \mathbf{p}_2 \tag{8.162b}$$

where

$$\begin{aligned}\mathbf{A}_{11} = {} & \mathbf{C}_{11}\mathbf{S},_{xx}^{(1)} + 2\mathbf{C}_{13}\mathbf{S},_{xy}^{(1)} + \mathbf{C}_{33}\mathbf{S},_{yy}^{(1)} \\ & + (\mathbf{C}_{11},_x + C_{13},_y)\mathbf{S},_x^{(1)} + (\mathbf{C}_{13},_x + \mathbf{C}_{33},_y)\mathbf{S},_y^{(1)}\end{aligned} \tag{8.163a}$$

$$\begin{aligned}\mathbf{A}_{12} = {} & \mathbf{C}_{13}\mathbf{S},_{xx}^{(2)} + (\mathbf{C}_{12} + \mathbf{C}_{33})\mathbf{S},_{xy}^{(2)} + \mathbf{C}_{23}\mathbf{S},_{yy}^{(2)} \\ & + (\mathbf{C}_{13},_x + \mathbf{C}_{33},_y)\mathbf{S},_x^{(2)} + (\mathbf{C}_{12},_x + \mathbf{C}_{23},_y)\mathbf{S},_y^{(2)}\end{aligned} \tag{8.163b}$$

$$\begin{aligned}\mathbf{A}_{21} = {} & \mathbf{C}_{13}\mathbf{S},_{xx}^{(1)} + (\mathbf{C}_{12} + \mathbf{C}_{33})\mathbf{S},_{xy}^{(1)} + \mathbf{C}_{23}\mathbf{S},_{yy}^{(1)} \\ & + (\mathbf{C}_{13},_x + \mathbf{C}_{12},_y)\mathbf{S},_x^{(1)} + (\mathbf{C}_{33},_x + \mathbf{C}_{23},_y)\mathbf{S},_y^{(1)}\end{aligned} \tag{8.163c}$$

$$\begin{aligned}\mathbf{A}_{22} = {} & \mathbf{C}_{33}\mathbf{S},_{xx}^{(2)} + 2\mathbf{C}_{23}\mathbf{S},_{xy}^{(2)} + \mathbf{C}_{22}\mathbf{S},_{yy}^{(2)} \\ & + (\mathbf{C}_{33},_x + \mathbf{C}_{23},_y)\mathbf{S},_x^{(2)} + (\mathbf{C}_{23},_x + \mathbf{C}_{22},_y)\mathbf{S},_y^{(2)}\end{aligned} \tag{8.163d}$$

$$\begin{aligned}\mathbf{p}_1 = {} & -\mathbf{f}_x - \Big[\mathbf{C}_{11}\mathbf{z},_{xx}^{(1)} + 2\mathbf{C}_{13}\mathbf{z},_{xy}^{(1)} + \mathbf{C}_{33}\mathbf{z},_{yy}^{(1)} \\ & + (\mathbf{C}_{11},_x + C_{13},_y)\mathbf{z},_x^{(1)} + (\mathbf{C}_{13},_x + \mathbf{C}_{33},_y)\mathbf{z},_y^{(1)}\Big] \\ & - \Big[\mathbf{C}_{13}\mathbf{z},_{xx}^{(2)} + (\mathbf{C}_{12} + \mathbf{C}_{33})\mathbf{z},_{xy}^{(2)} + \mathbf{C}_{23}\mathbf{z},_{yy}^{(2)} \\ & + (\mathbf{C}_{13},_x + \mathbf{C}_{33},_y)\mathbf{z},_x^{(2)} + (\mathbf{C}_{12},_x + \mathbf{C}_{23},_y)\mathbf{z},_y^{(2)}\Big]\end{aligned} \tag{8.163e}$$

$$\begin{aligned}\mathbf{p}_2 = {} & -\mathbf{f}_y - \Big[\mathbf{C}_{13}\mathbf{z},_{xx}^{(1)} + (\mathbf{C}_{12} + \mathbf{C}_{33})\mathbf{z},_{xy}^{(1)} + \mathbf{C}_{23}\mathbf{z},_{yy}^{(1)} \\ & + (\mathbf{C}_{13},_x + \mathbf{C}_{12},_y)\mathbf{z},_x^{(1)} + (\mathbf{C}_{33},_x + \mathbf{C}_{23},_y)\mathbf{z},_y^{(1)}\Big] \\ & - \Big[\mathbf{C}_{33}\mathbf{z},_{xx}^{(2)} + 2\mathbf{C}_{23}\mathbf{z},_{xy}^{(2)} + \mathbf{C}_{22}\mathbf{z},_{yy}^{(2)} \\ & + (\mathbf{C}_{33},_x + \mathbf{C}_{23},_y)\mathbf{z},_x^{(2)} + (\mathbf{C}_{23},_x + \mathbf{C}_{22},_y)\mathbf{z},_y^{(2)}\Big]\end{aligned} \tag{8.163f}$$

in which $\mathbf{C}_{11}, \mathbf{C}_{12}, \ldots \mathbf{C}_{11,x}, \ldots$ are $L \times L$ diagonal matrices containing the values of the respective quantities at the internal nodes. The $\mathbf{S},_{kl}^{(i)}$ and the vectors $\mathbf{z},_{kl}^{(i)}$ $(i = 1, 2 \;\; k, l = 0, x, y)$ are given by Eqs. (8.142). For the homogeneous isotropic body the stiffness matrix is given by Eqs. (8.158), and it is readily shown that Eqs. (8.163) reduce to (8.145).

Equations (8.162) are solved to give the vectors $\boldsymbol{\alpha}_1, \boldsymbol{\alpha}_2$. Then the displacements and their derivatives are obtained from Eqs. (8.141).

EXAMPLES

On the basis of the procedure presented in the Section "Inhomogenous Anisotrpic Elastic Body" a FORTRAN code has been written for the solution of Eqs. (8.153) under the boundary conditions (8.131). The employed RBFs ϕ_j are MQs. This computer program uses constant boundary elements and has been given the name AEMELBE. The electronic version of the program is given on this book's companion website. Certain examples problems are presented which demonstrate the efficiency and accuracy of the AEM for the solution of inhomogeneous anisotropic elasticity problems.

Example 8.6

In this example the deformation of the inhomogeneous orthotropic rectangular plane body of Fig. 8.24 with unit thickness ($h = 1$) is studied. The body is subjected to a uniform normal traction along the two opposite sides. The other two edges are free to move tangentially, while they are restraint in the normal direction.

The material parameters are given as

$$
\begin{gathered}
C_{11} = \frac{8(10 - 9\xi^2)}{8(10 - 9\xi^2)(1 + 9\xi^2) - 5} E \\
C_{22} = \frac{0.8(10 - 9\xi^2)^2(1 + 9\xi^2)}{8(10 - 9\xi^2)(1 + 9\xi^2) - 5} E \qquad (8.164) \\
C_{12} = 0.25 C_{11}, \quad C_{33} = \frac{2}{5(10 + 9\xi^2)} E, \quad C_{13} = C_{23} = 0
\end{gathered}
$$

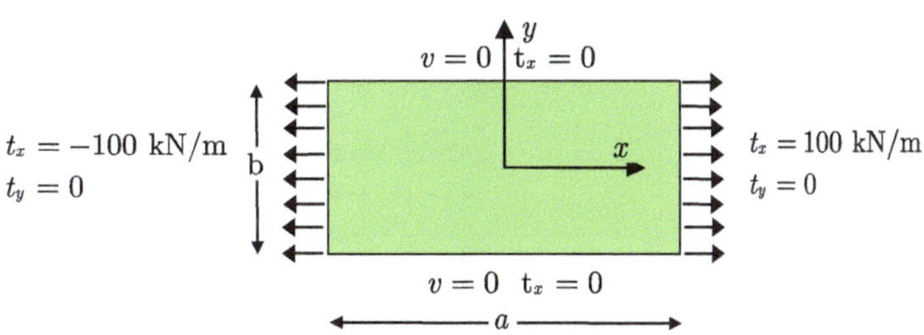

FIGURE 8.24 Plane inhomogeneous orthotropic body in Example 8.6.

where $\xi = 2x/a$. The problem admits the exact solution [25]

$$u_{exact}(x, y) = N_x \frac{a}{2E}\left(3\xi^3 + \xi - \frac{5}{48\sqrt{10}}\ln\frac{\sqrt{10}+3\xi}{\sqrt{10}-3\xi}\right) \tag{8.165}$$

Numerical results were obtained for $a = 2$ m, $b = 1$ m, $N_x = ht_x$, and $E = 100$ MPa using N = 298 boundary elements, L = 200 domain nodal points and $c = 0.1$. Fig. 8.25 shows the computed displacement $u(x, 0)$ versus the exact one.

Example 8.7

In this example the inhomogeneous orthotropic square plane body of Fig. 8.26 is studied. The body is fixed along the side $x = 0$ and subjected to the uniform normal traction along the side $x = a$. The other two edges are free to move tangentially, while they are restraint in the normal direction.

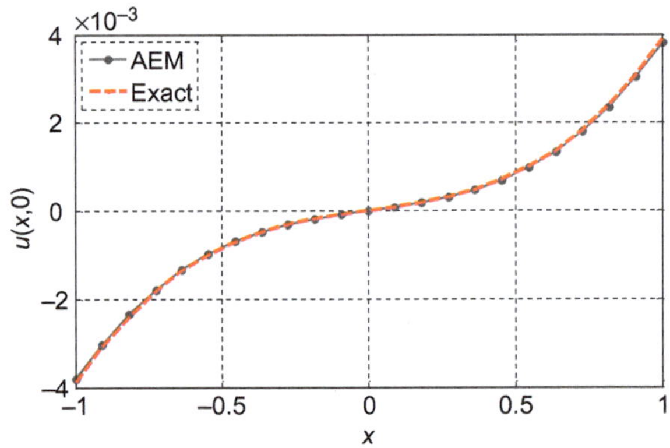

FIGURE 8.25 Displacement $u(x, 0)$ in the plane body in Example 8.6.

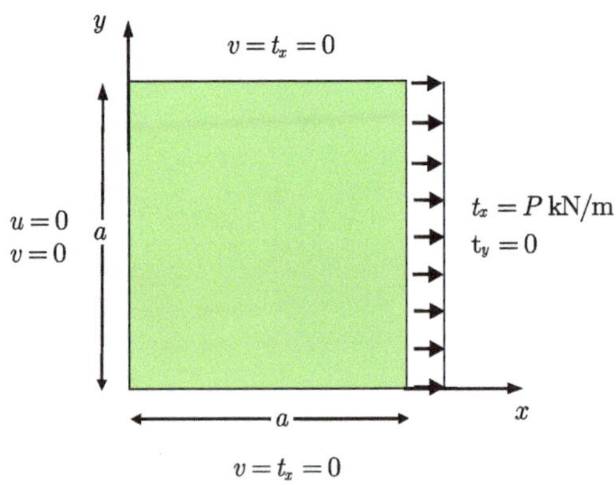

FIGURE 8.26 Plane inhomogeneous orthotropic square body in Example 8.7.

The material parameters are given as

$$
\begin{aligned}
&C_{11} = 6.14(1+x)^2, \quad C_{12} = 2.14(1+x)^2, \quad C_{13} = 0 \\
&C_{22} = 5.96(1+x)^2, \quad C_{33} = 1.64(1+x)^2 \\
&C_{13} = C_3 = 0
\end{aligned}
\tag{8.166}
$$

The problem admits the exact solution [26]

$$
u_{exact} = \frac{x}{6.14(1+x)} \tag{8.167}
$$

Numerical results were obtained for $a = 1$, $P = 1$ using N = 200 boundary elements, L = 81 domain nodal points and $c = 0.1$. The computed results are shown in Figs. 8.27 and 8.28 as compared with the exact ones.

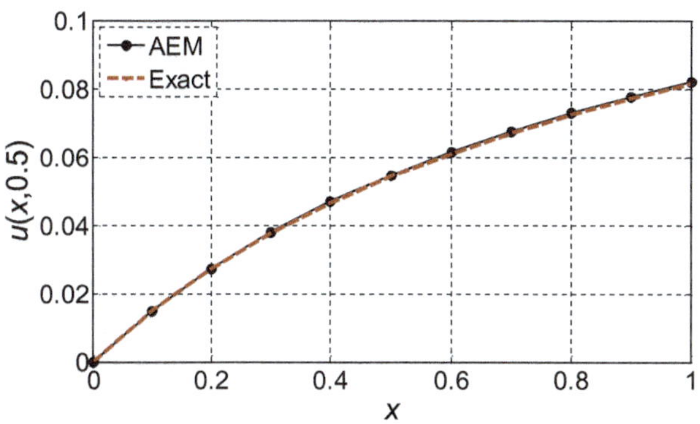

FIGURE 8.27 Displacement $u(x, 0.5)$ in the plane body in Example 8.7.

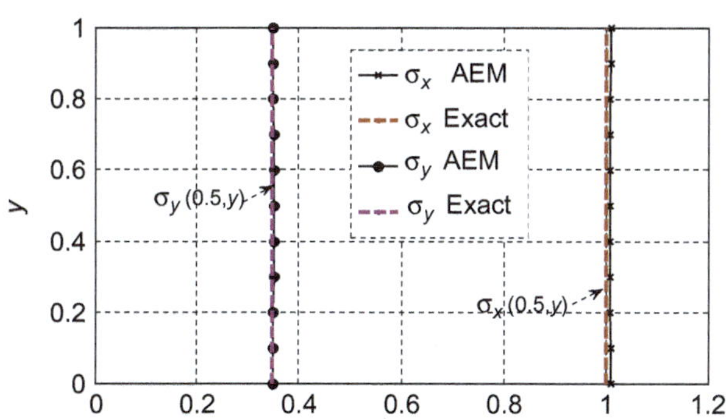

FIGURE 8.28 Stresses $\sigma_x(0.5, y)$ and $\sigma_y(0.5, y)$ in the plane body in Example 8.7.

8.6 REFERENCES

[1] Nardini D, Brebbia CA. New approach to free vibration analysis using boundary elements. In: Brebbia CA, editor. Boundary element methods in engineering. Southampton and Boston: Springer-Verlag, Berlin and Computational Mechanics Publications; 1982. p. 312–26.

[2] Partridge PW, Brebbia CA, Wrobel LC. The dual reciprocity boundary element method. Southampton: Computational Mechanics Publications; 1992.

[3] Ang WT, Clemens DC, Vahdati N. A dual reciprocity boundary element method for a class of elliptic boundary value problems for non-homogeneous anisotropic media. Eng Anal Bound Elem 2003;27:49–53.

[4] Tanaka M, Matsumoto T, Sueda Y. 2001. A dual reciprocity boundary element method applied to the steady-state state heat conduction problem of functionally gradient materials. Proc. BETEQ 2001, Rudgers University, USA.

[5] Nerantzaki MS, Kandilas CB. A boundary element method solution for anisotropic, nonhomogeneous elasticity. Acta Mech 2008;200:199–211.

[6] Katsikadelis JT, Nerantzaki MS. The boundary element method for nonlinear problems. Eng Anal Bound Elem 1999;23(5):365–73.

[7] Brebbia CA. In praise of John Katsikadelis. In: Sapountzakis EJ, editor. Recent developments in boundary element methods, 2010. Southampton and Boston: WIT Press; 2010. p. 1–16.

[8] Tyn Myint U, Debnath L. Linear partial differential equations for scientists and engineers. 4th ed. Boston: Birkhäuser; 2007.

[9] Dominguez J. Boundary elements in dynamics. Southampton, Boston: Computational Mechanics Publications; 1993.

[10] Gaul L, Kögl M, Wagner M. Boundary elements for engineers and scientists. Berlin: Springer-Verlag; 2003.

[11] Golberg MA, Chen CS. The theory of radial basis functions applied to the BEM for inhomogeneous partial differential equations. Bound Elem Comm 1994;5:57–61.

[12] Golberg MA. The numerical evaluation of particular solutions in the BEM–A review. Bound Elem Comm 1995;6:99–106.

[13] Partridge PW, Sensale B. Hybrid approximation functions in the dual reciprocity boundary element method. Comm Numer Methods Eng 1997;13(2):83–94.

[14] Golberg MA, Chen CS, Bowman H, Power H. Some comments on the use of radial basis functions in the dual reciprocity method. Comp Mech 1998;22:61–9.

[15] Partridge PW. Radial basis approximation functions in the boundary element dual reciprocity method. Boundary element technology XIII. Southampton, Boston: Computational Mechanics Publications; 1999. p. 325–34.

[16] Liu GR. Meshfree methods: Moving beyond the finite element method. 2nd Ed Boca Raton, US: CRC Press; 2009.

[17] Agnantiaris JP, Polyzos D, Beskos DE. Some studies on dual reciprocity BEM for elastodynamic analysis. Comp Mech 1996;17:270–7.

[18] Katsikadelis JT. The analog equation method. A powerful BEM-based solution technique for solving linear and nonlinear engineering problems. In: Brebbia CA, editor. Boundary element method XVI. Southampton, UK: Computational Mechanics Publications; 1994. p. 167–82. Proc. 16th International Boundary Element Method Conference (BEM XVI), July 12-15.

[19] Katsikadelis JT. The boundary element method for plate analysis. U.K.: Academic Press, Elsevier; 2014.

[20] Goldberg MA, Chen CS, Karur SR. Improved multiquadric approximation for partial differential equations. Eng Anal Bound Elem 1996;18:9–17.

[21] Fasshauer GE, Zhang JG. On choosing "optimal" shape parameters for RBF approximation. Numer Algor 2007;45:345–68.

[22] Katsikadelis JT. The 2D elastostatic problem in inhomogeneous anisotropic bodies by the meshless analog equation method (MAEM). Eng Anal Bound Elem 2008;32:997–1005.

[23] Katsikadelis JT. A generalized Ritz method for partial differential equations in domains of arbitrary geometry using global shape functions. Eng Anal Bound Elem 2008;32:353–67.

[24] Cameron AD, Casey JA, Simpson GB. Benchmark tests for thermal analysis. Glasgow: NAFEMS; 1986.
[25] Leissa AW, Martin AF. Some exact plane elasticity solutions for nonhomogeneous, orthotropic sheets. J Elasticity 1990;23:97–112.
[26] Azis MI, Clements DL. A boundary element method for anisotropic inhomogeneous elasticity. Int J Solids Struct 2001;38:5747–63.

PROBLEMS

8.1. Use the DRM and AEM to solve the boundary value problem

$$\nabla^2 u + u_{,x} + u_{,y} - 4u = -2[(1 + a^2 + x - x^2)(b^2 - y^2) + (1 + b^2 + y - y^2)(a^2 - x^2)] \quad \text{in } \Omega$$

$$u = 0 \text{ on } \Gamma$$

The domain Ω is the rectangle $-a \le x \le a$, $-b \le y \le b$. Compare the results with the exact solution $u_{exact} = (a^2 - x^2)(b^2 - y^2)$.

8.2. Solve the boundary value problem below for the domain of Fig. P8.2

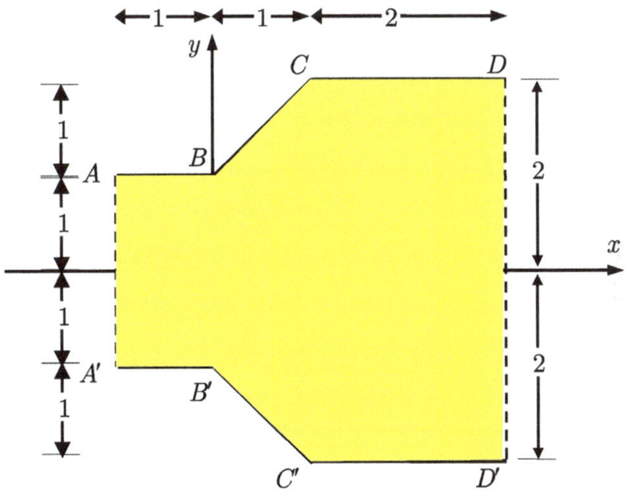

FIGURE P8.2

$$(1 + e^{2y})u_{xx} - 2xe^y u_{xy} + (4x^2 + 1)u_{yy} - xe^y u_x - e^y u_y - (x^2 + e^y)u = 2 + e^y - x^4 \quad \text{in } \Omega$$

$$(1 + e^{2y})u_n - xe^y u_t = -xe^{2y} - 2x \quad \text{on } AA' \qquad \text{(a)}$$

$$(1 + e^{2y})u_n - xe^y u_t = xe^{2y} + 2x \quad \text{on } DD' \qquad \text{(b)}$$

$$u = x^2 + e^y \quad \text{on } ABCD \text{ and } A'B'C'D' \qquad \text{(c)}$$

The problem admits an exact solution $u = x^2 + e^y$

8.3. A cantilever of length $L = 3.0$ m, height $b = 1.0$ m, and thickness $h = 0.1$ m (Fig. P8.3) is subjected to the uniform load $P = 1$ kN/m. Determine the displacements u and v, and the stresses σ_x, σ_y, and τ_{xy} along the cross-section $x = L/2$. Use $E = 2 \times 10^4$ kN/m^2, $\nu = 0.2$. Compare the results with those obtained from an analytical solution based on the technical theory of beams including shear deformation

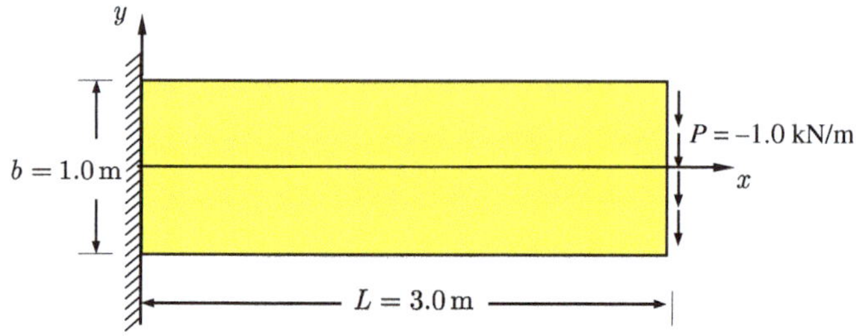

FIGURE P8.3

8.4. Determine the displacement field of the plane inhomogeneous body with unit thickness shown in Fig. P8.4 governed by the system of equations

$$\nabla^2 u + \frac{1+\nu}{1-\nu}(u_{,xx} + v_{,xy}) + \left(\frac{\lambda^*{}_{,x}}{\mu} + 2\frac{\mu_{,x}}{\mu}\right)u_{,x} + \frac{\mu_{,y}}{\mu}u_{,y} + \frac{\mu_{,y}}{\mu}v_{,x} + \frac{\lambda^*{}_{,x}}{\mu}v_{,y} = -\frac{b_x}{\mu}$$

$$\nabla^2 v + \frac{1+\nu}{1-\nu}(u_{,xy} + v_{,yy}) + \frac{\lambda^*{}_{,y}}{\mu}u_{,y} + \frac{\mu_{,x}}{\mu}u_{,y} + \frac{\mu_{x}}{\mu}v_{,x} + \left(\frac{\lambda^*{}_{,y}}{\mu} + 2\frac{\mu_{,y}}{\mu}\right)v_{,y} = -\frac{b_y}{\mu}$$

Consider mixed boundary conditions

$$u = xy + 0.2y^2, \quad v = xy + 0.2x^2, \quad \text{on } DEAB$$

$$t_x = E(\nu x + y)/(1 - \nu^2), \quad t_y = 0.7E(x + y)/(1 + \nu), \quad \text{on } BC$$

$$t_x = 0.7E(x + y)/(1 + \nu), \quad t_y = E(x + \nu y)/(1 - \nu^2), \quad \text{on } CD$$

Use the following data:

$$E = E_0(1 + x)^2, \quad E_0 = 2 \times 10^5 \text{kN/m}^2, \quad \nu = 0.2,$$

$$\lambda^* = \nu E_0(1 + x)^2/(1 - \nu^2), \quad \mu = E_0(1 + x)^2/2(1 + \nu).$$

$$b_x = E_0(1 + x)[-0.1(1 + x)(7 + 3\nu) - 2(\nu x + y)]/(1 - \nu^2)$$

$$b_y = E_0(1 + x)[-0.1(1 + x)(7 + 3\nu) - 1.4(x + y)(1 - \nu)]/(1 - \nu^2)$$

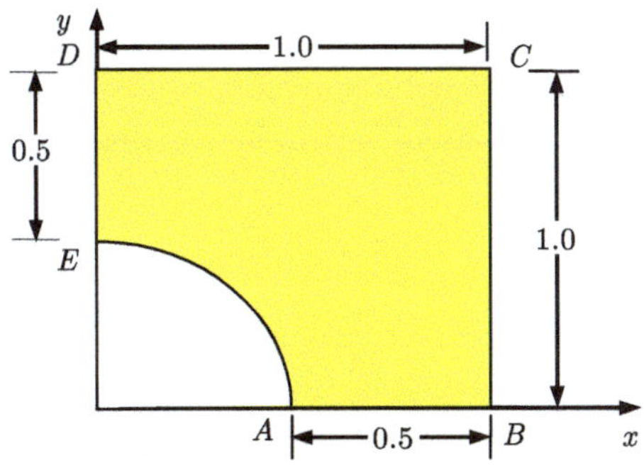

FIGURE P8.4 Inhomogeneous plane body.

Chapter | Nine

The BEM for Time Dependent Problems

CHAPTER OUTLINE

9.1 INTRODUCTION

In this chapter, the boundary element method (BEM) is developed for solving problems described by the general second order hyperbolic and parabolic partial differential equations with variable coefficients. Many problems in physics and engineering science are described by such equations, for example, vibrations and transient response of bounded inhomogeneous anisotropic bodies, wave propagation in inhomogeneous anisotropic media, transient heat conduction, etc. A reciprocal identity can be derived using the time dependent (dynamic) reciprocal theorem [1,2]. For some simple equations, a time dependent fundamental solution is available. This allows the derivation of the integral representation of the solution and hence the development of the BEM. However, for the general hyperbolic or parabolic equation, the establishment of a fundamental solution is out of question. Moreover, the derivation of the boundary integral equation and its numerical solution, even in cases where a fundamental solution is available [1,3], is a tedious task

The Boundary Element Method for Engineers and Scientists.

and requires special care. On the other hand these problems can be efficiently solved directly in the time domain using the dual reciprocity method (DRM), subject to the restrictions mentioned in Section 8.4, or the analog equation method (AEM). Since both methods include internal nodal points, they apply to bounded domains. For unbounded domains these methods can be applied by confining the solution to a finite domain and imposing artificial absorbing boundaries, namely boundaries which ensure the Sommerfeld's radiation condition [4]. The time dependent problems can be solved in the transformed domain, for example, using Laplace transform. However, the efficiency of this method diminishes as compared with a time domain solution. In this chapter only the AEM is presented as it is developed for the solution of time dependent problems in bounded domains in the time domain. This method produces linear systems of ordinary differential equations, that is, semidiscrete hyperbolic or parabolic equations, which are solved using the methods presented in [5,6]. The chapter is enriched by developing the BEM for the fractional wave-diffusion equation in inhomogeneous anisotropic bodies. The efficiency of the AEM is illustrated through well corroborated examples.

9.2 THE BEM FOR THE GENERAL SECOND ORDER HYPERBOLIC PARTIAL DIFFERENTIAL EQUATION

We consider the second order $t-$*hyperbolic* equation [7]

$$\rho\ddot{u}+\eta\dot{u}=Au_{,xx}+2Bu_{,xy}+Cu_{,yy}+Du_{,x}+Eu_{,y}+Fu+f(\mathbf{x},t),\ \mathbf{x}(x,y)\in\Omega,\ t>0 \tag{9.1}$$

subject to the boundary conditions

$$u=\alpha(\mathbf{x},t),\quad \mathbf{x}\in\Gamma_1 \tag{9.2a}$$

$$\nabla u\cdot\mathbf{m}=\gamma(\mathbf{x},t),\quad \mathbf{x}\in\Gamma_2 \tag{9.2b}$$

and to the initial conditions

$$u(\mathbf{x},0)=g_1(\mathbf{x}),\quad \dot{u}(\mathbf{x},0)=g_2(\mathbf{x}),\quad \mathbf{x}\in\Omega \tag{9.3a, b}$$

where $\Gamma_1\cup\Gamma_2=\Gamma$; $u=u(\mathbf{x},t)$ represents the unknown field function; $\rho(\mathbf{x})$, $\eta(\mathbf{x})$, $A(\mathbf{x}), B(\mathbf{x}),\ldots, F(\mathbf{x})$ are specified position dependent coefficients, whose physical meaning depends on the field function $u(\mathbf{x},t)$; $f(\mathbf{x},t)$ is the external source, and $g_1(\mathbf{x})$, $g_2(\mathbf{x})$ are functions specified in Ω. The quantity $\nabla u\cdot\mathbf{m}$ represents the flux in the direction of the *conormal* to the boundary $\mathbf{m}=(An_x+Bn_y)\mathbf{i}+(Bn_x+Cn_y)\mathbf{j}$. Apparently, the direction of the conormal coincides with the normal to the boundary, if $A=C$ and $B=0$. Finally, $\alpha(\mathbf{x},t)$ and $\gamma(\mathbf{x},t)$ are functions specified on boundary, which in general depend on time. The boundary Γ may be multiply-connected.

The initial boundary value problem (9.1)–(9.3) for suitable meaning of its coefficients occurs in many problems in engineering and mathematical physics, such as vibrations of membranes, acoustic vibrations, scalar waves in inhomogeneous anisotropic media, etc.

9.2.1 The AEM solution for the hyperbolic equation

Let $u = u(\mathbf{x}, t)$ be the sought solution to the initial boundary value problem (9.1)–(9.3). This function is two times continuously differentiable in Ω. Since Eq. (9.1) is of the second order, the operator $\hat{\mathrm{N}} = \nabla^2$, that is, the Laplace operator with respect to the space coordinates x, y, can be employed to produce the analog equation. This yields

$$\nabla^2 u = b(\mathbf{x}, t), \quad \mathbf{x} \in \Omega, \quad t > 0 \tag{9.4}$$

The time in Eq. (9.4) is a parameter, that is, the equation is quasi-static. This implies that the principle of the analog equation replaces the hyperbolic equation with a sequence of elliptic equations in time. Obviously, the AEM procedure employed in Section 8.4.2 for the elliptic equation can be employed to this case, too, provided that the unknown coefficients in Eq. (8.76) are treated as functions of time, $a_j = a_j(t)$. Hence, Eq. (8.76) reads

$$b \simeq \sum_{j=1}^{L} a_j(t)\phi_j \tag{9.5}$$

Discretizing the boundary using N elements, taking L internal nodes, and using the procedure presented in Section 8.4.2 give

$$\mathbf{Hu} - \mathbf{Gq} = \mathbf{T}\boldsymbol{\alpha} \tag{9.6a}$$

$$\mathbf{C}_1\mathbf{u} + \mathbf{C}_2\mathbf{q} = \mathbf{c}_3 \tag{9.6b}$$

$$\boldsymbol{\rho}\ddot{\mathbf{u}} + \boldsymbol{\eta}\dot{\mathbf{u}} = \mathbf{A}\,\mathbf{u}_{,xx} + 2\mathbf{B}\mathbf{u}_{,xy} + \mathbf{C}\mathbf{u}_{,yy} + \mathbf{D}\mathbf{u}_{,x} + \mathbf{E}\mathbf{u}_{,y} + \mathbf{F}\mathbf{u} + \mathbf{f} \tag{9.6c}$$

where $\boldsymbol{\rho}, \boldsymbol{\eta}$ are $L \times L$ diagonal matrices containing the values of the functions $\rho(\mathbf{x}), \eta(\mathbf{x})$ at the internal nodes. All other matrices and vectors involved in Eqs. (9.6) are given in Section 8.4.2. We should have in mind that the vector $\boldsymbol{\alpha}$ and in general $\mathbf{c}_3$ depend on time, that is,

$$\boldsymbol{\alpha} = \{\, a_1(t) \quad a_2(t) \quad \cdots \quad a_L(t) \,\}^T \tag{9.7a}$$

$$\mathbf{c}_3 = \{\, c_{31}(t) \quad c_{32}(t) \quad \ldots \quad c_{3\mathrm{N}}(t) \,\}^T \tag{9.7b}$$

Introducing Eqs. (8.97) in Eq. (9.6c) yields

$$\mathbf{M}\ddot{\boldsymbol{\alpha}} + \mathbf{C}\dot{\boldsymbol{\alpha}} + \mathbf{K}\boldsymbol{\alpha} = \mathbf{p}(t) \tag{9.8}$$

where

$$\mathbf{M} = \rho \mathbf{S} \tag{9.9a}$$

$$\mathbf{C} = \eta \mathbf{S} \tag{9.9b}$$

$$\mathbf{K} = -(\mathbf{A}\ \mathbf{S},_{xx} + 2\mathbf{B}\mathbf{S},_{xy} + \mathbf{C}\mathbf{S},_{yy} + \mathbf{D}\mathbf{S},_{x} + \mathbf{E}\mathbf{S},_{y} + \mathbf{S}) \tag{9.9c}$$

$$\mathbf{p}(t) = \mathbf{f}(t) - (\rho\ddot{\mathbf{z}} + \eta\dot{\mathbf{z}}) + (\mathbf{A}\ \mathbf{z},_{xx} + 2\mathbf{B}\mathbf{z},_{xy} + \mathbf{C}\mathbf{z},_{yy} + \mathbf{D}\mathbf{z},_{x} + \mathbf{E}\mathbf{z},_{y} + \mathbf{F}\mathbf{z}) \tag{9.9d}$$

Note that for time independent boundary conditions (no support excitations) it is $\dot{\mathbf{z}} = \ddot{\mathbf{z}} = 0$.

Equation (9.8) represents the set of semidiscrete equations of motion, which should be solved with initial conditions $\boldsymbol{\alpha}(0), \dot{\boldsymbol{\alpha}}(0)$ to yield $\boldsymbol{\alpha}(t)$. The matrices $\mathbf{M}, \mathbf{C}, \mathbf{K}$ represent in some sense mass, damping and stiffness matrices, respectively, while $\mathbf{p}(t)$ a load vector.

The initial conditions $\boldsymbol{\alpha}(0), \dot{\boldsymbol{\alpha}}(0)$ can be established from Eqs. (9.3a,b) as follows:

Solving Eq. (8.97f) for $\boldsymbol{\alpha}(t)$ and differentiating give

$$\boldsymbol{\alpha}(t) = \mathbf{S}^{-1}[\mathbf{u}(t) - \mathbf{z}(t)] \tag{9.10a}$$

$$\dot{\boldsymbol{\alpha}}(t) = \mathbf{S}^{-1}[\dot{\mathbf{u}}(t) - \dot{\mathbf{z}}(t)] \tag{9.10b}$$

which by virtue of Eqs. (9.3a,b) become

$$\boldsymbol{\alpha}(0) = \mathbf{S}^{-1}[\mathbf{g}_1 - \mathbf{z}(0)] \tag{9.11a}$$

$$\dot{\boldsymbol{\alpha}}(0) = \mathbf{S}^{-1}[\mathbf{g}_2 - \dot{\mathbf{z}}(0)] \tag{9.11b}$$

The initial value problem (9.8), (9.11a,b) can be solved using any of the established numerical methods for the direct time integration of the equations of motion, for example, Newmark's method. The numerical results in the examples of this chapter have been obtained using the method presented in [5]. The solution algorithm is given in Table 9.1. A Matlab program based on this algorithm is available on this book's companion website.

Finally, the obtained solution $\boldsymbol{\alpha}(t)$ is inserted in Eqs. (8.97) to evaluate the solution and its derivatives at the internal nodes. The solution and its derivatives at any point $\mathbf{x} \in \Omega$ can be evaluated using also Eqs. (8.97), but in this case the involved matrices must be evaluated with respect to the point $\mathbf{x}$.

TABLE 9.1 Algorithm for the Numerical Solution of the Semidiscretized Linear Equations of Motion

A. Data for the equation $\mathbf{M}\ddot{\mathbf{u}}+\mathbf{C}\dot{\mathbf{u}}+\mathbf{K}\mathbf{u}=\mathbf{p}(t)$

Read: $\mathbf{M}$, $\mathbf{C}$, $\mathbf{K}$, $\mathbf{u}_0$, $\dot{\mathbf{u}}_0$, $\mathbf{p}(t)$, t_{tot}

B. Initial computations

1. Choose: $h:=\Delta t$ and compute n_{tot}
2. Compute: $c_1:=h^2/2 \quad c_2:=h \quad \mathbf{q}_0:=\mathbf{M}^{-1}(\mathbf{p}_0-\mathbf{C}\dot{\mathbf{u}}_0-\mathbf{K}\mathbf{u}_0)$
3. Formulate $\mathbf{U}_0:=\{\mathbf{q}_0 \quad \dot{\mathbf{u}}_0 \quad \mathbf{u}_0\}^T$
4. Compute:

$$\mathbf{A}:=\begin{bmatrix}\mathbf{M} & \mathbf{C} & \mathbf{K}\\ \frac{c_1}{2}\mathbf{I} & -h\mathbf{I} & \mathbf{I}\\ -\frac{c_2}{2}\mathbf{I} & \mathbf{I} & \mathbf{0}\end{bmatrix}^{-1}\begin{bmatrix}\mathbf{0} & \mathbf{0} & \mathbf{0}\\ -\frac{c_1}{2}\mathbf{I} & \mathbf{0} & \mathbf{I}\\ \frac{c_2}{2}\mathbf{I} & \mathbf{I} & \mathbf{0}\end{bmatrix},\quad \mathbf{b}:=\begin{bmatrix}\mathbf{M} & \mathbf{C} & \mathbf{K}\\ \frac{c_1}{2}\mathbf{I} & -h\mathbf{I} & \mathbf{I}\\ -\frac{c_2}{2}\mathbf{I} & \mathbf{I} & \mathbf{0}\end{bmatrix}^{-1}\begin{Bmatrix}\mathbf{I}\\ 0\\ 0\end{Bmatrix}$$

C. Compute solution $\mathbf{U}_n=\{q_n \quad \dot{u}_n \quad u_n\}^T \; (q_n=\ddot{u}_n)$

for $n:=1$ to n_{tol}

$\mathbf{U}_n=\mathbf{A}\mathbf{U}_{n-1}+\mathbf{b}\mathbf{p}_n$

end

For free undamped vibrations it is $f(t)=0$, $c_3=0$, $\eta=0$, hence $\mathbf{C}=\mathbf{p}(t)=0$. Thus Eq. (9.8) becomes

$$\mathbf{M}\ddot{\boldsymbol{\alpha}}+\mathbf{K}\boldsymbol{\alpha}=\mathbf{0} \tag{9.12}$$

Assuming a solution of the form $\boldsymbol{\alpha}=\mathbf{a}e^{i\omega t}$, where $\mathbf{a}$ is a constant vector, and substituting it in Eq. (9.12) gives the generalized linear eigenvalue problem

$$(\mathbf{K}-\omega^2\mathbf{M})\mathbf{a}=\mathbf{0} \tag{9.13}$$

The solution of this problem gives the eigenfrequencies ω_i $(i=1,2,\ldots,L)$ and the corresponding eigenvectors $\mathbf{a}_i$. Then the mode shapes are obtained from Eq. (8.97f), for $\boldsymbol{\alpha}=\mathbf{a}_i$, $\mathbf{z}=0$

$$\mathbf{u}_i=\mathbf{S}\mathbf{a}_i, \quad i=1,2,\ldots,L \tag{9.14}$$

SOLUTION PROCEDURE STEPS

It is obvious that the steps from 1 to 7 described in the Section 8.4.2 apply to this problem, too. The additional work consists in computing the matrices $\mathbf{M},\mathbf{C},\mathbf{K}$ and the vector $\mathbf{p}(t)$ using Eqs. (9.9), and the initial conditions using Eqs. (9.11). Note that $\mathbf{K}=-\overline{\mathbf{K}}$, cf. Eqs. (8.100). Then Eq. (9.8) is solved to yield the vector $\boldsymbol{\alpha}(t)$, which is used in Eqs. (8.97) to obtain the field function and its derivatives at the internal nodes.

EXAMPLES

On the basis of the procedure presented in Section 9.2.1 a FORTRAN code has been written for the solution of the initial boundary value problem (9.1)–(9.3). The employed radial basis functions ϕ_j are multiquadrics (MQs). Certain example problems are presented which demonstrate the efficiency and accuracy of the AEM.

Example 9.1

The free vibrations of the rectangular membrane $(-a \leq x \leq a,\ -b \leq y \leq b)$ under a specified initial deflection is studied. The membrane is fixed along the boundary and stretched by the forces $N_x = (a^2 - x^2)$, $N_{xy} = 0$, $N_y = (b^2 - y^2)$. The dynamic response of the membrane is governed by the following initial boundary value problem

$$\rho \ddot{u} + \eta \dot{u} = N_x u_{,xx} + 2N_{xy}\ u_{,xy} + N_y\ u_{,xy} + f(\mathbf{x}, t) \quad \text{in } \Omega \tag{9.15}$$

$$u(\mathbf{x}, t) = 0 \quad \mathbf{x} \in \Gamma \tag{9.16}$$

$$u(\mathbf{x}, 0) = g_1(\mathbf{x}), \quad \dot{u}(\mathbf{x}, 0) = g_2(\mathbf{x}) \quad \text{in } \Omega \tag{9.17a, b}$$

The numerical results have been obtained using the following data: $a=6$, $b=4$; $\rho=5$, $\eta=2\rho\omega\xi$, $\omega^2=4/\rho$, $f(\mathbf{x}, t)=0$, $g_1(\mathbf{x})=10^{-3}(a^2-x^2)(b^2-y^2)$, $g_2(\mathbf{x})=-10^{-3}\omega\xi(a^2-x^2)(b^2-y^2)$. The problem admits an exact solution $u_{exact}(\mathbf{x}, t)=10^{-3}(a^2-x^2)(b^2-y^2)e^{-\xi\omega t}\cos\omega_D t$, $\omega_D=\omega\sqrt{1-\xi^2}$.

The time history of the solution at point (0,0) obtained with $N=200$ constant boundary elements and $L=225$ domain nodal points distributed as in Fig. 9.1 for $\xi=0$ and $\xi=0.03$ is shown in Fig. 9.2 as compared with the exact solution. Figs. 9.3–9.5 show the dependence of the error $e = u(0,0,t) - u_{exact}(0,0,t)$ on the damping parameter, the number of the domain nodal points, and time step, respectively.

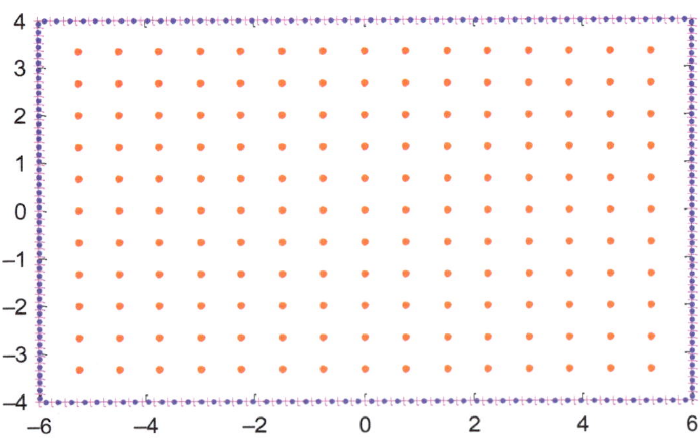

FIGURE 9.1 Boundary and domain nodal points in Example 9.1.

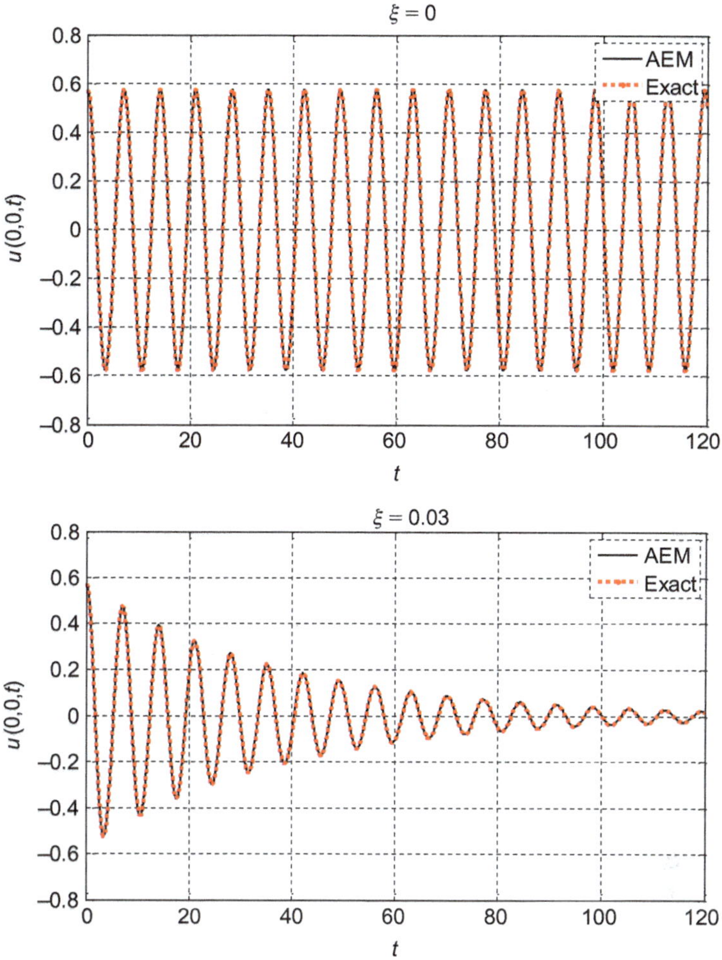

FIGURE 9.2 Time history of $u(0, 0, t)$ in Example 9.1.

9.2.2 The AEM solution for coupled hyperbolic equations

The method is illustrated by applying it to the solution of the vibration problem of the plane linear elastic body. The dynamic response is governed by the initial boundary value problem (cf. Eqs. (8.121) and (8.122))

$$\nabla^2 u + \frac{1+\nu}{1-\nu}(u_{,xx} + v_{,xy}) + \frac{1}{G} f_x - \frac{\rho}{G}\ddot{u} = 0 \quad \text{in } \Omega \tag{9.18a}$$

$$\nabla^2 v + \frac{1+\nu}{1-\nu}(u_{,xy} + v_{,yy}) + \frac{1}{G} f_y - \frac{\rho}{G}\ddot{v} = 0 \quad \text{in } \Omega \tag{9.18b}$$

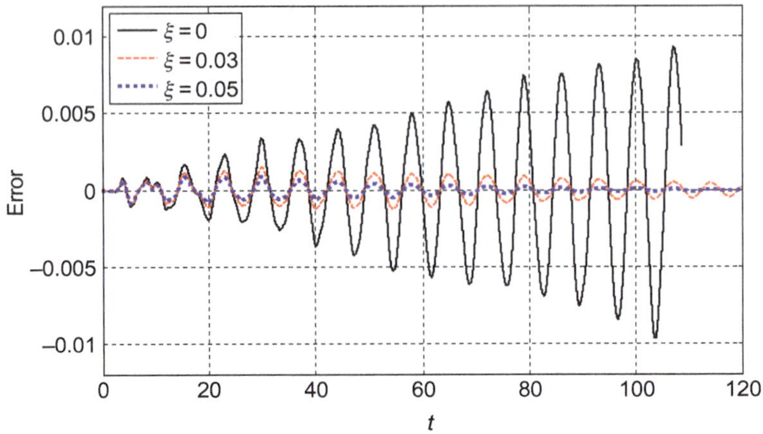

FIGURE 9.3 Dependence of the error $e = u(0, 0, t) - u_{exact}(0, 0, t)$ on the damping parameter (L = 315) in Example 9.1.

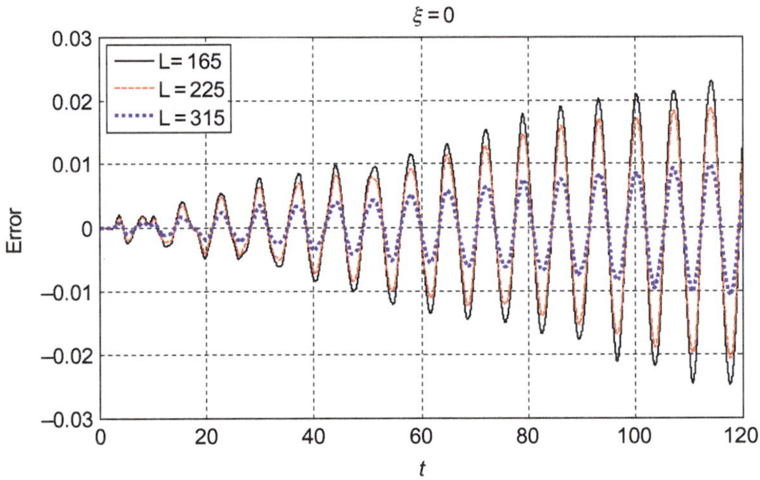

FIGURE 9.4 Dependence of the error $e = u(0, 0, t) - u_{exact}(0, 0, t)$ on the number of the domain nodal point in Example 9.1.

$$\gamma_1 u + \gamma_2 t_x = \gamma_3 \quad \text{on } \Gamma \tag{9.19a}$$

$$\delta_1 v + \delta_2 t_y = \delta_3 \quad \text{on } \Gamma \tag{9.19b}$$

$$u(\mathbf{x}, 0) = g_1(\mathbf{x}), \quad \dot{u}(\mathbf{x}, 0) = g_2(\mathbf{x}) \text{ in } \Omega \tag{9.20a, b}$$

$$v(\mathbf{x}, 0) = \overline{g}_1(\mathbf{x}), \quad \dot{v}(\mathbf{x}, 0) = \overline{g}_2(\mathbf{x}) \text{ in } \Omega \tag{9.20c, d}$$

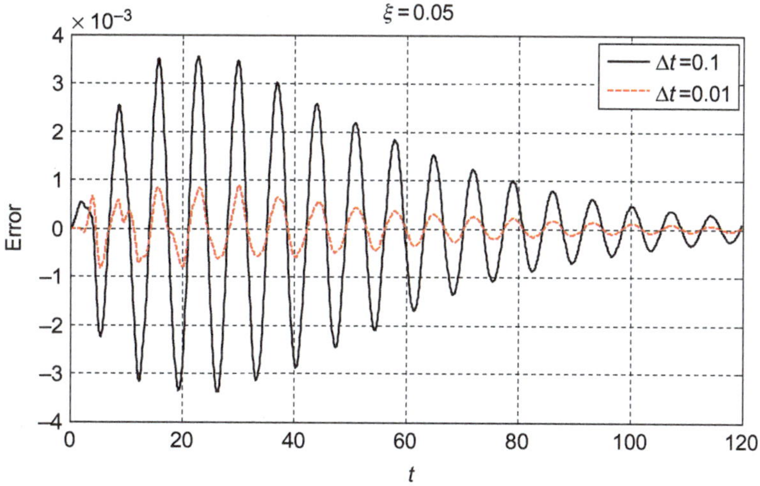

FIGURE 9.5 Dependence of the error $e = u(0, 0, t) - u_{exact}(0, 0, t)$ on the time step in Example 9.1.

where $u = u(\mathbf{x}, t)$, $v = v(\mathbf{x}, t)$ designate the displacement components, $f_x(\mathbf{x}, t)$, $f_y(\mathbf{x}, t)$ the body forces, and t_x, t_y the boundary tractions given by Eqs. (8.132), ρ the surface mass density. The quantities γ_i, δ_i $(i = 1, 2, 3)$ have the same meaning as in Eqs. (8.131). Note that γ_3, δ_3 may depend on time.

Since Eqs. (9.18) are of the second order, the operator $\hat{\mathrm{N}} = \nabla^2$, that is, the Laplace operator with respect to the space coordinates x, y, can be employed to produce the analog equations. This yields

$$\nabla^2 u = b^{(1)}(\mathbf{x}, t) \tag{9.21a}$$

$$\nabla^2 v = b^{(2)}(\mathbf{x}, t) \tag{9.21b}$$

Approximating the fictitious sources with radial basis function series

$$b^{(i)} \simeq \sum_{j=1}^{L} a_j^{(i)}(t)\phi_j, \quad i = 1, 2 \tag{9.22}$$

and applying the procedure described in Section 8.5, we obtain the following system of semidiscrete equations of motion

$$\mathbf{M}_1\ddot{\boldsymbol{\alpha}}_1 + \mathbf{A}_{11}\boldsymbol{\alpha}_1 + \mathbf{A}_{12}\boldsymbol{\alpha}_2 = \mathbf{p}_1(t) \tag{9.23a}$$

$$\mathbf{M}_2\ddot{\boldsymbol{\alpha}}_2 + \mathbf{A}_{21}\boldsymbol{\alpha}_1 + \mathbf{A}_{22}\boldsymbol{\alpha}_2 = \mathbf{p}_2(t) \tag{9.23b}$$

where $\mathbf{M}_1$, $\mathbf{M}_2$ represent in some sense mass matrices, given as

$$\mathbf{M}_1 = -\frac{\rho}{G}\mathbf{S}^{(1)} \tag{9.24a}$$

$$\mathbf{M}_2 = -\frac{\rho}{G}\mathbf{S}^{(2)} \tag{9.24b}$$

The quantities $\mathbf{A}_{11}, \mathbf{A}_{12}, \mathbf{A}_{21}, \mathbf{A}_{22}$ are defined by Eqs. (8.145) and $\mathbf{p}_1(t), \mathbf{p}_2(t)$ are given as

$$\mathbf{p}_1(t) = -\frac{1}{G}\mathbf{f}_x - \frac{1+\nu}{1-\nu}(\mathbf{z},_{xx}^{(1)} + \mathbf{z},_{xy}^{(2)}) + \frac{\rho}{G}\ddot{\mathbf{z}}^{(1)} \tag{9.25a}$$

$$\mathbf{p}_2(t) = -\frac{1}{G}\mathbf{f}_y - \frac{1+\nu}{1-\nu}(\mathbf{z},_{xy}^{(1)} + \mathbf{z},_{yy}^{(2)}) + \frac{\rho}{G}\ddot{\mathbf{z}}^{(2)} \tag{9.25b}$$

The initial conditions for the vectors $\boldsymbol{\alpha}_1, \boldsymbol{\alpha}_2$ result from Eqs. (8.141) and (9.20). Thus we have

$$\boldsymbol{\alpha}_1(0) = (\mathbf{S}^{(1)})^{-1}[\mathbf{g}_1 - \mathbf{z}^{(1)}(0)], \quad \dot{\boldsymbol{\alpha}}_1(0) = (\mathbf{S}^{(1)})^{-1}[\mathbf{g}_2 - \dot{\mathbf{z}}^{(1)}(0)] \tag{9.26a, b}$$

$$\boldsymbol{\alpha}_2(0) = (\mathbf{S}^{(2)})^{-1}[\bar{\mathbf{g}}_1 - \mathbf{z}^{(2)}(0)], \quad \dot{\boldsymbol{\alpha}}_2(0) = (\mathbf{S}^{(2)})^{-1}[\bar{\mathbf{g}}_2 - \dot{\mathbf{z}}^{(2)}(0)] \tag{9.26c, d}$$

EXAMPLES

On the basis of the procedure presented in Section 9.2.2 a FORTRAN code has been written for the solution of the initial boundary value problem (9.18)–(9.20). The employed radial basis functions ϕ_j are multiquadrics (MQs). The obtained numerical results validate the efficiency and accuracy of the AEM.

Example 9.2

In this example the dynamic response of the plane isotropic homogeneous body of Fig. 9.6 is studied. The body has a thickness $h = 0.02$ m and is subjected to a suddenly applied constant load $P = 1000$ kN/m along the side 3–4. The displacements are restricted along sides 1–4 and 2–3. The circular holes are centered at points (1.0,1.10) and (2.5,1.2). The surface mass density is $\rho = 54$ kg/m^2, while the elastic parameters are $E = 70 \times 10^6$ kN/m^2 and $\nu = 0.3$. The results were obtained using N = 545 boundary elements and L = 252 domain nodal points and shape parameter $c = 0.35$. The results are compared with those from the finite element method (FEM) solution (program NASTRAN) with 649 quadrilateral elements. Table 9.2 presents the first 10 eigenfrequencies, while Fig. 9.7 the first three modes of vibrations. Finally, Fig. 9.8 shows the time history of the transverse displacement v at point A(2,2).

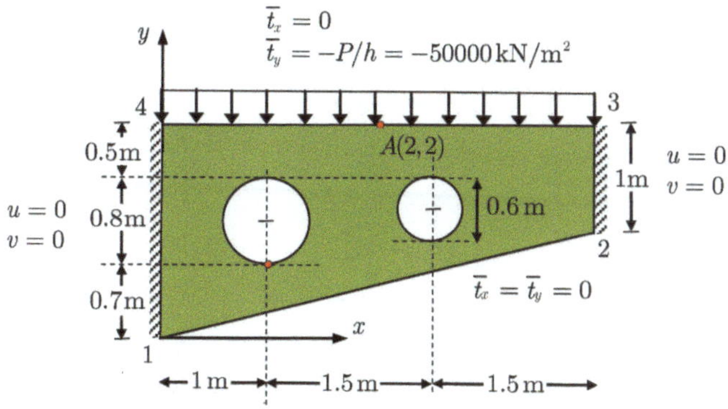

FIGURE 9.6 Plane elastic body in Example 9.2.

TABLE 9.2 Eigenfrequencies in Example 9.2

ω_i	AEM	FEM	Difference %
1	1595.8	1581.9	0.879
2	3592.6	3418.6	5.089
3	3687.6	3620.9	1.841
4	5608.3	5390.5	4.040
5	6072.0	6020.5	0.855
6	6388.3	6098.7	4.749
7	7341.9	7045.8	4.203
8	7858.3	7702.4	2.024
9	8603.5	8407.7	2.328
10	9469.3	9239.1	2.492

9.3 THE BEM FOR THE GENERAL SECOND ORDER PARABOLIC PARTIAL DIFFERENTIAL EQUATION

We consider the second order $t-$*parabolic* equation [7]

$$\eta\dot{u} = A\,u_{,xx} + 2Bu_{,xy} + Cu_{,yy} + Du_{,x} + Eu_{,y} + Fu + f(\mathbf{x}, t), \quad \mathbf{x}(x, y) \in \Omega, \quad t > 0 \tag{9.27}$$

subject to the boundary conditions

$$u = \alpha(\mathbf{x}, t), \quad \mathbf{x} \in \Gamma_1 \tag{9.28a}$$

$$\nabla u \cdot \mathbf{m} = \gamma(\mathbf{x}, t), \quad \mathbf{x} \in \Gamma_2 \tag{9.28b}$$

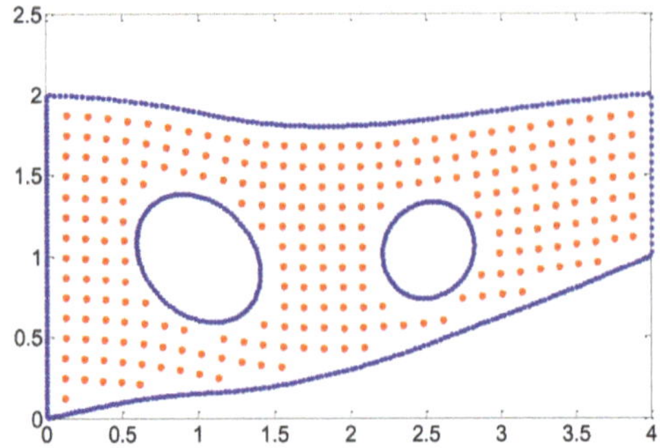

Mode 1, $\omega_1 = 1595.8$

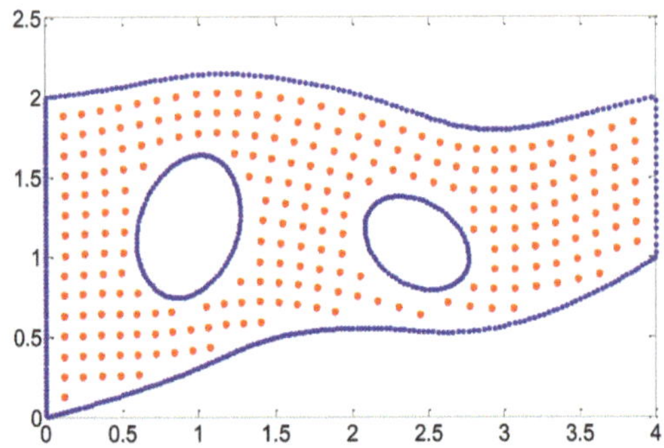

Mode 2, $\omega_2 = 3592.6$

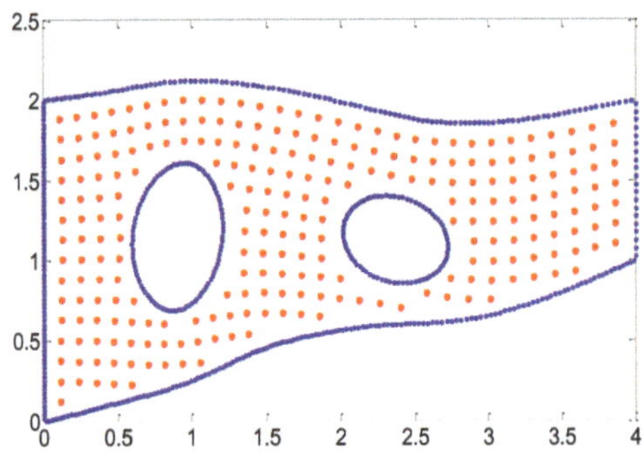

Mode 3, $\omega_3 = 3687.6$

FIGURE 9.7 Plane elastic body in Example 9.2.

and to the initial condition

$$u(\mathbf{x}, 0) = g(\mathbf{x}), \quad \mathbf{x} \in \Omega \tag{9.29}$$

where $\Gamma_1 \cup \Gamma_2 = \Gamma$; $u = u(\mathbf{x}, t)$ represents the unknown field function; $\eta(\mathbf{x})$, $A(\mathbf{x}), B(\mathbf{x}), \ldots, F(\mathbf{x})$ are specified position dependent coefficients, whose

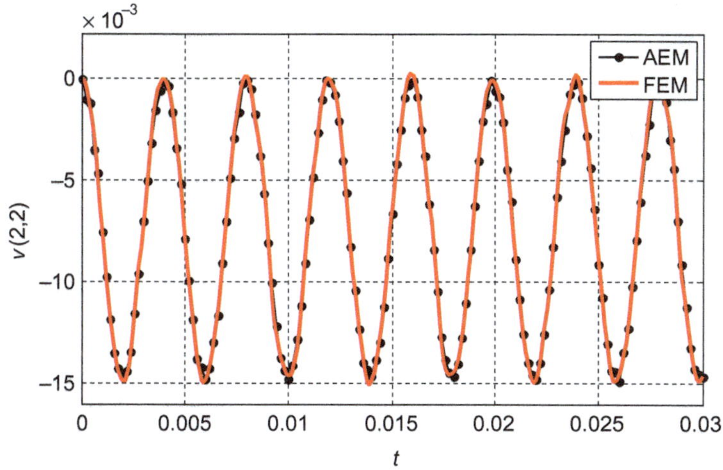

FIGURE 9.8 The time history of the transverse displacement v at point A(2,2) in Example 9.2.

physical meaning depends on that of the field function $u(\mathbf{x}, t)$; $f(\mathbf{x}, t)$ is the external source, and $g(\mathbf{x})$ a given function specified in Ω. The quantity $\nabla u \cdot \mathbf{m}$ represents the flux in the direction of the *conormal* to the boundary $\mathbf{m} = (An_x + Bn_y)\mathbf{i} + (Bn_x + Cn_y)\mathbf{j}$. Finally, $\alpha(\mathbf{x}, t)$ and $\gamma(\mathbf{x}, t)$ are given functions defined on the boundary, which in general depend on time. The boundary Γ may be multiply-connected.

The initial boundary value problem (9.27)–(9.29) for suitable meaning of its coefficients occurs in many problems in engineering and mathematical physics such as heat conduction problems and in general diffusion problems in two-dimensional inhomogeneous anisotropic media.

9.3.1 The AEM solution for the parabolic equation

The initial boundary value problem (9.27)–(9.29) is solved using the AEM. Eq. (9.4) is employed as the analog equation for the parabolic equation, too. Adhering to the procedure described in Section 9.2.1, we obtain the following initial value problem for the vector $\boldsymbol{\alpha}(t)$

$$\mathbf{C}\dot{\boldsymbol{\alpha}} + \mathbf{K}\boldsymbol{\alpha} = \mathbf{p}(t) \tag{9.30}$$

$$\boldsymbol{\alpha}(0) = \mathbf{S}^{-1}[\mathbf{g} - \mathbf{z}(0)] \tag{9.31}$$

The involved matrices and vectors in the above equations are the same as those derived in Section 9.2.1 with $\rho = 0$.

The initial value problem (9.30), (9.31) is solved using the time step method for the semidiscrete parabolic differential equations presented in [6]. The reason is that the coefficients matrices **C**, **K** in Eq. (9.30) resulting from

TABLE 9.3 Algorithm for the Numerical Solution of the Semidiscretized Linear Parabolic Equations

A. Data for the equation $\mathbf{C}\dot{\mathbf{u}} + \mathbf{K}\mathbf{u} = \mathbf{p}(t)$

Read: $\mathbf{C}$, $\mathbf{K}$, $\mathbf{u}_0$, $\mathbf{p}(t)$, t_{tot}

B. Initial computations

1. Choose: $h := \Delta t$ and compute n_{tot}
2. Compute: $\mathbf{q}_0 := \mathbf{C}^{-1}(\mathbf{p}_0 - \mathbf{K}\mathbf{u}_0)$
3. Formulate: $\mathbf{U}_0 := \{\mathbf{q}_0 \quad \mathbf{u}_0\}^T$
4. Compute: $\mathbf{A} := \begin{bmatrix} \mathbf{C} & \mathbf{K} \\ -\frac{h}{2}\mathbf{I} & \mathbf{I} \end{bmatrix}^{-1} \begin{bmatrix} \mathbf{0} & \mathbf{0} \\ \frac{h}{2}\mathbf{I} & \mathbf{I} \end{bmatrix}, \ \mathbf{b} := \begin{bmatrix} \mathbf{C} & \mathbf{K} \\ -\frac{h}{2}\mathbf{I} & \mathbf{I} \end{bmatrix}^{-1} \begin{bmatrix} \mathbf{I} \\ \mathbf{0} \end{bmatrix}$

C. Compute solution: $\mathbf{U}_n = \{q_n \quad u_n\}^T$, $(q_n = \dot{u}_n)$

for $n := 1$ to n_{tol}

$\mathbf{U}_n = \mathbf{A}\mathbf{U}_{n-1} + \mathbf{b}\mathbf{p}_n$

end

the spatial discretization are in general nonsymmetric and nonpositive definite. Hence, the well known generalized trapezoidal family or α-family of methods cannot be employed, because they are applicable if $\mathbf{C}$, $\mathbf{K}$ are symmetric, $\mathbf{C}$ positive definite, and $\mathbf{K}$ semipositive definite [8]. The solution algorithm is given in Table 9.3. A Matlab program based on this algorithm is available on this book's companion website.

SOLUTION PROCEDURE STEPS

Apparently, the solution procedure steps are the same as the solution procedure steps in Section 9.2.1. The only difference is that the matrix $\mathbf{M}$ does not appear and the second initial condition is dropped. The vector $\boldsymbol{\alpha}(t)$ results from the solution of the initial value problem Eqs. (9.30), (9.31). Then it is inserted in Eqs. (8.97) to obtain the field function and its derivatives at the internal nodes.

It should be emphasized here that the computer code for the hyperbolic (dynamic) problem can solve also the parabolic (diffusion) and the elliptic (static) problems as special cases, without modification, except for the solution of the respective semidiscrete equations, yielding the vector of the series coefficients $\boldsymbol{\alpha}(t)$. This is an important advantage of the AEM, because the computer code depends only on the order of the differential equation and not on its specific form.

EXAMPLES

The FORTRAN code employed for the hyperbolic equation has been adjusted to solve the initial boundary value problem (9.30), (9.31). Numerical results are presented to illustrate the efficiency of the AEM.

Example 9.3

In this example the heat conduction equation which describes the unsteady temperature distribution in a two-dimensional homogeneous orthotropic body occupying the domain Ω is studied. The temperature distribution is governed by the following initial value problem (cf. Eq. 8.117)

$$\rho(\mathbf{x})c\dot{u} = \nabla\cdot(\mathbf{D}\nabla u) + f(\mathbf{x}, t) \quad \text{in } \Omega \tag{9.32}$$

$$u = \overline{u} \quad \text{on } \Gamma_1 \tag{9.33a}$$

$$\nabla u\cdot\mathbf{m} = -\overline{q}_m \quad \text{on } \Gamma_2 \tag{9.33b}$$

$$\nabla u\cdot\mathbf{m} = a(u - u_a) \quad \text{on } \Gamma_3 \tag{9.33c}$$

and

$$u(\mathbf{x}, 0) = g_1(\mathbf{x}) \quad \text{in } \Omega \tag{9.34}$$

where $\Gamma = \Gamma_1 \cup \Gamma_2 \cup \Gamma_3$; $u(\mathbf{x}, t)$ is the temperature in the solid; $\mathbf{D}$ is the conductivity matrix; c the specific heat; ρ the mass density; $f(\mathbf{x}, t)$ the external heat flux; $\overline{q}_m$ the intensity of the heat input on Γ_2; $\overline{u}$ the specified temperature on Γ_1; a the heat transfer coefficient and u_a the ambient temperature. Finally, $g_1(\mathbf{x})$ is the initial temperature distribution; $\mathbf{m} = k_x\mathbf{i} + k_y\mathbf{j}$ the vector in the direction of the conormal with k_x, k_y being the specified thermal conductivities.

For the problem at hand the conductivity matrix is

$$\mathbf{D} = \begin{bmatrix} k_x & 0 \\ 0 & k_y \end{bmatrix} \tag{9.35}$$

and Eq. (9.32) reduces to

$$\rho c\dot{u} = k_x u_{,xx} + k_y u_{,yy} + f(\mathbf{x}, t) \quad \text{in } \Omega \tag{9.36}$$

The problem has been solved for the rectangular body $\Omega{:}\{a \times b\}$ with $f(\mathbf{x}, t) = 0$,

$$u(0, y, t) = u(x, 0, t) = u(a, y, t) = u(x, b, t) = 0 \quad \text{on } \Gamma_1 = \Gamma \tag{9.37a}$$

and

$$u(x, y, 0) = u_0, \quad \text{in } \Omega \tag{9.37b}$$

Equation (9.36) admits an analytic solution [9]

$$u_{ex}(x, y, t) = \sum_{n=1}^{\infty}\sum_{j=1}^{\infty} A_n \sin\frac{n\pi x}{a}\sin\frac{j\pi y}{b}\exp\left[-\left(\frac{k_x n^2\pi^2}{a^2} + \frac{k_y j^2\pi^2}{b^2}\right)t\right] \tag{9.38a}$$

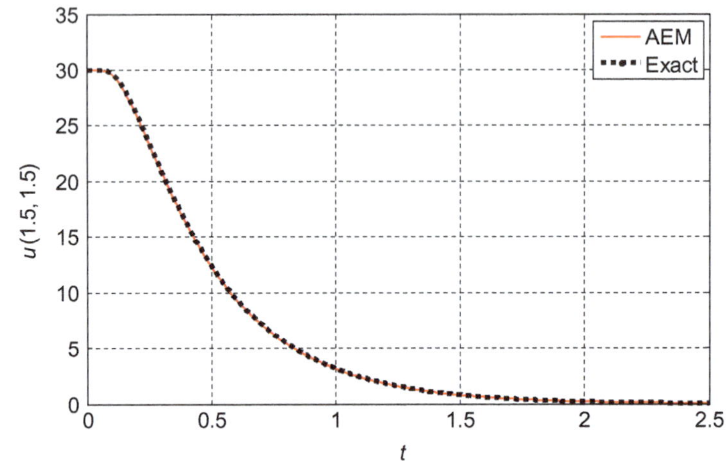

FIGURE 9.9 Time history of the temperature at point $(1.5, 1.5)$ in Example 9.3.

where

$$A_n = \frac{4u_0}{nj\pi^2}[(-1)^n - 1][(-1)^j - 1] \tag{9.38b}$$

The adopted numerical values are: $a = b = 3$, $k_x = k_y = 1.25$ $c = 1$, $\rho = 1$, and $u_0 = 30$.

The solution was computed using N = 200 constant boundary elements and L = 121 domain nodal points distributed uniformly using MQs with shape parameter $c = 0.2$. The time history of the temperature at point (1.5,1.5) is shown in Fig. 9.9, while Fig. 9.10 shows the relative error $[u(1.5, 1.5, t) - u_{ex}(1.5, 1.5, t)]/u_{ex}(1.5, 1.5, t)$ with $\Delta t = 0.005$. Moreover, Fig. 9.11 shows the contours of the temperature distribution at $t = 0.5$. The semidiscretized parabolic equations were solved using the algorithm given in Table 9.3.

Example 9.4

In this example the heat conduction in the inhomogeneous anisotropic plane body shown in Fig. 9.12 is studied. The conductivity matrix is given as

$$\mathbf{D} = \begin{bmatrix} (2x+y+2)^2 & (x-y)^2 \\ (x-y)^2 & (x+2y+2)^2 \end{bmatrix} \tag{9.39}$$

The transient response is described by Eq. (9.32), which by virtue of Eq. (9.39) becomes

$$\begin{aligned} \rho(\mathbf{x})c\dot{u} &= (2x+y+2)^2 u_{,xx} + 2(x-y)^2 u_{,xy} + (x+2y+2)^2 u_{,yy} \\ &\quad + (6x + 6y + 8)u_{,x} + (6x+6y+8)u_{,y} + f(x, y, t) \end{aligned} \tag{9.40}$$

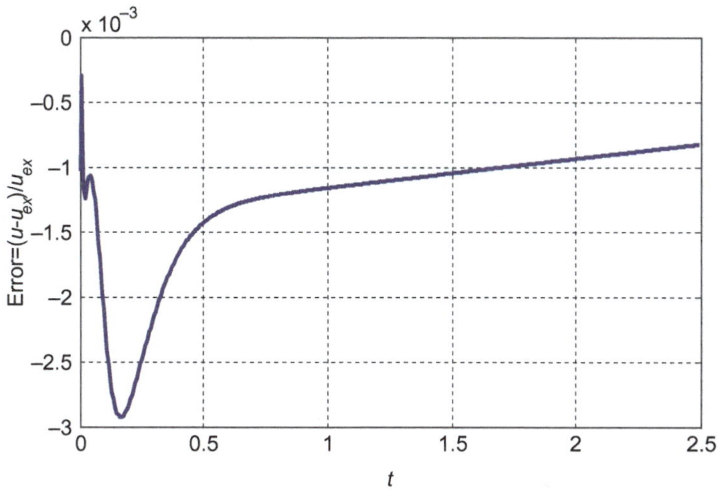

FIGURE 9.10 Relative error at point $(1.5, 1.5)$ in Example 9.3.

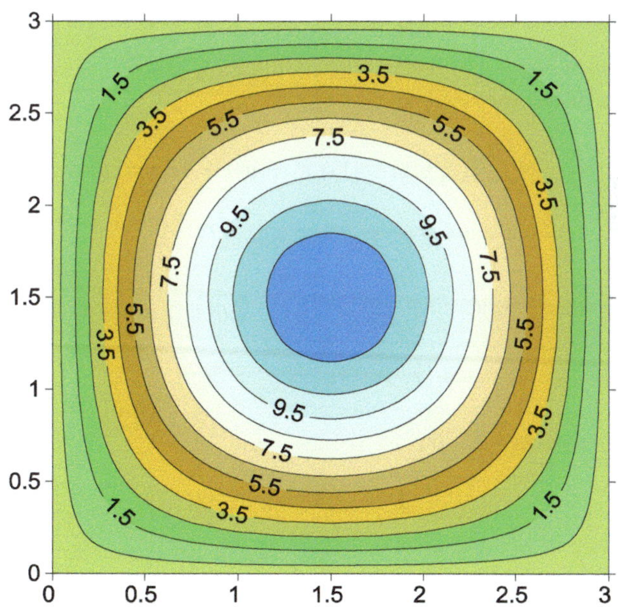

FIGURE 9.11 Contours of the temperature distribution at $t = 0.5$ in Example 9.3.

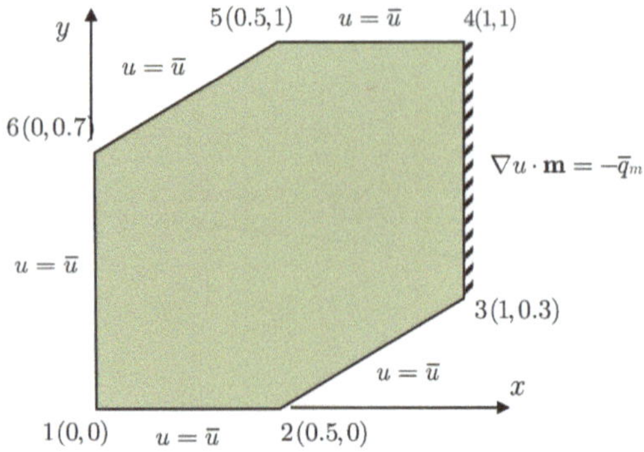

FIGURE 9.12 Geometry of the domain and boundary conditions in Example 9.4.

The considered boundary and initial conditions are (see Fig. 9.12)

$$u = \overline{u} \quad \text{on } \Gamma_u \tag{9.41a}$$

$$\nabla u \cdot \mathbf{m} = -\overline{q}_m \quad \text{on } \Gamma_q \tag{9.41b}$$

$$u(\mathbf{x}, 0) = g_1(\mathbf{x}) \quad \text{in } \Omega \tag{9.42}$$

Equation (9.40) for

$$c = 1,\ \rho = 1,\ f(x, y, t) = [17(x^2 + y^2) + 5xy - 16]\exp(-t)$$

$$\overline{u} = (x^2 + y^2 - 5xy)\exp(-t),$$

$$\overline{q}_m = [(2x - 5y)(2x + y + 2)^2 + (2x - 5y)(x - y)]\exp(-t), \text{ and}$$

$$g_1(x, y) = (x^2 + y^2 - 5xy),$$

admits the analytical solution

$$u_{exact} = (x^2 + y^2 - 5xy)\exp(-t) \tag{9.43}$$

The AEM solution was computed using N = 200 constant boundary elements and L = 103 domain nodal points distributed uniformly (Fig. 9.13) using MQs with shape parameter $c = 0.1$. The time history of the temperature at point (0.5,0.5) is shown in Fig. 9.14, while Fig. 9.15 shows the error $u(0.5, 0.5, t) - u_{exact}(0.5, 0.5, t)$ for $\Delta t = 0.005$. Moreover, Fig. 9.16 shows the contours of the temperature distribution at $t = 1$.

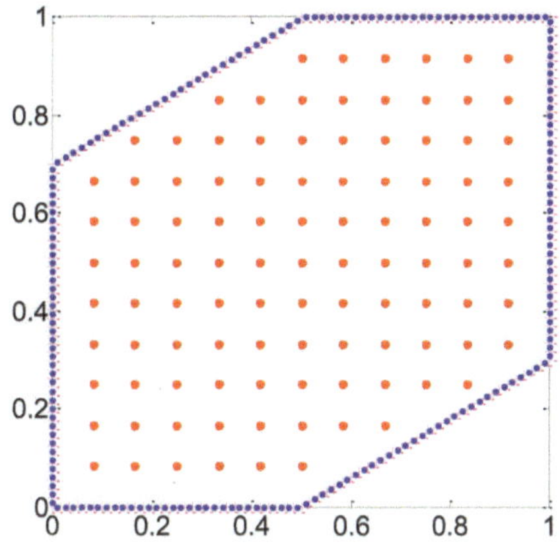

FIGURE 9.13 Boundary and domain nodal points.

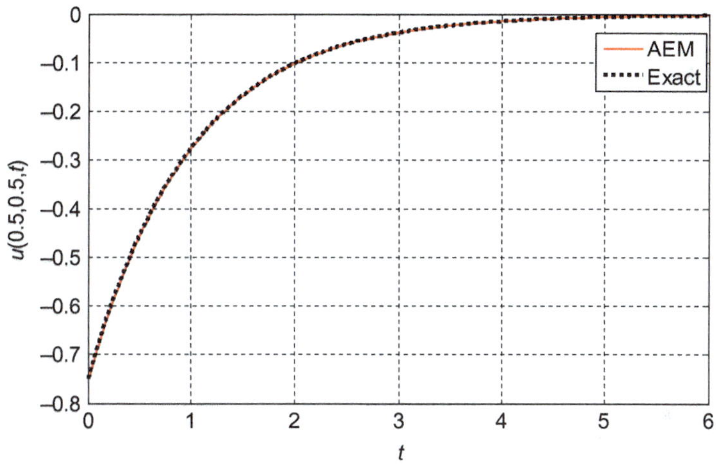

FIGURE 9.14 Time history of the temperature at point (0.5, 0.5) in Example 9.4.

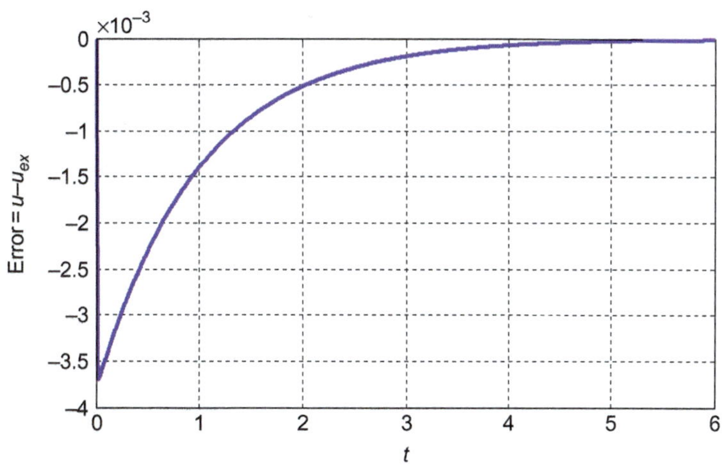

FIGURE 9.15 Error $u - u_{exact}$ at point (0.5, 0.5) in Example 9.4.

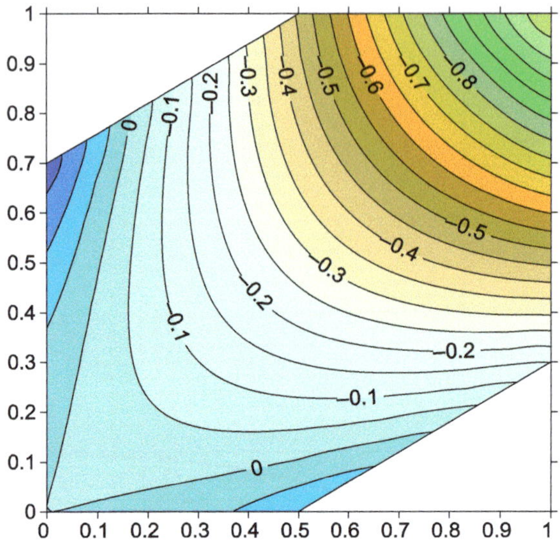

FIGURE 9.16 Contours of the temperature distribution at $t=1$.

9.4 THE FRACTIONAL WAVE-DIFFUSION EQUATION IN BOUNDED INHOMOGENEOUS ANISOTROPIC MEDIA

The fractional wave-diffusion equation for an inhomogeneous anisotropic plane body occupying the domain Ω bounded by the curve Γ reads [10]

$$\rho\frac{\partial^\beta u}{\partial t^\beta}+\eta\frac{\partial^a u}{\partial t^\alpha}=A\frac{\partial^2 u}{\partial x^2}+2B\,\frac{\partial^2 u}{\partial x\partial y}+C\,\frac{\partial^2 u}{\partial y^2}+D\frac{\partial u}{\partial x}+E\frac{\partial u}{\partial y}+Fu+f(\mathbf{x},t) \tag{9.44}$$

where $\mathbf{x}(x,y)\in\Omega$, $t>0$, $0<\alpha<\beta\leq 2$ and $\rho=\rho(\mathbf{x})$, $\eta=\eta(\mathbf{x}), A(\mathbf{x})$, $B(\mathbf{x}),\ldots,F(\mathbf{x})$ are specified functions, whose physical meaning depends on that of the field function $u(\mathbf{x},t)$; $f(\mathbf{x},t)$ is the external source and $\partial^\beta u/\partial t^\beta$ and $\partial^\alpha u/\partial t^\alpha$ are the time fractional derivatives of order β and α, respectively. The Caputo fractional derivative is employed, since it permits the application of initial conditions having direct physical significance. It is usually designated by D_c^α and is defined as [11]

$$D_c^\alpha u(t)=\frac{\partial^\alpha u}{\partial t^\alpha}=\begin{cases}\dfrac{1}{\Gamma(m-\alpha)}\displaystyle\int_0^t\frac{d^m u(\tau)/dt^m}{(t-\tau)^{\alpha+1-m}}d\tau, & m-1<\alpha<m,\\ d^m u(t)/dt^m, & \alpha=m\end{cases} \tag{9.45}$$

where m is a positive integer. Apparently, the classical diffusion equation is retrieved for $\rho=0$, $\alpha=1$, and the classical wave equation in presence of damping for $\beta=2$, $\alpha=1$.

Equation (9.44) is subjected to the boundary conditions

$$u = \alpha(\mathbf{x}, t), \quad \mathbf{x} \in \Gamma_1 \quad \text{and} \quad \nabla u \cdot \mathbf{m} = \gamma(\mathbf{x}, t), \quad \mathbf{x} \in \Gamma_2, \tag{9.46a, b}$$

and to the initial conditions

$$u(\mathbf{x}, 0) = g_1(\mathbf{x}), \quad \mathbf{x} \in \Omega \text{ if } \beta \leq 1 \tag{9.47a}$$

or

$$u(\mathbf{x}, 0) = g_1(\mathbf{x}), \quad \dot{u}(\mathbf{x}, 0) = g_2(\mathbf{x}), \quad \mathbf{x} \in \Omega \text{ if } \beta > 1 \tag{9.47b}$$

where $\Gamma_1 \cup \Gamma_2 = \Gamma$; $\mathbf{m} = (An_x + Bn_y)\mathbf{i} + (Bn_x + Cn_y)\mathbf{j}$ is a vector in the direction of the conormal on the boundary; $\alpha(\mathbf{x}, t), \gamma(\mathbf{x}, t)$ are functions specified on Γ_1 and Γ_2, respectively, and $g_i(\mathbf{x})$ $i = 1, 2$ are functions specified in Ω.

It has been shown that many of the universal electromagnetic, acoustic, and mechanical responses can be modeled accurately using the fractional wave-diffusion equation. Relaxation phenomena in complex viscoelastic materials are described by the fractional diffusion equation [12]. Propagation of stress waves in viscoelastic solids are successfully described by the fractional wave equation [13]. More information about the physical phenomena that can be described by the fractional wave-diffusion equation and in general about the fractional differential equations can be found in the relevant literature, for example, [10,14–16] and references cited there.

9.4.1 The AEM solution for wave-diffusion equation

Equation (9.4) is employed as the analog equation for the fractional wave-diffusion equation also in this case. Following the procedure described in Section 9.2.1, we obtain the following initial value problem for the vector $\boldsymbol{\alpha}(t)$

$$\mathbf{M} D_c^{\beta} \mathbf{a} + \mathbf{C}\, D_c^{\alpha} \mathbf{a} + \mathbf{K}\mathbf{a} = \mathbf{p}(t) \tag{9.48}$$

$$\boldsymbol{\alpha}(0) = \mathbf{S}^{-1}[\mathbf{g}_1 - \mathbf{z}(0)] \quad \text{if } \beta \leq 1 \tag{9.49a}$$

or

$$\boldsymbol{\alpha}(0) = \mathbf{S}^{-1}[\mathbf{g}_1 - \mathbf{z}(0)], \ \dot{\boldsymbol{\alpha}}(0) = \mathbf{S}^{-1}[\mathbf{g}_2 - \dot{\mathbf{z}}(0)] \quad \text{if } \beta > 1 \tag{9.49b}$$

The involved matrices and vectors in the above equations are the same as those derived in Section 9.2.1.

Equation (9.48) represents a system of linear time fractional differential equations, which can be solved using the method presented in [17].

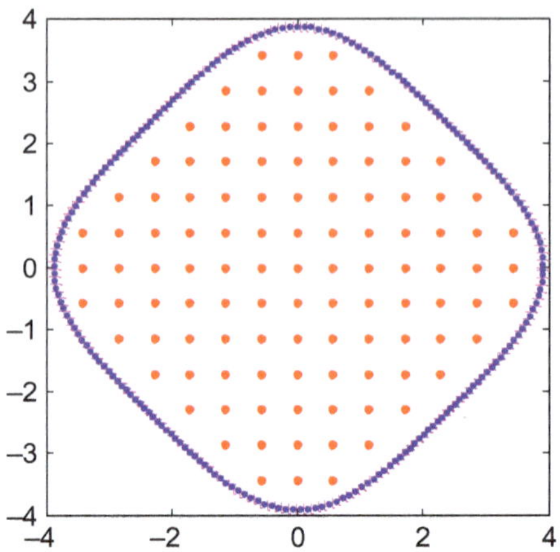

FIGURE 9.17 Geometry of the plane body with distribution of the nodal points in Example 9.5.

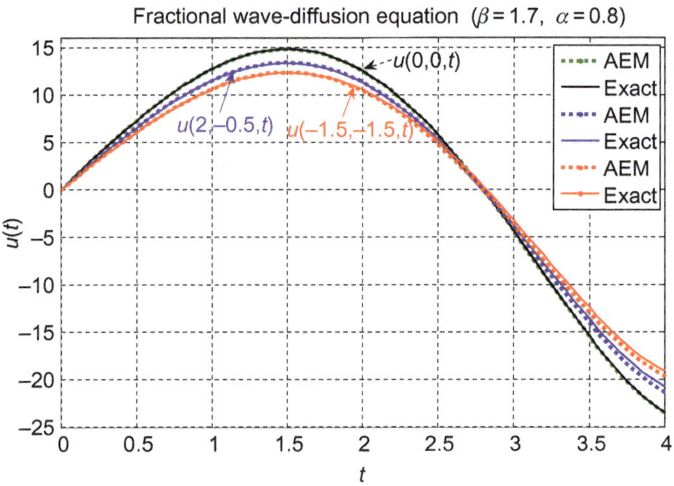

FIGURE 9.18. Time history of the solution $u(x, y, t)$ in Example 9.5.

Example 9.5

The wave-diffusion equation (9.44) in the inhomogeneous anisotropic plane body shown in Fig. 9.17 has been solved [10]. The boundary of the domain is defined by the curve

$$r=(ab)^{1/2}/\left[(\cos\theta/a)^2+(\sin\theta/b)\right]^{(1/4)}\left[(\cos\theta/b)^2+(\sin\theta/a)^2\right]^{(1/4)},\quad 0\leq\theta\leq 2\pi \tag{9.50}$$

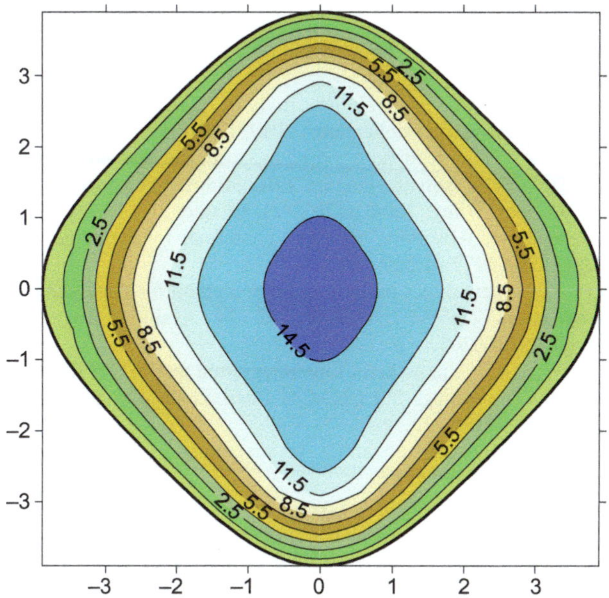

FIGURE 9.19 Contours of the solution $u(x, y, 1.5)$ in Example 9.5.

The numerical results have been obtained using the following data: $a = 3$, $b = 1.3$, $A = (y^2 - x^2 + 50)/50$, $B = 2xy/50$, $C = (x^2 - y^2 + 50)/50$, $D = E = 0$ $F = 0$; $\rho = 5\exp\left(-0.1\left\{|x| + |y|\right\}\right)$, $\eta = 0.4(x^2+y^2)^{1/2}$, $\beta = 1.7$, $\alpha = 0.8$; boundary condition $u(\mathbf{x}, t) = 0$, $\mathbf{x} \in \Gamma$, initial conditions $u(\mathbf{x}, 0) = 0$, $\dot{u}(\mathbf{x}, 0) = U(x, y)$, and external force

$$f(x, y, t) = \rho D_c^{1.7}T + \eta D_c^{0.8}T - (AU_{,xx} + 2BU_{,xy} + CU_{,yy})$$

where

$$U(x, y) = a^2b^2 - [(x/a)^2 + (y/b)^2][(x/b)^2 + (y/a)^2]$$

$$T(t) = t - t^3/6 + t^5/200$$

The problem admits an exact solution $u_{exact} = U(x, y)T(t)$. The time history of the solution at certain points obtained with N=210, L=109, $\Delta t = 0.005$ is shown in Fig. 9.18 as compared with the exact solution. Moreover, Fig. 9.19 shows the contours of the solution at $t = 1.5$.

9.5 REFERENCES

[1] Dominguez J. Boundary elements in dynamics. Southampton, Boston: Computational Mechanics Publications; 1993.
[2] Katsikadelis JT. *Dynamic analysis of structures. Theory and applications*. 2nd ed. Athens, Greece: Symmetria Publications; 2012, in Greek.
[3] Costabel, M. M. Time-dependent problems with the boundary integral equation method. In: Stein E, de Borst R, Hughes TJR, editors. Encyclopedia of computational mechanics. NJ: John Wiley; 2004. p. 703–21. Chapter 25.
[4] Engquist B, Majda A. Absorbing boundary conditions for the numerical simulation of waves. Math Comp 1977;31:629–51.
[5] Katsikadelis JT. A new direct time integration method for the equations of motion in structural dynamics. ZAMM 2014;94(9):757–74. Available from: http://dx.doi.org/10.1002/zamm.201200245.
[6] Katsikadelis JT. 2015. A new direct time integration method for the semi-discrete parabolic equation, Proceedings 8th GRACM International Congress on Computational Mechanics, Volos, 12 July 15.
[7] Mikhlin SG, editor. Linear equations of mathematical physics. New York: Holt, Rinehart and Winston, Inc; 1967.
[8] Hughes TJR. The finite element method. Englewood Cliffs, NJ, USA: Prentice Hall Inc; 1987.
[9] Bruch Jr. JC, Zyvoloski G. Transient two-dimensional heat conduction problems solved by the finite element method. Int J Numer Methods Eng 1974;8:481–94.
[10] Katsikadelis JT. The fractional wave-diffusion equation in bounded inhomogeneous anisotropic media. An AEM solution. In: Manolis GD, Polyzos D, editors. Advances in boundary element methods: a volume to honor Professor Dimitri Beskos. Dordrecht, Netherlands: Springer Science; 2009. p. 255–6.
[11] Podlubny I. Fractional differential equations. New York: Academic Press; 1999.
[12] Ginoa M, Cerbelli S, Roman HE. Fractional diffusion equation and relaxation in complex viscoelastic materials. Physica A 1992;191:449–53.
[13] Mainardi F, Paradisi P. Model of diffusive waves in viscoelasticity based on fractional calculus, In: Proceedings of the IEEE Conference on Decision and Control, Vol. 5, O. R. Gonzales, IEEE, New York; 1997. p. 4961–4966.
[14] Agrawal OP. Solution for a fractional diffusion-wave equation defined in a bounded domain. Nonlin Dyn 2002;29:145–55.
[15] Oldham KB, Spanier J. The fractional calculus. New York: Academic Press; 1974.
[16] Miller KS, Ross B. An introduction to the fractional calculus and fractional differential equations. New York: Wiley; 1993.
[17] Katsikadelis JT. Numerical Solution of Multi-term Fractional Differential Equations, ZAMM. Zeitschrift für Angewandte Mathematik und Mechanik 2009;89(7):593–608. Available from: http://dx.doi.org/10.1002/zamm.200900252.
[18] Tyn Myint U, Debnath L. Linear partial differential equations for scientists and engineers. 4th ed. Boston: Birkhäuser; 2007.

PROBLEMS

9.1. Determine the solution for the wave equation

$$\ddot{u} = c^2\nabla^2 u + xy \sin t \quad \text{in } \Omega \tag{a}$$

$$u=0 \quad \text{on } \Gamma \tag{b}$$

$$u(\mathbf{x},0)=0, \quad \dot{u}(\mathbf{x},0)=0 \quad \text{in } \Omega \tag{c,d}$$

The domain Ω is the rectangular $(0 \le x \le \pi) \times (0 \le y \le \pi)$. The problem admits an exact solution (see [18], Chapter 9)

9.2. Determine the temperature distribution in a plane rectangular body with radiation from the boundary. The temperature distribution is described by

$$\dot{u} = k(u_{,xx} + u_{,yy}) - h(u - u_0) \quad \text{in } \Omega \tag{a}$$

$$u = 0 \quad \text{on } \Gamma \tag{b}$$

$$u(\mathbf{x}, 0) = xy \quad \text{in } \Omega \tag{c}$$

The domain Ω is the rectangle $(-a \leq x \leq a) \times (-b \leq y \leq b)$; k, h, u_0 are constants designating the thermal conductivity, the heat transfer coefficient, and the ambient temperature, respectively.

9.3. Determine the solution of the initial boundary value problem

$$10\dot{u} = (a^2 - x^2)u_{,xx} + (b^2 - y^2)u_{,yy} + 4u \quad \text{in } \Omega \tag{a}$$

$$u = 0 \quad \text{on } \Gamma \tag{b}$$

$$u(\mathbf{x}, 0) = (a^2 - x^2)(b^2 - y^2) \quad \text{in } \Omega \tag{c, d}$$

The domain Ω is the rectangle $(-a \leq x \leq a) \times (-b \leq y \leq b)$. The problem admits an exact solution $u(\mathbf{x}, t) = (a^2 - x^2)(b^2 - y^2)\exp(-0.4t)$.

9.4. Study the dynamic response of an isotropic homogeneous viscoelastic plane body of uniform thickness. The geometry, the loading and the boundary conditions are shown in Fig. P.9.4. The viscoelastic response is described by the fractional standard solid viscoelastic model

$$\boldsymbol{\sigma} + p_1 {}_c D^{\alpha} \boldsymbol{\sigma} = q_0 \mathbf{C}\boldsymbol{\varepsilon} + q_1 \mathbf{C} {}_c D^{\alpha} \boldsymbol{\varepsilon} \quad \text{in } \Omega \tag{a}$$

where

$$p_1 = \frac{\eta b}{1+b}, \quad q_0 = \frac{b}{1+b}, \quad q_1 = \frac{\eta b}{1+b}, \quad \mathbf{C} = \frac{E}{1-\nu^2}\begin{vmatrix} 1 & \nu & 0 \\ \nu & 1 & 0 \\ 0 & 0 & \frac{1}{2}(1-\nu) \end{vmatrix}$$

Use $E = 2\ \text{N/m}^2$, $\nu = 0$, $\eta = 1$, $b = 1$, $a = 0.8$

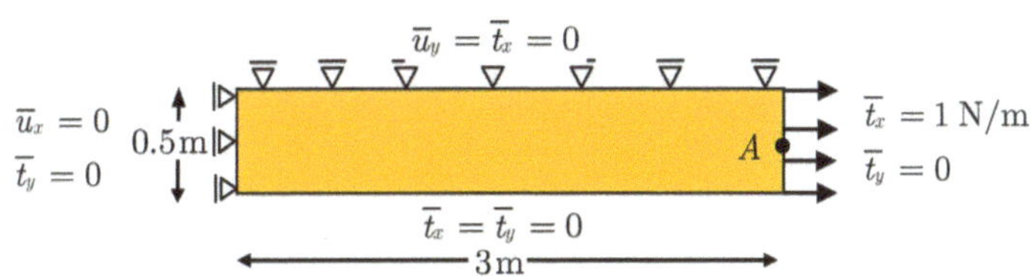

FIGURE P9.4

Chapter | Ten

The BEM for Nonlinear Problems

CHAPTER OUTLINE

10.1 INTRODUCTION

In this chapter, the boundary element method (BEM) is developed for solving nonlinear steady state and time dependent potential problems described by the general second order nonlinear partial differential equation as well as systems of them. Many problems in physics and engineering science are described by such equations [1], for example, the minimal surface problem [2], static and dynamic response of elastic and viscoelastic membranes undergoing large deflections [2–6], heat conduction in bodies with temperature dependent conductivity [7], etc. Apparently, the standard BEM cannot be applied to such problems, since no integral representation of the solution can be established. The dual reciprocity method (DRM) can treat such problems but only for specific forms of equations [8,9] as it is subjected to the restrictions mentions in Section 8.3. Therefore, the analog equation method (AEM) proves to be to the most appropriate method to solve these problems without any restrictions. The AEM, based on the principle of the analog equation, converts a nonlinear equation into a simple linear one with known fundamental solution, allowing

The Boundary Element Method for Engineers and Scientists.

thus the application of the standard BEM [10]. The AEM has been efficiently employed to solve many problems of engineering and science described by second orders nonlinear partial differential equations and systems of them as well. Among them, the soap bubble problem [11], static and dynamic analysis of elastic and viscoelastic membranes [10,12,13], space membranes [14], etc. In the following the AEM is developed as applied to the solution of nonlinear problems. Several example problems are presented, which validate the efficiency of the BEM to solve difficult nonlinear problem in engineering analysis and mathematical physics.

10.2 THE NONLINEAR WAVE EQUATION

We develop the AEM for the equation describing the nonlinear time dependent problems (dynamic, diffusion). The steady state problem (static) will result as a special case. Thus, we consider first the following initial boundary value problem

$$\rho\ddot{u} + \eta\dot{u} = \mathrm{N}(u_{xx}, u_{xy}, u_{yy}, u_x, u_y, u) + f(\mathbf{x}, t), \quad \mathbf{x}(x, y) \in \Omega \tag{10.1}$$

$$u = \alpha(\mathbf{x}, t), \quad \mathbf{x} \in \Gamma_1 \tag{10.2a}$$

$$u_{,n} = \gamma(\mathbf{x}, t), \quad \mathbf{x} \in \Gamma_2 \tag{10.2b}$$

and

$$u(\mathbf{x}, 0) = g_1(\mathbf{x}), \qquad \dot{u}(\mathbf{x}, 0) = g_2(\mathbf{x}), \quad \mathbf{x} \in \Omega \tag{10.3a, b}$$

where $\Gamma_1 \cup \Gamma_2 = \Gamma$; $u(\mathbf{x}, t)$ is the unknown field function and N is a nonlinear operator, that is, a nonlinear function of its arguments; $\rho = \rho(\mathbf{x})$, $\eta = \eta(\mathbf{x})$ are functions specified in Ω; $f(\mathbf{x}, t)$ is the external source, and $g_1(\mathbf{x})$, $g_2(\mathbf{x})$ are functions specified in Ω. The quantities $\alpha(\mathbf{x}, t)$ and $\gamma(\mathbf{x}, t)$ are functions specified on the boundary. The boundary Γ may be multiply-connected.

The initial boundary value problem (10.1)–(10.3) for suitable meaning of the field function $u(\mathbf{x}, t)$ occurs in many problems in engineering and mathematical physics, such as the minimal surface problem, large deflections of elastic membranes, nonlinear vibrations of membranes, heat flow in bodies with temperature dependent, etc.

10.2.1 The AEM solution

Let $u = u(\mathbf{x}, t)$ be the sought solution to the initial boundary value problem (10.1)–(10.3). This function is two times continuously differentiable in Ω. Since Eq. (10.1) is of the second order, the operator $\hat{N} = \nabla^2$, that is, the Laplace operator with respect to the space coordinates x, y, can be employed to produce the analog equation. This yields

$$\nabla^2 u = b(\mathbf{x}, t), \quad \mathbf{x} \in \Omega, \quad t > 0 \tag{10.4}$$

Note that the time in Eq. (10.4) is a parameter, that is, the equation is quasi-static. This implies that the principle of the analog equation replaces the time dependent equation (10.1) with a sequence of elliptic equations in time. Obviously, the AEM procedure employed in Section 8.4.2 for the elliptic equation can be employed to this case, too, provided that the unknown coefficients in Eq. (8.76) are treated as functions of time, $a_j = a_j(t)$. Hence, Eq. (8.76) reads

$$b \simeq \sum_{j=1}^{L} a_j(t)\phi_j \tag{10.5}$$

Discretizing the boundary using N elements, taking L internal nodes, and using the procedure presented in Section 8.4.2 give

$$\mathbf{Hu} - \mathbf{Gq} = \mathbf{T}\boldsymbol{\alpha} \tag{10.6a}$$

$$\mathbf{c}_1\mathbf{u} + \mathbf{c}_2\mathbf{u}_n = \mathbf{c}_3 \tag{10.6b}$$

$$\boldsymbol{\rho}\ddot{\mathbf{u}} + \boldsymbol{\eta}\dot{\mathbf{u}} = \mathbf{N}(\mathbf{u}_{,xx}, \mathbf{u}_{,xy}, \mathbf{u}_{,yy}, \mathbf{u}_{,x}, \mathbf{u}_{,y}, \mathbf{u}) + \mathbf{f} \tag{10.6c}$$

where $\mathbf{N}(\mathbf{u}_{,xx}, \mathbf{u}_{,xy}, \mathbf{u}_{,yy}, \mathbf{u}_{,x}, \mathbf{u}_{,y}, \mathbf{u})$ represents an $L \times 1$ vector whose elements are nonlinear functions of the components of the involved vectors; $\boldsymbol{\rho}, \boldsymbol{\eta}$ are $L \times L$ diagonal matrices containing the values of the functions $\rho(\mathbf{x}), \eta(\mathbf{x})$ at the internal nodes. All other matrices and vectors involved in Eqs. (10.6) are given in Section 8.4.2. We should have in mind that the vector $\boldsymbol{\alpha}$ and in general the vector $\mathbf{c}_3$ depend on time, that is,

$$\boldsymbol{\alpha} = \{a_1(t) \quad a_2(t) \quad \cdots \quad a_L(t)\}^T \tag{10.7a}$$

$$\mathbf{c}_3 = \{c_{31}(t) \quad c_{32}(t) \quad \ldots \quad c_{3N}(t)\}^T \tag{10.7b}$$

Introducing Eqs. (8.97) in Eq. (10.6c) yields

$$\mathbf{M}\ddot{\boldsymbol{\alpha}} + \mathbf{C}\dot{\boldsymbol{\alpha}} + \mathbf{N}(\boldsymbol{\alpha}) = \mathbf{p}(t) \tag{10.8}$$

where

$$\mathbf{M} = \boldsymbol{\rho}\mathbf{S} \tag{10.9a}$$

$$\mathbf{C} = \boldsymbol{\eta}\mathbf{S} \tag{10.9b}$$

$$\mathbf{p}(t) = \mathbf{f}(t) - (\boldsymbol{\rho}\ddot{\mathbf{z}} + \boldsymbol{\eta}\dot{\mathbf{z}}) \tag{10.9c}$$

and N(α) is a vector, whose elements are nonlinear functions of the components of α; the vector $\mathbf{z}$ is given by Eq. (8.98b).

Note that for time independent boundary conditions, it is $\dot{\mathbf{z}} = \ddot{\mathbf{z}} = 0$.

The initial conditions $\alpha(0), \dot{\alpha}(0)$ can be established from Eqs. (10.3a,b) as follows:

Solving Eq. (8.97f) for $\alpha(t)$ and differentiating give

$$\alpha(t) = \mathbf{S}^{-1}[\mathbf{u}(t) - \mathbf{z}(t)] \tag{10.10a}$$

$$\dot{\alpha}(t) = \mathbf{S}^{-1}[\dot{\mathbf{u}}(t) - \dot{\mathbf{z}}(t)] \tag{10.10b}$$

which by virtue of Eqs. (10.3a,b) become

$$\alpha(0) = \mathbf{S}^{-1}[\mathbf{g}_1 - \mathbf{z}(0)] \tag{10.11a}$$

$$\dot{\alpha}(0) = \mathbf{S}^{-1}[\mathbf{g}_2 - \dot{\mathbf{z}}(0)] \tag{10.11b}$$

The initial value problem (10.8), (10.11a,b) can be solved using any of the established numerical methods for the direct time integration of the nonlinear equations of motion. The numerical results in the examples of this chapter have been obtained using the method presented in [15].

Finally, the obtained solution $\alpha(t)$ is inserted in Eqs. (8.97) to evaluate the solution and its derivatives at the internal nodes at time t. The solution and its derivatives at any point $\mathbf{x} \in \Omega$ can be evaluated using also Eqs. (8.97), but in this case the involved matrices must be evaluated with respect to the point $\mathbf{x}$.

10.3 THE NONLINEAR DIFFUSION EQUATION

The nonlinear diffusion problem is described by the initial boundary value problem

$$\eta \dot{u} = \mathrm{N}(u_{,xx}, u_{,xy} u_{,yy}, u_{,x}, u_{,y}, u) + f(\mathbf{x}, t), \quad \mathbf{x}(x, y) \in \Omega \tag{10.12}$$

$$u = \alpha(\mathbf{x}, t), \quad \mathbf{x} \in \Gamma_1 \tag{10.13a}$$

$$u_{,n} = \gamma(\mathbf{x}, t), \quad \mathbf{x} \in \Gamma_2 \tag{10.13b}$$

and

$$u(\mathbf{x}, 0) = g_1(\mathbf{x}), \quad \mathbf{x} \in \Omega \tag{10.14}$$

Apparently, the AEM solution is obtained from Eq. (10.8) by setting $\rho = \mathbf{0}$ and considering only the first initial condition, that is, Eq. (10.11a).

Thus the RBF series coefficient vector $\boldsymbol{\alpha}(t)$ is computed from the solution of the following initial value problem

$$\mathbf{C}\dot{\boldsymbol{\alpha}} + \mathbf{N}(\boldsymbol{\alpha}) = \mathbf{p}(t) \tag{10.15a}$$

$$\boldsymbol{\alpha}(0) = \mathbf{S}^{-1}[\mathbf{g}_1 - \mathbf{z}(0)] \tag{10.15b}$$

The initial value problem (10.15a,b) is solved using the method for the semidiscrete parabolic equation presented in [16].

10.4 THE NONLINEAR POTENTIAL EQUATION

The nonlinear potential problem is described by the boundary value problem

$$N(u_{,xx}, u_{,xy} u_{,yy}, u_{,x}, u_{,y}, u) + f(\mathbf{x}) = 0, \quad \mathbf{x}(x, y) \in \Omega \tag{10.16}$$

$$u = \alpha(\mathbf{x}), \quad \mathbf{x} \in \Gamma_1 \tag{10.17a}$$

$$u_{,n} = \gamma(\mathbf{x}), \quad \mathbf{x} \in \Gamma_2 \tag{10.17b}$$

Apparently, the AEM solution is obtained from Eqs. (10.8)–(10.11) by setting $\boldsymbol{\rho} = \mathbf{0}$, $\boldsymbol{\eta} = \mathbf{0}$. Thus we have

$$\mathbf{N}(\boldsymbol{\alpha}) = \mathbf{f} \tag{10.18}$$

Apparently, the coefficient vector $\boldsymbol{\alpha}$ does not now depend on time. Hence Eq. (10.18) is a nonlinear algebraic equation, which can be solved to obtain the coefficients $\boldsymbol{\alpha}$. Any of the available methods can be used to solve such equations. The function *fsolve* of Matlab or subroutine NEQNF of the IMSL has been employed to obtain the numerical results in the examples of this chapter.

10.4.1 Examples

On the basis of the procedure presented in Section 10.2.1 a FORTRAN code has been written for the solution of the initial boundary value problem (10.1), (10.2). The employed radial basis functions ϕ_j are multiquadrics (MQs). Apparently the same program can be employed to solve the problems (10.12), (10.13a,b) and (10.16), (10.17a,b) with only difference the solution of the discretized equations. Certain examples problems are presented which demonstrate the efficiency and accuracy of the AEM.

EXAMPLE 10.1

In this example the steady state heat transfer in a plane body with temperature dependent conductivity $k_x = k_y = k = k(u)$, $k_{xy} = 0$ in absence of internal heat sources is studied. In this case Eq. (9.32) for $c = 0$, $f(x, t) = 0$ becomes

$$(ku_{,x})_{,x} + (ku_{,y})_{,y} = 0 \quad \text{in } \Omega \tag{10.19}$$

Assuming that k varies according to the law

$$k = k_0\left(1+\beta\frac{u-u_0}{u_0}\right) \tag{10.20}$$

where u_0 is the initial temperature and k_0 and β are material constants, Eq. (10.19) becomes

$$k\nabla^2 u+\frac{k_0}{u_0}\beta(u_{,x}^2+u_{,y}^2)=0 \tag{10.21}$$

The nonlinear Eq. (10.21) is solved for a square plane body with unit side length, $0\le x, y\le 1$, under mixed boundary conditions

$$u(0,y)=300, \quad u(1,y)=400, \quad u_{,n}(x,0), \quad u_{,n}(x,1)=0, \tag{10.22}$$

for $k_0 = 1$, $\beta = 3$, $u_0 = 300$

Numerical results are presented in Table 10.1 as compared with those obtained using DRM and the Kirchhoff's transformation [9]. The results were obtained using N = 200 constant boundary elements and $L = 49, 121, 225$ internal nodal points and shape parameter $c = 0.1$. The contours of the temperature distribution are shown in Fig. 10.1

EXAMPLE 10.2

This problem concerns the determination of the surface bounded by one or more nonintersecting space curves and has a minimal area. The physical analog of the minimal surface is that of the soap bubble constrained between bounding contours (Plateau's problem [17]). The minimal surface is obtained as the solution of the following boundary value problem [10,11]

$$(1+u_{,y}^2)u_{,xx}-2u_{,x}u_{,y}u_{,xy}+(1+u_{,x}^2)u_{,yy}=0 \quad \text{in } \Omega \tag{10.23}$$

$$u=\overline{u} \quad \text{on } \Gamma \tag{10.24}$$

TABLE 10.1 Numerical Results for the Temperature u in Example 10.1

Position		DRM	Kirchhoff	AEM		
x	y			L = 49	L = 121	L = 225
0.1	0.5	314.15	314.00	314.01	314.01	314.01
0.3	0.5	338.34	337.82	337.88	337.85	337.84
0.5	0.5	358.49	358.11	358.17	358.13	358.12
0.7	0.5	376.27	376.08	376.11	376.08	376.07
0.9	0.5	392.43	392.36	392.37	392.36	392.35

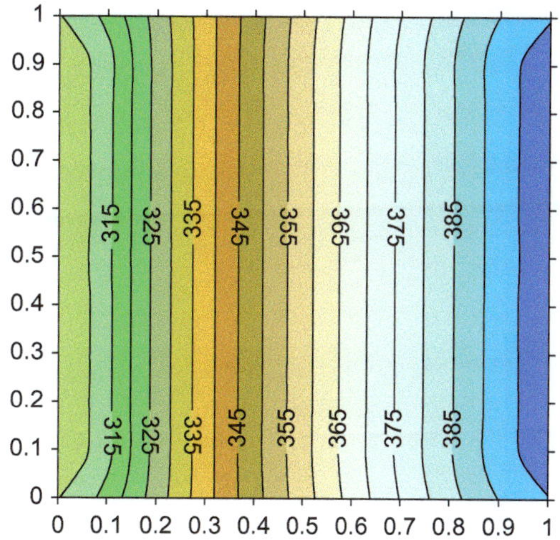

FIGURE 10.1 Contours of the temperature distribution in Example 10.1.

The problem has been solved for the square domain $1 \leq x \leq 5$, $-2 \leq y \leq 2$. The surface is supported on the space curve $\overline{u} = \tan^{-1}(y/x)$, $(x, y) \in \Gamma$. The problem adopts an exact solution $u = \tan^{-1}(y/x)$. The numerical solution has been obtained with $\mathrm{N} = 200$ constant boundary elements, $\mathrm{L} = 49$ and shape parameter $c = 0.1$. The contours and the 3D view of the computed minimal surface are shown in Fig. 10.2. Moreover, Table 10.2 presents results as compared with those obtained from the analytical solution. It should be emphasized that the results are in excellent agreement not only for the field function but also for its derivatives.

EXAMPLE 10.3

In this example the surface of a soap bubble (minimal surface), whose plan form has the shape of a rounded cross (Fig. 10.3), has been established. The soap bubble passes through the space curve which is the intersection of the cylindrical surface $r = 5(\sin^4\theta + \cos^4\theta)$, $0 \leq \theta \leq 2\pi$, $z \geq 0$, and the sphere $x^2 + y^2 + z^2 = R^2$, $R = 5$. The surface was obtained from the solution of the boundary value problem (10.23), (10.24) using $\mathrm{N} = 315$ constant boundary elements, $\mathrm{L} = 109$ domain nodal points and shape parameter $c = 0.1$. Fig. 10.3 shows the distribution of the nodal points. The contours and the 3D view of the soap bubble surface are shown in Fig. 10.4.

EXAMPLE 10.4

In this example we consider the transient heat conduction in a plane body with temperature dependent conductivity $k = k_0(1 + \beta u)$. The temperature distribution $u(\mathbf{x}, t)$ at time t when the temperature on the boundary is kept zero is described by the following initial boundary value problem

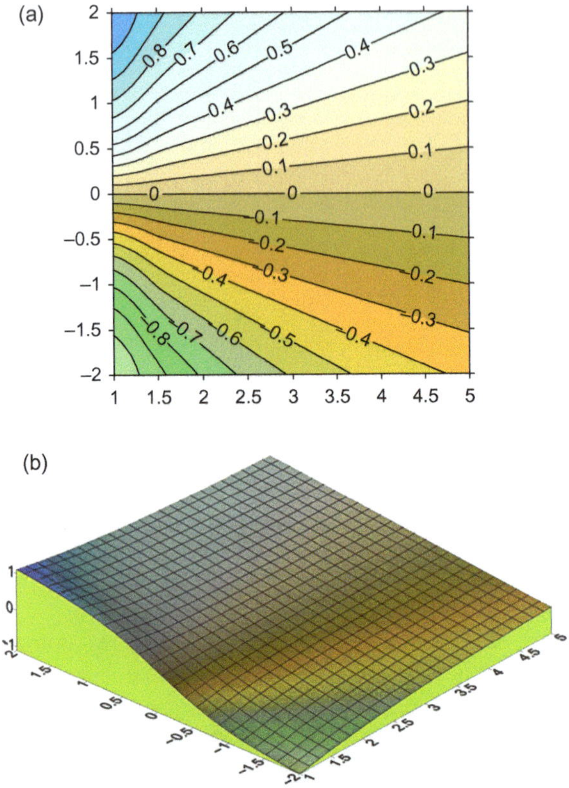

FIGURE 10.2 Computed minimal surface in in Example 10.2. (a) Contours; (b) 3D View.

$$\eta \dot{u} = k\nabla^2 u + k_0 \beta(u,_x^2 + u,_y^2) + f(\mathbf{x}, t) \quad \text{in } \Omega \tag{10.25}$$

$$u = 0 \quad \text{on } \Gamma \tag{10.26}$$

$$u(\mathbf{x}, 0) = [1 - (x^2 + y^2)] \quad \text{in } \Omega \tag{10.27}$$

where $f(\mathbf{x}, t)$ is the internal source, and k_0, β are material constants.

The problem has been solved for a circular domain Ω of unit circle with $\eta = 1/2$, $k_0 = 1$, $\beta = 3$, $f = (3 + x^2 + y^2)\exp(-2t) + 12[1 - 2(x^2 + y^2)]\exp(-4t)$. The results were obtained using N = 200 constant boundary elements and L = 129 domain nodal points uniformly distributed as shown in Fig. 10.5 with shape parameter $c = 0.1$. Moreover, Fig. 10.6 shows the time variation of the computed temperature u and its derivatives $u,_x$, $u,_{xx}$ at point $(0, 0)$ as compared with the exact ones, while Fig. 10.7 shows the respective relative errors for $\Delta t = 0.001$. Finally, Fig. 10.8 presents the contours and the 3D view of temperature distribution u at $t = 0.25$.

TABLE 10.2 Solution u and Its Derivatives Along the Line $y = 1$ in Example 10.2. Upper value: computed; lower value: exact

x	u	$u_{,x}$	$u_{,y}$	$u_{,xx}$	$u_{,xy}$	$u_{,yy}$
1.5	0.5879	−0.3076	0.4616	0.2840	−0.1184	−0.2840
	0.5880	−0.3077	0.4616	0.2840	−0.1183	−0.2840
2.0	0.4636	−0.2000	0.4000	0.1600	−0.1200	−0.1600
	0.4636	−0.2000	0.4000	0.1600	−0.1200	−0.1600
2.5	0.3805	−0.1379	0.3448	0.0951	−0.0999	−0.0951
	0.3805	−0.1379	0.3448	0.0951	−0.0999	−0.0951
3.0	0.3217	−0.1000	0.3000	0.0600	−0.0800	−0.0600
	0.3217	−0.1000	0.3000	0.0600	−0.0800	−0.0600
3.5	0.2783	−0.0755	0.2642	0.0399	−0.0641	−0.0399
	0.2783	−0.0755	0.2642	0.0399	−0.0641	−0.0399
4.0	0.2449	−0.0588	0.2353	0.0277	−0.0519	−0.0277
	0.2449	−0.0588	0.2353	0.0277	−0.0519	−0.0277
4.5	0.2186	−0.0471	0.2118	0.0199	−0.0426	−0.0199
	0.2186	−0.0471	0.2118	0.0199	−0.0426	−0.0199

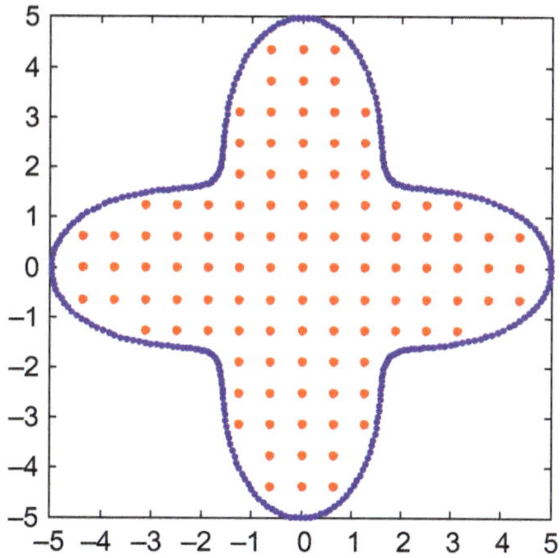

FIGURE 10.3 Plan form of the soap bubble with boundary and domain nodal points in Example 10.3 $\{r = 5(\sin^4\theta + \cos^4\theta)\}$.

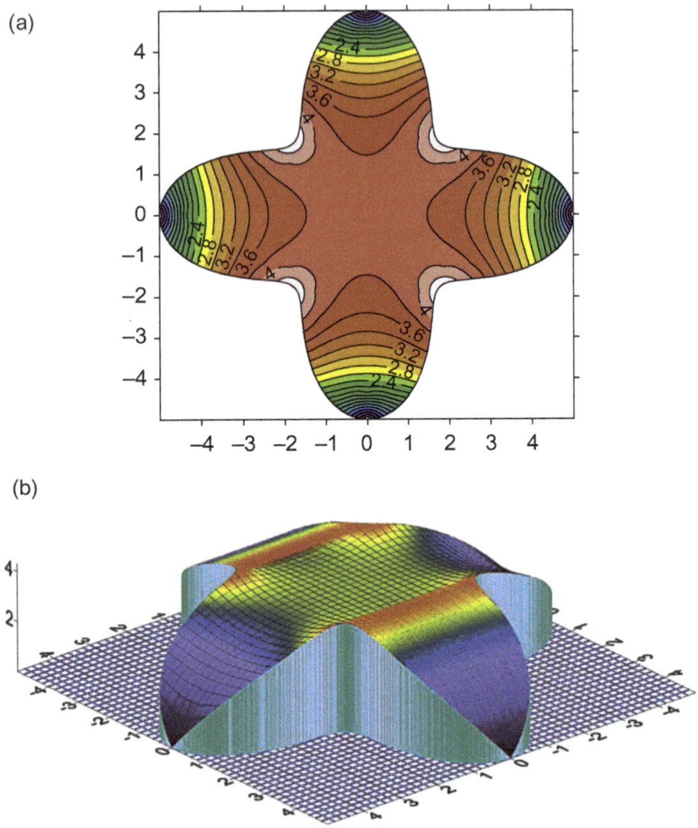

FIGURE 10.4 Computed surface of the soap bubble in Example 10.3. (a) Contours; (b) 3D view.

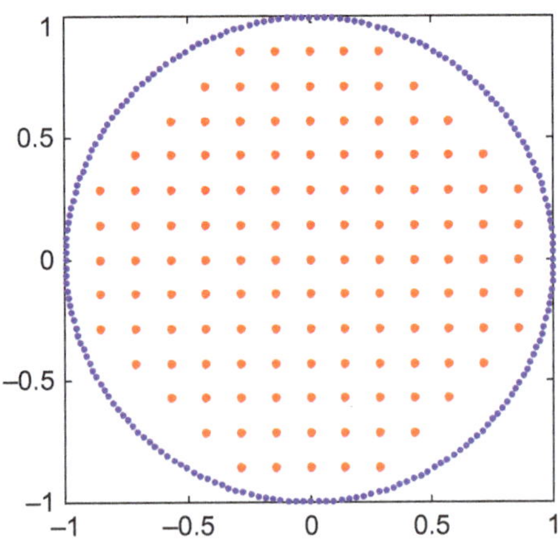

FIGURE 10.5 Boundary and domain nodal points in Example 10.4.

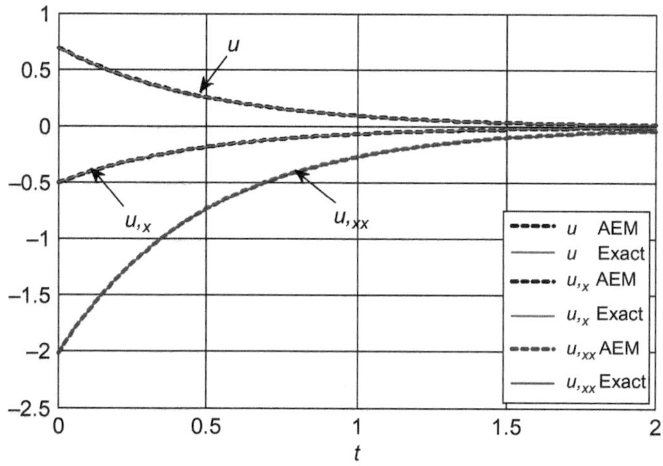

FIGURE 10.6 Time variation of u and its derivatives $u_{,x}, u_{,xx}$ at point $(0, 0)$ in Example 10.4.

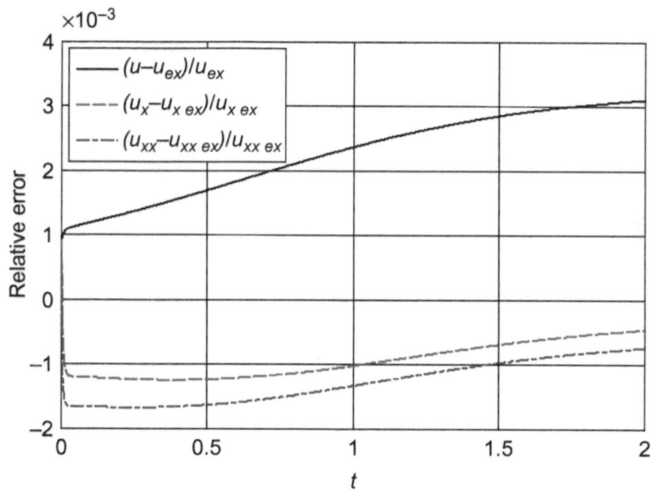

FIGURE 10.7 Relative error at point $(0, 0)$ with $\Delta t = 0.001$ in Example 10.4.

10.5 COUPLED NONLINEAR EQUATIONS

The AEM solution for coupled nonlinear second order differential equations is illustrated by analyzing elastic membranes undergoing large deflections. The dynamic response of a membrane, which is prestressed and fixed on its boundary, is governed by the following nonlinear initial boundary value problem [18]

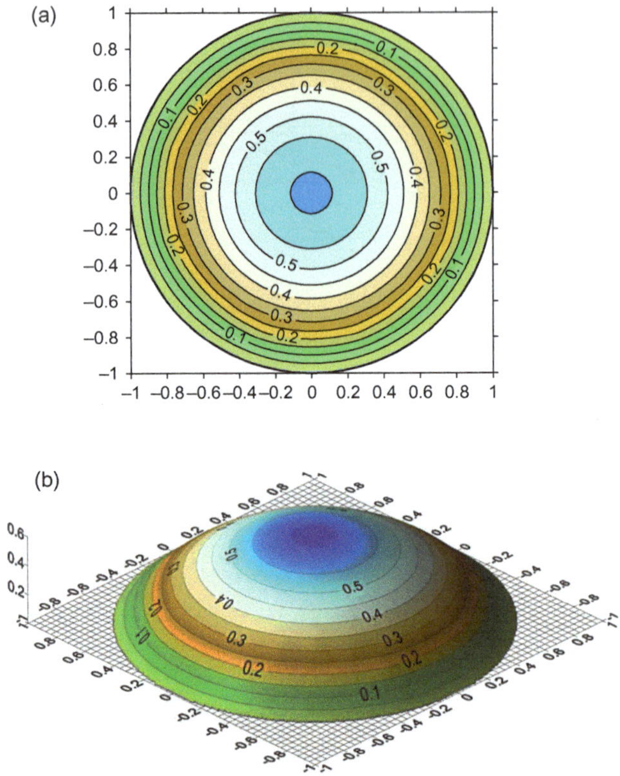

FIGURE 10.8 Contours and 3D view of the temperature distribution at $t = 0.25$ in Example 10.4. (a) Contours; (b) 3D view.

$$N_x w_{,xx} + 2N_{xy} w_{,xy} + N_y w_{,yy} - \rho h \ddot{w} + \rho h \ddot{u} w_{,x} + \rho h \ddot{v} w_{,y} = -f \quad \text{in } \Omega \tag{10.28a}$$

$$\begin{aligned} &\nabla^2 u + \frac{1+\nu}{1-\nu}\left(u_{,x} + v_{,y}\right)_{,x} + w_{,x}\left(\frac{2}{1-\nu} w_{,xx} + w_{,yy}\right) \\ &+ \frac{1+\nu}{1-\nu} w_{,xy} w_{,y} - \frac{\rho h}{G}\ddot{u} = 0 \end{aligned} \quad \text{in } \Omega \tag{10.28b}$$

$$\begin{aligned} &\nabla^2 v + \frac{1+\nu}{1-\nu}\left(u_{,x} + v_{,y}\right)_{,y} + w_{,y}\left(\frac{2}{1-\nu} w_{,yy} + w_{,xx}\right) \\ &+ \frac{1+\nu}{1-\nu} w_{,xy} w_{,x} - \frac{\rho h}{G}\ddot{v} = 0 \end{aligned} \quad \text{in } \Omega \tag{10.28c}$$

with boundary conditions(*)

$$w=0, \quad u_n=u_n^*, \quad u_t=u_t^*, \quad \text{on } \Gamma \tag{10.29a, b, c}$$

and initial conditions

$$w(\mathbf{x},0)=g_1(\mathbf{x}), \quad \dot{w}(\mathbf{x},0)=h_1(\mathbf{x}) \quad \mathbf{x}\in\Omega \tag{10.30a}$$

$$u(\mathbf{x},0)=g_2(\mathbf{x}), \quad \dot{u}(\mathbf{x},0)=h_2(\mathbf{x}) \quad \mathbf{x}\in\Omega \tag{10.30b}$$

$$v(\mathbf{x},0)=g_3(\mathbf{x}), \quad \dot{v}(\mathbf{x},0)=h_3(\mathbf{x}) \quad \mathbf{x}\in\Omega \tag{10.30c}$$

where $w(\mathbf{x}, t)$ is the transverse displacement and $u(\mathbf{x}, t)$, $v(\mathbf{x}, t)$ the membrane displacements; h is the thickness of the membrane; ρ the mass density; G the shear modulus; u_n, u_t are the normal and tangential displacements on the boundary; $g_i(\mathbf{x})$, $h_i(\mathbf{x})$ $(i=1,2,3)$ are specified functions and $f=f(\mathbf{x},\mathrm{t})$ is the transverse surface load. Finally, quantities N_x, N_y, N_{xy} represent the membrane forces produced in the interior of the membrane and are given by the relations (cf. [18])

$$N_x=C\left[u_{,x}+\frac{1}{2}w_{,x}^2+\nu\left(v_{,y}+\frac{1}{2}w_{,y}^2\right)\right] \tag{10.31a}$$

$$N_y=C\left[v_{,y}+\frac{1}{2}w_{,y}^2+\nu\left(u_{,x}+\frac{1}{2}w_{,x}^2\right)\right] \tag{10.31b}$$

$$N_{xy}=C\frac{1-\nu}{2}(u_{,y}+v_{,x}+w_{,x}w_{,y}) \tag{10.31c}$$

where $C=Eh/(1-\nu^2)$ is the membrane stiffness with E and ν being the elastic constants of the material.

It was shown (see [18]) that the influence of the membrane inertia forces is negligible and thus can be omitted. Obviously, the static problem results by neglecting all inertia forces.

10.5.1 The AEM solution

Since Eqs. (10.28) are of the second order, the operator $\hat{N}=\nabla^2$, that is, the Laplace operator with respect to the space coordinates x, y, can be employed to produce the analog equations. This yields

$$\nabla^2 w=b^{(1)}(\mathbf{x}, t) \tag{10.32a}$$

(*) These boundary conditions are valid when the prestress is due to boundary displacements, which are imposed simultaneously with the transverse load f [18].

$$\nabla^2 u = b^{(2)}(\mathbf{x}, t) \tag{10.32b}$$

$$\nabla^2 v = b^{(3)}(\mathbf{x}, t) \tag{10.32c}$$

Approximating the fictitious sources with RBF series

$$b^{(i)} \simeq \sum_{j=1}^{L} a_j^{(i)}(t)\phi_j, \quad i = 1, 2 \tag{10.33}$$

and using the procedure presented in Section 8.4.2, we obtain the system of $3 \times L$ nonlinear semidiscretized equations

$$\mathbf{M}_1\ddot{\boldsymbol{\alpha}}_1 + \mathbf{F}_1(\boldsymbol{\alpha}_1, \ddot{\boldsymbol{\alpha}}_1, \boldsymbol{\alpha}_2, \ddot{\boldsymbol{\alpha}}_2, \boldsymbol{\alpha}_3, \ddot{\boldsymbol{\alpha}}_3) = \mathbf{p}_1(t) \tag{10.34a}$$

$$\mathbf{M}_2\ddot{\boldsymbol{\alpha}}_2 + \mathbf{A}_{11}\boldsymbol{\alpha}_2 + \mathbf{A}_{12}\boldsymbol{\alpha}_3 + \mathbf{F}_2(\boldsymbol{\alpha}_1) = \mathbf{p}_2 \tag{10.34b}$$

$$\mathbf{M}_3\ddot{\boldsymbol{\alpha}}_3 + \mathbf{A}_{21}\boldsymbol{\alpha}_2 + \mathbf{A}_{22}\boldsymbol{\alpha}_3 + \mathbf{F}_3(\boldsymbol{\alpha}_1) = \mathbf{p}_3 \tag{10.34c}$$

where $\mathbf{M}_k$ $(k=1,2,3)$ represent $L \times L$ known mass matrices, given as

$$\mathbf{M}_1 = -\frac{\rho h}{G}\mathbf{S}^{(1)}, \quad \mathbf{M}_2 = -\frac{\rho h}{G}\mathbf{S}^{(2)} \quad \mathbf{M}_3 = -\frac{\rho h}{G}\mathbf{S}^{(3)} \tag{10.35a, b, c}$$

The quantities $\mathbf{A}_{11}, \mathbf{A}_{12}, \mathbf{A}_{21}, \mathbf{A}_{22}$ are defined by Eqs. (8.145) and

$$\mathbf{p}_1(t) = \mathbf{f}(t) \tag{10.36a}$$

$$\mathbf{p}_2(t) = -\frac{1+\nu}{1-\nu}(\mathbf{z},_{xx}^{(1)} + \mathbf{z},_{xy}^{(2)}) \tag{10.36b}$$

$$\mathbf{p}_3(t) = -\frac{1+\nu}{1-\nu}(\mathbf{z},_{xy}^{(1)} + \mathbf{z},_{yy}^{(2)}) \tag{10.36c}$$

When the membrane inertia force are neglected, it is $\mathbf{M}_2 = \mathbf{M}_3 = 0$, and Eqs. (10.34) reduce to

$$\mathbf{M}_1\ddot{\boldsymbol{\alpha}}_1 + \mathbf{F}_1(\boldsymbol{\alpha}_1, \ddot{\boldsymbol{\alpha}}_1, \boldsymbol{\alpha}_2, \boldsymbol{\alpha}_3) = \mathbf{p}_1(t) \tag{10.37a}$$

$$\mathbf{A}_{11}\boldsymbol{\alpha}_2 + \mathbf{A}_{12}\boldsymbol{\alpha}_3 + \mathbf{F}_2(\boldsymbol{\alpha}_1) = \mathbf{p}_2 \tag{10.37b}$$

$$\mathbf{A}_{21}\boldsymbol{\alpha}_2 + \mathbf{A}_{22}\boldsymbol{\alpha}_3 + \mathbf{F}_3(\boldsymbol{\alpha}_1) = \mathbf{p}_3 \tag{10.37c}$$

Equation (10.37a) is a differential equation, while Eqs. (10.37b,c) are algebraic equations, which can be viewed as constraint equations. Since the latter equations are linear with respect to $\boldsymbol{\alpha}_2, \boldsymbol{\alpha}_3$, they can be solved for these vectors. Then substitution in Eq. (10.37a) yields a single equation of motion for $\boldsymbol{\alpha}_1$. In this case the initial conditions for Eq. (10.37a) result from Eqs. (8.97), (10.29a), and (10.30a). Thus we have

$$\boldsymbol{\alpha}_1(0) = (\mathbf{S}^{(1)})^{-1}\mathbf{g}_1, \quad \dot{\boldsymbol{\alpha}}_1(0) = (\mathbf{S}^{(1)})^{-1}\mathbf{h}_1 \tag{10.38a, b}$$

EXAMPLES

On the basis of the procedure presented in previous Section, a FORTRAN code has been written for the solution of the initial boundary value problem (10.37), (10.38). The employed radial basis functions ϕ_j are multiquadrics (MQs). The example below demonstrates the efficiency of the AEM.

Example 10.5

The dynamic response of the membrane of Fig. 10.9 is studied in this example. The boundary of the domain is defined by the curve $r=(ab)^{1/2}/\left[(\cos\theta/a)^2+(\sin\theta/b)^2\right]^{(1/4)}/\left[(\cos\theta/b)^2+(\sin\theta/a)^2\right]^{(1/4)}$, $0\leq\theta\leq2\pi$. The membrane is prestressed by the imposed displacement $u_n = 0.2\ m$ normal to the boundary, while the tangential displacement is zero, $u_t=0\ m$. The employed data are $a=3$ m, $b=1.3$ m; $h=0.002$ m, $\rho=5000$ kg/m^3, $\nu=0.3$, $E=1.1\times10^5$ kN/m^2. The membrane is subjected to the suddenly applied uniform transverse load $f=2$ kN/m^2. The results were obtained using N = 210 constant boundary elements and L = 101 domain nodal points distributed as shown in Fig. 10.9 with shape parameter $c=0.1$. Fig. 10.10 shows the

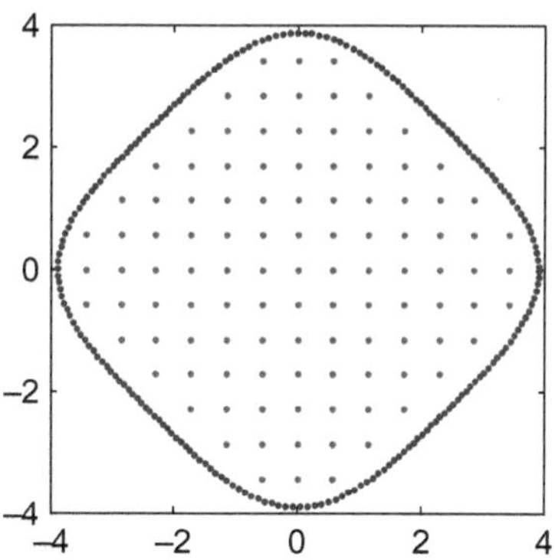

FIGURE 10.9 Boundary and domain nodal points of the membrane in Example 10.5.

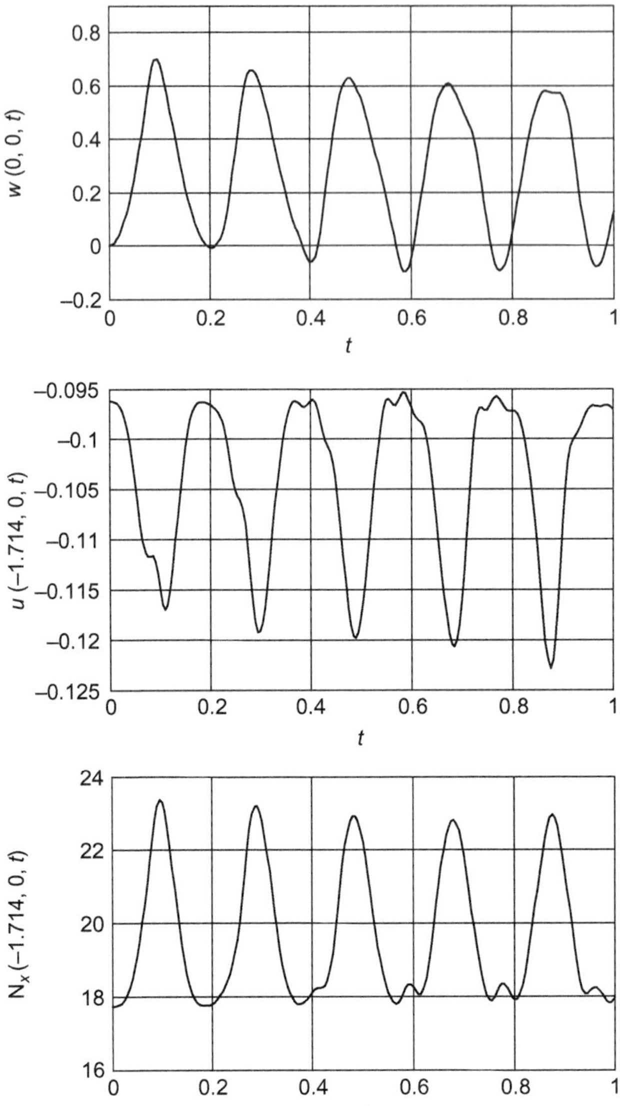

FIGURE 10.10 Time history of the displacements $w(0, 0, t)$, $u(-1.741, 0, t)$, and membrane force $N_x(-1.741, 0, t)$ in Example 10.5.

time history of the transverse displacement $w(0, 0, t)$, the membrane displacement $u(-1.714, 0, t)$ and the membrane force $N_x(-1.714, 0, t)$. Finally, Figs. 10.11–10.14 present the contours and the 3D view of the transverse displacement w and the membrane forces N_x, N_y, N_{xy} at $t = 0.1$. The solution of the nonlinear system of the semidiscretized equation of motion was obtained using the method presented in [15].

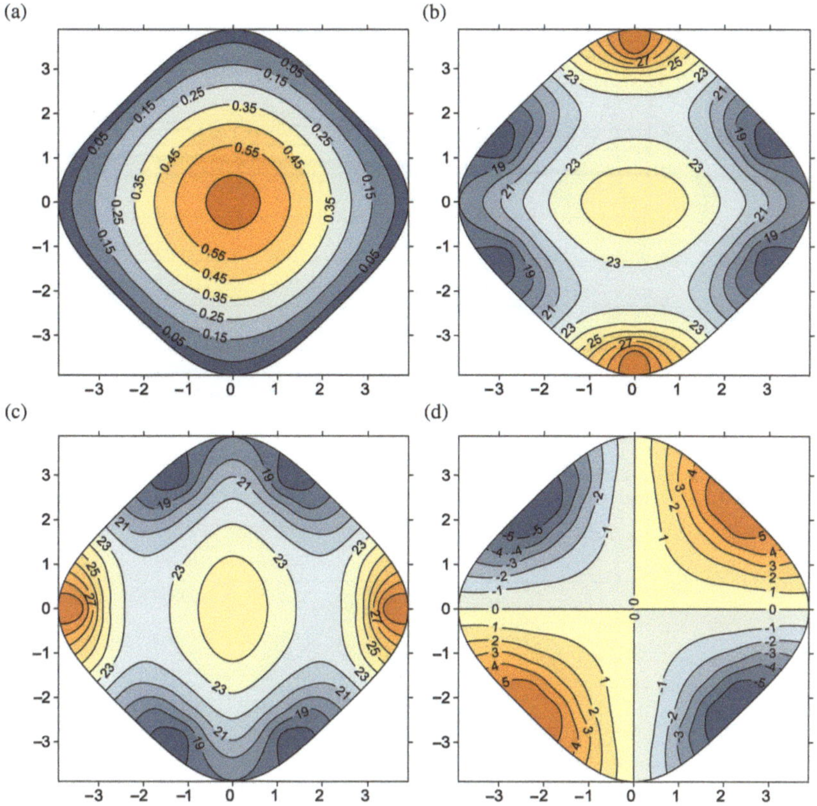

FIGURE 10.11 Contours of the deflection w and membrane forces N_x, N_y, N_{xy} at $t = 0.1$ in Example 10.5. (a) w; (b) N_x; (c) N_y; (d) N_{xy}.

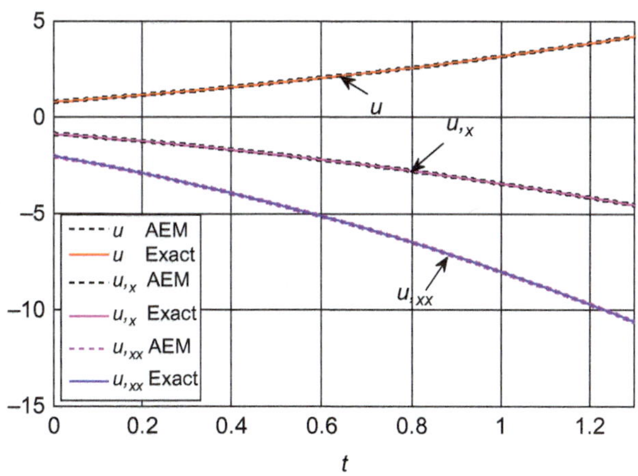

FIGURE 10.12 Time variation of the displacement u and its derivatives $u_{,x}, u_{,xx}$ at point $(0.428, 0.142)$ in Example 10.6.

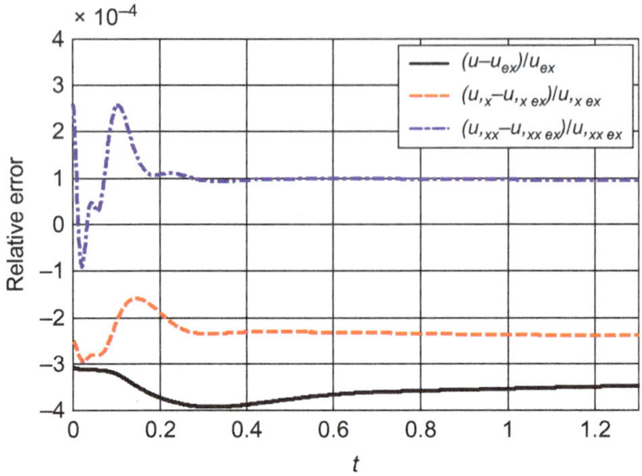

FIGURE 10.13 Relative error at point $(0.428, 0.142)$ with $\Delta t = 0.001$ in Example 10.6.

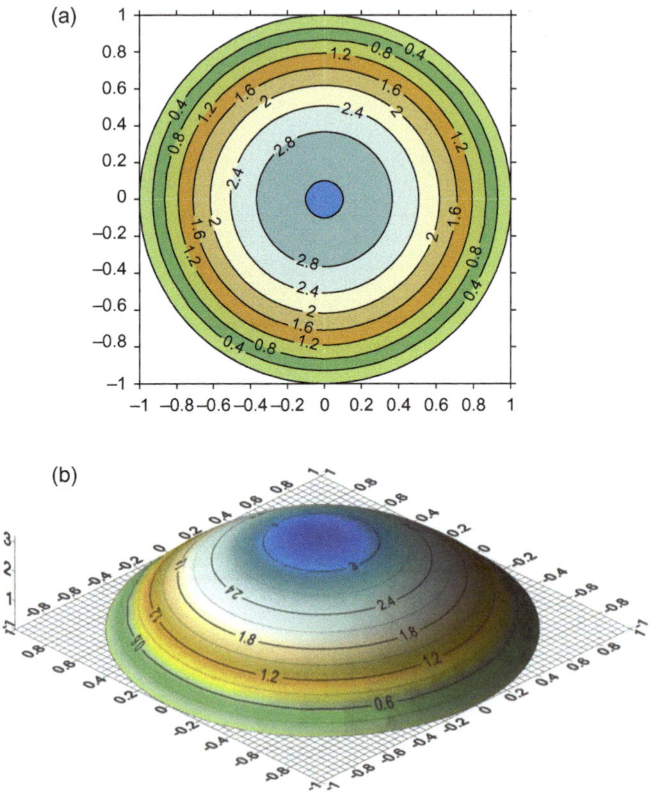

FIGURE 10.14 Contours and 3D view of the solution $u(x, y)$ at $t = 0.80$ in Example 10.6. (a) Contours; (b) 3D view.

10.6 THE NONLINEAR FRACTIONAL WAVE-DIFFUSION EQUATION

The nonlinear fractional wave-diffusion equation for a plane body occupying the domain Ω bounded by the curve Γ reads [19]

$$\rho \frac{\partial^\beta u}{\partial t^\beta} + \eta \frac{\partial^a u}{\partial t^\alpha} = \mathrm{N}(u_{,xx}, u_{,xy} u_{,yy}, u_{,x}, u_{,y}, u,) + f(\mathbf{x}, t) \tag{10.39}$$

where $\mathbf{x}(x, y) \in \Omega$, $t > 0$, $0 < \alpha < \beta \leq 2$ and $\rho = \rho(\mathbf{x})$, $\eta = \eta(\mathbf{x})$ are specified functions, whose physical meaning depends on the field function $u(\mathbf{x}, t)$; $f(\mathbf{x}, t)$ is the external source and $\partial^\beta u / \partial t^\beta$ and $\partial^\alpha u / \partial t^\alpha$ are the time fractional derivatives of order β and a. Apparently, the classical diffusion equation is retrieved for $\rho = 0$, $\alpha = 1$, and the classical wave equation in presence of damping for $\beta = 2$, $\alpha = 1$.

Equation (10.39) is subjected to the boundary conditions

$$u = \alpha(\mathbf{x}, t), \quad \mathbf{x} \in \Gamma_1, \tag{10.40a}$$

$$u_{,n} = \gamma(\mathbf{x}, t), \quad \mathbf{x} \in \Gamma_2 \tag{10.40b}$$

and to the initial conditions

$$u(\mathbf{x}, 0) = g_1(\mathbf{x}) \quad \text{if } \beta \leq 1 \tag{10.41a}$$

or

$$u(\mathbf{x}, 0) = g_1(\mathbf{x}), \quad \dot{u}(\mathbf{x}, 0) = g_2(\mathbf{x}) \quad \text{if } \beta > 1 \tag{10.41b}$$

where $\alpha(\mathbf{x}, t)$, $\gamma(\mathbf{x}, t)$ are functions specified on Γ_1 and Γ_2, respectively, and $g_i(\mathbf{x})$, $i = 1, 2$ functions specified in Ω.

Apparently, the AEM solution when applied as in Section (10.2) (see also [19]) leads to the following initial value problem for the evaluation of the series coefficients $\boldsymbol{\alpha}(t)$

$$\mathbf{M}\mathbf{D}_c^\beta \boldsymbol{\alpha} + \mathbf{C}\mathbf{D}_c^\alpha \boldsymbol{\alpha} + \mathbf{N}(\boldsymbol{\alpha}) = \mathbf{p}(t) \tag{10.42}$$

$$\boldsymbol{\alpha}(0) = \mathbf{S}^{-1}[\mathbf{g}_1 - \mathbf{z}(0)] \quad \text{if } \beta \leq 1 \tag{10.43a}$$

$$\boldsymbol{\alpha}(0) = \mathbf{S}^{-1}[\mathbf{g}_1 - \mathbf{z}(0)], \quad \dot{\boldsymbol{\alpha}}(0) = \mathbf{S}^{-1}[\mathbf{g}_2 - \dot{\mathbf{z}}(0)] \quad \text{if } \beta > 1 \tag{10.43b}$$

Equation (10.42) represents a multi-term nonlinear ordinary fractional differential equation with respect to time. The method presented in [20] has been used to obtain numerical results in the example below.

EXAMPLES

On the basis of the procedure presented in previous, a FORTRAN code has been written for the solution of the initial boundary value problem (10.39)–(10.41).

The employed radial basis functions ϕ_j are multiquadrics (MQs). The example that follows demonstrates the efficiency of the AEM.

EXAMPLE 10.6

In this example we consider the following initial boundary value problem for the nonlinear wave-diffusion equation.

$$\rho \frac{\partial^\beta u}{\partial t^\beta} + \eta \frac{\partial^a u}{\partial t^\alpha} = k\nabla^2 u + k_0 e(u_{,x}^2 + u_{,y}^2) + f(\mathbf{x}, t) \quad \text{in } \Omega \tag{10.44}$$

$$u = 0 \quad \text{on } \Gamma \tag{10.45}$$

$$u(\mathbf{x}, 0) = [1-(x^2+y^2)]\,T(0), \qquad \dot{u}(\mathbf{x}, 0) = [1-(x^2+y^2)]\,\dot{T}(0) \quad \text{in } \Omega \tag{10.46a, b}$$

where $f(\mathbf{x}, t)$ is the internal source, and $k = k_0(1+eu)$ with k_0 and e being material constants.

For a circular domain Ω with unit radius and $\rho = 1$, $\eta = 0.2$, $k_0 = 1$, $e = 3$, $T(t) = (1+t)^2, \alpha = 0.5, \beta = 1.5$ $\;f(t) = (2.7081\sqrt{t} + 0.3009t^{3/2})(1-x^2-y^2)+ 4(1+t)^2 + 12(1+t)^4(1-2x^2-2y^2)$the problem admits an analytic solution $u_{exact} = [1-(x^2+y^2)]\,T(t)$. The results were obtained using N = 200 constant boundary elements and L = 145 domain nodal points uniformly distributed (see Fig. 10.5) with shape parameter $c = 0.1$. Fig. 10.12 shows the time variation of the computed solution u and its derivatives $u_{,x}$, $u_{,xx}$ at point $(0.428, 0.142)$ as compared with the exact ones, while Fig. 10.13 shows the respective relative errors for $\Delta t = 0.001$. Finally, Fig. 10.14 presents the contours and the 3D view of $u(x, y)$ at $t = 0.80$.

10.7 REFERENCES

[1] Debnath L. Nonlinear partial differential equations for scientists and engineers. Boston: Birkhäuser; 1997.

[2] Nitsche JCC. Lectures on minimal surfaces, vol. 1. New York: Cambridge University Press; 1989.

[3] Green AE, Zerna W. Theoretical elasticity. London: Oxford University Press; 1968.

[4] Leonard JW. Tension structures. Behavior and analysis. New York: MacGraw Hill; 1988.

[5] Ivovich VA, Pokrovskii LN. Dynamic analysis of suspended roofs. Rotterdam: A.A. Balkema; 1991.

[6] Katsikadelis JT. Nonlinear dynamic analysis of viscoelastic membranes described with fractional differential models. J Theor Appl Mech 50th Anniversary Issue 2012;50(3):743–53.

[7] Bialecki R, Nowak AJ. Boundary value problems in heat conduction with nonlinear material and nonlinear boundary conditions. Appl Math Model 1981;5(6):417–21.

[8] Wrobel LC, Brebbia CA, editors. Boundary element methods in heat transfer. UK: Computational Mechanics Publications and Elsevier Applied Science; 1992.

[9] Partridge PW, Brebbia CA, Wrobel LC. The dual reciprocity boundary element method. Southampton: Computational Mechanics Publications; 1992.

[10] Katsikadelis JT, Nerantzaki MS. The boundary element method for nonlinear problems. Eng Anal Bound Elem 1999;23(5):365–73.

[11] Katsikadelis JT, Nerantzaki MS. A boundary element solution to the soap bubble problem. Comput Mech 2001;27(2):154–9.
[12] Katsikadelis JT, Nerantzaki MS, Tsiatas GC. The analog equation method for large deflection analysis of membranes. A boundary only solution. Comput Mech 2001;27(6):513–23.
[13] Katsikadelis JT. Dynamic analysis of nonlinear membranes by the analog equation method. A boundary-only solution. Comput Mech 2002;29(2):170–7.
[14] Tsiatas CG, Katsikadelis JT. Large deflection analysis of elastic space membranes. Int J Numer Methods Eng 2006;65(2):264–94.
[15] Katsikadelis JT. A new direct time integration method for the equations of motion in structural dynamics. ZAMM 2014;94(9):757–74. Available from: http://dx.doi.org/10.1002/zamm.201200245.
[16] Katsikadelis JT. 2015. A new direct time integration method for the semi-discrete parabolic equation, Proc. 8th GRACM International Congress on Computational Mechanics, Volos, 12 July.
[17] Rado T. On the problem of Plateau. Berlin: Springer-Verlag; 1932.
[18] Katsikadelis JT. The boundary element method for plate analysis. Oxford, UK: Academic Press, Elsevier; 2014.
[19] Katsikadelis JT. The BEM for numerical solution of partial fractional differential equations. Comput Math Appl 2011;62:891–901.
[20] Katsikadelis JT. Numerical solution of multi-term fractional differential equations. ZAMM 2009;89(7):593–608.

PROBLEMS

10.1. The surface $u = u(x, y)$ bounded by a space curve $z = \overline{u}(x, y)$ of plan form Γ and having constant mean curvature κ is obtained from the solution of the following boundary value problem

$$(1+u,_y^2)u,_{xx} - 2u,_x u,_y u,_{xy} + (1+u,_x^2)u,_{yy} - \kappa(1+u,_x^2+u,_y^2) = 0 \quad \text{in } \Omega \tag{a}$$

$$u = \overline{u} \quad \text{on } \Gamma \tag{b}$$

Solve the problem when the domain Ω is the square $0 \le x, y \le 5$ under the boundary conditions:

$$u(0, y) = (50 - y^2)^{1/2}, \quad u(5, y) = (25 - y^2)^{1/2} \tag{c, d}$$

$$u(x, 0) = (50 - x^2)^{1/2}, \quad u(x, 5) = (25 - x^2)^{1/2} \tag{e, f}$$

and mean curvature $\kappa = -\sqrt{2}/5$.

Show that the problem admits an exact solution $u_{exact} = [50 - (x^2 + y^2)]^{1/2}$ and compare the computed results.

10.2. Determine the minimal surface which is supported on the circumference of the two concentric circles $R = 5,\ z = 0$ and $R = 4,\ z = 3$.

10.3. Determine the surface of the soap bubble supported on the space curve, which is the intersection of the cylindrical surface $x^2/2.25 + y^2/1 = 1$, $z \ge 0$ and the sphere $x^2 + y^2 + z^2 = 25$.

10.4. Determine the solution of the boundary value for the Burger equation

$$\nabla^2 u + u_{,x} u = 0 \quad \text{in } \Omega \tag{a}$$

$$u(0, y) = 2, \; u(1, y) = 3, \quad u(x, 0) = u(x, 1) = 2 + x \quad \text{on } \Gamma \tag{b}$$

The domain Ω is the square $0 \leq x, y \leq 1$.

10.5. Determine the solution of the initial boundary value problem

$$10\dot{u} = (a^2 - x^2)u_{,xx} + (b^2 - y^2)u_{,yy} + 4u \quad \text{in } \Omega \tag{a}$$

$$u = 0 \quad \text{on } \Gamma \tag{b}$$

$$u(\mathbf{x}, 0) = (a^2 - x^2)(b^2 - y^2) \tag{c}$$

The domain Ω is the rectangle $(-a \leq x \leq a) \times (-b \leq y \leq b)$. The problem admits an exact solution $u(\mathbf{x}, t) = (a^2 - x^2)(b^2 - y^2)\exp(-0.4t)$.

10.6. Using the procedure described in Section 10.5.1 write a computer code and analyze the static response of the square membrane with side length a, Ω:$\{-a/2 \leq x, y \leq a/2\}$, under a uniform load f. The prestress is imposed by boundary displacements, whose distribution is shown in Fig. P10.6. Determine the deflection surface $w(x, y)$, the membrane forces N_x, N_y, N_{xy} and the boundary reactions $N_{\text{n}}, N_{nt}, V_n$ using the following data: $a = 5.0\text{m}$; $h = 0.002\text{m}$, $f = 10\ \text{kN/m}^2$, $E = 1.1 \times 10^5\ \text{kN/m}^2$, $\nu = 0.3$; $\overline{u} = \overline{v} = 0.10\text{m}$, and $\overline{w} = 0$.

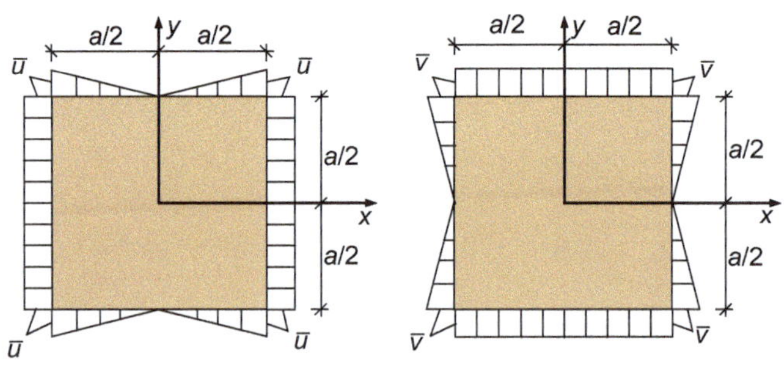

FIGURE P10.6

10.7. Using the procedure described in Section 10.5.1 write a computer code and analyze the dynamic response of the square membrane of Fig. P10.7. The membrane is prestressed by imposed displacements $u_n^* = 0.10\text{m}$ in the direction normal to the boundary, while $u_t^* = 0$ in the tangential direction. Data: $a = 10.0$ m, $h = 0.002$ m $E = 1.1 \times 10^5$ kN/m^2 and $\nu = 0.3$, $f = 10$ kN/m^2, $R = a/5$, $\rho = 5000$ kg/m^3.

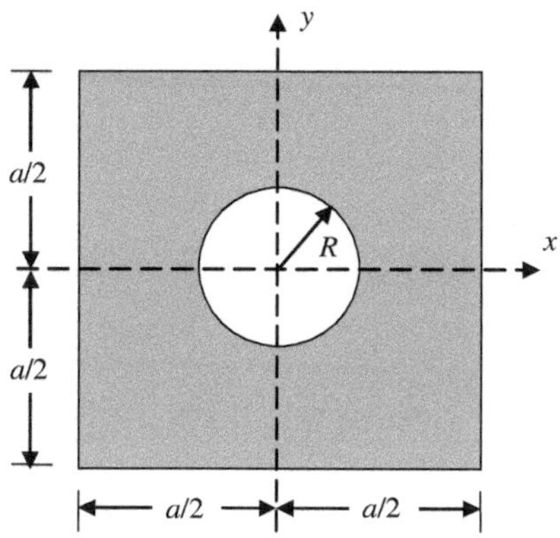

FIGURE P10.7 Square membrane with a hole.

10.8. Derive the differential equations governing the dynamic response of an isotropic viscoelastic membrane with constitutive equations described by the three parameter (solid) fractional differential model

$$\begin{Bmatrix} \sigma_x \\ \sigma_y \\ \tau_{xy} \end{Bmatrix} + \frac{\eta b}{1+b} \begin{Bmatrix} {}_c^{}D_t^a \sigma_x \\ {}_c^{}D_t^a \sigma_y \\ {}_c^{}D_t^a \tau_{xy} \end{Bmatrix} = \frac{b}{1+b} \frac{E}{1-\nu^2} \begin{bmatrix} 1 & \nu & 0 \\ \nu & 1 & 0 \\ 0 & 0 & \dfrac{1-\nu}{2} \end{bmatrix} \begin{Bmatrix} \varepsilon_x + \eta\, {}_c^{}D_t^a \varepsilon_x \\ \varepsilon_y + \eta\, {}_c^{}D_t^a \varepsilon_y \\ \gamma_{xy} + \eta\, {}_c^{}D_t^a \gamma_{xy} \end{Bmatrix}$$

and develop the AEM procedure for its numerical analysis (see [18]).

Appendix A: Derivatives of *r*

A.1 DERIVATIVES OF *r*

In this Appendix, we present certain relations which facilitate the differentiation of the kernels of the integral equations. Points inside the domain Ω are denoted by upper case letters, for example, $P(x, y)$, while points on the boundary by lower case letters, for example, $q(\xi, \eta)$. The angle between the x-axis and the vector r is denoted by α and the angle between the x-axis and the unit vector $\mathbf{n}$ normal to the boundary at point q by β (see Fig. A.1). Using these two angles, we also define the angle ϕ as

$$\phi = \text{angle}(\mathbf{r}, \mathbf{n}) = \beta - \alpha \tag{A.1}$$

Referring to Fig. A.1, it is

$$\cos\alpha = \frac{\xi - x}{r} \tag{A.2}$$

$$\sin\alpha = \frac{\eta - y}{r} \tag{A.3}$$

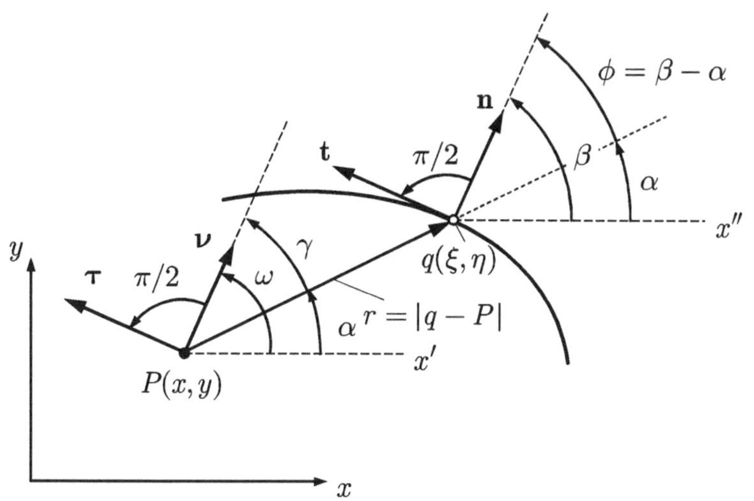

FIGURE A.1 Geometric definitions related to the relative position of a field point *P* and a boundary point *q*.

where

$$r=\sqrt{(\xi-x)^2+(\eta-y)^2} \tag{A.4}$$

Differentiation of Eq. (A.4) yields

$$r,_x=-r,_\xi=-\frac{\xi-x}{r}=-\cos\alpha \tag{A.5}$$

$$r,_y=-r,_\eta=-\frac{\eta-y}{r}=-\sin\alpha \tag{A.6}$$

where the subscripts x, y, ξ, and η preceded by a comma denote differentiation with respect to the corresponding independent variable.

Noting that

$$\begin{aligned}\cos\beta&=n_x\\ \sin\beta&=n_y\end{aligned}$$

we can derive the following expressions for the derivatives of r

$$\begin{aligned} r,_n&=r,_\xi\, n_x+r,_\eta\, n_y\\ &=r,_\xi\cos\beta+r,_\eta\sin\beta\\ &=\cos\alpha\cos\beta+\sin\alpha\sin\beta\\ &=\cos(\beta-\alpha)\\ &=\cos\phi\end{aligned} \tag{A.7}$$

$$\begin{aligned} r,_t&=-r,_\xi\, n_y+r,_\eta\, n_x\\ &=-r,_\xi\sin\beta+r,_\eta\cos\beta\\ &=-\cos\alpha\sin\beta+\sin\alpha\cos\beta\\ &=-\sin(\beta-\alpha)\\ &=-\sin\phi\end{aligned} \tag{A.8}$$

$$\begin{aligned} r,_{xx}&=-\left(\frac{\xi-x}{r}\right),_x\\ &=-\frac{(\xi-x),_x\, r-(\xi-x)\, r,_x}{r^2}\\ &=\frac{(\eta-y)^2}{r^3}\\ &=\frac{r,_y^2}{r}\end{aligned} \tag{A.9}$$

Similarly, we obtain

$$r,_{yy} = \frac{r,_x^2}{r} \tag{A.10}$$

$$r,_{xy} = -\frac{r,_x r,_y}{r} \tag{A.11}$$

$$r,_{\xi\xi} = \frac{r,_\eta^2}{r} \tag{A.12}$$

$$r,_{\eta\eta} = \frac{r,_\xi^2}{r} \tag{A.13}$$

$$r,_{\xi\eta} = -\frac{r,_\xi r,_\eta}{r} \tag{A.14}$$

$$r,_{\xi x} = -r,_{xx} = -\frac{r,_y^2}{r} \tag{A.15}$$

$$r,_{\xi y} = -r,_{xy} = \frac{r,_x r,_y}{r} \tag{A.16}$$

$$r,_{\eta x} = -r,_{yx} = \frac{r,_x r,_y}{r} \tag{A.17}$$

$$r,_{\eta y} = -r,_{yy} = -\frac{r,_x^2}{r} \tag{A.18}$$

$$\begin{aligned} r,_{nx} &= (r,_\xi \cos\beta + r,_\eta \sin\beta),_x \\ &= r,_{\xi x} \cos\beta + r,_{\eta x} \sin\beta \\ &= \frac{r,_y}{r}(r,_\eta \cos\beta - r,_\xi \sin\beta) \\ &= \frac{r,_t r,_y}{r} \end{aligned} \tag{A.19}$$

$$\begin{aligned} r,_{ny} &= (r,_\xi \cos\beta + r,_\eta \sin\beta),_y \\ &= r,_{\xi y} \cos\beta + r,_{\eta y} \sin\beta \\ &= \frac{r,_x}{r}(-r,_\eta \cos\beta + r,_\xi \sin\beta) \\ &= -\frac{r,_t r,_x}{r} \end{aligned} \tag{A.20}$$

$$\begin{aligned} r,_{tx} &= (-r,_\xi \sin\beta + r,_\eta \cos\beta),_x \\ &= -r,_{\xi x}\sin\beta + r,_{\eta x}\cos\beta \\ &= -\frac{r,_y}{r}(r,_\xi \cos\beta + r,_\eta \sin\beta) \\ &= -\frac{r,_y r,_n}{r} \end{aligned} \tag{A.21}$$

$$\begin{aligned} r,_{ty} &= (-r,_\xi \sin\beta + r,_\eta \cos\beta),_y \\ &= -r,_{\xi\eta}\sin\beta + r,_{\eta y}\cos\beta \\ &= \frac{r,_x}{r}(r,_\xi \cos\beta + r,_\eta \sin\beta) \\ &= \frac{r,_x r,_n}{r} \end{aligned} \tag{A.22}$$

$$\begin{aligned} r,_{nt} &= (r,_{t\xi}\cos\beta + r,_{t\eta}\sin\beta) \\ &= \frac{r,_n}{r}(r,_\eta \cos\beta - r,_\xi \sin\beta) \\ &= \frac{r,_n r,_t}{r} \end{aligned} \tag{A.23}$$

$$\begin{aligned} r,_\nu &= r,_x \cos\omega + r,_y \sin\omega \\ &= -(\cos\alpha\cos\omega + \sin\alpha\sin\omega) \\ &= \cos(\omega-\alpha) \end{aligned} \tag{A.24}$$

$$\begin{aligned} r,_\tau &= (-r,_x \sin\omega + r,_y \cos\omega) \\ &= -(-\cos\alpha\sin\omega + \sin\alpha\cos\omega) \\ &= \sin(\omega-\alpha) \end{aligned} \tag{A.25}$$

$$\begin{aligned} r,_{\nu x} &= (r,_x \cos\omega + r,_y \sin\omega),_x \\ &= r,_{xx}\cos\omega + r,_{yx}\sin\omega \\ &= \frac{r,_y}{r}(r,_y \cos\omega - r,_x \sin\omega) \\ &= \frac{r,_y r,_\tau}{r} \end{aligned} \tag{A.26}$$

$$\begin{aligned} r,_{\nu y} &= (r,_x \cos\omega + r,_y \sin\omega),_y \\ &= r,_{xy}\cos\omega + r,_{yy}\sin\omega \\ &= \frac{r,_x}{r}(-r,_y \cos\omega + r,_x \sin\omega) \\ &= -\frac{r,_x r,_\tau}{r} \end{aligned} \tag{A.27}$$

$$\begin{aligned} r,_{\tau x} &= (-r,_x \sin\omega + r,_y \cos\omega),_x \\ &= -r,_{xx} \sin\omega + r,_{yx} \cos\omega \\ &= -\frac{r,_y}{r}(r,_y \sin\omega + r,_x \cos\omega) \\ &= \frac{r,_y r,_\nu}{r} \end{aligned} \tag{A.28}$$

$$\begin{aligned} r,_{\tau y} &= (-r,_x \sin\omega + r,_y \cos\omega),_y \\ &= -r,_{xy} \sin\omega + r,_{yy} \cos\omega \\ &= \frac{r,_x}{r}(r,_y \sin\omega + r,_x \cos\omega) \\ &= \frac{r,_x r,_\nu}{r} \end{aligned} \tag{A.29}$$

$$\begin{aligned} r,_{\nu\nu} &= (r,_x \cos\omega + r,_y \sin\omega),_\nu \\ &= r,_{\nu x} \cos\omega + r,_{\nu y} \sin\omega \\ &= \left(\frac{r,_y r,_\tau}{r} \cos\omega - \frac{r,_x r,_\tau}{r} \sin\omega\right) \\ &= \frac{r,_\tau^2}{r} \end{aligned} \tag{A.30}$$

$$\begin{aligned} r,_{\tau\tau} &= (-r,_x \sin\omega + r,_y \cos\omega),_\tau \\ &= (-r,_{\tau x} \sin\omega + r,_{\tau y} \cos\omega) \\ &= \left(\frac{r,_y r,_\nu}{r} \sin\omega + \frac{r,_x r,_\nu}{r} \cos\omega\right) \\ &= \frac{r,_\nu^2}{r} \end{aligned} \tag{A.31}$$

Appendix B: Gauss Integration

B.1 GAUSS INTEGRATION OF A REGULAR FUNCTION

The success of the boundary element method as a computational method is contingent upon accurately evaluating line and domain integrals. Any numerical integration method can be utilized, for example, trapezoidal rule, Simpson rule, Newton–Cotes rule, etc. The numerical methods approximate an integral with a sum of products of the values of the integrand at specific points of the integration interval, usually equidistant, multiplied by known coefficients, the *weight factors* (or *weights*), resulting from the employed integration rule. Hence, an essential criterion for choosing the integration rule is the achievement of the desired degree of accuracy in the approximation of the integral by utilizing the lowest possible number of integration points. The Gauss integration method satisfies this criterion. In this method, the points are not equidistant but they arrange themselves in the optimal pattern. The weight factors are also adjusted in this optimization process. The concept behind the Gauss integration is quite simple, and will be explained in the sequel.

Let us consider the integral

$$I = \int_{-1}^{1} f(\xi)\, d\xi \tag{B.1}$$

In the Gauss integration, the integral is approximated by the sum

$$I \approx I_n = \sum_{k=1}^{n} f(\xi_k)\, w_k \tag{B.2}$$

where $f(\xi_k)$ $(k = 1, 2, \ldots, n)$ are the values of the function $f(\xi)$ at the n points ξ_k $(-1 < \xi_k < +1)$ and w_k the associated *weights*. The points ξ_k, which are called *Gauss integration points* or simply *Gauss points*, are not equidistant within the integration interval, but, for a given n, their location and the associated weights are chosen so that the error $E_n = I - I_n$ becomes minimum.

We first study the trivial case where the function $f(\xi)$ is a linear polynomial

$$f(\xi) = a_0 + a_1\,\xi \tag{B.3}$$

The exact value of the integral is

$$\begin{aligned} I &= \int_{-1}^{+1} (a_0 + a_1\xi)\, d\xi \\ &= \left[a_0\,\xi + a_1 \frac{\xi^2}{2} \right]_{-1}^{+1} \\ &= 2a_0 \end{aligned}$$

while the approximate value obtained from Eq. (B.2) for $n = 1$ is

$$I_1 = (a_0 + a_1\xi_1)\, w_1$$

Therefore, the error is

$$E_1(a_0, a_1) = I - I_1 = 2a_0 - (a_0 + a_1\xi_1)\, w_1$$

and is minimized for

$$\frac{\partial E_1}{\partial a_0} = 2 - w_1 = 0 \tag{B.4a}$$

$$\frac{\partial E_1}{\partial a_1} = -w_1\,\xi_1 = 0 \tag{B.4b}$$

which yield

$$\xi_1 = 0, \;\; w_1 = 2$$

and $E_1(a_0, a_1) = 0$.

We notice that Eq. (B.2) gives

$$\begin{aligned} \int_{-1}^{1} f(\xi)\, d\xi &= 2f_0 \\ &= 2a_0 \end{aligned} \tag{B.5}$$

where $f_0 = f(0)$.

Namely, the value of the integral is independent of the slope a_1 of the straight line representing the function. Hence, the definite integrals on the interval $[-1, +1]$ of all the linear functions passing through the point $(0, f_0)$ have the same value (see Fig. B.1).

Consider now the case of a function $f(\xi)$ which is a quadratic polynomial. The exact value of the integral is

$$\begin{aligned} I &= \int_{-1}^{1} (a_0 + a_1\,\xi + a_2\,\xi^2)\,d\xi \\ &= \left[a_0\,\xi + a_1 \frac{\xi^2}{2} + a_2 \frac{\xi^3}{3} \right]_{-1}^{+1} \\ &= 2a_0 + \frac{2}{3} a_2 \end{aligned}$$

while the approximate one for two Gauss points ($n = 2$) is

$$\begin{aligned} I_2 &= f(\xi_1) w_1 + f(\xi_2) w_2 \\ &= (a_0 + a_1\,\xi_1 + a_2\,\xi_1^2) w_1 + (a_0 + a_1\,\xi_2 + a_2\,\xi_2^2) w_2 \end{aligned}$$

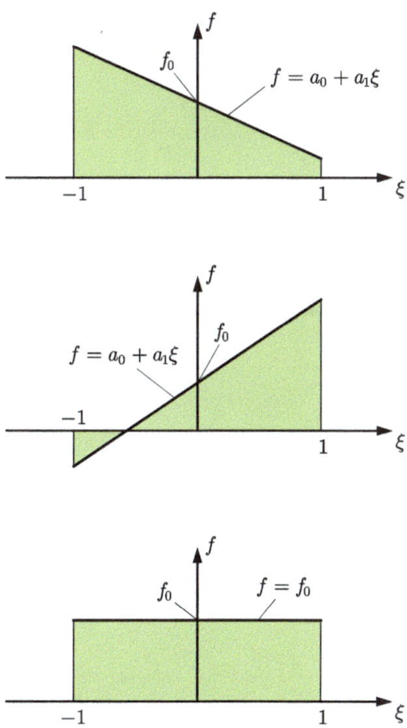

FIGURE B.1 Typical linear polynomials.

The resulting error is

$$
\begin{aligned}
E_2(a_0, a_1, a_2) &= I - I_2 \\
&= 2a_0 + \frac{2}{3}a_2 - (a_0 + a_1\,\xi_1 + a_2\,\xi_1^2)w_1 \\
&= -(a_0 + a_1\,\xi_2 + a_2\,\xi_2^2)w_2
\end{aligned}
\tag{B.6}
$$

which is minimized under the following conditions

$$\frac{\partial E_2}{\partial a_0} = 2 - w_1 - w_2 = 0 \tag{B.7}$$

$$\frac{\partial E_2}{\partial a_1} = -\xi_1\,w_1 - \xi_2\,w_2 = 0 \tag{B.8}$$

$$\frac{\partial E_2}{\partial a_2} = \frac{2}{3} - \xi_1^2\,w_1 - \xi_2^2\,w_2 = 0 \tag{B.9}$$

The above three simultaneous equations include four unknowns, w_1, w_2, ξ_1, and ξ_2. Therefore, one of them can be chosen arbitrarily, and since it is convenient to place the points ξ_1 and ξ_2 symmetrically with respect to the origin, we set

$$\xi_2 = -\xi_1 \tag{B.10}$$

From Eq. (B.8), we obtain $w_1 = w_2$ and from Eq. (B.7) we find

$$w_1 = w_2 = 1 \tag{B.11}$$

Finally, Eq. (B.9) gives

$$\xi_1 = \frac{1}{\sqrt{3}} = 0.577350269189626 \tag{B.12}$$

Introducing the above values of ξ_1, ξ_2, w_1, and w_2 into Eq. (B.6), the error becomes $E_2(a_0, a_1, a_2) = 0$. The approximate value of the integral is computed from the sum

$$I_2 = f(\xi_1)\,w_1 + f(\xi_2)\,w_2 \tag{B.13}$$

where ξ_1 and ξ_2 are given by Eqs. (B.10) and (B.12) and w_1, w_2 by Eq. (B.11).

We consider now the case where the function $f(\xi)$ is a cubic polynomial and we apply the same procedure. The exact value is

$$
\begin{aligned}
I_3 &= \int_{-1}^{1} (a_0 + a_1\,\xi + a_2\,\xi^2 + a_3\,\xi^3)\,d\xi \\
&= 2\,a_0 + \frac{2}{3}\,a_2
\end{aligned}
$$

Using again two Gauss points ($n = 2$), the expression for the error becomes

$$
\begin{aligned}
E_3(a_0, a_1, a_2, a_3) = \left(2\,a_0 + \frac{2}{3}\,a_2\right) &- (a_0 + a_1\,\xi_1 + a_2\,\xi_1^2 + a_3\,\xi_1^3)w_1 \\
&- (a_0 + a_1\,\xi_2 + a_2\,\xi_2^2 + a_3\,\xi_2^3)w_2
\end{aligned}
$$

which is minimized if the following conditions are satisfied

$$\frac{\partial E_3}{\partial a_0} = 2 - w_1 - w_2 = 0 \tag{B.14a}$$

$$\frac{\partial E_3}{\partial a_1} = -\xi_1\,w_1 - \xi_2\,w_2 = 0 \tag{B.14b}$$

$$\frac{\partial E_3}{\partial a_2} = \frac{2}{3} - \xi_1^2\,w_1 - \xi_2^2\,w_2 = 0 \tag{B.14c}$$

$$\frac{\partial E_3}{\partial a_3} = -\xi_1^3\,w_1 - \xi_2^3\,w_2 = 0 \tag{B.14d}$$

The solution $w_1 = w_2 = 1$ and $\xi_1 = -\xi_2 = 1/\sqrt{3}$, obtained when $f(\xi)$ is a second-order polynomial, is also a solution of the simultaneous equations (B.14), and yields $E_3(a_0, a_1, a_2, a_3) = 0$. Hence, Eq. (B.2) renders exact values for the integrals of the second- and third-order polynomials using the same Gauss integration rule, that is, the same integration points and weights.

This procedure can also be applied for higher order polynomials. In general, it can be proved that the formula given by Eq. (B.2) with n terms gives the exact value for the integral of a polynomial $f(\xi)$ of order less than or equal to $2n - 1$. However, this procedure can hardly be employed to determine the coordinates of the Gauss points and the weight factors, because it becomes complicated and laborious for the increasing degree of the polynomial. The Gauss points can be obtained by a simpler method, which uses Legendre's orthogonal polynomials [1,2] to represent the function $f(\xi)$ and that is why it is known as *Gauss–Legendre integration.*

Legendre's polynomials are defined as

$$P_n(\xi) = \frac{1}{2^n n!} \frac{d^n}{d\xi^n} \left(\xi^2 - 1\right)^n \tag{B.15}$$

For example, the first-, second-, and third-order Legendre's polynomials are

$$\text{for} \quad n = 1: \quad P_1(\xi) = \xi \tag{B.16a}$$

$$\text{for} \quad n = 2: \quad P_2(\xi) = \frac{1}{2}(3\xi^2 - 1) \tag{B.16b}$$

$$\text{for} \quad n = 3: \quad P_3(\xi) = \frac{1}{2}(5\xi^3 - 3\xi) \tag{B.16c}$$

It is proven that the coordinates ξ_k of the integration points are the zeros of these polynomials. The weight factors w_k are computed by the expression

$$w_k = \frac{2(1 - \xi_k^2)}{n^2 \left[P_{n-1}(\xi_k)\right]^2} \tag{B.17}$$

Table B.1 provides the coordinates of the integration points and the weights for various values of n.

If $f(\xi)$ is not a polynomial, the integral can be evaluated approximately as

$$\int_{-1}^{1} f(\xi)\, d\xi \approx \sum_{k=1}^{n} f(\xi_k)\, w_k \tag{B.18}$$

where the function in the integrand is actually approximated by a polynomial of degree $2n - 1$. The error associated with the Gauss–Legendre integration is given by Lanczos in the following form [1]

$$E_n = I - I_n = \frac{1}{2n + 1} \left[f(1) + f(-1) - I_n - \sum_{k=1}^{n} w_k\, \xi_k\, f'(\xi_k) \right] \tag{B.19}$$

in which n is the number of integration points. This estimation is very good for smooth functions $f(\xi)$.

Usually, the interval over which a function is integrated differs from the interval $[-1, +1]$. Suppose that we have to evaluate the integral

$$I = \int_{a}^{b} f(x)\, dx \tag{B.20}$$

TABLE B.1 Abscissas and Weights for the Gauss–Legendre Integration

n	$I = \int_{-1}^{1} f(\xi)\, d\xi \approx \sum_{k=1}^{n} f(\xi_k)\, w_k$ $\pm \xi_k$	w_k
2	0.577350269189626	1.000000000000000
3	0.000000000000000	0.888888888888889
	0.774596669241483	0.555555555555556
4	0.339981043584856	0.652145154862546
	0.861136311594053	0.347854845137454
5	0.000000000000000	0.568888888888889
	0.538469310105683	0.478628670499366
	0.906179845938664	0.236926885056189
6	0.238619186083197	0.467913934572691
	0.661209386466265	0.360761573048139
	0.932469514203152	0.171324492379170
7	0.000000000000000	0.417959183673469
	0.405845151377397	0.381830050505119
	0.741531185599394	0.279705391489277
	0.949107912342759	0.129484966168870
8	0.183434642495650	0.362683783378362
	0.525532409916329	0.313706645877887
	0.796666477413627	0.222381034453374
	0.960289856497536	0.101228536290376
9	0.000000000000000	0.330239355001260
	0.324253423403809	0.312347077040003
	0.613371432700590	0.260610696402935
	0.836031107326636	0.180648160694857
	0.968160239507626	0.081274388361574
10	0.148874338981631	0.295524224714753
	0.433395394129247	0.269266719309996
	0.679409568299024	0.219086362515982
	0.865063366688985	0.149451349150581
	0.973906528517172	0.066671344308688
12	0.125233408511469	0.249147045813403
	0.367831498998180	0.233492536538355
	0.587317954286617	0.203167426723066
	0.769902674194305	0.160078328543346
	0.904117256370475	0.106939325995318
	0.981560634246719	0.047175336386512

Continued

TABLE B.1 continued

n	$I = \int_{-1}^{1} f(\xi)\, d\xi \approx \sum_{k=1}^{n} f(\xi_k)\, w_k$ $\pm \xi_k$	w_k
16	0.095012509837637	0.189450610455068
	0.281603550779258	0.182603415044923
	0.458016777657227	0.169156519395002
	0.617876244402643	0.149595988816576
	0.755404408355003	0.124628971255533
	0.865631202387831	0.095158511682492
	0.944575023073232	0.062253523938647
	0.989400934991649	0.027152459411754

Using the transformation

$$x = \frac{b-a}{2}\,\xi + \frac{b+a}{2} \tag{B.21}$$

the interval of Eq. (B.20) is mapped onto the interval $[-1, +1]$. Thus,

$$I = \int_a^b f(x)\, dx = \frac{b-a}{2}\int_{-1}^{+1} f(\xi)\, d\xi \tag{B.22}$$

EXAMPLE B.1

Determine the value of the integral

$$I = \int_1^3 \frac{\sin^2 x}{x}\, dx$$

The transformation $x = \xi + 2$ yields the expression of the above integral in the interval $[-1, +1]$, which is of the following form

$$I = \int_{-1}^{+1} \frac{\sin^2(\xi + 2)}{\xi + 2}\, d\xi$$

The integral is computed using Gaussian integration. Results for various values of n are presented in Table B.2. We notice that a five-point Gaussian quadrature is sufficient to approximate the value of the integral to seven significant figures.

TABLE B.2 Computed Values of the Integral $I = \int_1^3 \frac{\sin^2 x}{x}\, dx$ for Various Numbers of Gauss Points, n

n	ξ_k	w_k	$\int_1^3 \frac{\sin^2 x}{x}\, dx$
1	$\xi_1 = 0$	$w_1 = 2.0000000000$	0.82682181
2	$\xi_1 = -0.5773502692$	$w_1 = 1.0000000000$	0.79856002
	$\xi_2 = -\xi_1$	$w_2 = w_1$	
3	$\xi_1 = -0.7745966692$	$w_1 = 0.5555555556$	0.79465269
	$\xi_2 = 0$	$w_2 = 0.8888888889$	
	$\xi_3 = -\xi_1$	$w_3 = w_1$	
4	$\xi_1 = -0.8611363116$	$w_1 = 0.3478548451$	0.79482835
	$\xi_2 = -0.3399810436$	$w_2 = 0.6521451549$	
	$\xi_3 = -\xi_2$	$w_3 = w_2$	
	$\xi_4 = -\xi_1$	$w_4 = w_1$	
5	$\xi_1 = -0.9061798459$	$w_1 = 0.2369268851$	0.79482516
	$\xi_2 = -0.5384693101$	$w_2 = 0.4786286705$	
	$\xi_3 = 0$	$w_3 = 0.5688888889$	
	$\xi_4 = -\xi_2$	$w_4 = w_2$	
	$\xi_5 = -\xi_1$	$w_5 = w_1$	
Exact value			0.79482518

B.2 INTEGRALS WITH A LOGARITHMIC SINGULARITY

The Gaussian integration gives good results when a function $f(\xi)$ varies smoothly within the integration interval. If the integrand is singular the Gauss–Legendre integration cannot be employed and for this purpose special integration rules have been developed which depend on the type of singularity. Integrals whose integrand includes a logarithmic singularity are approximated by the formula [3]

$$\int_0^1 f(\xi) \ln\xi \, d\xi \approx -\sum_{k=1}^{n} f(\xi_k)\, w_k \tag{B.23}$$

The integration points ξ_k and the weights w_k are given in Table B.3.

B.3 DOUBLE INTEGRALS OF A REGULAR FUNCTION

The Gaussian integration can also be employed to evaluate double integrals. In this case the integration rule depends on the geometry of the domain over which the integration is performed. Various Gauss integration rules have

TABLE B.3 Abscissas and Weights for the Gaussian Quadrature of a Function With Logarithmic Singularity

	$\int_0^1 f(\xi)\ln\xi \, d\xi \approx -\sum_{k=1}^{n} f(\xi_k)\, w_k$	
n	ξ_k	w_k
2	0.112008806166976	0.718539319030384
	0.602276908118738	0.281460680969615
3	0.0638907930873254	0.513404552232363
	0.368997063715618	0.391980041201487
	0.766880303938941	0.0946154065661491
4	0.0414484801993832	0.383464068145135
	0.245474914320602	0.386875317774762
	0.556165453560278	0.190435126950142
	0.848982394532985	0.0392254871299598
5	0.0291344721519720	0.297893471782894
	0.173977213320897	0.349776226513224
	0.411702520284902	0.234488290044052
	0.677314174582820	0.0989304595166331
	0.894771361031008	0.0189115521431957
6	0.0216340058441169	0.238763662578547
	0.129583391154950	0.308286573273946
	0.314020449914765	0.245317426563210
	0.538657217351802	0.142008756566476
	0.756915337377402	0.0554546223248862
	0.922668851372120	0.0101689586929322
7	0.0167193554082581	0.196169389425248
	0.100185677915675	0.270302644247272
	0.246294246207930	0.239681873007690
	0.433463493257033	0.165775774810432
	0.632350988047766	0.0889432271376579
	0.811118626740105	0.0331943043565710
	0.940848166743347	0.0593278701512592
8	0.0133202441608924	0.164416604728002
	0.0797504290138949	0.237525610023306
	0.197871029326188	0.226841984431919
	0.354153994351909	0.175754079006070
	0.529458575234917	0.112924030246759
	0.701814529939099	0.0578722107177820
	0.849379320441106	0.0209790737421329
	0.953326450056359	0.00368640710402761

been developed for domains of specific geometry, such as rectangles, triangles, circles as well as domains that can be transformed into one of those geometries. Efforts have been made to develop Gauss integration rules for domains of arbitrary geometry, for example, the method of finite sectors [4]. In the sequel, we will limit our discussion to Gaussian quadratures for rectangular and triangular domains, which are the common geometries for two-dimensional discretizations.

B.3.1 Gauss integration for rectangular domains

For domains resulting as Cartesian products of lower dimensions, for example, squares, cubes, cylinders, etc., integration rules can always be formulated by *multiplying* rules of lower dimensions [2,5]. For example, if

$$\int_{-1}^{+1} f(\xi)\, d\xi \approx \sum_{k=1}^{n} f(\xi_k)\, w_k \tag{B.24}$$

is a one-dimensional rule, then

$$\int_{-1}^{+1}\int_{-1}^{+1} f(\xi,\eta)\, d\xi\, d\eta \approx \sum_{j=1}^{n}\sum_{i=1}^{n} f(\xi_j,\eta_i)\, w_i\, w_j \tag{B.25}$$

is the corresponding two-dimensional rule. These rules, however, are not necessarily on the side of economy, with regard to the number of integration points. For square domains Ω: $|x| < h$, $|y| < h$, the following simple rule can be employed [2]

$$\frac{1}{4h^2}\iint_{\Omega} f(x,y)\, dx\, dy \approx \sum_{k=1}^{n} f(x_k, y_k)\, w_k \tag{B.26}$$

The coordinates of the Gauss points x_k and y_k and the associated weight factors are given in Table B.4. The order of the error is $E = O(h^4)$ for the first two cases and $E = O(h^6)$ for the third one.

B.3.2 Gauss integration for triangular domains

The integration over a triangular domain is simplified by employing a natural coordinate system, which here is referred to as *triangular coordinate system* (see Fig. B.2). The sides a_1, a_2, a_3 are designated by the number of the opposite vertices. The triangular coordinates ξ_i $(i = 1,\ 2,\ 3)$ for an interior point P are defined as the ratios of the areas A_i to the total area A of the triangle,

$$\xi_1 = \frac{A_1}{A},\ \xi_2 = \frac{A_2}{A},\ \xi_3 = \frac{A_3}{A} \tag{B.27}$$

TABLE B.4 Coordinates and Weights for Gauss Integration on Square Domains of Side $2h$

Number and Position of the Gauss Points	x_k	y_k	w_k
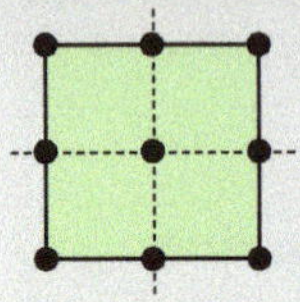	0	0	$4/9$
	$\pm h$	$\pm h$	$1/36$
	$\pm h$	0	$1/9$
$n = 9$	0	$\pm h$	$1/9$
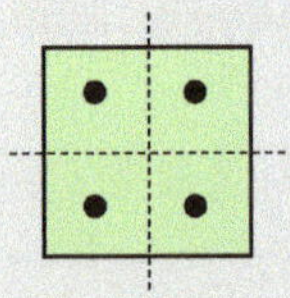 $n = 4$	$\pm h\frac{\sqrt{3}}{3}$	$\pm h\frac{\sqrt{3}}{3}$	$1/4$
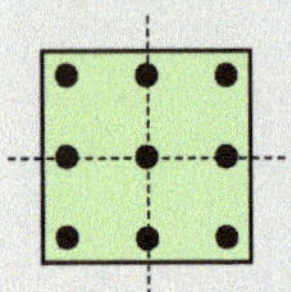	0	0	$16/81$
	$\pm h\sqrt{\frac{3}{5}}$	$\pm h\sqrt{\frac{3}{5}}$	$25/324$
	0	$\pm h\sqrt{\frac{3}{5}}$	$10/81$
$n = 9$	$\pm h\sqrt{\frac{3}{5}}$	0	$10/81$

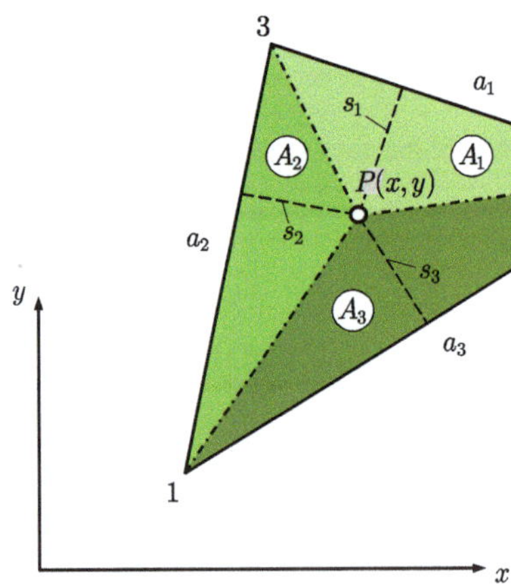

FIGURE B.2 Triangular coordinate systems.

Since the sum of the three areas is

$$A_1 + A_2 + A_3 = A$$

it is evident that

$$\xi_1 + \xi_2 + \xi_3 = 1 \tag{B.28}$$

The Cartesian coordinates x and y of point P are related to the triangular coordinates by

$$x = \xi_1 x_1 + \xi_2 x_2 + \xi_3 x_3 \tag{B.29a}$$

$$y = \xi_1 y_1 + \xi_2 y_2 + \xi_3 y_3 \tag{B.29b}$$

where x_i, y_i $(i = 1, 2, 3)$ are the coordinates of the triangle's vertices. Equations (B.29a,b) can be checked at a few specific points. For example, at the centroid of the triangle it is:

$$A_1 = A_2 = A_3 = \frac{A}{3},$$

$$x = \frac{x_1 + x_2 + x_3}{3}, \text{ and } y = \frac{y_1 + y_2 + y_3}{3}$$

while at point 2 (see Fig. B.2): $A_2 = A$, $A_1 = A_3 = 0$, $x = x_2$, and $y = y_2$.

Equations (B.28) and (B.29) yield the relation between rectangular and orthogonal coordinates which can be written in matrix form as

$$\begin{Bmatrix} 1 \\ x \\ y \end{Bmatrix} = \begin{bmatrix} 1 & 1 & 1 \\ x_1 & x_2 & x_3 \\ y_1 & y_2 & y_3 \end{bmatrix} \begin{Bmatrix} \xi_1 \\ \xi_2 \\ \xi_3 \end{Bmatrix} \tag{B.30}$$

The integration over the triangular domain can be performed by considering the infinitesimal parallelogram shown in Fig. B.3. Denoting by s_1, s_2, and

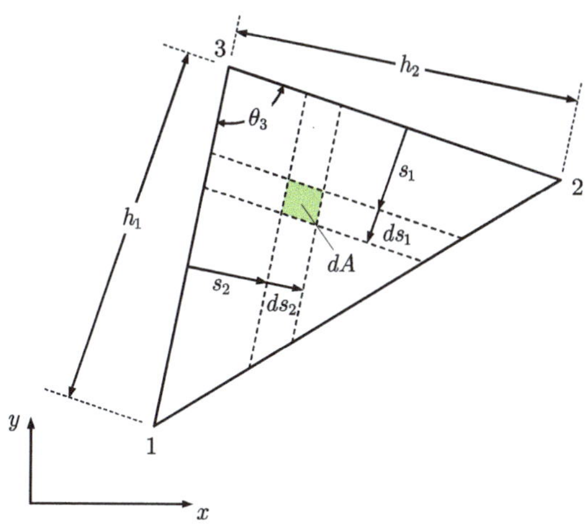

FIGURE B.3 Integration over a triangular domain.

s_3 the distances of point P from the corresponding opposite sides and by h_1, h_2, and h_3 the respective heights of the triangle, we can write

$$A = \frac{a_i h_i}{2} \quad \text{and} \quad A_i = \frac{a_i s_i}{2} \quad (i = 1, 2, 3)$$

which yield according to Eq. (B.27)

$$\frac{s_i}{h_i} = \frac{A_i}{A} = \xi_i$$

and hence

$$s_i = h_i \, \xi_i$$

So, the surface element dA can be expressed as

$$\begin{aligned} dA &= \frac{ds_1 \, ds_2}{\sin\theta_3} \\ &= \frac{(h_1 \, d\xi_1)(h_2 \, d\xi_2)}{\sin\theta_3} \\ &= 2A \; d\xi_1 \; d\xi_2 \end{aligned} \tag{B.31}$$

Consequently, a domain integral may be written as

$$\begin{aligned} \int_A f(x, y) \; dA &= 2A \int_0^1 \int_0^{1-\xi_1} f(\xi_1, \xi_2, \xi_3) \; d\xi_2 \; d\xi_1 \\ &= 2A \int_0^1 \int_0^{1-\xi_1} f[\xi_1, \xi_2, (1 - \xi_1 - \xi_2)] \; d\xi_2 \; d\xi_1 \end{aligned} \tag{B.32}$$

If $f(\xi_1, \xi_2, \xi_3)$ is a polynomial term of the form $\xi_1^a \, \xi_2^b \, \xi_3^c$, where a, b, and c are nonnegative integers, then we have [6]

$$\int_A \xi_1^a \, \xi_2^b \, \xi_3^c \; dA = \frac{a! \, b! \, c!}{(a + b + c + 2)!} \; 2A \tag{B.33}$$

A comprehensive introduction to triangular coordinates can be found in the book by Gallagher [7].

Gauss integration rules over triangles have been derived by means of triangular coordinates and have the form

$$\begin{aligned} I &= \int_0^1 \left[\int_0^{1-\xi_1} f(\xi_1, \xi_2, \xi_3) \; d\xi_2 \right] d\xi_1 \\ &\approx \sum_{k=1}^{n} f(\xi_1^k, \xi_2^k, \xi_3^k) \; w_k \end{aligned} \tag{B.34}$$

The coordinates ξ_1^k, ξ_2^k, ξ_3^k and the weight factors w_k for integration rules which are accurate for polynomials of order 1–5 are given in Table B.5. These values have been derived by Hammer et al. [8].

TABLE B.5 Coordinates and Weights for Gauss Integration Over Triangular Domains

Number and Position of the Gauss Points	k	ξ_1^k	ξ_2^k	ξ_3^k	w_k/A
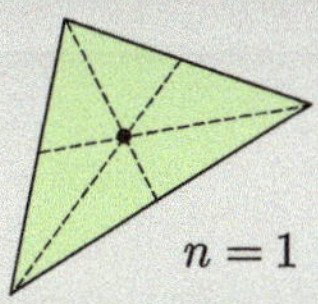 $n=1$ Degree of accuracy: 1	1	1/3	1/3	1/3	1
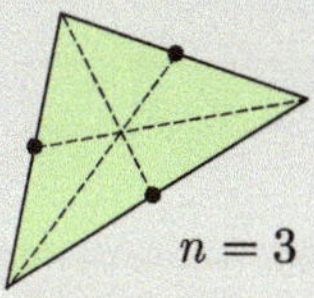 $n=3$ Degree of accuracy: 2	1	1/2	1/2	0	1/3
	2	0	1/2	1/2	1/3
	3	1/2	0	1/2	1/3
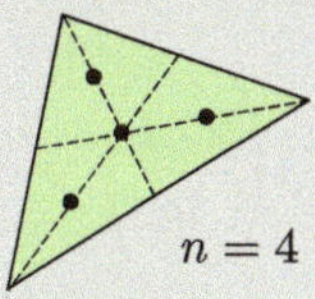 $n=4$ Degree of accuracy: 3	1	1/3	1/3	1/3	−27/48
	2	3/5	1/5	1/5	25/48
	3	1/5	3/5	1/5	25/48
	4	1/5	1/5	3/5	25/48
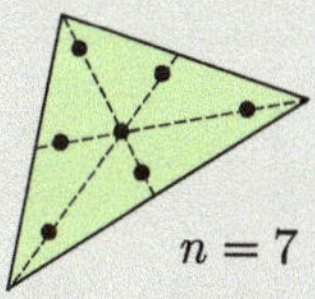 $n=7$ Degree of accuracy: 5	1	0.33333333	0.33333333	0.33333333	0.22500000
	2	0.79742699	0.10128651	0.10128651	0.12593918
	3	0.10128651	0.79742699	0.10128651	0.12593918
	4	0.10128651	0.10128651	0.79742699	0.12593918
	5	0.05971587	0.47014206	0.47014206	0.13239415
	6	0.47014206	0.05971587	0.47014206	0.13239415
	7	0.47014206	0.47014206	0.05971587	0.13239415

EXAMPLE B.2

The integrals below are computed using a 4-point Gaussian integration over the triangle shown in Fig. B.4.

Equations (B.29a,b) applied to the triangle of Fig. B.4 become

$$x = 2\xi_1 + 3\xi_2 + \xi_3$$
$$y = \xi_1 + 2\xi_2 + 3\xi_3$$

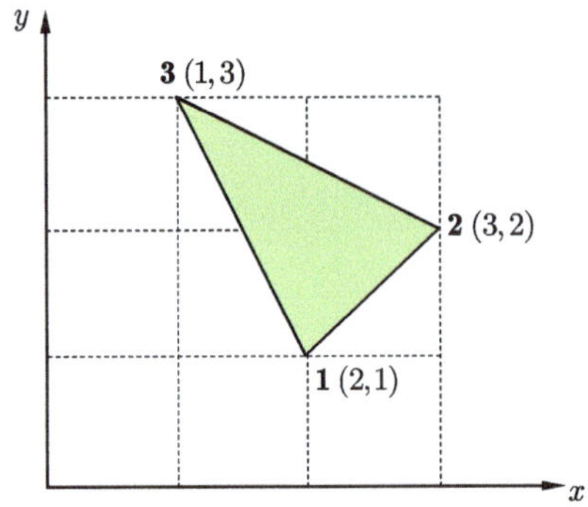

FIGURE B.4 Triangular domain of integration.

and formula (B.34) yields

$$\int_A (1)\, dA = \sum_{k=1}^{4} (1)_k \, w_k$$

$$= (1)\left(-\frac{27A}{48}\right) + (1)\left(\frac{25A}{48}\right) + (1)\left(\frac{25A}{48}\right) + (1)\left(\frac{25A}{48}\right) = A$$

$$\int_A x\, dA = \sum_{k=1}^{4} (2\xi_1^k + 3\xi_2^k + \xi_3^k)\, w_k$$

$$= \left(2\times\frac{1}{3} + 3\times\frac{1}{3} + \frac{1}{3}\right)\left(-\frac{27}{48}A\right) + \left(2\times\frac{3}{5} + 3\times\frac{1}{5} + \frac{1}{5}\right)\left(\frac{25}{48}A\right)$$

$$+ \left(2\times\frac{1}{5} + 3\times\frac{3}{5} + \frac{1}{5}\right)\left(\frac{25}{48}A\right) + \left(2\times\frac{1}{5} + 3\times\frac{1}{5} + \frac{3}{5}\right)\left(\frac{25}{48}A\right)$$

$$= 2A$$

$$\int_A y^2\, dA = \sum_{k=1}^{4} (\xi_1^k + 2\xi_2^k + 3\xi_3^k)^2\, w_k$$

$$= \left(\frac{1}{3} + 2\times\frac{1}{3} + 3\times\frac{1}{3}\right)^2\left(-\frac{27}{48}A\right) + \left(\frac{3}{5} + 2\times\frac{1}{5} + 3\times\frac{1}{5}\right)^2\left(\frac{25}{48}A\right)$$

$$+ \left(\frac{1}{5} + 2\times\frac{3}{5} + 3\times\frac{1}{5}\right)^2\left(\frac{25}{48}A\right) + \left(\frac{1}{5} + 2\times\frac{1}{5} + 3\times\frac{3}{5}\right)^2\left(\frac{25}{48}A\right)$$

$$= \frac{25}{6}A$$

The computed values are exact because the 4-point integration rule is exact for polynomials up to the third degree.

B.4 DOUBLE SINGULAR INTEGRALS

The domain discretization method to compute the integrals (4.30) for Poisson's equation or (7.105) and (7.116) for plane elasticity problems requires the evaluation of double integrals, whose integrand behaves like $\ln r$, $1/r$, or $1/r^2$. When the integration is performed over the element on which the field point lies, these integrals become singular or hypersingular. Several special methods have been developed for their evaluation [9]. The method presented here is relatively simple and effective. It was developed in [10] for the fundamental solution of the Laplace equation and here it is extended also to that of Navier's equations.

B.4.1 Domain Integrals of the Fundamental Solution for the Laplace Equation

Consider the domain integral on the e-th element

$$\int_{\Omega^e} v \, d\Omega \tag{B.35}$$

where $v(r)$ is the fundamental solution for the Laplace equation

$$v = \frac{1}{2\pi}\ln r, \; r = |Q - P|, \; (P, Q \in \Omega^e) \tag{B.36}$$

The domain integral (B.35) can be evaluated by converting it into a regular line integral on the boundary Γ^e. This is achieved using Green's identity

$$\int_{\Omega^e} (u\,\nabla^2 U - U\,\nabla^2 u) d\Omega = \int_{\Gamma^e} \left(u\frac{\partial U}{\partial n} - U\frac{\partial u}{\partial n} \right) ds \tag{B.37}$$

which for functions u and U defined as

$$u = 1 \quad \text{and} \quad \nabla^2 U = v^* \tag{B.38}$$

where

$$v^* = \frac{1}{2\pi}(\ln r + 1) \tag{B.39}$$

yields

$$\int_{\Omega^e} v^* \, d\Omega = \int_{\Gamma^e} \frac{\partial U}{\partial n} ds \tag{B.40}$$

Combining Eqs. (B.36) and (B.39), we find that

$$\begin{aligned}\int_{\Omega^e} v\, d\Omega &= \int_{\Omega^e} v^*\, d\Omega - \frac{A^e}{2\pi} \\ &= \int_{\Gamma^e} U_n\, ds - \frac{A^e}{2\pi}\end{aligned} \tag{B.41}$$

where A^e is the area of the e-th element.

The function U may be established by expressing the second of Eqs. (B.38) in polar coordinates, namely,

$$\frac{1}{r}\frac{d}{dr}\left(r\frac{dU}{dr}\right) = \frac{1}{2\pi}(\ln r + 1)$$

which gives after two consecutive integrations

$$U = \frac{1}{8\pi} r^2 \ln r \tag{B.42}$$

In the case that the domain integral involves derivatives of v, the function U in Eq. (B.41) is replaced with the respective derivatives, namely,

$$\int_{\Omega^e} v_{,m}\, d\Omega = \int_{\Gamma^e} U_{,mn}\, ds \qquad (m = x,\, y,\, xx,\, xy,\, yy) \tag{B.43}$$

where subscripts preceded by a comma denote differentiation.

The derivatives $U_{,n}$ and $U_{,mn}$ are evaluated from the relations

$$U_{,n} = \frac{1}{8\pi}(2\ln r + 1)\;\; r\; r_{,n} \tag{B.44a}$$

$$U_{,xn} = -\frac{1}{8\pi}[(2\ln r + 1)\;\cos(\alpha + \phi) + 2\cos\alpha\;\;\cos\phi] \tag{B.44b}$$

$$U_{,yn} = -\frac{1}{8\pi}[(2\ln r + 1)\;\sin(\alpha + \phi) + 2\sin\alpha\;\;\cos\phi] \tag{B.44c}$$

$$U_{,xxn} = \frac{1}{4\pi r}(\cos\phi - \sin 2\alpha\;\;\sin\phi) \tag{B.44d}$$

$$U_{,xyn} = \frac{1}{4\pi r}\cos 2\alpha\;\;\sin\phi \tag{B.44e}$$

$$U_{,yyn} = \frac{1}{4\pi r}(\cos\phi + \sin 2\alpha\;\;\sin\phi) \tag{B.44f}$$

where $\alpha = \text{angle}(\mathbf{x}, \mathbf{r})$ and $\phi = \text{angle}(\mathbf{r}, \mathbf{n})$ (see Fig. A.1).

B.4.2 Domain integrals of the fundamental solution for the Navier equations

Consider the domain integral on the e-th element

$$\int_{\Omega^e} v^* \, d\Omega \tag{B.45}$$

where

$$v^* = U_{\xi x} \text{ or } U_{\eta x} \text{ or } U_{\xi y} \text{ or } U_{\eta y} \tag{B.46}$$

The integral of Eq. (B.45) can be evaluated by converting it into a regular line integral on the boundary Γ^e. This is achieved using Eqs. (7.139) and (7.140). For example, choosing $V = \xi$ as the potential function in Eqs. (7.135) and (7.139) results in

$$\int_{\Omega^e} U_{\xi x} \, d\Omega = \frac{2(1-\overline{\nu})}{8\pi G} \int_{\Gamma^e} \left[\xi \left(\frac{\partial \phi_{,\xi}}{\partial n} \right) - \phi_{,\xi} \, n_x \right] ds + \int_{\Gamma^e} (U_{\xi x} \, n_x + U_{\eta x} \, n_y) ds \tag{B.47}$$

The domain integrals of the kernels $T_{\xi x}$, $T_{\eta x}$, $T_{\xi y}$, $T_{\eta y}$, $\sigma_{x\xi}$, $\sigma_{x\eta}$, ..., $\tau_{xy\eta}$, $\overline{\sigma}_{x\xi}$, $\overline{\sigma}_{x\eta}$, ..., $\overline{\tau}_{xy\eta}$ are expressed in terms of the derivatives of (B.45), and are evaluated after they have been converted to line integrals.

REFERENCES

[1] Scheid F. Numerical analysis, Schaum's outline series. New York: McGraw-Hill Book Company; 1968.
[2] Abramowitz M, Stegun I, editors. Handbook of mathematical functions. 10th ed. New York: Dover Publications; 1972.
[3] Stroud AH, Secrest D. Gaussian quadrature formulas. Englewood Cliffs, New Jersey: Prentice Hall, Inc; 1966.
[4] Katsikadelis JT. A boundary element solution to the vibration problem of plates. J Sound Vibration 1990;141:313–22.
[5] Davis P, Rabinowitz P. Methods of numerical integration. New York: Academic Press; 1975.
[6] Eisenberg MA, Malvern LE. On finite element integration in natural coordinates. Int J Numer Methods Eng 1973;7:574–5.
[7] Gallagher RH. Finite element analysis fundamentals. Englewood Cliffs, New Jersey: Prentice Hall; 1975.
[8] Hammer PC, Marlowe OJ, Stroud AH. Numerical integration over simplexes and cones. Math Tables Other Aids Comput 1956;Vol. X:130–7.
[9] Sladek V, Sladek J, editors. Singular integrals in boundary element methods. Southampton: Computational Mechanics Publications; 1998.
[10] Katsikadelis JT. The analog equation method. In: Brebbia CA, editor. A powerful BEM-based solution technique for linear and nonlinear engineering problems. Bound Elem 1994;XVI:167–183.

Appendix C: Answers to Selected Problems

CHAPTER 2

2.1. **i.** $\int_\Gamma \frac{x^2}{2} n_x\, ds$ or $\int_\Gamma xyn_y\, ds$

ii. $\int_\Gamma xyn_x\, ds$ or $\int_\Gamma \frac{y^2}{2} n_y\, ds$

iii. $\int_\Gamma \frac{x^2 y}{2} n_x\, ds$ or $\int_\Gamma \frac{xy^2}{2} n_y\, ds$

iv. $\int_\Gamma \frac{x^3}{3} n_x\, ds$ or $\int_\Gamma x^2 yn_y\, ds$

v. $\int_\Gamma xy^2 n_x\, ds$ or $\int_\Gamma \frac{y^3}{3} n_y\, ds$

vi. $\int_\Gamma \frac{1}{3}(x^3 n_x + y^3 n_y)\, ds$ or $\int_\Gamma xy(yn_x + xn_y)\, ds$

vii. $\int_\Gamma \sin x\; n_x\, ds$ or $\int_\Gamma y \cos x\; n_y\, ds$

2.2. *Hint*

a. Use the transformation relations from Cartesian to polar coordinates

$$x = r\cos\theta \text{ and } y = r\sin\theta$$

b. Apply the chain rule of differentiation to express the derivatives in polar coordinates

$$\frac{\partial}{\partial x} = r_{,x}\frac{\partial}{\partial r} + \theta_{,x}\frac{\partial}{\partial \theta}, \qquad \frac{\partial}{\partial y} = r_{,y}\frac{\partial}{\partial r} + \theta_{,y}\frac{\partial}{\partial \theta}$$

$$r_{,x} = \cos\theta,\ r_{,y} = \sin\theta, \qquad \theta_{,x} = -\frac{\sin\theta}{r},\ \theta_{,y} = \frac{\cos\theta}{r}$$

c. Repeat the differentiation to obtain the second derivatives and add them to derive the Laplacian in polar coordinates

$$\nabla^2 = \frac{\partial^2}{\partial r^2} + \frac{1}{r}\frac{\partial}{\partial r} + \frac{1}{r^2}\frac{\partial^2}{\partial \theta^2}$$

which for the axisymmetric function, $v = v(r)$, becomes

$$\nabla^2 = \frac{\partial^2}{\partial r^2} + \frac{1}{r}\frac{\partial}{\partial r} = \frac{1}{r}\frac{\partial}{\partial r}\left(r\frac{\partial}{\partial r}\right)$$

d. Establish a particular solution of the equation

$$\nabla^2 U = \ln r, \text{ which gives} \qquad U = \frac{1}{4}r^2(\ln r - 1)$$

e. Apply Green's identity (2.16) for $u = 1$ and U to obtain

$$\int_\Omega \ln r \, d\Omega = \int_\Gamma \frac{\partial U}{\partial n} ds$$

2.3. **i.** $\int_a^b \delta(x - x_0)\, dx = 1$

ii. $\int_a^b \delta(kx)f(x)\, dx = \frac{f(0)}{|k|}$ (*hint:* use appropriately Eq. 2.40)

iii. $\int_a^b \delta(-x)\, dx = 1$

iv. $\int_a^b \delta^{(n)}(x)\, \phi(x)\, dx = (-1)^n\, \phi^{(n)}(x_0)$

2.4. *Hint:* Use appropriately Eq. (2.40).

2.5. *Hint:* Apply Eq. (2.40) for $x = r\cos\theta$, $y = r\sin\theta$ to obtain

$$\delta(P - P_0) = \delta(x - x_0)\delta(y - y_0) = \frac{\delta(r - r_0)\delta(\theta - \theta_0)}{r}$$

2.6. *Hint:* Apply repeatedly the Gauss–Green theorem for all terms in the integral

$$\int_\Omega v\, L(u)\, d\Omega$$

until all derivatives of u are eliminated and group appropriately the boundary terms.

2.7. *Hint:* Apply Green's identity for u and $v = 1$.

2.8. **iii.** $\int_\Omega [vL(u) - uL^*(v)]d\Omega = \int_\Gamma \left[v\frac{\partial u}{\partial n} - u\frac{\partial v}{\partial n} + uv(\boldsymbol{a}\cdot\boldsymbol{n})\right] ds$

CHAPTER 3

3.1. $u = \frac{1}{12}xy(x^2y + xy^2)$

3.2. $\int_\Omega vf\, d\Omega = \frac{1}{4}R^2(2\ln R - 1)\, f(x_o, y_o), \qquad f(x_o, y_o) = \alpha_0 + \alpha_1\, x_o + \alpha_2\, y_o$

3.3. **i.** *Hint:* use Eq. (3.49) and differentiate it with respect to x. See also Section B.4 of Appendix B.
ii. *Hint:* Establish a function F satisfying the equation

$$\nabla^2 F = x^2 + y^2$$

then apply Green's identity for $v = \ln r/2\pi$ and $u = F$, differentiate it with respect to x and use Eq. (2.42).

3.4. *Hint:* Show first that the integral representation (3.18) can be written as

$$u(P) = -\int_\Gamma \left\{ v(P,q)\frac{\partial u(q)}{\partial n_q} - [u(q) - u(p)]\frac{\partial v(P,q)}{\partial n_q} \right\} ds_q + u(p)$$

where p is a point on Γ. Then, to obtain $\partial u/\partial n_p$, differentiate with respect to a direction n and find the limit as $P \to p \in \Gamma$ choosing n to coincide with the normal to the boundary at point $p \in \Gamma$. The limiting procedure is similar to that of Section 3.3. Work in the same way to obtain $\partial u/\partial t_p$. For points p where the boundary is smooth, we find

$$\frac{1}{2}\frac{\partial u(p)}{\partial n_p} = -\int_\Gamma \left\{ \frac{\partial v(p,q)}{\partial n_p}\frac{\partial u(q)}{\partial n_q} - [u(q) - u(p)]\frac{\partial^2 v(p,q)}{\partial n_p \partial n_q} \right\} ds_q$$

$$\frac{1}{2}\frac{\partial u(p)}{\partial t_p} = -\int_\Gamma \left\{ \frac{\partial v(p,q)}{\partial t_p}\frac{\partial u(q)}{\partial n_q} - [u(q) - u(p)]\frac{\partial^2 v(p,q)}{\partial t_p \partial n_q} \right\} ds_q$$

3.5. *Hint:* For $p \in \Gamma$ use the integral representations of $\partial u/\partial n_p$ and $\partial u/\partial t_p$ to derive the expressions for $\partial u/\partial x$ and $\partial u/\partial y$.

CHAPTER 4

4.1. *Hint:* See subroutine DERIV of program FLUIDCON in Chapter 6

4.2. *Hint:*

a. Evaluate the matrices $[H]^1$, $[G]^1$, $[H]^2$ and $[G]^2$.

Domain Ω_1 is isotropic. The boundary integral equation is given by Eq. (3.29), which after discretization yields

$$[H]^1\{u\}^1 = [G]^1\{q_n\}^1,\ \{q_n\}^1 = \{u_{,n}\}^1\ (u_{,n} = \nabla u \cdot \mathbf{n})$$

The matrices $[H]^1$, $[G]^1$ are evaluated as described in Section 4.3.

Domain Ω_2 is orthotropic. The boundary integral equation is given by Eq. (3.79), which after discretization yields

$$[H]^2\{u\}^2 = [G]^2\{q_n\}^2,\ \{q_n\}^2 = \{u_{,m}\}^2\ (u_{,m} = \nabla u \cdot \mathbf{m})$$

where

$$H_{ij} = \int_{\Gamma_j} \frac{1}{2\pi\sqrt{|D|}} \frac{r_{,m}}{r} ds_q - \frac{1}{2}\delta_{ij}$$

$$G_{ij} = \int_{\Gamma_j} \frac{1}{2\pi\sqrt{|D|}} \ln r ds_q$$

in which

$$r(q, p_i) = \sqrt{\frac{(\xi - x_i)^2}{k_{xx}} + \frac{(\eta - y_i)^2}{k_{yy}}},$$

$$q{:}(\xi, \eta) \in \Gamma_j \text{ and } p_i{:}(x_i, y_i) \in \Gamma_i.$$

The elements of these matrices can be evaluated in a way similar to that presented in Section 4.3.

b. Apply the given boundary conditions on the outer boundaries and the interface continuity conditions to obtain the matrices $[A]$ and $\{B\}$ as presented in Section 4.9.

CHAPTER 5

5.1. **i.** $\psi_1 = -\frac{1}{16} + \frac{1}{16}\xi + \frac{9}{16}\xi^2 - \frac{9}{16}\xi^3$, $\psi_2 = \frac{9}{16} - \frac{27}{16}\xi - \frac{9}{16}\xi^2 + \frac{27}{16}\xi^3$,
$\psi_3 = \frac{9}{16} + \frac{27}{16}\xi - \frac{9}{16}\xi^2 - \frac{27}{16}\xi^3$, $\psi_4 = -\frac{1}{16} - \frac{1}{16}\xi + \frac{9}{16}\xi^2 + \frac{9}{16}\xi^3$

ii. $\psi_1 = -\frac{1}{6} + \frac{1}{6}\xi + \frac{2}{3}\xi^2 - \frac{2}{3}\xi^3$, $\psi_2 = \frac{2}{3} - \frac{4}{3}\xi - \frac{2}{3}\xi^2 + \frac{4}{3}\xi^3$,
$\psi_3 = \frac{2}{3} + \frac{4}{3}\xi - \frac{2}{3}\xi^2 - \frac{4}{3}\xi^3$, $\psi_4 = -\frac{1}{6} - \frac{1}{6}\xi + \frac{2}{3}\xi^2 + \frac{2}{3}\xi^3$

5.2. The exact value for the area of the circular sector is given as

$$A_{\text{exact}} = \frac{\theta_0}{2} R^2 \Rightarrow A_{\text{exact}} = \frac{3\pi}{8}$$

i. For a linear element approximation: $A_{\text{linear}} = \frac{1}{2} R^2 \sin\theta_0$

ii. For a quadratic element approximation follow the steps:

a. Use polar coordinates to evaluate the area

$$A = \frac{1}{2}\int_0^{\theta_0} r^2\, d\theta$$

b. Transform the integral on the interval $-1 \le \xi \le 1$

$$A = \frac{1}{2}\int_{-1}^{+1} \sqrt{x(\xi)^2 + y(\xi)^2}\ \left|J(\xi)\right| d\xi$$

where $x(\xi)$ and $y(\xi)$ are obtained from Eq. (5.66) and the Jacobian $\left|J(\xi)\right|$ from Eq. (5.68).

c. Evaluate the integral using a four-point Gauss integration.

iii. For a cubic element approximation establish first

$$x(\xi) = \sum_{k=1}^{4} x_k \psi_k(\xi) \quad \text{and} \quad y(\xi) = \sum_{k=1}^{4} y_k \psi_k(\xi)$$

from the known coordinates of the four nodal points, which are

$$x_k = R\cos\left[(k-1)\frac{\theta_0}{3}\right] \quad \text{and} \quad y_k = R\sin\left[(k-1)\frac{\theta_0}{3}\right]$$

where $k = 1, 2, 3, 4$. The shape functions $\psi_k(\xi)$ are those derived in Problem 5.1-(i). Then, follow the same procedure as in case (ii) to evaluate the area integral.

The exact value of the area along with the computed values for the three types of elements are given in the following table.

	Exact	Linear	Parabolic	Cubic
Area	1.178097	1.164686	1.178086	1.178099
Error (%)	0	1.14	9.34×10^{-6}	-1.70×10^{-6}

5.3. The integrand has a peak at $x = 0.25$ (see its graph in the figure). The interval is divided into subintervals as in the table below and Gauss integration is applied on each of them

Number of Subintervals	Subintervals	Number of Gauss Points	$\int_{-1}^{+1} \frac{dx}{\left[(x-0.25)^2+0.05\right]^4}$
4	(−1.00, + 0.10)	8	35124.17
	(+0.10, + 0.25)	6	
	(+0.25, + 0.40)	6	
	(+0.40, + 1.00)	8	
Exact value			35123.22

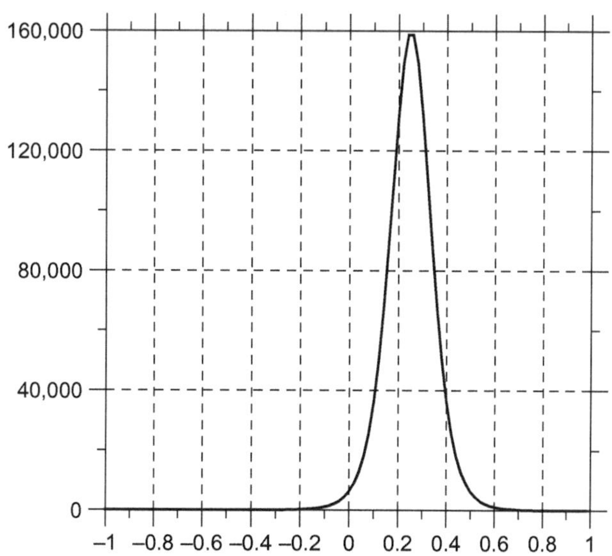

5.4. *Hint:* Transform the integral on the interval $-1 \leq \xi \leq +1$ and use Gauss integration to evaluate it. It is suggested to follow the steps:

a. Express the coordinates of the points on the parabolic element in terms of ξ using Eq. (5.63), namely

$$x(\xi) = 4.30\ \psi_1(\xi) + 4.10\ \psi_2(\xi) + 3.80\ \psi_3(\xi)$$

$$y(\xi) = 2.50\ \psi_1(\xi) + 2.90\ \psi_2(\xi) + 3.20\ \psi_3(\xi)$$

b. Find the expressions of $r(\xi)$ and $|J(\xi)|$ to obtain the integrand in the form

$$\begin{aligned} f(\xi) &= \psi_1(\xi)\ \ln\ [r(\xi)]\ |J(\xi)| \\ &= -\frac{1}{2}\xi\,(1-\xi)\ln\left\{\sqrt{[x(\xi)-4.15]^2 + [y(\xi)-2.65]^2}\right\} \\ &\times\ \sqrt{[x'(\xi)]^2 + [y'(\xi)]^2} \end{aligned}$$

c. Check whether the integral is singular or near-singular. For this purpose evaluate $\xi = \xi_0$ for which $r(\xi)$ becomes minimum as well as the values of $r(\xi_0)$ and $f(\xi_0)$ (answer: $\xi_0 = -0.56595534$, $r(\xi_0) = 0.08357687$, $f(\xi_0) = -0.49520518$). Hence the integral is neither singular nor near-singular. This also becomes evident from the graph of the integrand shown in the figure below.

d. Evaluate the integral using Gauss integration

$$\int_{-1}^{+1} f(\xi)d\xi \simeq \sum_{k=1}^{n} f(\xi_k)\ w_k = I_n \Rightarrow \begin{cases} I_4 = -0.34181408 \\ I_6 = -0.35966759 \\ I_8 = -0.35957404 \end{cases}$$

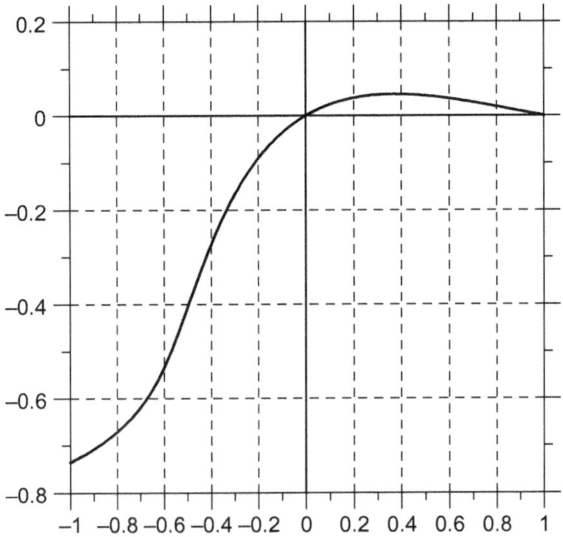

5.5. *Hint:*

a. Evaluate the length of the j-th element,

$$\ell_j = \frac{2}{\kappa}\sqrt{(x_2 - x_1)^2 + (y_2 - y_1)^2} = 1.166190$$

b. Use Eq. (5.27) together with Eqs. (5.32) and (5.35) to evaluate g_1^{ij}.

c. Use Eq. (5.36) together with Eqs. (5.37) and (5.38) to evaluate g_2^{ij}.

The influence coefficients, when the source node i coincides with nodes 1 and 2 (local numbering) of the integration element j, are

for node$-i$ at $\xi_1 = -0.5$: $g_1^{1j} = -0.215355$, $g_2^{1j} = -0.039519$

for node$-i$ at $\xi_2 = +0.5$: $g_1^{2j} = -0.046087$, $g_2^{2j} = -0.184665$

CHAPTER 6

6.1. Use program TORSCON to evaluate the required quantities for the three cross-sections. The computed answers along with the exact values or those obtained with other approximate solutions are given below in tabular form.

Cross-Section	Solution Method	$\frac{I_t}{h^4}$	$\frac{\max\tau_{tz}}{G\theta h}$
	BEM (N = 250)	0.028593	0.464960
	Analytical[a]	0.028585	0.465030
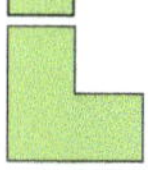	BEM (N = 250)	0.053527	0.779140
	Approximate[b]	0.051157	
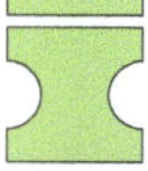	BEM (N = 250)	0.053780	0.677690
	Approximate[b]	0.057911	

[a] *See Example 6.2.*
[b] *See chapter "Applications," Hirsch C.* Numerical computation of internal and external flows, *vol. 1 and 2. Chichester: John Wiley & Sons; 1988.*

6.2. *Hint:* Seek the solution as a sum of the homogeneous solution and a particular one

$$u = u_0 + u_1$$

a. The particular solution is obtained employing the procedure described in Section 3.4.2, hence

$$\nabla^2 u_1 = -\frac{1}{30}x$$

which yields

$$u_1 = -\frac{x(x^2 + y^2)}{240}$$

b. The homogeneous solution is obtained from the boundary value problem

$$\nabla^2 u_0 = 0 \text{ in } \Omega$$

$$u_0 = \frac{x(x^2+y^2)}{240} \quad \text{on } \Gamma$$

using the program LABECON.

Answer: $u(a/2, b/3) = 0.018$ m employing N = 200 boundary elements.

6.3. *Hint:* Use program LABECONMU.

6.4. *Hint:* At the insulated parts of the boundary it is $T_n = 0$.

Answer: Values of the temperature T at selected points on the axis of symmetry of the cross-section are given in the table below as obtained using LABECON with N = 240.

y	0.05	0.25	0.45	0.65	0.85
T	14.54	32.67	50.82	67.35	77.86

6.5. *Hint:* Modify program LABECON, according to the procedure presented in Section 4.9 (method of subdomains), so that it can be used for the analysis of composite domains.

6.6. *Hint:* Use the program developed for the requirements of Problem 4.2.

6.7. *Hint:* Use program FLUIDCON.

Answer: Values of the velocity component v_n at selected points of the outlet cross-section are given in the following table as obtained using FLUIDCON with N = 240.

y	1.70	1.30	0.90	0.50	0.10
v_n	0.484	0.492	0.503	0.512	0.516

CHAPTER 7

7.1. Given the initial strains $\{\varepsilon_0\}$ use Eq. (7.13) to establish the corresponding stresses $\{\sigma_0\}$. Then substitute Eq. (7.26) into the equilibrium equation (7.18) with $b_x = b_y = 0$ to obtain

$$\nabla^2 u + \frac{1}{1-2\nu}\left(\frac{\partial^2 u}{\partial x^2} + \frac{\partial^2 v}{\partial x \partial y}\right) + \frac{1}{G} b_x^0 = 0$$

$$\nabla^2 v + \frac{1}{1-2\nu}\left(\frac{\partial^2 u}{\partial x \partial y} + \frac{\partial^2 v}{\partial y^2}\right) + \frac{1}{G} b_y^0 = 0$$

where

$$b_x^0 = -\left(\frac{\partial \sigma_x^0}{\partial x} + \frac{\partial \tau_{xy}^0}{\partial y}\right)$$

$$b_y^0 = -\left(\frac{\partial \tau_{xy}^0}{\partial x} + \frac{\partial \sigma_y^0}{\partial y}\right)$$

The boundary tractions due to the initial strains are obtained by substituting Eq. (7.26) into Eq. (7.22) and taking into account that

$$t_x = t_x^t - t_x^0 = 0$$
$$t_y = t_y^t - t_y^0 = 0$$

where t_x^t and t_y^t denote the total boundary tractions. The derived expressions are

$$t_x^t = t_x^0 = -\left(\sigma_x^0\, n_x + \tau_{xy}^0\, n_y\right)$$
$$t_y^t = t_y^0 = -\left(\tau_{xy}^0\, n_x + \sigma_y^0\, n_y\right)$$

Therefore, the displacement field u_0 and v_0 due to initial strains is the one produced by the effective body forces b_x^0, b_y^0 in Ω and the prescribed boundary tractions t_x^0 and t_y^0 on Γ. In this case, the boundary integral equations result from Eq. (7.90) by replacing b_x, b_y and t_x, t_y with b_x^0, b_y^0 and t_x^0, t_y^0, respectively. In solving this problem using program ELBECON, attention should be paid in restraining the rigid body motion, so that the matrix $[H]$ can be inverted.

7.2. The initial stress distribution caused by the temperature change $\Delta T(x, y)$ is given by Eq. (7.28) and the effective body forces (see answer for Problem 7.1) become

$$b_x^0 = -\frac{\overline{E}\alpha}{1-\overline{\nu}}\,\frac{\partial \Delta T}{\partial x} = 0 \quad \text{and} \quad b_y^0 = -\frac{\overline{E}\alpha}{1-\overline{\nu}}\,\frac{\partial \Delta T}{\partial y} = 0$$

while the boundary tractions become

$$t_x^0 = \frac{\overline{E}\alpha\Delta T}{1-\overline{\nu}}\, n_x = 222.22\, n_x \quad \text{and} \quad t_y^0 = \frac{\overline{E}\alpha\Delta T}{1-\overline{\nu}}\, n_y = 222.22\, n_y$$

The solution of the problem is obtained following three steps:

a. Evaluation of the displacements produced by the temperature change. Due to the double symmetry of the cross-section the solution can be limited to the lower left quadrant (see figure below). This problem is solved using program ELBECON with the boundary conditions shown in the figure.

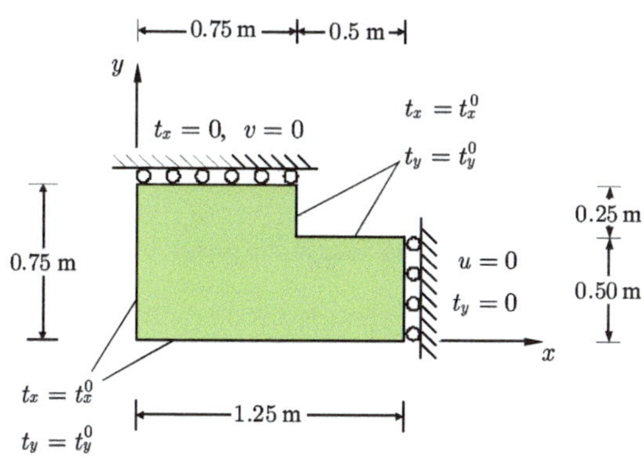

b. Solve the problem for the whole domain with the following boundary conditions:

$$\text{outer boundary:}\quad \begin{array}{ll} u(0,y) = -u^0(0,y), & t_y(0,y)=0 \\ v(x,0) = -v^0(x,0), & t_x(x,0)=0 \\ u(2.5,y) = -u^0(2.5,y), & t_y(2.5,y)=0 \\ t_x(x,1.5)=0, & t_y(x,1.5)=0 \end{array}$$

inner boundary: $t_x = 0, \quad t_y = 0$ (traction-free)

c. Superimpose the solutions of steps (a) and (b).

Numerical results obtained for N = 348 boundary elements are given in the following table.

y	$x=0$		$x=0.425$				
	$v \times 10^3$	t_x	$u \times 10^3$	$v \times 10^3$	σ_x	σ_y	τ_{xy}
0.3125	0.366	145.71	0.040	0.371	− 166.29	15.90	21.50
0.5125	0.577	120.34	0.115	0.595	− 118.07	2.93	40.49
0.9875	1.018	119.13	0.126	− 0.616	− 111.11	− 1.93	− 27.24
1.1875	1.216	150.09	0.043	− 0.477	− 164.88	8.29	− 12.87

7.3. *Hint:*

a. Establish the displacement and stress fields for temperature change working in two stages as in Problem 7.2, including, however, the body forces

$$b_x^0 = -\frac{\overline{E}\alpha}{1-\overline{\nu}}\frac{\partial T}{\partial x},\quad b_y^0 = -\frac{\overline{E}\alpha}{1-\overline{\nu}}\frac{\partial T}{\partial y}$$

which do not vanish in this case. Nevertheless, the domain integrals involving the body forces can be converted to boundary line integrals by means of Eqs. (7.139) and (7.140), since they are derived from a potential T ($\nabla^2 T = 0$). For the needs of this problem, program ELBECON must be modified to include the vector $\{F\}$ of Eq. (7.111) resulting from the presence of body forces.

b. Superimpose the solution of part (a) to that for internal pressure in order to obtain the total deformation and stresses in the pipe.

7.4. *Hint:* Determine a particular solution u_1, v_1 using the relevant expressions derived in Example 7.1. Then employ program ELBECON after modifying appropriately the boundary conditions as indicated in Eq. (7.124).

7.5. Answer:

$$u_G = u_H = 7.3699 \times 10^{-3}\,\text{m}$$

$$K_{el} = \frac{P_{total}}{u_G} = \frac{2 \times 750\ \text{kN/m} \times 0.60\text{m}}{7.3699 \times 10^{-3}\,\text{m}} = 1.22118 \times 10^5\ \text{kN/m}$$

It is worth mentioning that, modeling the given frame with three beam elements (accounting also for shear deformation) and considering that the frame is clamped at the bottom of the two columns, the lateral displacement and corresponding stiffness are found to be

$$u_{beam} = 7.6 \times 10^{-3}\,\text{m}$$
$$K_{beam} = 1.18421 \times 10^{5}\,\text{kN/m}$$

Index

Note: Page numbers followed by "*f*" and "*t*" refer to figures and tables, respectively.

C

F

O

P

U

V

W